Statistical Outliers and Related Topics

Editors

Mir Masoom Ali
Distinguished Professor Emeritus
Ball State University, Muncie, Indiana, USA

Rahmatullah Imon
Professor, Ball State University
Muncie, Indiana, USA

Irfan Ali
Aligarh Muslim University
Department of Statistics & Operation Research
Aligarh, India

Haitham M. Yousof
Department of Statistics
Mathematics and Insurance, Benha University, Egypt

CRC Press
Taylor & Francis Group
Boca Raton London New York

CRC Press is an imprint of the
Taylor & Francis Group, an **informa** business
A SCIENCE PUBLISHERS BOOK

First edition published 2025
by CRC Press
2385 NW Executive Center Drive, Suite 320, Boca Raton FL 33431

and by CRC Press
4 Park Square, Milton Park, Abingdon, Oxon, OX14 4RN

CRC Press is an imprint of Taylor & Francis Group, LLC

Library of Congress Cataloging-in-Publication Data (applied for)

ISBN: 978-1-032-46056-7 (hbk)
ISBN: 978-1-032-46059-8 (pbk)
ISBN: 978-1-003-37988-1 (ebk)

DOI: 10.1201/9781003379881

Typeset in Palatino Linotype
by Prime Publishing Services

Preface

Data collected in a broad range of applications frequently contains one or more typical observations, called outliers. The concept of outliers in a data set is considered as old as the subject of statistics. Even before the formal development of the statistical methods, arguments ranged over whether, and on what basis, we should discard observations from a set of data on the grounds that they are unrepresentative or outliers. It is now evident that a routine data set typically contains about 1-10% outliers and even the highest quality data set cannot be guaranteed free of outliers. Outliers occur mostly because of gross measurement and recording errors. However, there are some other sources as well and for this reason; they are not necessarily wrong information. If the data are not homogeneous then the conclusions extracted from the data are likely to be incorrect. One immediate consequence of the presence of outliers is that they may cause apparent non-Normality and the entire classical inferential procedure might break down in the presence of outliers. For this reason, the identification of outliers before performing any statistical analysis is necessary to obtain accurate conclusions from the analysis.

The proposed book is a collection of different aspects of outliers and related topics written by researchers in that particular area of interest. That includes the definition of outliers, their sources, consequences, identification, computational and robustness issues, handling of outliers in diversified areas of statistics such as univariate and multivariate data, linear and generalized linear models, time series, linear functional and structural models, circular data, spatial data, big data, high dimensional data, multi-view data. Some chapters of this book emphasized how important outliers are for data scientists. This feature alone make the book appealing to workers in Data Mining; an area which can be described as the currently fastest-growing area of statistics. Data mining applications typically deal with data sets that consist of thousands and sometimes millions of observations. Big data poses big challenges in every aspect of statistics including the detection and handling of outliers and this book deals with this issue very seriously. The proposed book includes the outlier detection methods that are applicable to very large data sets, data sets of hundreds of thousands of observations, which is what practitioners need. The book presents methods that are applicable to very large data sets, which makes it suitable for Data Mining applications.

This book will be helpful for all those who are interested in learning new developments in Statistical outliers and their applications in regression analysis, sampling theory, non-parametric statistics, quality control, life testing and survival analysis, for example and many other applied areas of statistics. This book is intended mainly for practitioners and students (graduates and upper-level undergraduates) who do not necessarily have a strong or sophisticated mathematical and statistical background.

Editors
Mir Masoom Ali
Rahmatullah Imon
Irfan Ali
Haitham M. Yousof

Contents

Chapter 1

A Brief History of Statistical Outliers

M. Masoom Ali*[1], A.H.M. Rahmatullah Imon[1] and Irfan Ali[2]

[1]Department of Mathematical Sciences, Ball State University, Muncie, IN 47306 USA

[2]Department of Statistics and Operations Research, Aligarh Muslim University, Aligarh 202002, India

1.1 CLASSICAL STATISTICS AND OUTLIERS

The word 'outlier' appeared in English in the sixteenth century as another word for outsider, and the concept of outlier came from Astrophysics even before the formal development of statistics and statistical techniques (Stigler 1973). This word had much use in astrophysics to denote planets which were wrongly recorded as members of our solar system which in fact belonged to other solar systems. According to the Oxford English Dictionary an outlier is *"an individual whose origins, beliefs, or behavior place him or her outside a particular establishment or community"*. The roots of 'outlier' are as simple as its pronunciation: it is just a combination of "to lie" with 'out', carrying the sense of someone rooted outside the norms of a given community. In classical statistics, the norm of the community is the data that come from a target population. At the time

*For Correspondence: Masoom Ali (mali@bsu.edu)

of the development of statistics there was a general belief among the statisticians that the dataset should follow a normal distribution. To quote Pearson (1905), "... *towards the end of the nineteenth century not all were convinced of the need for curves other than normal.*" Hence, any observation that does not belong to a normal distribution was traditionally called outliers although its meaning and definitions have been changed over the years. Awareness of outliers in some form or another has existed for at least several hundred years. The concept of outliers in a dataset is considered to be as old as the subject of statistics. To quote Peirce (1852), "*In almost every true series of observations, some are found, which differ so much from the others as to indicate some abnormal source of error not contemplated in the theoretical discussions, and the introduction of which into the investigations can only serve ... to perplex and mislead the inquirer.*" Even before the formal development of the statistical method, argument raged over whether, and on what basis, we should discard observations from a set of data on the grounds that they are 'unrepresentative' (Barnett and Lewis, 1994). These issues are still alive after so many years and although much work has been done in this area of research, there are still many problems unresolved.

An excellent review of different aspects of outliers is available in Hawkins (1980), Barnett and Lewis (1994), and Hadi, Imon and Werner (2009). It is now generally believed that the existence of outliers is inevitable in statistical data. Based on a survey conducted by the Princeton University group on various well-known US data bases, Hampel et al. (1986) claim that a routine dataset typically contains about 1–10% outliers, and even the highest quality dataset cannot be guaranteed free of outliers. Although datasets (especially large ones) often (if not always) contain outliers and subgroups, classical statistical analyses customarily assume that the data are homogeneous and free from outliers. According to Chatterjee and Hadi (1988) there exists an implicit assumption in classical statistics that says that "*All observations are equally reliable and should have an equal role in determining the results and influencing conclusions.*" The presence of outliers destroys this implicit assumption. There are several sources of outliers. It may occur due to inherent variability. The natural feature of a population could be uncontrollable. Another prime source of outliers could be gross measurement and recording error. Outliers may also occur due to execution error. Imperfect collection of data could be responsible for this. We may inadvertently choose a biased sample or include individuals not truly representative of the population we aimed to sample.

The presence of outliers in statistical data and their notorious effects on classical statistical methods have been recognized for hundreds of years. All classical statistical methods can be severely distorted by outliers. It is therefore essential for data analysts to be able to identify the outliers in the data before they proceed with classical analyses. This is a huge topic and warrants a separate discussion. Sometimes it is a matter of subjective judgment on the part of the observer whether or not some observations are genuine members of the main population. If they are contaminants (arising from some other distribution), they may frustrate attempts to draw inferences about the original population. Of course, any contaminants which occur in the midst of the dataset will not be conspicuous. We call contaminants to be outliers when they appear surprisingly extreme. Outliers may or may not be contaminants; contaminants may or may not be outliers. When we order the sample, the smallest and the largest ordered observations are known as extremes. Whether we declare either of them to be an outlier depends on consideration of how they appear in relation to the postulated model. Extreme values may or may not be outliers. Attitudes towards the handling of outliers vary from one extreme to another: "*... from the view that we should never sully the sanctity of the data to 'if in doubt, throw it out* et al. (1994). Outliers do not inevitably 'perplex' or 'mislead'; they are not necessarily 'bad' or 'erroneous', and the experimenter may be tempted in some situations not to reject an outlier but to welcome it as an indication of some unexpectedly useful industrial treatment or surprisingly successful agricultural variety. The ozone depletion history is a classic example of it. Farman et al. (1985) were puzzled by data gathered by the British Antarctic Survey showing that ozone levels for Antarctica had dropped 10% below normal levels. The question arises why did the Nimbus 7 satellite, which had instruments aboard for recording ozone levels, not record similarly low ozone concentrations? After a long investigation it was revealed that the ozone concentrations recorded by the satellite were so low they were being treated as outliers by a computer program used in Nimbus 7 and were discarded!

1.2 OUTLIERS IN UNIVARIATE DATA

The term statistical outlier drew a great deal of attention in the middle of the last century. According to Barnett and Lewis (1994) in 1949, in the case of Hadlum vs Hadlum, Major Hadlum appealed against the failure of an earlier petition of divorce. His claim was based on an alleged adultery by Mrs. Hadlum, the evidence which consisted of the fact that Mrs. Hadlum gave birth to a child which was 349 days later than when Major Hadlum had left the country to serve the nation during World War II. The appeal judge rejected the appeal. But in a similar case the judge approved divorce petition. Since there was no guideline and medical evidence for paternity like DNA test was unavailable at that time, the judges passed their verdicts based on their own judgment and they ended up with lots of conflicting conclusions. In 1951, the House of Lords had ruled that the limit is 360 days based on a huge survey conducted by the British Medical Association for a sample of 13,634 British births. In this study, the mean gestation period was 40 weeks with a standard deviation of 4 weeks. The histogram of this data looked pretty normal and the cut off date was determined by the three-sigma rule. The three-sigma rule is still a very popular choice for the detection of outliers in univariate statistical data. Since extreme observations are suspect outliers, Dixon (1950) developed several detection rules for univariate outliers based on order statistics. Another popular choice is the box plot where the outliers are identified based on the median and the interquartile range.

The diagnostic methods mentioned above seem to be successful for the identification of a single outlier and it is now generally believed that locating and testing for the significance of a single outlier has been largely resolved (Hawkins, Bradu and Kass 1984). Although much work has been done on the identification of multiple outliers, there are many unresolved issues. The main reason is that single outlier detection techniques have been proved to be ineffective to detect a potential outlier in the presence of multiple outliers. The repeated application of diagnostic tools for the identification of outliers one by one is not effective as well. When a sample contains more than one outlier, Pearson and Chandra Sekar (1936) pointed out the existence of what has come to be termed the masking effect which makes outliers appear inlying. Because of this effect after the deletion of one or more influential points, another observation may emerge as extremely influential, which was not visible at first. The opposite effect is known as the swamping effect (Barnett and Lewis 1994), which makes inliers appear outlying. A group deletion technique may be useful but there is no general agreement among statisticians which methods would be most reliable.

Robust methods are designed specifically to be resistant to outliers. This is usually done by downweighting suspected or potential outliers. Robust methods are very effective in the identification of the outliers but, unfortunately, they are computationally very intensive and at times practically infeasible. This may explain why robust statistics have not been widely implemented in statistical packages although they have been proposed in the literature for so many years. Outlier detection methods are designed to identify outliers and then leave the choice to the data analysts as to what to do with the detected outliers. The data analysts, for example, can set the outliers aside and proceed with the analysis using classical methods on the rest of the data. In robust statistics language, this is equivalent to giving outliers zero weight and all other observations full weight. Most of the robust techniques are based on the most compact dataset (nearly half of the observations). Those techniques indeed may be very robust and may possess very high breakdown, but they have a tendency to identify innocent observations as outliers. Robust estimators of scale are often seriously underestimated. This gives a spurious precision to subsequent analysis resulting so that the detection procedures may identify observations as outliers which are actually not. Cook and Hawkins (1990) noted that even considering a sample of 20 five-dimensional $N(0, 1)$ data the robust techniques identify 6 cases as outliers. They described this situation as a case of *"outliers everywhere"*. The problem with many of the conventional outlier diagnostic methods is that they are often based on classical estimators and, hence they are not effective in the identification of outliers. A combination of the above two approaches has been recently (during the last 20 years) proposed in the literature. This new approach attempts to find methods which are reasonably not expensive to compute and, at the same time, are highly reliable in the identification of the outliers. One of these approaches, called diagnostic-robust, has been proposed in the literature (Hadi and Simonoff 1993, Imon 2008) which is a combination of these two approaches. This approach broadly consists of three steps: In Step 1, we find all suspect outliers by robust techniques; in Step 2, we apply diagnostic techniques to verify whether the suspected cases are genuine outliers are not; in Step 3, innocent observations (if any) are put back into the dataset if they are wrongly identified by robust techniques.

Another important thing we need to mention is that in the classical statistics literature the population was assumed to be normal. Observations that do not match with the normal distribution were called outliers. This attitude has changed in recent years. We start believing that normal is not the only standard distribution, the inherent pattern of the data could be uniform, exponential, or other standard distributions. In this situation, any observation that fails to match with the target distribution is called an outlier. Different aspects of outliers in uniform, exponential and other standard statistical distributions are studied by D'Agostino and Stephens (1986), Balasooriya (1989), Woo and Ali (1996), Ali, Woo and Yoon (2000), Dixit, Ali and Woo (2003), Ali and Woo (2004), and others.

1.3 OUTLIERS IN MULTIVARIATE DATA

Detection of outliers is more complicated in multivariate data. Here outliers may occur in different positions for different variables. Moreover, the individual effect of outliers may be different from their joint effects. For univariate outlier graphical methods could be very useful for locating outliers, but which is not the case for multivariate outliers. Only for a bivariate data we can use a Bagplot proposed by Rousseeuw, Ruts and Tukey (1999) which is basically a bivariate box plot.

For multivariate data there are two general approaches to outlier identification: methods based on projection pursuit, and methods based on robust Mahalanobis distances. Projection pursuit was introduced by Friedman and Tukey (1974) as an attempt to find interesting structure in the data that might not be obvious at first glance. Projection pursuit assigns a numerical index to one- or two-dimensional projections of the data to characterize the amount of structure present in that projection and finds the most 'interesting' possible projections in multidimensional data. In this process it can focus on the unexpected characteristics of the data and can identify outliers. A well-known special case of projection pursuit is principal component analysis (PCA) in which the index to be maximized is the variance within each component. Another popular projection pursuit method is the Stahel-Donoho estimator which was popularized by Maronna and Yohai (1995). This method is computationally extensive. To reduce the extreme computational burden Peña and Prieto (2001) propose a method called Kurtosis 1 that projects the data onto a set of $2p$ directions, where p is the dimension of the data. Outliers are searched along the projections that maximize and minimize the kurtosis coefficient. A variety of PCA methods are in use for the identification of multivariate outliers. Among them the spatial sign covariance matrix proposed by Locantore et al. (1999), robust PCA (ROBPCA) proposed by Hubert et al. (2005), outliers for high dimensions proposed by Filzmoser et al. (2008) have wide applications.

An alternative to the projection pursuit strategy is the approach based on the Mahalanobis distance. The classical Mahalanobis distance contains components like mean vector and variance-covariance matrix, both of them can be severely distorted in the presence of outliers and hence this method cannot be readily applied for the detection of outliers. Attempts have been made to robustify Mahalanobis distance by replacing the mean vector by the median vector and the variance-covariance matrix by the minimum volume ellipsoid (MVE) or by the minimum covariance determinants (MCD) proposed by Rousseeuw (1985). Several modifications of this method are found in the literature. The most important modification, the FAST-MCD developed by Rousseeuw and van Driessen (1999), is almost universally accepted as the best successor to the original MCD today. An alternative method based on robust Mahalanobis distance was proposed by Rocke and Woodruff (1996).

1.4 OUTLIERS IN STATISTICAL MODELS

Modelling is another important and interesting issue regarding the identification of outliers. In univariate data outliers are always extreme values in the sample. But this is not necessarily true for linear or other statistical models. Let us consider the following example. If we inspect the values of X and Y in the following data, there is nothing unusual in it. The following box plots, as shown in Figure 1.1, support this statement as well. But if we try to establish a linear relationship of Y on X, the scatterplot (see Figure 1.2) exhibits that there exists a point that is off from the line of best fit. This circled point definitely looks like an outlier as far as the linear relationship is concerned.

X	1	7	10	8	7	10	0	7	5	3
Y	11	12	39	32	31	40	9	28	25	19

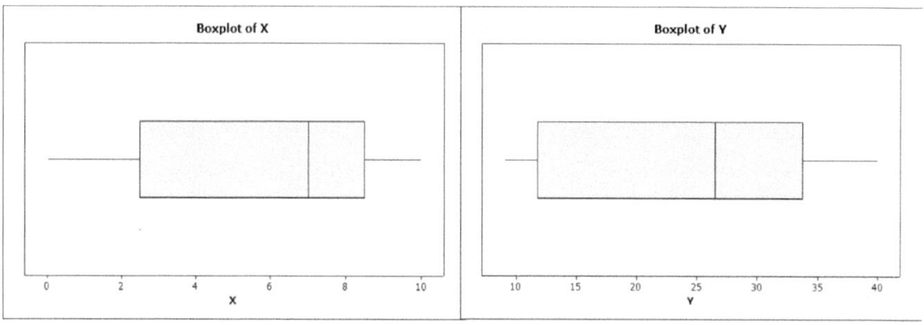

Figure 1.1 Box plot of X and Y.

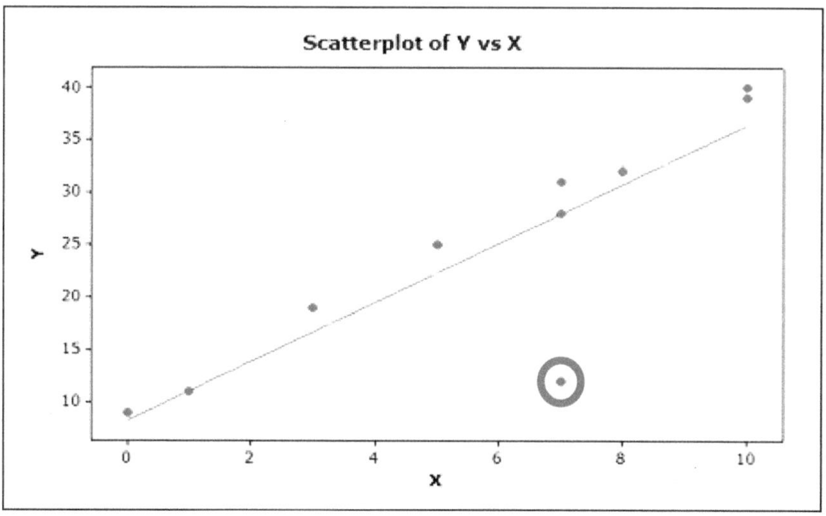

Figure 1.2 Scatter plot of Y on X.

In a regression problem, there are at least three ways of outliers.

X-Outlier: This is a point that is outlying in regard to the x-coordinate. In the literature, an X-outlier is more popularly known as a high leverage point.

Y-Outlier: This is a point that is outlying only because its y-coordinate is extreme.

X- and Y-Outlier: A point that is outlying in both x and y coordinates is known as X- and Y-outlier.

In addition to these, a regression outlier is a point that deviates from the linear relationship determined from the other points, or at least from the majority of those points. In a regression problem, observations are judged as outliers on the basis of how unsuccessful the fitted regression equation is in accommodating them and that is why observations corresponding to excessively large residuals are treated as outliers. Srikantan (1961) led the way of identifying outliers in linear regression using scaled versions of residuals followed by Daniel and Wood (1971), Ellenberg (1976), Cook and Weisberg (1982), and many others. Thes e methods are successful in the identification of a single outlier, but they are ineffective in the presence of multiple outliers. Robust regression methods are very successful in the identification of multiple outliers. The least median of squares proposed by Rousseeuw (1984), the least trimmed squares and the reweighted least squares, proposed by Rousseeuw and Leroy (1987), are very effective. The only issue of robust regression method is that sometimes they are too sensitive and identify too many observations as outliers. To overcome this problem, the diagnostic-robust generalized Studentized residual (GSR) proposed by Imon (2005) could be very useful.

For the identification of X-outliers, Hoaglin and Welsch (1978), and Vellman and Welsch (1981) suggested some methods which are very effective in the identification of a single X-outlier or high leverage point, but not in the presence of multiple X-outliers. Hadi (1992) proposed a method, called potentials, based on a single case deletion. Extending his idea to a group deletion study Imon (2002) developed a diagnostic-robust approach of identifying multiple X-outliers called generalized potentials (GP). Some graphical displays are suggested for the simultaneous identification of outliers and high leverage points. Among them the leverage-residual plot proposed by Gray (1986) and the potential-residual plot proposed by Hadi (1992) have become popular. A diagnostic-robust version of this type of graphical display, called GP-GSR plot, was proposed by Davies, Imon and Ali (2004). A numerical method for the simultaneous identification of multiple outliers and high leverage points was proposed by Imon and Ali (2005). The simultaneous identification of outliers and collinearity is studied by Chatterjee and Hadi (1988), Bagheri, Midi and Imon (2012), and others.

Another important type of unusual observations that warrant extra attention are known as influential observations. To quote Belsley, Kuh and Welsch (1980), *"An influential observation is one which, either individually or together with several other observations, has a demonstrably larger impact on the calculated values of various estimates... than is the case for most of the other observations.*" In this situation parameter estimates or predictions may depend more on the influential observations than on the majority of the data and their omission from the data may result in substantial changes to important features of an analysis. Cook (1977) suggested considering both the location of the point and the response variable in measuring influence. He proposed use of the distance measure which is known in the statistical literature as Cook's distance. Another very popular diagnostics used in measuring influence is difference in fits (DFFITS) proposed by Belsley, Kuh and Welsch (1980). Likewise, the outliers and leverage measures, Cook's distance and DFFITS are not effective in the identification of multiple influential observations. Some robust versions of these two methods are available in the literature, but they have some issues too. The diagnostic-robust generalized DFFITS proposed by Imon (2005) has become very effective.

Outlier detection methods described for the linear regression models can be applied to the experimental design models (Montgomery 2019). Most of them are also readily applicable to the generalized linear model (GLM) problems. The key features of a GLM is here the random errors can come from any probability distributions (not only normal) and instead of the least squares method parameters are estimated by the maximum likelihood method. Pearson residuals, analogous to the Studentized residuals in a linear model, are frequently used for the identification of outliers. For multiple outliers, diagnostic-robust generalized Studentized Pearson residuals (GSPR) proposed by Imon and Hadi (2008) have become very useful. Leverage measures used for a linear regression model are not very effective in the identification of multiple *X*-outliers in a GLM (Hosmer and Lemeshow 2000). The deletion distance from the median (DDM) proposed by Imon and Hadi (2013) have become very useful in this regard. Cook's distance and DFFITS can be used for the identification of an influential observation for a GLM, but in case of multiple influential cases they are not very effective. The generalized difference in fits (GDFFITS) proposed by Nurunnabi et al. (2010) is very useful in this regard.

Outlier detection methods used for the linear regression models are also applicable to the time series models Montgomery et al 2008). Fox (1972) introduced four special types of outliers for time series data. They are additive outlier (AO), innovational outlier (IO), level shifts outlier (LSO) and transitory change outliers (TCO). Fox (1972) developed methods for the identification of these four types of outliers. Few of them were further modified by Tsay (1988). A large body of literature is now available to study the impact of outliers in the estimation of parameters of ARIMA (Denby and Martin 1979, Peña 1987, 1990,

Peña and Maravall 1991, Peña et al. 2001, Bianco et al. 2001, McQuarrie and Tsai 2003). Outliers can seriously distort the autocorrelation structure of the time series (Chernick et al. 1982). If the outliers are ignored and left in the time series, they may seriously bias the autocorrelation function (ACF) and partial autocorrelation function (PACF) of the series (Mills and Prasad 1992). Not much attention has been focused on the issues of the identification of outliers in non linear time series data until quite recently. However, an excellent review of outliers in non linear time series is available in Gounder et al. (2007). A very popular approach is, however, to use the robustified estimation techniques of linear models or linear time series models to identify outliers in non linear time series data (Rousseuw and Leroy 1987, Barnett and Lewis 1994, Gabr 1998). Some other useful techniques are developed recently for ARCH, GARCH, and other similar models (Franses and van Dijk 2000, Peña et al. 2001, Battaglia and Orfei 2005, Tsay 2002, Charles and Darne 2005, Doornik and Ooms 2005) to handle this problem. Imon et al. (2007) introduced a new type of observations, called ARCH-influential observations, which are responsible for inducing or breaking the ARCH effect in the data.

1.5 OUTLIERS IN BIG DATA

This is the era of big data and big data poses big challenges including the understanding of the concept of outliers and their identifications. In statistical data, the concept of outliers is global. Outliers are the observations which fail to match with a global pattern (parent distribution). But most of the times big data does not obey any common global pattern and observations may come in clusters where the concept of majority and minority simply do not work as shown in Figure 1.3 taken from Imon and Hadi (2021). In this figure, it is not

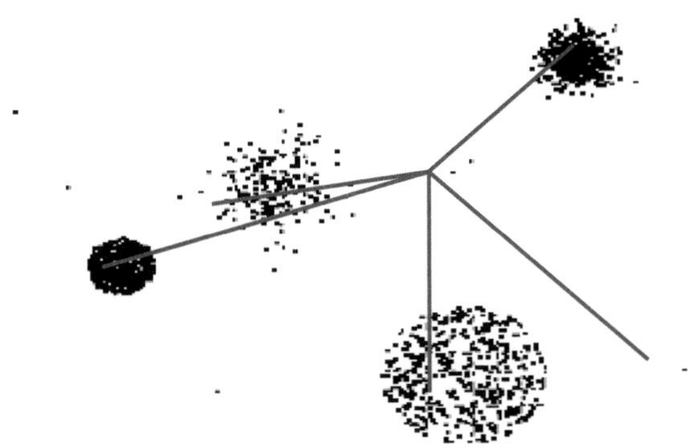

Figure 1.3 Outliers in big data.

easy to tell which portion of the data is major. However, there exists four major clusters here and in addition to those four clusters, there exists few scatter data points which fail to accommodate any of the four clusters.

Given the massive datasets that have recently become available, it is not surprising that the machine learning community has also taken an interest in outlier detection. An extensive comparative review is provided by Hodge and Austin (2004). Most of these methods can be classified as distance- and density-based methods. Cluster based outlier detection methods are widely used nowadays (Aggarwal and Yu 2001). Among the cluster-based outliers, the DB(ε, π)-outliers and grid-based outliers, index-based outliers, nested-loop-based outliers proposed by Knorr and Ng (1997, 1998), the local outlier factor (LOF) proposed by Breunig et al. (1999), *k*–nearest neighborhood approach proposed by Ramaswamy et al. (2000), resolution-based outlier factor (ROF) proposed by Fan et al. (2006) have become popular. Aggarwal and Yu (2001) pointed out that the distance between any pair of data points in high-dimensional space is so similar that either each data point or none of the data points can be viewed as an outlier if the concept of proximity is used to define outliers. As a result, using the traditional Euclidean distance function one cannot effectively get outliers in a high-dimensional dataset due to the averaging behavior of the noisy and irrelevant dimensions. In addition to the methods described above there are several tree based unsupervised learning algorithms like "isolation forest" proposed by Liu et al. (2008) to identify both univariate and multivariate outliers. Since the concept of spatial outliers is local in nature, most of the big data outlier detection techniques can be readily applied to spatial data (Hadi and Imon 2019). The issue of robustness of spatial outlier methods in the presence of multiple outliers is discussed by Filzmoser et al. (2014). Support vector machine (SVM) analysis is a popular machine learning tool for classification and regression, first introduced by Vapnik (1995). A special type of SVM designed for a regression problem is referred to as support vector regression (SVR). In a linear regression, we try to obtain the "line of best fit" by minimizing the sum of squares of the differences between the observed and fitted observation which we call errors. This fitting method is popularly known as the "method of least squares (of errors)". In SVR, our main objective is to fit the best possible 'hyperplane' through the observed points. Observations that lie on the margin of this hyperplane form a vector, called a support vector. Thus, an SVR is the method of fitting the best hyperplane through the data points using support vectors. Dhhan et al. (2017) proposed a method of identifying outliers using the support vector regression.

Another wide area of applications of big data is the gene expression data that are useful for cancer study. Tomlins et al. (2005) led the way by developing the cancer outlier profile analysis (COPA) method for detecting cancer genes which are differentially expressed in a subset of disease samples. COPA statistic uses a fixed sample percentile, which is determined by users. This limitation was overcome by the outlier sums (OS) statistic defined by Tibshirani and Hastie (2007). Wu (2007) modified the OS statistic by changing the definition of the upper fence (Third quartile + 1.5 × Interquartile Range) used in a box plot for the identification of outliers. His proposed method is known as outliers' robust *t* (ORT) statistic. Lian (2008) argued that OS and ORT statistics used arbitrary outliers and proposed the maximum ordered subset *t* (MOST) statistic which considers all possible values for outlier thresholds. Wang and Rekaya (2010) proposed a new method named least sum of ordered subset square *t*-statistic (LSOSS). In LSOSS, mean expression values in normal and cancer samples were considered instead of median expression values. Farazi and Imon (2021) proposed expressed robust *t* (ERT) method for the same.

1.6 CONCLUSIONS

Successful identification of outliers has a very close connection with how outliers are defined. For regular data the concept of outlier is global and we can employ a variety of diagnostic tools to identify outliers. But most of the diagnostic tools become ineffective in the presence of multiple outliers. Classical estimators such as the mean and standard deviation are not suitable for data containing outliers and can cause the statistical analysis to produce results exactly opposite to the correct conclusions. Thus, robust statistical techniques that can be computed in a reasonable time should be used if outliers are thought to be present. Although robust techniques can successfully identify genuine outliers, they are too prone to declare observations as outliers when they are not. To overcome this issue, a diagnostic-robust approach could be very useful. For multivariate data, projection pursuit aims to find the best projections that reveal the outliers in a highly visible placement. Such approaches can find outliers in a wide variety of configurations since the original placement of outliers is transformed to more informative projections. However, such methods tend to be very computationally intensive and are not currently suitable for big data. Distance-based methods are based on obtaining robust estimates of the mean and covariance matrix so that a robust Mahalanobis distance can be computed for each point. Principal components are generally recommended for high dimensional data, but other approaches may be promising too. In big data, the concept of outlier is local and we need to find new strategies to identify them. Handling of outliers is another important issue. Most of the statistics practitioners use data as it appears, they do not even check for outliers. Those who do some outlier analysis, usually discard all suspect outliers. But outliers are not necessarily wrong information, rather they could be the most informative part of the dataset. Instead of discarding outliers we can use some inferential method which are less sensitive to outliers. Wrong choice of model can be responsible for outliers as well. Proper selection of models can reduce the risk of outliers as well. Current research in this area is the detection and handling of outliers in big data with computationally less extensive methods.

REFERENCES

Ali, M.M. and Woo, J. 2004. Effects of an outlier for estimation in a uniform distribution. Journal of Information and Optimization Sciences. 25(3): 521–532.

Ali, M.M., Woo, J. and Yoon, G. 2000. Effects of an unidentified outlier for parametric estimation in a power function distribution. Journal of Statistical Research. (A Special Volume in Honor of the 80th Birthday of Professor C.R. Rao). 34(2): 91–98.

Aggarwal, C.C. and Yu, P.S. 2001. Outlier detection for high dimensional data, In: Proceedings of the 2001 ACM SIGMOD International Conference on Management of Data. 37–46.

Bagheri, A., Habshah, M. and Imon, A.H.M.R. 2012. A novel collinearity-influential observation diagnostic measure based on a group deletion approach. Communications in Statistics—Simulation and Computation. 41(8): 1379–1396.

Balasooriya, U. 1989. Detection of outliers in the exponential distribution based on prediction. Communications in Statistics—Theory and Methods. 18: 711–720.

Barnett, V. and Lewis, T.B. 1994. Outliers in Statistical Data, 3rd Edn. Wiley, New York.

Belsley, D.A., Kuh, E. and Welsch, R.E. 1980. Regression Diagnostics: Identifying Influential Data and Sources of Collinearity. Wiley, New York.

Battagila, F. and Orfei, L. 2005. Outlier detection and estimation in nonlinear time series. Journal of Time Series Analysis. 26: 107–121.

Bianco, A.M., Ben, M.G., Martinez, E.J. and Yohai, V.J. 2001. Outlier detection in regression models with ARIMA errors using robust estimates. Journal of Forecasting. 20: 565–579.

Breunig, M.M., Kriegel, H.P., Ng, R. and Sander, J.R. 1999. OPTICS-OF: Identifying local outliers. In: Proceedings of the Third European Conference on Principles of Data Mining and Knowledge Discovery. 262–270.

Charles, A. and Darne, O. 2005. Outliers and GARCH models in financial data. Economics Letters. 86: 347–352.

Chatterjee, S. and Hadi, A.S. 1988. Sensitivity Analysis in Linear Regression. Wiley, New York.

Chernick, M.R., Downing, D.J. and Pike, D.H. 1982. Detecting outliers in time series data. Journal of the American Statistical Association. 77: 743–747.

Cook, R.D. 1977. Detection of influential observations in linear regression. Technometrics. 19: 15–18.

Cook, R.D. and Hawkins, D.M. 1990. Comment on "Unmasking Multivariate Outliers and Leverage Points," by P.J. Rousseeuw and B.C. van Zomeren. Journal of the American Statistical Association. 85: 640–644.

Cook, R.D. and Weisberg, S. 1982. Residuals and Influence in Regression. Chapman and Hall, London.

D'Agostino, R.B. and Stephens, M.A. (Eds.) 1986. Goodness-of-fit Techniques, Marcel Dekker, New York.

Daniel, C. and Wood, F.S. 1971. Fitting Equations to Data. Wiley, New York.

Davies, P., Imon, A.H.M.R. and Ali, M.M. 2004. A Conditional expectation method for improved residual estimation and outlier identification in linear regression. International Journal of Statistical Sciences. 3: 191–208.

Denby, L. and Martin, R.D. 1979. Robust estimation of the first order autoregressive parameter. Journal of the American Statistical Association. 74: 140–146.

Dhhan, W., Rana, S., Alshaybawee, T. and Midi, H. 2017. The Single-index support vector regression model to address the problem of high dimensionality. Communications in Statistics—Simulation and Computation. 47: 2792–2799.

Dixit, U.J., Ali, M.M. and Woo, J. 2003. Efficient estimation of parameters of a uniform distribution in the presence of outliers. Soochow Journal of Mathematics. 29(4): 363–369.

Dixon, W.J. 1950. Analysis of extreme values. The Annals of Mathematical Statistics. 21: 488–506.

Doornik, J.A. and Ooms, M. 2005. Outlier detection in GARCH models, 2005-W24 In: Nuffield Economics Working Papers. Nuffield College, Oxford.

Ellenberg, J.H. 1976. Testing for a single outlier from a general regression. Biometrics. 32: 637–645.

Fan, H., Zaïane, O.R., Foss, A. and Wu, J. 2006. A nonparametric outlier detection for efficiently discovering top-n outliers from engineering data. In: Proceedings of Pacific-Asia Conference on Knowledge Discovery and Data Mining. Singapore. 557–566.

Farazi, M.M.R. and Imon, A.H.M.R. 2021. Identification of outliers in gene expression data, Chapter 11. In: Data Science and SDGs: Challenges, Opportunities and Realities, Sinha, B.K. and Mollah, M.N.H. (Eds.), Springer. 135–145.

Farman, J.C., Gardinar, B.G. and Shanklin, J.D. 1985. Large losses of total ozone in antarctica reveal seasonal ClOx/NOx interaction. Nature. 315: 207–210.

Filzmoser, P., Maronna, R.A. and Werner, M. 2008. Outlier identification in high dimensions. Computational Statistics and Data Analysis. 52: 1694–1711.

Filzmoser, P., Ruiz-Gazen, A. and Thomas-Agnan, C. 2014. Identification of local multivariate outliers. Statistical Papers. 55: 29–47.

Fox, A.J. 1972. Outliers in time series. Journal of the Royal Statistical Society, Series–B. 34: 350–363.

Franses, P.H. and van Dijk, D. 2000. Nonlinear Time Series Models in Empirical Finance. Cambridge University Press, Cambridge.

Friedman, J.H. and Tukey, J.W. 1974. A projection pursuit algorithm for exploratory data analysis. In: IEEE Transactions on Computers. 23(9): 881–889.

Gabr, M.M. 1998. Robust estimation of bilinear time series models. Communications in Statistics–Theory and Methods. 27: 41–53.

Gounder, M.K., Shitan, M. and Imon, A.H.M.R. 2007. Detection of outliers in non linear time series: A review. Festschrifts in Honor of Professor Mir Masoom Ali, Department of Mathematical Sciences, Ball State University, USA. 213–224.

Gray, J.B. 1986. A simple graphic for assessing influence in regression. Journal of Statistical Computation and Simulation. 24: 121–134.

Hadi, A.S. 1992. A new measure of overall potential influence in linear regression. Computational Statistics and Data Analysis. 14: 1–27.

Hadi, A.S. and Imon, A.H.M.R. 2019. Identification of multiple outliers in spatial data. International Journal of Statistical Sciences. 16: 87–96.

Hadi, A.S. and Simonoff, J.S. 1993. Procedures for the identification of outliers in linear models. Journal of the American Statistical Association. 88: 1264–1272.

Hadi, A.S., Imon, A.H.M.R. and Werner, M. 2009. Detection of outliers. Wiley Interdisciplinary Reviews, Computational Statistics. 1: 57–70.

Hampel, F.R., Ronchetti, E.M., Rousseeuw, P.J. and Stahel, P.J. 1986. Robust Statistics: The Approach Based on Influence Function. Wiley, New York.

Hawkins, D.M. 1980. Identification of Outliers, Chapman and Hall, London.

Hawkins, D.M., Bradu, D. and Kass, G.V. 1984. Location of several outliers in multiple regression data using elemental sets. Technometrics. 26: 197–208.

Hoaglin, D.C. and Welsch, R.E. 1978. The hat matrix in regression and ANOVA. The American Statisticians. 32: 17–22.

Hodge, V.J. and Austin, J. 2004. A survey of outlier detection methodologies. Artificial Intelligence Reviews. 22: 85–126.

Hosmer, D.W. and Lemeshow, S. 2000. Applied Logistic Regression, 2nd Edn. Wiley, New York.

Hubert, M., Rousseeuw, P.J. and Branden, K.V. 2005. ROBPCA: A new approach to robust principal components analysis. Technometrics. 47: 64–79.

Imon, A.H.M.R. 2002. Identifying multiple high leverage points in linear regression. Journal of Statistical Studies, Special Volume in Honor of Professor Mir Masoom Ali. 22: 207–218.

Imon, A.H.M.R. 2005. Identifying multiple influential observations in linear regression. Journal of Applied Statistics. 32(9): 929–946.

Imon, A.H.M.R. 2008. Diagnostic robust approach of outlier detection in regression. Journal of Statistical Research. 42(2): 105–120.

Imon, A.H.M.R. and Ali, M.M. 2005. Simultaneous identification of multiple outliers and high leverage points in linear regression. Journal of Korean Data Information. Science. Society. 16(2): 429–444.

Imon, A.H.M.R., Doula, M.S. and Hamzah, N.A. 2007. On the detection of ARCH effect in time series data. In: Proceedings of an International Conference on Mathematical Sciences on Integrating Mathematical Sciences within Society. Bangi–Putrajaya, Malaysia. 783–789.

Imon, A.H.M.R. and Hadi, A.S. 2008. Identification of multiple outliers in logistic regression. Communications in Statistics—Theory and Methods. 37(11): 1697–1709.

Imon, A.H.M.R. and Hadi, A.S. 2013. Identification of multiple high leverage points in logistic regression. Journal of Applied Statistics. 40(12): 2601–2616.

Imon, A.H.M.R. and Hadi, A.S. 2021. Identification of multiple unusual observations in spatial regression. Journal of Statistics and Applications. 18(2): 155–162.

Knorr, E. and Ng, R. 1997. A unified notion of outliers: properties and computation. In: Proceedings of the International Conference on Knowledge Discovery and Data Mining. 219–222.

Knorr, E. and Ng, R. 1998. Algorithms for mining distance-based outliers in large datasets. In: Proceedings of 24th International Conference on Very Large Data Bases. 392–403.

Lian, H. 2008. MOST: Detecting cancer differential gene expression. Biostatistics. 9: 411–418.

Liu, F.T., Ting, K.M. and Zhou, Z. 2008. Isolation forest. In: Eighth IEEE International Conference on Data Mining. 413–422.

Locantore, N., Marron, J.S., Simpson, D.G., Tripoli, N. and Zhang, J.T. 1999. Robust principal component analysis for functional data. Test. 8(1): 1–73.

Maronna, R.A. and Yohai, V.J. 1995. The behavior of the Stahel-Donoho robust multivariate estimator. Journal of the American Statistical Association. 90(429): 330–341.

McQuarrie, A.D. and Tsai, C-L. 2003. Outlier detections in autoregressive models. Journal of Computational and Graphical Statistics. 12: 450–471.

Mills, J.A. and Prasad, K. 1992. A comparison of model selection criteria. Econometric Reviews. 11: 201–233.

Montgomery, D.C. 2019. Design and Analysis of Experiments, 10th Edn. Wiley, New York.

Montgomery, D.C., Jennings, C. and Kulachi, M. 2008. Introduction to Time Series Analysis and Forecasting. Wiley, New York.

Nurunabi, A.A.M., Imon, A.H.M.R. and Nasser, M. 2010. Identification of multiple influential observations in logistic regression. Journal of Applied Statistics. 37(10): 1605–1624.

Pearson, E.S. and Chandra Sekar, C. 1936. The efficiency of statistical tools and a criterion of rejection of outlying observations. Biometrika. 28: 308–319.

Pearson, K. 1905. On the general theory of skew correlation and non-linear regression. Biometrika. 4: 171–212.

Peirce, B. 1852. Criterion for the rejection of doubtful observations. Astronomical Journal. 45(2): 161–163.

Peña, D. 1987. Measuring the importance of outliers in ARIMA models. In: New Perspective in Theoretical and Applied Statistics, Vilaplana, J.P., Wretz, W. and Puri, M.L. (Eds.), Wiley, New York. 109–118.

Peña, D. 1990. Influential observations in time series. Journal of Business and Economic Statistics. 8: 241.

Peña, D. and Maravall, A. 1991. Interpolations, outliers and inverse autocorrelations. Communications in Statistics—Theory and Methods. 20: 3175–3186.

Peña, D., Tiao, C. and Tsay, S. 2001. A Course in Time Series Analysis. Wiley, New York.

Peña, D. and Prieto, F. 2001. Multivariate outlier detection and robust covariance matrix estimation. Technometrics. 43(3): 286–310.

Ramaswamy, S., Rastogi, R. and Shim, K. 2000. Efficient algorithms for mining outliers from large datasets. In: Proceedings of the ACM SIGMOD International Conference on Management of Data. 427–438.

Rocke, D.M. and Woodruff, D.L. 1996. Identification of multivariate outliers. Journal of the American Statistical Association. 91(435): 1047–1061.

Rousseeuw, P.J. 1984. Least median of squares regression. Journal of the American Statistical Association. 79: 871–880.

Rousseeuw, P.J. 1985. Multivariate estimation with high breakdown point. In: Grossmann, W., Pflug, G., Vincze, I. and Wertz, W. (Eds.), Mathematical Statistics and Applications, Vol. B. Reidel Publishing Company, Dordrecht. 283–297.

Rousseeuw, P.J. and Leroy, A. 1987. Robust Regression and Outlier Detection. Wiley, New York.

Rousseeuw, P.J. and van Driessen, K. 1999. A fast algorithm for the minimum covariance determinant estimator. Technometrics. 41(3): 212–223.

Rousseeuw, P.J., Ruts, I. and Tukey, J.W. 1999. The Bagplot: A bivariate box plot. The American Statisticians. 53: 382–387.

Srikantan, K.S. 1961. Testing for the single outlier in a regression model. Sankhya, A. 23: 251–260.

Stigler, S.M. 1973. Simon Newcombe, Percy Daniell, and the history of robust estimation 1885–1920. Journal of the American Statistical Association. 68: 872–879.

Tibshirani, R. and Hastie, T. 2007. Outlier sums for differential gene expression analysis. Biostatistics. 8: 2–8.

Tomlins, S.A., Rhodes, D.R. and Perner, S. 2005. Recurrent fusion of TMPRSS2 and ETS transcription factor genes in prostate cancer. Science. 310: 644–648.

Tsay, R.S. 1988. Outliers, level shifts and variance changes in time series. Journal of Forecasting. 7: 1– 20.

Tsay, R.S. 2002. Analysis of Financial Time Series. Wiley, New York.

Vapnik, V. 1995. The Nature of Statistical Learning Theory. Springer, New York.

Velleman, P.F. and Welsch, R.E. 1981. Efficient computing of regression diagnostics. The American Statistician. 35: 234–242.

Wang, Y. and Rekaya, R. 2010. LSOSS: Detection of cancer outlier differential gene expression. Biomarker Insights. 5: 69–78.

Woo, J. and Ali, M.M. 1996. Parametric estimation of two-parameter exponential model in the presence of unidentified outliers. Journal of Information and Optimization. Sciences. 17: 57–63.

Wu, B. 2007. Cancer outlier differential gene expression detection. Biostatistics. 8: 566–575.

Chapter **2**

Some Recent Developments in the Identification of Outliers in Spatial Data and Spatial Regression

Ali S. Hadi[1] and A.H.M. Rahmatullah Imon*[2]

[1]Department of Mathematics and Actuarial Science, The American University in Cairo, Egypt

[2]Department of Mathematical Sciences, Ball State University, Muncie, Indiana, USA

2.1 INTRODUCTION

Awareness of outliers in some form or another has existed for at least several hundred years. There are numerous definitions of outliers in the statistical literatures (Barnett and Lewis 1994, Hadi et al. 2009). A commonly used definition is that outliers are the minority of observations in a dataset that have different patterns than the majority of observations. The assumption here is that there is a core of at least 50% of observations in a dataset that are homogeneous (that is, represented by a common pattern) and the remaining observations (hopefully few) have patterns that are inconsistent with this common pattern. It is now evident that the presence of outliers can lead to wrong inferences. Although outliers

*For Correspondence: A.H.M. Rahmatullah Imon (rimon@bsu.edu)

could be easily identified in univariate, bivariate, or even trivariate data through graphical examination of the data, but this approach is very subjective. Moreover, visual inspection does not usually work for more than three dimensions. Automated identification of outliers is tricky. Even the repeated use of this test may not work and in the end the genuine outliers may be left undetected (a problem known as masking) and/or some of the non outlying observations may be incorrectly declared as outliers (a problem known as swamping). Identification of outlying data points is often by itself the primary goal, without any intention of fitting a statistical model. The outliers themselves are points of primary interest, drawing attention to unknown aspects of the data, or especially if unexpected, leading to new discoveries.

Detection and handling of anomalous spatial patterns or spatial outliers have received a great deal of attention in recent years. The identification of spatial outliers is important because it can reveal hidden but valuable knowledge in many applications such as identifying aberrant genes or tumour cells, discovering highway traffic congestion points, locating extreme meteorological events such as tornadoes, hurricanes, etc. Conceptually spatial outliers are very different from classical outliers. According to Shekhar et al. (2002) spatial outliers are those observations whose characteristics are markedly different from their spatial neighbours. Thus, in spatial data the concept of outlier is local, not global and hence warrants different identification techniques. An excellent review of different aspects of spatial outliers is available in Shekhar et al. (2002) and Hadi and Imon (2019). It is now evident that likewise the regular outliers, spatial outliers may be affected by masking and swamping in the presence of multiple outliers. Things could be more cumbersome in regression models where outliers can occur along the y-dimension (outlier), or along the x-dimension (high leverage points) or both. Moreover, these methods are not readily applicable for spatial regression. For example, temperatures and amount of rainfall of different regions may vary due to their distances from sea or mountain. Once we fit this relationship by regression we may observe not only strange temperature or rainfall patterns, the distance factor may also be unusual. Attempts have been made to identify outliers based on residuals but it only focuses on the outliers in y, but not in x or both and the whole concept is rather global than local. In this book chapter we have discussed some recently developed techniques to overcome these problems. These techniques are the identification of multiple spatial outliers in general developed by Hadi and Imon (2019) and the identification of multiple outliers in spatial regression developed by Imon and Hadi (2021).

2.2 IDENTIFICATION OF OUTLIERS IN SPATIAL DATA

A good number of spatial outlier detection methods are now available in the literature. These methods can be generally grouped into two categories, namely graphic approaches and quantitative tests. An excellent review of graphic approaches for the identification of spatial outliers is available in Shekhar et al. (2002). Graphic approaches are based on visualization of spatial data which highlights spatial outliers. These methods include variogram clouds (Haslett et al. 1991), pocket plots (Panatier 1996), spatial scatter plot (Haining 1993), and Moran scatter plot (Anselin 1996). The commonly used '*k*-means' clustering method may not work here. But some other distance-based methods such as DB(ε, π)-outliers and grid-based outliers, index-based outliers, nested-loop-based outliers (Knorr and Ng 1997, 1998), *k*-nearest neighborhood approach (Ramaswamy et al. 2000), resolution-based outlier factor (ROF) (Fan et al. 2006) techniques can be applied there. Distance-based outlier detection models have problems with different densities. These methods cannot compare the neighborhood of points from areas of different densities. To overcome this problem some density-based outlier detection methods are suggested. Among them, local outlier factor (LOF) proposed by Breunig et al. (1999) have become very popular. Following this method Huang and Qin (2004) proposed spatial outlier factor (SOF) which is designed to identify multifactor spatial outliers. For detecting outliers with multiple attributes, traditional outlier detection approaches could not be used properly due to the sparsity of the data objects in high- dimensional data space. Aggarwal and Yu (2001) show that the distance between any pair of data points in high-dimensional space is so similar that either each data point or no data point can be viewed as an outlier if the concept of proximity is used to define outliers. As a result, using traditional Euclidean distance function cannot effectively get outliers in high-dimensional dataset due to the averaging behavior of the noisy and irrelevant dimensions. The issue of robustness of spatial outlier methods in the presence of multiple outliers is discussed by Lu et al. (2004) and Filzmoser et al. (2014). These two approaches are based on Mahalanobis or robust distances computed in each neighborhood using a common estimation of the covariance matrix.

All the methods discussed above are effective in the identification of single spatial outlier but they may suffer a setback in the presence of multiple spatial outliers. We present a dataset in Table 2.1 taken from Hadi and Imon (2019). Although this data is artificial in nature, the use of this type of data is very common in the outlier detection literature (Rousseeuw and Leroy 1987, Hadi et al. 2009) because here we definitely know which observations are genuine outliers. For real data with multiple outliers due to masking and swamping there could always be lots of disagreements regarding which observations are genuine outliers or not.

Table 2.1 Multiple spatial outliers data

Index	Location	Attributes	Index	Location	Attributes
1	1.0	2.0	16	7.1	5.0
2	2.0	3.0	17	7.3	4.4
3	2.1	3.2	18	7.5	4.3
4	2.6	**7.0 C**	19	7.7	**6.9 E**
5	3.0	4.0	20	8.0	2.8
6	3.8	5.0	21	8.4	2.1
7	3.9	5.6	22	9.0	1.0 B
8	4.0	5.7	23	9.2	2.1
9	4.2	**1.6 D**	24	10.0	2.7
10	4.5	6.0	25	10.1	3.2
11	5.0	6.2	26	11.0	4.0
12	6.0	8.0 A	27	15.0	4.1
13	6.2	6.3	28	17.0	4.2
14	6.4	6.1	29	19.0	4.3
15	6.7	5.5	30	20.0	4.4

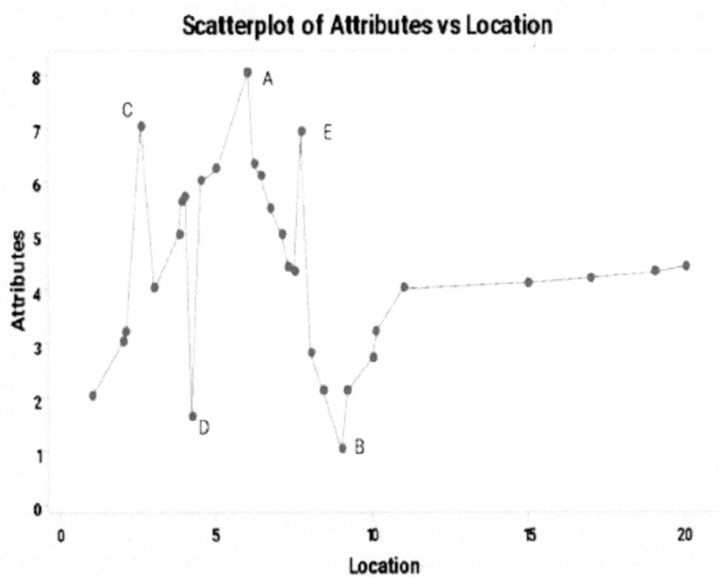

Figure 2.1 Plot of attribute values against locations for multiple outliers.

Attribute values are plotted against their locations in Figure 2.1. For global outliers, traditional statistics will essentially look at the attribute values in the y axis and the largest point such as A and/or the smallest point such as B would be suspect outliers. Although A and B are the largest and smallest observations respectively they are not too far from their neighbouring points. However, there are few other points such as C, D, and E which are too far from their neighbors but they are not global outliers because their values are moderate. Now in quest of checking the neighborhood stability we take the first-order difference of attributes. These differences are presented in Table 2.2 and they are plotted against their location in Figure 2.2.

Table 2.2 First order difference in attributes for multiple spatial outliers data

Index	Difference in Attributes	Index	Difference in Attributes	Index	Difference in Attributes
1	*	11	0.2	21	0.7
2	1.0	12	1.8 A	22	1.1 B
3	0.2	13	1.7 A	23	1.1 B
4	**3.8 C**	14	0.2	24	0.6
5	**3.0 C**	15	0.6	25	0.5
6	1.0	16	0.5	26	0.8
7	0.6	17	0.6	27	0.1
8	0.1	18	0.1	28	0.1
9	**4.1 D**	19	**2.6 E**	29	0.1
10	**4.4 D**	20	**4.1 E**	30	0.1

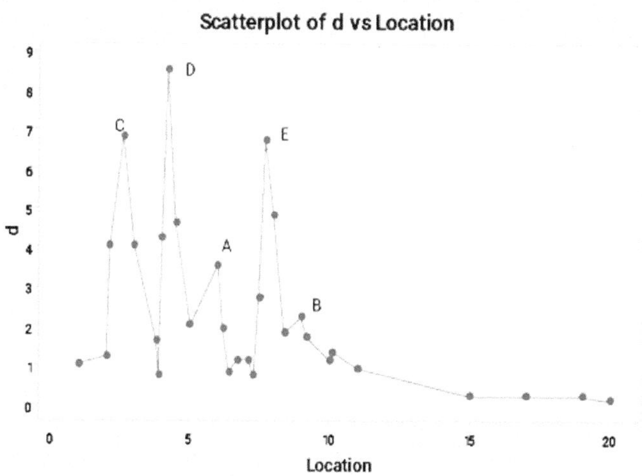

Figure 2.2 Plot of differenced attribute values against locations for multiple outliers.

We observe from Table 2.2 and Figure 2.2 that the first-order difference values for the largest and the smallest data, denoted by A and B, respectively, are not spatial outliers because the magnitudes of their difference from the neighbors are small. In contrast to that, the spatial outliers are like the spikes associated with points C, D, and E which are very different from their neighbors. Clearly C, D, and E look like spatial outliers because they violate the law of geography that the nearby things should be very similar. However, graphical methods are very subjective in nature. For this reason, we need to employ a quantitative test to confirm our suspicion regarding spatial outliers.

Shekhar et al. (2002) proposed a z test on the difference of attributes that we call a spatial z test. For a set of observations $x_1, x_2, \ldots, x_n$ let us define a set of differenced attributes (in absolute values) $D_1, D_2, \ldots, D_n$ where

$$D_i = |x_{i+1} - x_i| \text{ for } i = 1, 2, \ldots, n. \tag{1}$$

Under a normal assumption, a single value may be considered as an outlier if it falls outside a certain range of the standard deviation. A traditional measure of the 'outlyingness' of an observation D_i with respect to a sample is the ratio between its distance to the sample mean and the sample standard deviation

$$z_i = \frac{D_i - \underline{D}}{s_D} \text{ for } i = 1, 2, \ldots, n. \tag{2}$$

z-values defined in (2) will be called spatial z scores since they are calculated on spatial differences. Observations with $|z_i| > 3$ are traditionally deemed as suspicious (the three-sigma rule), based on the fact that they would be very unlikely under normality, since $P(|z| > 3) = 0.003$ for a random variable z with a standard normal distribution.

Table 2.3 Spatial z scores for multiple spatial outliers

Index	Spatial z_i	Index	Spatial z_i	Index	Spatial z_i
1	*	11	−0.75343	21	−0.38927
2	−0.17078	12	0.41188	22	−0.09795
3	−0.75343	13	0.33905	23	−0.09795
4	1.86851	14	−0.75343	24	−0.46211
5	1.28586	15	−0.46211	25	−0.53494
6	−0.17078	16	−0.53494	26	−0.31644
7	−0.46211	17	−0.46211	27	−0.82627
8	−0.82627	18	−0.82627	28	−0.82627
9	2.08701	19	0.99453	29	−0.82627
10	2.30551	20	2.08701	30	−0.82627

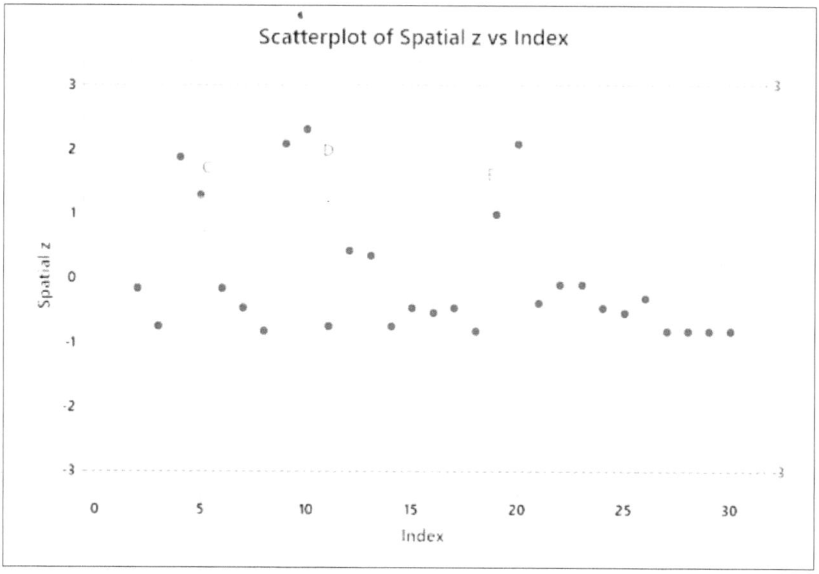

Figure 2.3 Index plot of spatial z scores for multiple spatial outliers.

For the above example we calculate the spatial z scores for the difference of attributes and those are presented in Table 2.3 and Figure 2.3. Although there are three suspect spatial outliers in this dataset, all z scores are within ±3. Thus we fail to identify any spatial outlier for this data. Since z scores are computed on the differences of attribute values, each point appears in pairs and if it is a genuine outlier the pairs should exhibit unusual patterns. Figure 2.3 shows that neither of the suspect spatial outliers is identified.

2.3 Identification of Spatial Outliers Using Robust Spatial z Test

In this section, we discuss a method developed by Hadi and Imon (2019) to overcome the problems associated with the spatial z scores as defined in (1). They redefined the first-order differences as

$$d_i \begin{cases} |x_{i+1} - x_i| & i = 1 \\ |x_i - x_{i-1}| + |x_i - x_{i+1}| & i = 2, 3, \ldots, n-1 \\ |x_i - x_{i-1}| & i = n \end{cases} \tag{3}$$

The main advantage of this type of distances is that a distance d_i is computed for all n observations, not just for $n - 1$ observations. Another advantage of the use of the distance between one observation and its two spatial neighbours is that, with some human intervention or interpretation, it can be used for the detection of clusters of outliers. A cluster of contiguous outliers can be detected if between two outliers with large distances, the observations in between will have very small distances. This indicates that the two end observations (with large distances) and the ones in between (with very small distances) are all clustered outliers. This can be decided on by the analyst from the index plot of the distances.

The z test is designed for the identification of a single outlier and it is now evident that they often fail to identify multiple outliers. Not only that, this test contains components like mean and standard deviation which can be severely affected in the presence of a single outlier and this distortion could be so huge that the z test may fail to identify even a single outlier. Here we replace mean and standard deviation by their robust counterparts. Sample median is a very effective robust measure of location. For the measure of dispersion, we can use the median absolute deviation (MAD). For the set of differenced attributes d_i as defined in (3) we compute the MAD defined as

$$MAD\ (d_i) = \frac{Median\left\{\left|d_i - Median\, d_i)\right|\right\}}{0.6745},\ \text{for } i = 1, 2, \ldots, n. \tag{4}$$

Here the correction factor 0.6745 is used to make the MAD comparable to the standard deviation in terms of efficiency. The robust spatial z-like statistic is defined as

$$z_i' = \frac{d_i - Median\,(d_i)}{MAD(d_i)},\ \text{for } i = 1, 2, \ldots, n. \tag{5}$$

Observations with $|z_i'| > 3$ will be identified as spatial outliers.

We calculate robust spatial z scores for the difference of attributes for the data given in Table 2.1 and these results are presented in Table 2.4

Table 2.4 Robust spatial z scores for multiple spatial outliers

Index	z_i'	Index	z_i'	Index	z_i'
1	−0.4615	11	0.2485	21	0.1065
2	−0.3195	12	1.3135	22	0.3905
3	1.6685	13	0.1775	23	0.0355
4	**3.6565 C**	14	−0.6035	24	−0.3905
5	1.6685	15	−0.3905	25	−0.2485
6	−0.0355	16	−0.3905	26	−0.5325
7	−0.6745	17	−0.6745	27	−1.0295
8	1.8105	18	0.7455	28	−1.0295
9	**4.8635 D**	19	**3.5855 E**	29	−1.0295
10	2.0945	20	2.2365	30	−1.1005

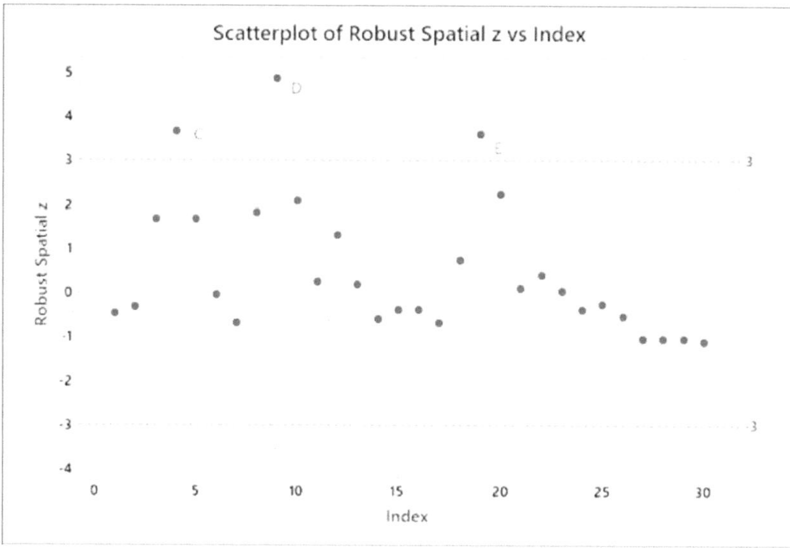

Figure 2.4 Index plot of robust spatial z scores for multiple spatial outliers.

and Figure 2.4. As we mentioned before we have two suspect outliers in this data. Although all of them were undetected by the spatial z scores, results presented in Table 4 clearly show that their corresponding robust spatial z scores exceed the cut off value 3 and hence are identified as outliers. Similar remarks apply with Figure 2.4 where these 3 spatial outliers C, D, E are correctly identified. Thus we can conclude that robust spatial z scores can successfully identify multiple outliers where the traditional spatial z scores fail to do so.

2.4 IDENTIFICATION OF MULTIPLE UNUSUAL OBSERVATIONS IN SPATIAL REGRESSION

Outlier detection methods discussed in Sections 2.2 and 2.3 are designed for regular spatial data, but modeling is an additional issue and therefore we need different outlier detection techniques for spatial regression. Let us assume that we have n pairs of spatial observations (u_i, v_i), $i = 1, 2, \ldots, n$. We further assume that V depends on U and we are interested to investigate their nature of relationship. In order to understand whether spatial observations are stable in their neighborhood, for both U and V we take the first-order differences defined as

$$x_i = u_i - u_{i-1},\ y_i = v_i - v_{i-1};\ i = 2, 3, \ldots, n \tag{6}$$

Based on the differenced observations obtained in (6), let us consider a standard regression model

$$Y = X\beta + \varepsilon \tag{7}$$

where Y is a vector of observed responses of order $(n - 1)$, X is an $(n - 1) \times 2$ matrix of explanatory variables including the constant, β is a vector of unknown finite parameters of order 2 and ε is an n-vector of random disturbances with $E(\varepsilon) = 0$ and $V(\varepsilon) = \sigma^2 I$. The traditionally used ordinary least squares (OLS) estimator of β is $\hat{\beta} = (X^T X)^{-1} X^T Y$ and the vector of fitted values is $\hat{Y} = X\hat{\beta} = HY$. The matrix

$$H = X(X^T X)^{-1} X^T \tag{8}$$

is often referred to as weight or leverage matrix whose diagonal elements h_{ii} are termed leverages. The OLS residual vector $\hat{\varepsilon}$ is defined as $\hat{\varepsilon} = Y - \hat{Y}$. Observations corresponding to exceptionally large $\hat{\varepsilon}$ values are termed outliers. However, the question still remains how large is large? For this reason we often consider the standardized version of residuals. One very popular choice is deleted spatial Studentized residuals (SDSR) defined as

$$t_i = \frac{y_i - x_i^T \hat{\beta}^{(-i)}}{\hat{\sigma}_{(i)} \sqrt{(1 - h_{ii})}}, \; i = 2, 3, \ldots, n \tag{9}$$

where $\hat{\beta}^{(-i)}$ and $\hat{\sigma}_{(i)}$ are the OLS estimates of β and σ respectively with the i-th observation deleted. We call an observation outlier when its corresponding deleted Studentized residual value exceeds 3 in absolute value. Observations corresponding to exceptionally large h_{ii} values are termed high leverage points which are essentially outliers in the X-space. However, since residuals are also functions of leverages, it is better if we identify both outliers and high leverage points simultaneously rather than separately. Gray (1986) proposed the Leverage-Residual (L-R) plot where the leverage value h_{ii} for each observation i, is plotted against the square of a normalized form of its corresponding residual. The bulk of the cases will be associated with low leverage and small residuals so that they cluster near the origin (0, 0). The unusual cases will have either high leverages or large residual components and so will tend to be separated from the bulk of the data. High leverage cases will be located in the upper area of the plot and observations with large residuals will be located in the area to the right.

The L-R plot may be effective in the identification of a single outlier but it may be ineffective in the presence of multiple outliers unless we remove a group of suspect outliers prior to fitting the model. Here we discuss a method suggested by Imon and Hadi (2021). Let us denote a set of cases 'remaining' in the analysis by R and a set of cases 'deleted' by D. Also suppose that R contains $(n - 1 - d)$ cases after $d < (n - 3)$

cases in D are deleted. Without loss of generality, assume that these observations are the last d rows of X and Y so that we can partition the matrices as

$$X = \begin{bmatrix} X_R \\ X_D \end{bmatrix}, \quad Y = \begin{bmatrix} Y_R \\ Y_D \end{bmatrix}, \quad H = \begin{bmatrix} H_R & H_{RD} \\ H_{DR} & H_D \end{bmatrix} \tag{10}$$

where $H_R = X_R(X^TX)^{-1}X_R^T$ and $H_D = X_D(X^TX)^{-1}X_D^T$ are symmetric matrices of order $(n - 1 - d)$ and d respectively, and $H_{RD} = X_R(X^TX)^{-1}X_D^T$ is an $(n - 1 - d) \times d$ matrix. However, $(X_R^TX_R)^{-1}$ can be expressed as

$$(X_R^TX_R)^{-1} = (X^TX - X_D^TX_D)^{-1} = (X^TX)^{-1} + (X^TX)^{-1}X_D^T(I_D - U_D)^{-1}(X^TX)^{-1} \tag{11}$$

where I_D is an identity matrix of order d and $U_D = X_D(X_D^TX_D)^{-1}X_D^T$. Using Eq. (6), Imon (2002) defined a group deleted version of high leverage points called spatial generalized potentials (SGP) defined as

$$p_{ii}^* = \begin{cases} \dfrac{h_{ii}^{(-D)}}{1 - h_{ii}^{(-D)}} & i \in R \\ h_{ii}^{(-D)} & i \in D \end{cases} \tag{12}$$

where $h_{ii}^{(-D)} = x_i^T(X_R^TX_R)^{-1}x_i$, $i = 2, 3, \ldots, n$. In other words, $h_{ii}^{(-D)}$ is the i-th diagonal element of $X(X_R^TX_R)^{-1}X^T$ matrix. The vector of estimated parameters after the deletion of d observations, denoted by $\hat{\beta}^{(-D)}$, is obtained using Eq. (11) as

$$\hat{\beta}^{(-D)} = (X_R^TX_R)^{-1}X_R^TY_R = \hat{\beta} - (X^TX)^{-1}X_D^T(I_D - U_D)^{-1}\hat{\varepsilon}_D \tag{13}$$

where $\hat{\varepsilon}_D = X_D\hat{\beta}$. Using Eqs. (11), (12) and (13), Imon (2005) introduced a group deleted version of residuals called spatial generalized Studentized residuals (SGSR) defined as

$$t_{ii}^* = \begin{cases} \dfrac{y_i - \hat{y}_i^{(-D)}}{\hat{\sigma}^{(-D)}\sqrt{1 - h_{ii}^{(-D)}}} & i \in R \\ \dfrac{y_i - \hat{y}_i^{(-D)}}{\hat{\sigma}^{(-D)}\sqrt{1 - h_{ii}^{(-D)}}} & i \in D \end{cases} \tag{14}$$

where $\hat{y}_i^{(-D)} = x_i^T\hat{\beta}^{(-D)}$ and $\hat{\sigma}^{(-D)}$ are the fitted values of y and the scale parameter σ respectively after the omission of the suspected outlier group indexed by D. Although the expression of spatial generalized potentials is available for any arbitrary set of deleted cases, D, the choice of such a set is clearly important since the omission of this group determines the weights for the whole set. We call an observation outlier when its corresponding spatial generalized Studentized residual value exceeds 3 in absolute value. No such value exists for generalized potentials. We follow Hadi (1992) to declare an observation as a high leverage point if its corresponding p_{ii}^* exceeds a threshold given as

$$p_{ii}^* > Median\ (p_{ii}^*) + 3\ MAD\ (p_{ii}^*) \tag{15}$$

where MAD stands for the median absolute deviation as defined in (4).

These above results enable us to define a simple graphical display of classifying group deleted leverages and residuals for possible identification of them. Generalized potentials are used as leverages and the generalized Studentized residuals as deletion residuals in a "spatial generalized potentials—spatial generalized Studentized residuals (SGP-SGSR)" plot. Since the high leverage points need not to be outliers and outliers may not be points of high leverage we may expect different deletion sets D from the computation of these two quantities. Since D is the group of suspected outliers we prefer to include all observations considered to be suspect either along the y dimension or along the x dimension. We employ the blocked adaptive computationally-efficient outlier nominators (BACON) proposed by Billor et al. (2000) as a classifier. Another possibility could be the application of support vector regression for the same, especially when the data is big. The main advantage of the GP-GSR plot is that it is suitable for the data where masking (false negative) and/or swamping (false positive) make single case diagnostic plots misleading. This plot, unlike the L-R plot, retains the signs of residuals, which can be very important when their interpretation is concerned. Since the bulk of the cases will be associated with low leverage and small residuals, most of the pairs (t_{ii}^*, p_{ii}^*) will cluster near the origin (0, 0). The unusual cases will have either high leverages or large residual components and will tend to be separated from the bulk of the cases. High leverage cases will be located at the right corner of the plot and observations with large residuals will be located either at the upper or lower corner of the plot depending on their signs; large positive outliers will be plotted at the upper corner and large negative outliers will be located at the bottom corner of the plot.

Now we would like to revisit the data presented in Table 2.1 to demonstrate how different methods work in the classification of spatial regression outliers in both x and y dimensions. In quest of spatial outlier, we looked at the first-order differences in attributes in the previous sections. But now we would like to investigate how location affects the attribute values. We compute the first-order differences in location and the results are presented in Table 2.5.

Table 2.5 First order differences in location for spatial outlier data

Index	Diff_Location	Index	Diff_Location	Index	Diff_Location
1	*	11	0.5	21	0.4
2	1.0	12	1.0	22	0.6
3	0.1	13	0.2	23	0.2
4	0.5	14	0.2	24	0.8
5	0.4	15	0.3	25	0.1
6	0.8	16	0.4	26	0.9
7	0.1	17	0.2	27	**4.0 F**
8	0.1	18	0.2	28	**2.0 G**
9	0.2	19	0.2	29	**2.0 H**
10	0.3	20	0.3	30	1.0

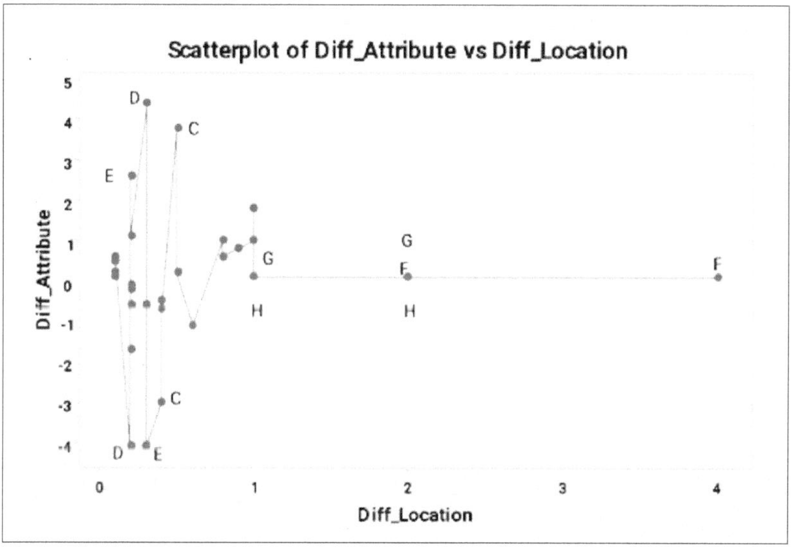

Figure 2.5 Scatter plot of differences in attributes vs differences in location for spatial outlier data.

In Sections 2.2 and 2.3, we observe that there exist 3 outliers, marked by C, D, and E along the y-axis. But we did not search for outliers along the x-axis. When we carefully look at the results given in Table 2.5, we observe that the point F has a marked difference from its neighbors. Points G and H look unusual too. This difference is visible more clearly when we look at Figure 2.5 where the first-order difference in attributes are plotted against the first-order difference in locations. Point F now clearly looks like a high leverage point or an outlier along the x-space and points G and H look more extreme as well.

Now we run a spatial regression of attributes on locations. Since our interest is to understand the neighborhood instability we consider the first-order difference of both attributes and locations as given in Tables 2.2 and 2.5. We then run a regression of differences in attributes on differences in location and the resulting deleted spatial Studentized residuals, leverages, SGSR and spatial generalized potentials are presented in Table 2.6.

Table 2.6 Residuals and leverages for the spatial regression data

Index	SDSR	Leverage	SGSR	SGP
1	*	*	*	*
2	0.45678	0.040885	1.09925	0.06658
3	0.11424	0.051079	0.20420	0.06290
4	2.15139	0.035779	**5.31765 C**	0.03590
5	-1.69711	0.037989	**-4.20445 C**	0.03835
6	0.47421	0.035612	1.13209	0.04571
7	0.32834	0.051079	0.72348	0.06290
8	0.06085	0.051079	0.07645	0.06290
9	-2.41622	0.045639	**-6.03394 D**	0.05185
10	2.61223	0.041275	**6.49375 D**	0.04367
11	0.07639	0.035779	0.07645	0.03590
12	0.89298	0.040885	0.13783	0.06658
13	-0.92108	0.045639	2.34566	0.05185
14	-0.10826	0.045639	-2.56908	0.05185
15	-0.33030	0.041275	-0.32208	0.04367
16	-0.28573	0.037989	-0.85865	0.03835
17	-0.32172	0.045639	-0.74085	0.05185
18	-0.05503	0.045639	-0.84428	0.05185
19	1.43538	0.045639	**3.58453 E**	0.05185
20	-2.42202	0.041275	**-6.02091 E**	0.04367
21	-0.39244	0.037989	-1.00843	0.03835
22	-0.62533	0.034647	-1.61273	0.03631
23	0.58737	0.045639	1.39415	0.05185
24	0.26077	0.035612	0.59882	0.04571
25	0.27470	0.051079	0.59176	0.06290
26	0.35834	0.037710	0.84471	0.05471
27	-0.49040	**0.636897 F**	-0.91806	**1.75404 F**
28	-0.12121	0.131865	-0.24012	**0.34271 G**
29	-0.12121	0.131865	-0.24012	**0.34271 H**
30	-0.02276	0.040885	-0.06854	0.06658

Although SDSR is very popular outlier measure it fails to identify even a single observation as an outlier. Here the cut off for the leverage is 0.2 and it can identify F as a high leverage point. The SGSR can successfully identify 3 spatial outliers C, D, and E since their values are higher than 3. It is worth mentioning that the cut off value for SGP is 0.1 based on equation (15). Thus SGP can successfully identify 3 high leverage points F, G, and H.

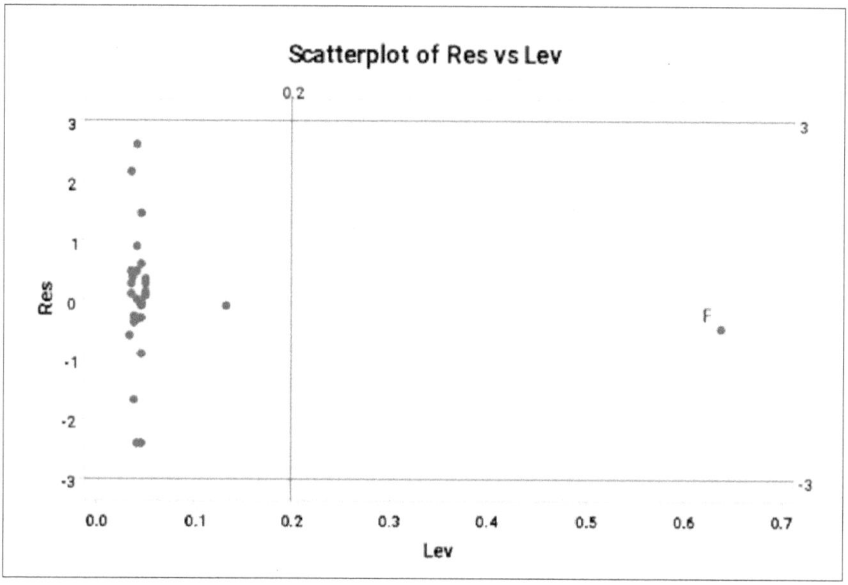

Figure 2.6 Leverage-residual plot for the spatial regression data.

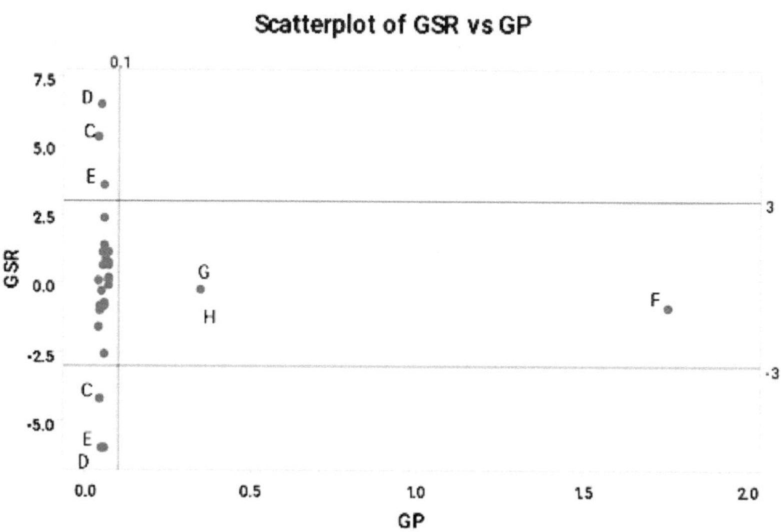

Figure 2.7 SGP-SGSR plot for the spatial regression data.

We see exactly the same picture in the L-R plot and the SGP-SGSR as shown in Figures 2.6 and 2.7. The L-R plot fails to identify even a single outlier and identifies only one high leverage point F out of three. Figure 2.7 clearly exhibits the merit of SGSR and SGP methods. They can successfully identify 3 spatial outliers (C, D, and E) and 3 spatial high leverage points (F, G, H).

2.5 CONCLUSIONS

Spatial *z* scores have been in use for the identification of spatial outliers. Since these scores contain some components which breakdown in the presence of outliers and consequently they suffer from masking. As a remedy to this problem, we recommend using robust spatial *z* scores especially when multiple outliers are present in spatial data. Likewise the regular data, outlier diagnostics for spatial regression may also break down in the presence of multiple unusual observations. This study suggests that instead of traditional spatial Studentized residuals, we should use the SGSR for the identification of multiple spatial outliers. We recommend using SGPs instead of traditional leverage measures for the identification of multiple high leverage points in spatial regression.

REFERENCES

Aggarwal, C.C. and Yu, P.S. 2001. Outlier detection for high dimensional data. In: Proceedings of the 2001 ACM SIGMOD International Conference on Management of Data, 37–46.

Anselin, L. 1996. The moran scatter plot as an esda tool to assess local instability in spatial association. In: Spatial Analytical Perspectives on GIS, M. Fischer, H. Scholten and D. Unwin (Eds.), Taylor and Francis, London. 111–125.

Barnett, V. and Lewis, T.B. 1994. Outliers in Statistical Data. Wiley, New York.

Billor, N., Hadi, A.S. and Velleman, P.F. 2000. BACON: Blocked adaptive computationally efficient outlier nominators. Computational Statistics and Data Analysis. 34: 279–298.

Breunig, M.M., Kriegel, H.P., Ng, R.T. and Sander, J.R. 1999. OPTICS-OF: Identifying local outliers. In: Proceedings of the Third European Conference on Principles of Data Mining and Knowledge Discovery. 262–270.

Fan, H., Zaïane, O.R., Foss, A. and Wu, J. 2006. A nonparametric outlier detection for efficiently discovering top-n outliers from engineering data. In: Proceedings of Pacific-Asia Conference on Knowledge Discovery and Data Mining. Singapore. 557–566.

Filzmoser, P., Ruiz-Gazen, A. and Thomas-Agnan, C. 2014. Identification of local multivariate outliers. Statistical Papers. 55: 29–47.

Gray, J.B. 1986. A simple graphic for assessing influence in regression. Journal of Statistical Computation and Simulation. 24: 121–134.

Hadi, A.S. 1992. A new measure of overall potential influence in linear regression. Computational Statistics and Data Analysis. 14: 1–27.

Hadi, AS. and Imon, A.H.M.R. 2019. Identification of multiple outliers in spatial data. International Journal of Statistical Sciences. 16: 87–96.

Hadi, A.S., Imon, A.H.M.R. and Werner, M. 2009. Detection of outliers. Wiley Interdisciplinary Reviews: Computational Statistics. 1: 57–70.

Haining, R. 1993. Spatial Data Analysis in the Social and Environmental Sciences. Cambridge University Press, London.

Haslett, J., Brandley, R., Craig, P., Unwin, A. and Wills, G. 1991. Dynamic graphics for exploring spatial data with applications to locating global and local anomalies. The American Statisticians. 45: 234–242.

Huang, T. and Qin, X. 2004. Detecting outliers in spatial database. In: Proceedings of the Third International Conference on Image and Graphics. Hong Kong. 556–559.

Imon, A.H.M.R. 2002. Identifying multiple high leverage points in linear regression. Journal of Statistical Studies. 22: 207–218.

Imon, A.H.M.R. 2005. Identifying multiple influential observations in linear regression. Journal of Applied Statistics. 32: 929–946.

Imon, A.H.M.R. and Hadi, A.S. 2021. Identification of multiple unusual observations in spatial regression. Journal of Statistics and Applications. 18: 155–162.

Knorr, E. and Ng, R. 1997. A unified notion of outliers: properties and computation. In: Proceedings of the International Conference on Knowledge Discovery and Data Mining. 219–222.

Knorr, E. and Ng, R. 1998. Algorithms for mining distance-based outliers in large datasets. In: Proceedings of 24th International Conference on Very Large Data Bases. 392–403.

Lu, C., Chen, D. and Kou, Y. 2004. Multivariate spatial outlier detection. International Journal of Artificial Intelligence Tools. 13: 801–812.

Panatier, Y. 1996. Variowin. Software for Spatial Data Analysis in 2D. Springer-Verlag, New York.

Ramaswamy, S., Rastogi, R. and Shim, K. 2000. Efficient algorithms for mining outliers from large datasets. In: Proceedings of the ACM SIGMOD International Conference on Management of Data. 427–438.

Rousseeuw, P.J. and Leroy, A. 1987. Robust Regression and Outlier Detection, Wiley, New York.

Shekhar, S., Lu, C. and Zhang, P. 2002. Detecting graph-based spatial outlier. Intelligent Data Analysis. 6: 451–468.

Chapter **3**

Identification of Multiple Unusual Observations in Generalized Linear Models

A.H.M. Rahmatullah Imon

Department of Mathematical Sciences,
Ball State University, Muncie, Indiana, USA

3.1 INTRODUCTION

The generalized linear model (GLM) is a flexible generalization of linear model. The idea of GLM generated by Nelder and Wedderburn (1972) seeks to extend the domain of applicability of the linear model by unifying various other statistical models via a link function and relaxing the normality assumption. In building statistical models we often assume that the response variable is quantitative, continuous, and normally distributed. However, in practise we may come across a variety of different situations. An excellent collection of examples is presented by Madsen and Thyregod (2011), Dobson and Barnett (2018), and others. The response variable could be binary or quantal such as developmental toxicity in chemical compounds or cancer where the data follow binomial distribution, count data such as accidents (Poisson distribution), count data containing waiting times (Negative binomial distribution), count data of multiple categories such as 'very satisfied', 'satisfied', 'neutral',

*For Correspondence: A.H.M. Rahmatullah Imon (rimon@bsu.edu)

'unsatisfied', 'very unsatisfied' (Multinomial distribution), continuous but positively skewed (Gamma distribution), and we need a unified way to model all these problems.

In particular, GLM can be used to model the relationship between the explanatory variable, X, and a function of the means, μ_i, of continuous or discrete responses. More precisely, GLM assumes that

$$(\mu_i) = \eta_i = \sum_{j=0}^{p} x_{ij}\beta_j,$$

where $x_{i0} = 1$ for $i = 1, 2, \ldots, n$ and $\beta = (\beta_0, \beta_1, \beta_2, \ldots, \beta_p)^T$ is a vector of unknown parameters of order $k = p + 1$ and $g(\cdot)$ is the link function which determines the scale on which linearity is assumed. Models of this type include logistic and probit regression, Poisson regression, linear regression with known variance, certain models for lifetime data, and many other models. Specifically, let $Y_1, Y_2, \ldots, Y_n$ be n independent random variables drawn from the exponential family with density (or probability function)

$$f(y_i;\theta_i,\phi) = \exp\left\{\frac{y_i\theta_i - b(\theta_i)}{a(\phi)} + c(y_i, \phi)\right\}$$

for some specific functions $a(\cdot)$, $b(\cdot)$ and $c(\cdot,\cdot)$. Here, $E(Y_i) = \mu_i = b'(\theta_i)$ and $Var(Y_i) = b''(\theta_i)\,a(\phi)$ with usual notation of derivative.

The most common method of estimating the unknown parameters, β, is that of maximum likelihood estimation (MLE) or quasi-likelihood methods (QMLE), which are equivalent if $g(\cdot)$ is the canonical link such as the logit function for the logistic regression, the log function for Poisson regression or the identity function for the Normal regression. That is, when $g(\mu_i) = \theta_i$, the MLE and QMLE estimator of β are the solutions of the $k = p + 1$ system of equations

$$\sum_{j=0}^{p} (y_i - \mu_i) x_{ij}\beta_j = 0, \quad j = 0, 1, 2, \ldots, p \qquad (1)$$

the estimator defined by (1) can be viewed as an M-estimator with score function

$$\psi(y_i; \beta) = (y_i - \mu_i)x_i$$

where $x_i = (1, x_{i1}, x_{i2}, \ldots, x_{ip})^T$.

3.2 DIAGNOSTIC MEASURES IN GLM

The linear structure of the model combined with error distributions within the exponential family results in a simple form for the likelihood and the maximum likelihood estimation of β can be achieved by iterative weighted least squares (McCullagh and Nelder 1989). In most diagnostic approaches, the MLE is first employed and subsequently diagnostic tools are used to identify potential unusual observations, known as outliers, high leverage points, and influential observations. Because of the relationship with least squares, analogous linear regression diagnostics are available. Pregibon (1981) gives a development of diagnostic quantities for the generalized linear model with example of single deletion diagnostics for the logistic model for binary data. Details of diagnostics for some generalized linear models are discussed by Cook and Weisberg (1982), McCullagh and Nelder (1989), Johnson (1985), Pierce and Schafer (1986), Williams (1987), Thomas and Cook (1990), Adimari and Ventura (2001), and others.

Although the general structure of these diagnostics procedures for generalized linear models is similar to that for the linear regression model, there are some complications. In particular, because the estimation method is iterative, there is the choice between one-step and fully iterated procedures, just as there was for nonlinear least squares. Another problem is that residuals and leverages which form the basis of all diagnostics are not uniquely defined. While these techniques have been quite successful in identifying individual unusual observation, its generalization to multiple unusual observations cannot guarantee success due to the so-called 'masking' (false negative) and 'swamping' (false positive) effects. The development of robust method in the early 1980's provides an option which offers automatic protection against anomalous data. Recent trend in diagnostics research is: (a) to detect unusual observations after deploying the robust method first and then apply the classical methods (Imon and Hadi 2008, 2013, Nurunnabi et al. 2010), or (b) use the robust method in any case (Cantoni and Ronchetti 2001, Maronna et al. 2006, Serigne and Ronchetti 2009).

Likewise, the linear models unusual observations can occur in either the design space X or response Y or both in a GLM. Observations are judged as 'outliers' on the basis of how unsuccessful the fitted regression equation is in accommodating them and that is why observations corresponding to "excessively large residuals" are treated as outliers. The leverage values are proportional to the distance from the mean of the data in the X-space. The points that are far from the mean may have considerable influence on the values of the estimated parameters. These points are known as "high leverage points". Observations that have a demonstratively larger effect in the estimation of covariates will be called "influential observations".

3.2.1 Identification of Outliers

After estimating the model by the ML method, let $\hat{\beta}$ denote the vector of estimated coefficients. Thus, the fitted values for the GLM are calculated for each covariate pattern and depend on the estimated probability for that covariate pattern. Let us consider a binary logistic regression model where

$$E(Y_i) = \mu_i = \pi_i = \frac{\exp\left(\sum_{j=0}^{p} x_{ij}\beta_j\right)}{1+\exp\left(\sum_{j=0}^{p} x_{ij}\beta_j\right)}$$

with $g(\mu_i) = \ln\left(\frac{\pi_i}{1-\pi_i}\right) = \sum_{j=0}^{p} x_{ij}\beta_j$. We denote the fitted value of the i-th covariate pattern as

$$\hat{y}_i = \hat{\pi}_i \text{ for all } i.$$

In linear regression, the i-th residual is defined as the difference between the observed and fitted value $(y_i - \hat{y}_i)$. In a similar fashion, we define the i-th residual as

$$\hat{\varepsilon}_i = y_i - \hat{\pi}_i, \quad i = 1, 2, \ldots, n$$

In linear regression, the hat matrix plays an extremely important role in the analysis. This is the matrix that provides the fitted values as the projection of the outcome variable into the covariate space. The linear regression residuals, $(Y - \hat{Y})$, are often expressed in terms of the hat matrix and this forms basis of many diagnostics. Using weighted least squares linear regression as a model, Pregibon (1981) derived a linear approximation to the fitted values, which yields a hat matrix in the form

$$H = V^{1/2}X(X^T VX)^{-1}X^T V^{1/2} \tag{2}$$

where V is an $n \times n$ diagonal matrix with general element $v_i = V(\hat{y}_i)$. For a binary logistic regression model we obtain

$$v_i = \hat{\pi}_i (1 - \hat{\pi}_i) \tag{3}$$

In general, the diagonal elements of the hat matrix h_i's as defined in Eq. (2) are called the leverage values.

In GLM, the residuals measure the extent of ill-fitted factor/covariate patterns. Hence, the observations possessing large residuals are suspect outliers. But at this stage a natural question comes to our mind, how big is big? The residuals defined in (2) are unscaled, that is why they are not readily applicable in detecting outliers. Now let us introduce some scaled version of the above residuals that are commonly used in diagnostics for the identification of outliers. The Pearson residuals are elements of the Pearson chi-square that can be used to detect ill-fitted factor/covariate patterns. In linear regression, a key assumption is that the error variance does not depend on the conditional mean $E(y_i|x_i) = \pi_i$. However, in GLM we may have errors where the error variance is a function of the conditional mean, i.e.,

$$V(y_i \mid x_i) = v_i = f(\hat{\pi}_i)$$

The Pearson residual defined for the i-th factor/covariate pattern is given by

$$r_i = \frac{y_i - \hat{\pi}_i}{\sqrt{v_i}}, \; i = 1, 2, \ldots, n$$

We can obtain a special form of Pearson residuals for a logistic regression model using (3) as

$$r_i = \frac{y_i - \hat{\pi}_i}{\sqrt{\hat{\pi}_i(1-\hat{\pi}_i)}}, \; i = 1, 2, \ldots, n$$

We call an observation outlier if its corresponding Pearson residual exceeds a quantity c in absolute term. Since Pearson residuals are scaled residuals a reasonable choice for c could be 3 (Ryan 1997), that matches with the 3σ distance rule used in the normal theory. For a linear regression model we obtain

$$\hat{\varepsilon}_i = (1 - h_i)y_i$$

If we use the Pregibon (1981) linear regression-like approximation for the residual for the i-th covariance pattern, the variance of the residual is given by

$$V(\hat{\varepsilon}_i) = v_i(1 - h_i)$$

which suggests that the Pearson residuals do not have variance equal to 1. For this reason, we use the standardized Pearson residuals (SPR) defined as

$$r_{si} = \frac{y_i - \hat{\pi}_i}{\sqrt{v_i(1-h_i)}}, \; i = 1, 2, \ldots, n \tag{4}$$

For a binary logistic regression it becomes

$$r_{si} = \frac{y_i - \hat{\pi}_i}{\sqrt{\hat{\pi}_i(1-\hat{\pi}_i)(1-h_i)}}, \; i = 1, 2, \ldots, n$$

We may declare the i-th observation as an outlier if $|r_{si}| > 3$. The standardized Pearson residuals have become most popular outlier diagnostics and they are readily available in most of the statistical software. However, they often suffer from masking and swamping problems and for this reason a single case deleted version of it is often used for the identification of outliers. The i-th deletion standardized Pearson residual (DSPR) is defined by

$$r_{si}^{(-i)} = \frac{y_i - \hat{\pi}_{(i)}}{\sqrt{v_{(i)}(1 - h_i)}}, \; i = 1, 2, \ldots, n \tag{5}$$

where $\hat{\pi}_{(i)}$ is the estimate of the i-th covariate and $v_{(i)}$ is the variance after the i-th observation was omitted from the fit. Observation possessing $|r_{si}^{(-i)}| > 3$ are identified as outliers. Some other outlier diagnostics available in the literature are Anscombe (1953) residuals and deviance residuals (Pierce and Schafer 1986).

3.2.2 Identification of High Leverage Points

Leverage values play extremely important role in generalized linear model. These are the quantities that provide the fitted values as the projection of the outcome variable into the covariate space. So it is really important to know which observations in the X space are exerting too much influence on covariates and we call these observations high leverage points. For linear regression, the well-known measure of leverage is given by the diagonal elements of the weight (or hat) matrix

$$H = X(X^T X)^{-1} X^T$$

For a perfect balanced design, the i-th diagonal element of H can be written as

$$h_i = \frac{1}{n} + \frac{(x_{i1} - \bar{x}_1)^2}{\sum (x_{i1} - \bar{x}_1)^2} + \frac{(x_{i2} - \bar{x}_2)^2}{\sum (x_{i2} - \bar{x}_2)^2} + \cdots + \frac{(x_{ip} - \bar{x}_p)^2}{\sum (x_{ip} - \bar{x}_p)^2}$$

Here h_i is the Euclidean distance between the i-th vector and the center of gravity $(1/n, \bar{x}_1, \bar{x}_2, \ldots, \bar{x}_p)$ for all vectors. We have already seen that a linear approximation to the fitted values suggested by Pregibon (1981) helps us to obtain a hat matrix H for a GLM as defined in Eq. (2). Let the quantity h_i denote the i-th diagonal element of this matrix. It is easy to show that

$$h_i = v_i x_i^T (X^T V X)^{-1} x_i \tag{6}$$

where $x_i = (1, x_{i1}, x_{i2}, \ldots, x_{ip})^T$ is the $1 \times k$ vector of observations corresponding to the i-th case. In linear regression, the leverages values are proportional to the distance from x_i to the mean of the data. The points that are far from the mean may have considerable influence on the values of the estimated parameters. These points are known as high leverage points. When we have n distinct values of x, the average value of h_i is k/n. Hoaglin and Welsch (1978) considered observations unusual when h_i exceeded $2k/n$ which is also known as *twice-the-mean* (2M) rule. Vellman and Welsch (1981) suggest considering the *thrice-the-mean* (3M) rule where h_i is considered as large when it exceeds $3k/n$. Other popular methods for the detection of high leverage points are the method suggested by Huber (1981), the method based on Mahalanobis distance (Rousseeuw and Leroy 1987) and the method based on potentials as suggested by Hadi (1992). Although the above methods are designed to detect high leverage points in linear regression, many of them are frequently used Hosmer and Lemeshow 2000) in logistic regression as well. Jennings (1986) demonstrates the fact that the leverage value of the i-th observation is closely related to the estimated logistic probability value $\hat{\pi}_i$. He suggested considering observations as high leverage points if their corresponding $\hat{\pi}$ values lie within the interval [0.1, 0.3] and/or within the interval [0.7, 0.9]. Another cut-off value for h_i, available in the literature is $1/k$ (Hosmer and Lemeshow 2000).

In linear regression, the leverage value is a monotonic increasing function of the distance of a covariance pattern from the mean. But in GLM this is not necessarily true. The diagonal elements of the hat matrix as defined in (6) can be re-expressed as

$$h_i = v_i b_i$$

where

$$b_i = x_i^T (X^T V X)^{-1} x_i \tag{7}$$

For a logistic regression model Eq. (6) becomes

$$h_i = \hat{\pi}_i (1 - \hat{\pi}_i) b_i \tag{8}$$

Referring to this result, Hosmer and Lemeshow (2000) point out that in logistic regression the most extreme points in the covariate space may have the smallest leverage. If we look at the leverage value as defined in Eq. (8), we will observe two components: a quantity that does increase with the distance from the mean (DM) is b_i as defined in Eq. (7) and another quantity $\hat{\pi}_i (1 - \hat{\pi}_i)$. For an extreme data point in the X-space, it is expected that the quantity $\hat{\pi}_i$ should be very close to 0 or 1 which automatically implies that the product $\hat{\pi}_i (1 - \hat{\pi}_i)$ should be very close to 0. Hence, even if the quantity b_i is large, its corresponding h_i could be very small making the procedure of identifying high leverage points on the magnitude of h_i very cumbersome. Hosmer and Lemeshow

(2000) suggest focusing on b_i if we are only interested in measuring the distance, however, they did not suggest any identification method based on these quantities. Imon (2006) suggested that it may not be easy to derive a theoretical distribution of b_i, but it does not make any problem to obtain a suitable confidence bound type cut off point for them. He considers b_i to be large if

$$b_i > Median\ (b_i) + 3\ MAD\ (b_i)$$

This type of form, analogous to a confidence bound for a location parameter, which was first introduced by Hadi (1992) in regression diagnostics and then used by many others.

3.2.3 Identification of Influential Observations

Influential observations are those whose presence or absence could make a huge difference in the fitting of the model. In linear regression, Cook's distance proposed by Cook (1977) and difference in fits (DFFITS) proposed by Belsley et al. (1980) are very commonly used measures of the influence of observations. These statistics are also very popular in the identification of influential observation in GLM as well (Atkinson and Riani 2000). For a GLM, we define the i-th Cook's distance as

$$CD_i = \frac{(\hat{\beta} - \hat{\beta}^{(-i)})^T X^T X(\hat{\beta} - \hat{\beta}^{(-i)})}{kv_i(1-h_i)},\ i = 1, 2, \ldots, n \tag{9}$$

where $\hat{\beta}^{(-i)}$ is the estimated parameter of β with the i-th observation deleted. The i-th Cook's distance can be expressed in terms of the i-th standardized Pearson residual and leverage as

$$CD_i = \frac{1}{k} r_{si}^2 \left(\frac{h_i}{1-h_i} \right)$$

We call an observation influential if its corresponding CD_i value is greater than 1. Another well-known measure for influence of observations is DFFITS. We can use the i-th DFFITS value to measure the influence of the i-th observation by

$$DFFITS_i = \frac{y_i - \hat{\pi}_{(i)}}{\sqrt{\hat{\pi}_{(i)}(1-\hat{\pi}_{(i)})(1-h_i)}},\ i = 1, 2, \ldots, n \tag{10}$$

Observation possessing DFFITS value bigger than $3\sqrt{k/n}$ is termed as an influential observation.

3.3 IDENTIFICATION OF MULTIPLE UNUSUAL OBSERVATIONS IN GLM

Unfortunately, most of the outlier detection methods discussed above suffer from masking and/or swamping when multiple outliers are present in the data. When multiple outliers are present in a dataset one might think of applying a single case diagnostic successively to identify all of them. However, this problem is not that simple. A group of outliers may distort the fitting of a model in such a way that outliers may have artificially very small residuals so that they may appear as inliers. This problem is known as masking (false negative) in the literature (Rousseeuw and Leroy 1987). The opposite effect of masking is known as swamping (false positive) (Barnett and Lewis 1994) for which innocent observations may appear as outliers.

3.3.1 Outliers

Imon and Hadi (2008) propose a group-deleted version of residuals and leverages to develop effective diagnostics for the identification of multiple outliers. Although they developed a rule for the identification of multiple outliers in logistic regression, but their formulae are general and can be readily applied to do the same in generalized linear models. Let us assume that d observations among a set of n observations are omitted before the fitting of the model. Let us denote a set of cases 'remaining' in the analysis by R and a set of cases 'deleted' by D. Hence, R contains (n-d) cases after d cases in D are deleted. Without loss of generality, assume that these observations are the last of d rows of X, Y and V so that

$$X = \begin{bmatrix} X_R \\ X_D \end{bmatrix} \quad Y = \begin{bmatrix} Y_R \\ Y_D \end{bmatrix} \quad V = \begin{bmatrix} V_R & 0 \\ 0 & V_D \end{bmatrix} \tag{11}$$

Let $\hat{\beta}^{(-D)}$ be the corresponding vector of estimated coefficients when a group of observations indexed by D is omitted. Thus, the corresponding fitted covariate values are $\hat{\pi}_i^{(-D)}$. Here we define the i-th deletion residual as

$$\hat{\varepsilon}_i^{(-D)} = y_i - \hat{\pi}_i^{(-D)}, \; i = 1, 2, \ldots, n$$

We also define the respective deletion variances and deletion leverages for the entire dataset as

$$v_i^{(-D)} = \hat{\pi}_i^{(-D)}\left(1 - \hat{\pi}_i^{(-D)}\right) \tag{12}$$

$$h_i^{(-D)} = v_i^{(-D)} x_i^T \left(X_R^T V_R X_R\right)^{-1} x_i \tag{13}$$

Using the above results and also using linear-regression like approximation, Imon and Hadi (2008) define the i-th generalized standardized Pearson residual (GSPR) as

$$r_{si}^{(-D)} = \begin{cases} \dfrac{y_i - \hat{\pi}_i^{(-D)}}{\sqrt{v_i^{(-D)}\left(1 - h_i^{(-D)}\right)}} & i \in R \\ \dfrac{y_i - \hat{\pi}_i^{(-D)}}{\sqrt{v_i^{(-D)}\left(1 + h_i^{(-D)}\right)}} & i \in D \end{cases} \tag{14}$$

This type of residuals is analogous to the residuals suggested by Hadi and Simonoff (1993) and Imon (2005) in linear regression. We call an observation outlier if its corresponding GSPR value is excessively large (say 3 or more) in absolute term. Although the expression for generalized standardized Pearson residual is available for any arbitrary set of deleted cases, D, the choice of such a set is the most important part of the study since the omission of this group determines the residuals for both the D set and the R set. For the detection of multiple outliers the initial deletion set must contain all suspect outliers because if any outlier remains in the R set, the entire GSPR values will be in faulty. In a search of suspect outliers, Imon and Hadi (2008) suggest a diagnostic-robust approach where suspect outliers (if at all) will be identified first by a suitable robust method and then diagnostic tools are applied to the resulting residuals to check whether the suspect outliers are genuine outliers or not. In their study Imon and Hadi (2008) used the block adaptive computationally-efficient outlier nominator (BACON) algorithm proposed by Billor et al. (2000) to find all suspect outliers. After the omission of suspect cases the GLM is fitted with the remaining observations and the generalized standardized Pearson residuals are computed for the entire dataset. Observations possessing GSPR values 3 or more are finally declared as outliers.

3.3.2 High Leverage Points

All the methods of leverage diagnostics discussed in Section 3.2.2 are designed for the identification of a single high leverage point and hence may be ineffective when multiple high leverage points are present in the data. Imon and Hadi (2013) propose a new method for the identification of multiple high leverage points in logistic regression. But this can be readily applied to a GLM as well. Partitioning the data matrix as it was done in (11), Imon and Hadi (2013) define deletion distance from the mean (DDM) for the entire dataset after the deletion of a group of suspect high leverage points indexed by D, as

$$b_i^{(-D)} = x_i^T (X_R^T V_R X_R)^{-1} x_i, \; i = 1, 2, \ldots, n \tag{15}$$

It should be noted that $b_i^{(-D)}$ is the i-th diagonal element of $X(X_R^T V_R X_R)^{-1} X^T$ matrix. Like b_i there exists no finite upper bound for DDMs. Hence, one could consider DDM to be large if

$$b_i^{(-D)} > Median\ (b_i^{(-D)}) + 3\ MAD\ (b_i^{(-D)}) \tag{16}$$

Although the expression for DDM is available for any arbitrary set of deleted cases, D, the choice of such a set is very important since the omission of this group determines leverages for the whole set. Imon and Hadi (2013) suggest a two-step procedure for identifying multiple high leverage points in logistic regression. Any suitable robust multivariate technique such as the robust Mahalanobis distance based on minimum covariance determinants can be employed at first to identify suspect cases. After the omission of suspect cases the model is refitted with the rest of the data and the DDM values are computed for the entire dataset. Observations corresponding to DDM values satisfying rule in Eq. (16) are declared as high leverage points.

3.3.3 Influential Observations

The problem of identifying a single influential observation has been now largely resolved. But there are many unresolved issues in the identification of multiple influential observations where masking and swamping makes the identification procedure much complicated. Nurunnabi et al. (2010) suggest generalized difference in fits (GDFFITS) to identify multiple influential observations in logistic regression that can be readily applied to a GLM. For a dataset containing D suspect influential observation, the GDFFITS for the entire dataset defined as

$$GDFFITS_i = \begin{cases} \dfrac{\hat{\pi}_i^{(-D)} - \hat{\pi}_i^{(-D-i)}}{\sqrt{v_i^{(-D)} h_i^{(-D)}}} & i \in R \\[2ex] \dfrac{\hat{\pi}_i^{(-D)} - \hat{\pi}_i^{(-D-i)}}{\sqrt{v_i^{(-D)} \left(\dfrac{h_i^{(-D)}}{1 - h_i^{(-D)}} \right)}} & i \in D \end{cases} \tag{17}$$

where $\hat{\pi}_i^{(-D)}$ and $h_{ii}^{(-D)}$ are defined in Eqs. (12) and (13), respectively. Nurunnabi et al. (2010) suggest declaring observations to be influential if their corresponding

$$|GDFFITS_i| \geq 3\sqrt{k/(n-d)} \tag{18}$$

Since all measures, irrespective of whether they belong to the deletion subset D or the remaining subset R are measured in a similar scale, this choice should not matter too much, but we still want to determine *D* as precise as possible. We can employ a variety of robust procedures in this regard. Our experience suggests that the block adaptive computationally-efficient outlier nominator (BACON) proposed by Billor et al. (2000) and robust Mahalanobis distance are very effective in this regard.

3.4 EXAMPLES

In this section we consider several examples which have frequent applications in GLM especially logistic regression diagnostics for the identification of unusual observations.

3.4.1 Brown Cancer Data

Here we consider a dataset given by Brown (1980). Here the main objective was to see whether an elevated level of acid phosphatase (A.P.) in the blood serum together with another continuous variable (age of the patients) and three dichotomized variables (X-ray, Stage and Grade) would be of value for predicting whether or not prostate cancer patients also had lymph node involvement (L.N.I). The data on the 53 patients are given in Table 3.1. It is worth mentioning that in this table we have presented only one explanatory variable, i.e., A.P. Here the dependent variable is nodal involvement, with 1 denoting the presence of nodal involvement and 0 indicating the absence of such involvement. This dataset has been extensively analyzed by many authors (Ryan 1997). Ryan (1997) pointed out that the original data on the 53 patients contains a single outlier (observation number 24).

Table 3.1 Brown cancer data

Index	L.N.I.	A.P.	Index	L.N.I.	A.P.	Index	L.N.I.	A.P.
1	0	48	19	0	83	37	0	63
2	0	56	20	0	98	38	0	102
3	0	50	21	0	52	39	0	76
4	0	52	22	0	75	40	0	95
5	0	50	23	1	99	41	0	66
6	0	49	**24**	**0**	**187**	42	1	84
7	0	46	25	1	136	43	1	81
8	0	62	26	1	82	44	1	76
9	1	56	27	0	40	45	1	70
10	0	55	28	0	50	46	1	78
11	0	62	29	0	50	47	1	70
12	0	71	30	0	40	48	1	67
13	0	65	31	0	55	49	1	82
14	1	67	32	0	59	50	1	67
15	0	47	33	1	48	51	1	72
16	0	49	34	1	51	52	1	89
17	0	50	35	1	49	53	1	126
18	0	78	36	0	48			

In linear regression, t–statistics are used in assessing the value of individual regressors, when other regressors are in the model. In GLM, we generally use the statistic

$$W = \frac{\widehat{\beta}_i}{se(\widehat{\beta}_i)}$$

which is called the Wald statistic. It should be noted that W does not have a t – distribution, even though it does have the same form as a t statistic, rather, W is assumed to be asymptotically normal.

Table 3.2 Effect of outliers on wald statistic for brown cancer data

Status of the Model	Value of W	p-value
With outliers	0.4521	0.360
Without outliers	2.6026	0.013

Table 3.2 shows how an outlier can breakdown the entire inferential procedure of GLM. In the presence of even a single outlier the effect of acid phosphatase becomes insignificant though it has a high significant impact when this outlier is removed from the data. This example re-emphasizes our concern that employing diagnostic measures is a must before reaching to an inference in generalized linear models.

When the response variable is plotted against acid phosphatase as shown in Figure 3.1, the outlying observation is easily identified.

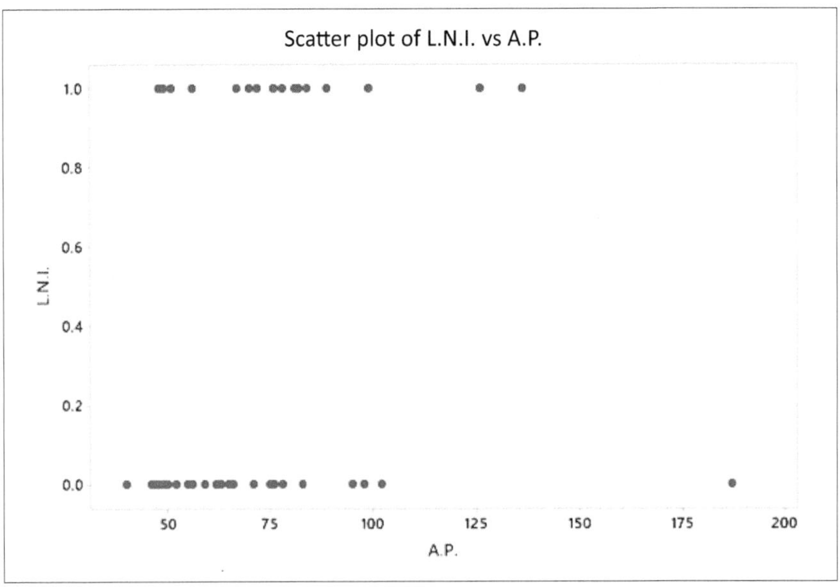

Figure 3.1 Scatter plot of acid phosphatase versus L.N.I. for brown (1980) data.

Observation 24 also looks like a high leverage point since its *X* value is very high. In addition to this observation, observations 25 and 53 have relatively higher *X* values. Hence, these two observations are suspect high leverage points. Since observation 24 is outlier and high leverage point there is a very strong suspicion that this observation should be an influential observation.

Table 3.3 Unusual observations diagnostics for brown cancer data

Index	SPR (±3)	DSPR (±3)	Leverage (0.0755, 0.1132)	Cook D (1.0)	DFFITS (±0.5828)
1	−0.82263	−0.81726	0.031893	0.00659	−0.11485
2	−0.87792	−0.87367	0.024567	0.00589	−0.10855
3	−0.83613	−0.83107	0.029807	0.00639	−0.11307
4	−0.84984	−0.84507	0.027881	0.00620	−0.11139
5	−0.83613	−0.83107	0.029807	0.00639	−0.11307
6	−0.82936	−0.82414	0.030831	0.00649	−0.11395
7	−0.80934	−0.80363	0.034124	0.00681	−0.11666
8	−0.92183	−0.91811	0.021160	0.00569	−0.10672
9	1.54254	1.54149	0.024567	0.02829	0.23788
10	−0.87081	−0.86645	0.025324	0.00596	−0.10917
11	−0.92183	−0.91811	0.021160	0.00569	−0.10672
12	−0.99215	−0.98855	0.020242	0.00653	−0.11431
13	−0.94465	−0.94107	0.020256	0.00579	−0.10756
14	1.43707	1.43514	0.019978	0.01821	0.19084
15	−0.81596	−0.81042	0.032991	0.00670	−0.11575
16	−0.82936	−0.82414	0.030831	0.00649	−0.11395
17	−0.83613	−0.83107	0.029807	0.00639	−0.11307
18	−1.05097	−1.04680	0.023510	0.00881	−0.13275
19	−1.09540	−1.09044	0.028128	0.01179	−0.15353
20	−1.24178	−1.23303	0.053295	0.03196	−0.25282
21	−0.84984	−0.84507	0.027881	0.00620	−0.11139
22	−1.02529	−1.02144	0.021660	0.00761	−0.12335
23	1.17132	1.16163	0.055533	0.02837	0.23822
24	−2.33948	−2.52228	**0.258612**	**1.55394**	**−1.76292**
25	0.92393	0.89308	**0.164472**	0.05048	0.31773
26	1.30485	1.30093	0.027050	0.01842	0.19195
27	−0.77062	−0.76385	0.041532	0.00744	−0.12200
28	−0.83613	−0.83107	0.029807	0.00639	−0.11307
29	−0.83613	−0.83107	0.029807	0.00639	−0.11307
30	−0.77062	−0.76385	0.041532	0.00744	−0.12200
31	−0.87081	−0.86645	0.025324	0.00596	−0.10917
32	−0.89960	−0.89565	0.022611	0.00574	−0.10716

(Contd.)

Index	SPR (±3)	DSPR (±3)	Leverage (0.0755, 0.1132)	Cook D (1.0)	DFFITS (±0.5828)
33	1.62320	1.62349	0.031893	0.04390	0.29630
34	1.59260	1.59226	0.028823	0.03708	0.27233
35	1.61296	1.61302	0.030831	0.04149	0.28805
36	−0.82263	−0.81726	0.031893	0.00659	−0.11485
37	−0.92937	−0.92571	0.020796	0.00571	−0.10685
38	−1.28426	−1.27435	0.062619	0.04156	−0.28830
39	−1.03377	−1.02983	0.022200	0.00797	−0.12623
40	−1.21088	−1.20300	0.046972	0.02615	−0.22867
41	−0.95239	−0.94884	0.020084	0.00585	−0.10818
42	1.28826	1.28389	0.029284	0.01923	0.19614
43	1.31323	1.30952	0.026049	0.01807	0.19010
44	1.35607	1.35322	0.022200	0.01692	0.18395
45	1.40952	1.40734	0.020070	0.01721	0.18554
46	1.33875	1.33559	0.023510	0.01725	0.18572
47	1.40952	1.40734	0.020070	0.01721	0.18554
48	1.43707	1.43514	0.019978	0.01821	0.19084
49	1.30485	1.30093	0.027050	0.01842	0.19195
50	1.43707	1.43514	0.019978	0.01821	0.19084
51	1.39146	1.38909	0.020486	0.01688	0.18374
52	1.24786	1.24208	0.036211	0.02179	0.20875
53	0.98701	0.96189	**0.133024**	0.04650	0.30495

From the diagnostics values presented in Table 3.3, we observe that neither SPR nor DSPR identify any observation as an outlier. However both twice the mean rule and thrice the mean rule identify observations 24, 25, and 53 are identified as high leverage points. Both Cook's distance and DFFITS identify observation 24 as influential. Now we employ the group deletion diagnostics and results are presented in Table 3.4.

We observe from the results presented in Table 3.4 that GSPR can successfully identify observation 24 as an outlier. The cut-off values for DDM and |GDFFITS| are 0.6741 and 0.588, respectively. The DDM identifies observations 24, 25, and 53 as high leverage points. GDFFITS identifies observation 24 as influential.

3.4.2 Modified Brown Data

We observe in the previous example that although there was a single outlier in the data, but the commonly used outlier diagnostic fail to identify that outlier. The GSPR can successfully identify that outlier. To judge the performance of different diagnostics we want a more

Table 3.4 Group deletion diagnostics for brown cancer data

Index	GSPR (±3)	DDM (0.6742)	GDFFITS (0.588)	Index	GSPR (±3)	DDM (0.6742)	GDFFITS (0.588)
1	−0.505	0.23449	−0.097	28	−0.529	0.20668	−0.098
2	−0.607	0.14208	−0.102	29	−0.529	0.20668	0.098
3	−0.529	0.20668	−0.098	30	−0.420	0.37708	−0.089
4	−0.554	0.18201	−0.100	31	−0.593	0.15089	−0.101
5	−0.529	0.20668	−0.098	32	−0.651	0.12037	−0.105
6	−0.517	0.22019	−0.098	33	2.054	0.23449	0.394
7	−0.482	0.26543	−0.095	34	1.911	0.19396	0.350
8	−0.697	0.10571	−0.108	35	2.005	0.22019	0.379
9	1.695	0.14208	0.285	36	−0.505	0.23449	−0.097
10	−0.593	0.15089	−0.101	37	−0.714	0.10239	−0.110
11	−0.697	0.10571	−0.108	38	−1.851	0.58462	−0.581
12	−0.862	0.10408	−0.137	39	−0.973	0.13062	−0.171
13	−0.748	0.09811	−0.114	40	−1.557	0.41024	−0.439
14	1.305	0.09696	0.199	41	−0.766	0.09714	−0.117
15	−0.493	0.24957	−0.096	42	0.885	0.21385	0.193
16	−0.517	0.22019	−0.098	43	0.947	0.17676	0.190
17	−0.529	0.20668	−0.098	44	1.060	0.13062	0.186
18	−1.021	0.14672	−0.189	45	1.217	0.10112	0.190
19	−1.155	0.20070	−0.245	46	1.013	0.14672	0.188
20	−1.677	0.48027	−0.498	47	1.217	0.10112	0.190
21	−0.554	0.18201	−0.100	48	1.305	0.09696	0.199
22	−0.950	0.12374	−0.163	49	0.926	0.18834	0.191
23	0.635	0.50518	0.191	50	1.305	0.09696	0.199
24	**−12.873**	**5.76796**	**−2.121**	51	1.162	0.10782	0.187
25	0.267	**1.97812**	0.086	52	0.792	0.29136	0.196
26	0.926	0.18834	0.191	**53**	0.340	**1.47417**	0.116
27	−0.420	0.37708	−0.089				

challenging example. For this reason we modify the Brown data by inserting two more outliers, high leverage points, and influential observations. These two additional observations are observation 54 for which L.N.I. is 0 and A.P. is 200, and observation 55 for which LN.I. is 0 and A.P. is 220. The scatter plot of L.N.I. against acid phosphates as shown in Figure 3.2 clearly exhibits that observations 24, 54, and 55 may severely distort the covariate pattern. They may be considered as outliers and high leverage points at the same time and thus should produce very high influences. In all this, the dataset contains 3 possible outliers (cases 24, 54, and 55), 5 possible high leverage points (cases 24, 25, 53, 54, and 55), and 3 influential observations (cases 24, 54, and 55).

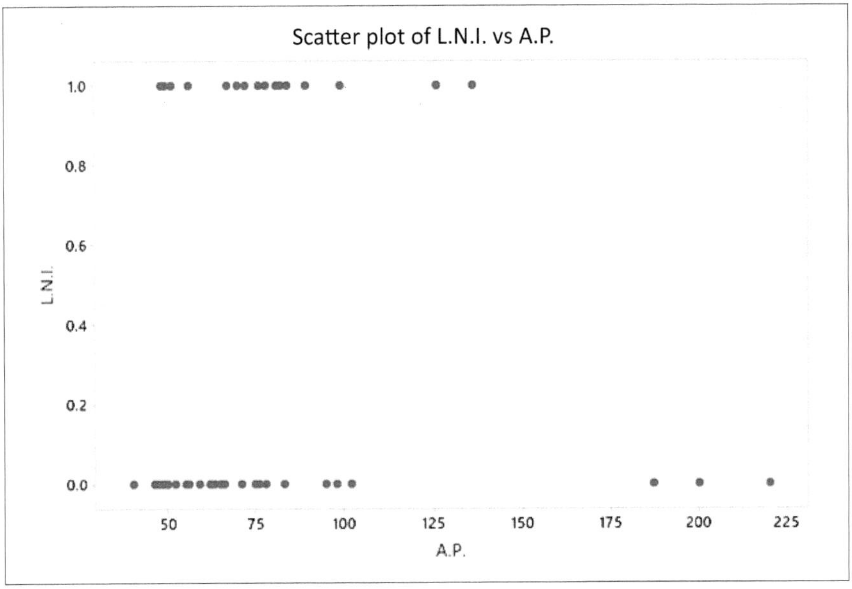

Figure 3.2 Scatter plot of acid phosphatase versus L.N.I. for modified brown data.

Table 3.5 Unusual observations diagnostics for modified brown data

Index	SPR (±3)	GSPR (±3)	Leverage (0.0727, 0.1091)	DDM (0.6742)	CD (1.00)	DFFITS (0.572)	GDFFITS (0.588)
1	−0.732	−0.505	0.027226	0.23449	0.01	−0.119	−0.097
2	−0.741	−0.607	0.022707	0.14208	0.01	−0.110	−0.102
3	−0.734	−0.529	0.025953	0.20668	0.01	−0.117	−0.098
4	−0.736	−0.554	0.024776	0.18201	0.01	−0.114	−0.100
5	−0.734	−0.529	0.025953	0.20668	0.01	−0.117	−0.098
6	−0.733	−0.517	0.026578	0.22019	0.01	−0.118	−0.098
7	−0.730	−0.482	0.028592	0.26543	0.01	−0.122	−0.095
8	−0.747	−0.697	0.020331	0.10571	0.01	−0.105	−0.108
9	1.382	1.695	0.022707	0.14208	0.02	0.207	0.285
10	−0.739	−0.593	0.023188	0.15089	0.01	−0.111	−0.101
11	−0.747	−0.697	0.020331	0.10571	0.01	−0.105	−0.108
12	−0.758	−0.862	0.018438	0.10408	0.01	−0.101	−0.137
13	−0.751	−0.748	0.019475	0.09811	0.01	−0.103	−0.114
14	1.354	1.305	0.019029	0.09696	0.02	0.186	0.199
15	−0.731	−0.493	0.027897	0.24957	0.01	−0.121	−0.096
16	−0.733	−0.517	0.026578	0.22019	0.01	−0.118	−0.098
17	−0.734	−0.529	0.025953	0.20668	0.01	−0.117	−0.098
18	−0.767	−1.021	0.018382	0.14672	0.01	−0.103	−0.189

(Contd.)

Index	SPR (±3)	GSPR (±3)	Leverage (0.0727, 0.1091)	DDM (0.6742)	CD (1.00)	DFFITS (0.572)	GDFFITS (0.588)
19	−0.774	−1.155	0.019117	0.20070	0.01	−0.106	−0.245
20	−0.796	−1.677	0.025282	0.48027	0.01	−0.128	−0.498
21	−0.736	−0.554	0.024776	0.18201	0.01	−0.114	−0.100
22	−0.763	−0.950	0.018252	0.12374	0.01	−0.102	−0.163
23	1.287	0.635	0.025908	0.50518	0.02	0.211	0.191
24	−1.015	**−12.873**	**0.191181**	**5.76796**	0.12	−0.498	**−2.121**
25	1.237	**0.267**	0.068522	**1.97812**	0.06	0.341	0.086
26	1.320	0.926	0.018918	0.18834	0.02	0.182	0.191
27	−0.725	−0.420	0.033249	0.37708	0.01	−0.131	−0.089
28	−0.734	−0.529	0.025953	0.20668	0.01	−0.117	−0.098
29	−0.734	−0.529	0.025953	0.20668	0.01	−0.117	0.098
30	−0.725	−0.420	0.033249	0.37708	0.01	−0.131	−0.089
31	−0.739	−0.593	0.023188	0.15089	0.01	−0.111	−0.101
32	−0.744	−0.651	0.021409	0.12037	0.01	−0.107	−0.105
33	1.404	2.054	0.027226	0.23449	0.03	0.232	0.394
34	1.395	1.911	0.025353	0.19396	0.02	0.222	0.350
35	1.401	2.005	0.026578	0.22019	0.03	0.229	0.379
36	−0.732	−0.505	0.027226	0.23449	0.01	−0.119	−0.097
37	−0.748	−0.714	0.020021	0.10239	0.01	−0.104	−0.110
38	−0.802	−1.851	0.027947	0.58462	0.01	−0.136	−0.581
39	−0.764	−0.973	0.018270	0.13062	0.01	−0.102	−0.171
40	−0.791	−1.557	0.023568	0.41024	0.01	−0.122	−0.439
41	−0.752	−0.766	0.019239	0.09714	0.01	−0.103	−0.117
42	1.316	0.885	0.019343	0.21385	0.02	0.184	0.193
43	1.322	0.947	0.018745	0.17676	0.02	0.181	0.190
44	1.333	1.060	0.018270	0.13062	0.02	0.180	0.186
45	1.347	1.217	0.018548	0.10112	0.02	0.182	0.190
46	1.329	1.013	0.018382	0.14672	0.02	0.180	0.188
47	1.347	1.217	0.018548	0.10112	0.02	0.182	0.190
48	1.354	1.305	0.019029	0.09696	0.02	0.186	0.199
49	1.320	0.926	0.018918	0.18834	0.02	0.182	0.191
50	1.354	1.305	0.019029	0.09696	0.02	0.186	0.199
51	1.342	1.162	0.018353	0.10782	0.02	0.181	0.187
52	1.306	0.792	0.020864	0.29136	0.02	0.190	0.196
53	1.248	**0.340**	0.053215	**1.47417**	0.04	0.300	0.116
54	−1.066	**−17.558**	**0.234341**	**7.06019**	0.18	**−0.594**	**−2.376**
55	−1.162	**−28.234**	**0.309937**	**9.30700**	0.30	**−0.782**	**−2.759**

Table 3.5 presents influence diagnostics for the modified Brown data. These results clearly show the advantage of using group deletion approaches as suggested by Imon and Hadi (2008, 2013) and Nurunnabi

et al. (2010). We have 3 outliers in this data, but the commonly used outlier diagnostic DSPR fails to identify even a single outlier. The GSPR, developed by Imon and Hadi (2008) can successfully identify all 3 genuine outliers without swamping any good case. Out of 5 high leverage points, the commonly used leverage measures can successfully identify 3 observations (cases 24, 54, and 55) but fail to identify the other 2 (cases 25, and 53). The DDM, developed by Imon and Hadi (2013) can successfully identify all 5 genuine high leverage points without swamping any good case. In case of the 3 influential observations, Cook's distance even a single influential observation. DFFITS can identify cases 54 and 55 correctly but masks case 24. But the GDFFITS developed by Nurunnabi et al. (2010) can successfully identify all 3 genuine influential observations without swamping any good case.

3.5 CONCLUSIONS

Here, we briefly discuss some commonly used diagnostics for the identification of multiple outliers, high leverage points, and influential observations in generalized linear models. Numerical examples show that all of them suffer from masking. Then we discuss some recently developed group deletion diagnostics which are developed by combining robust and diagnostic approaches together and observe that they could be very effective in the identification of multiple unusual cases in GLM.

REFERENCES

Adimari, G. and Ventura, L. 2001. Robust inference for generalized linear models with application to logistic regression. Statistics and Probability Letters. 55: 413–419.

Anscombe, F.J. 1953. Contribution to the discussion of H. Hotelling's paper. Journal of the Royal Statistical Society, Series B. 15: 229–230.

Atkinson, A.C. and Riani, M. 2000. Robust Diagnostic Regression Analysis. Springer Verlag, New York.

Barnett, V. and Lewis, T.B. 1994. Outliers in Statistical Data, 3rd Edn. Wiley, New York.

Belsley, D.A., Kuh, E. and Welsch, R.E. 1980. Regression Diagnostics: Identifying Influential Data and Sources of Collinearity. Wiley, New York.

Billor, N., Hadi, A.S. and Velleman, F. 2000. BACON: Blocked adaptive computationally-efficient outlier nominator. Computational Statistics and Data Analysis. 34: 279–298.

Brown, B.W., Jr. 1980. Prediction Analysis for Binary Data in Biostatistics Casebook, In: Miller, R.G., Jr., Efron, B., Brown, B.W., Jr., and Moses, L.E. (Eds.). Wiley, New York.

Cantoni, E. and Ronchetti, E. 2001, Robust inference for generalized linear models. Journal of the American Statistical Association. 96: 1022–1030.

Cook, R.D. 1977. Detection of influential observations in linear regression. Technometrics. 19: 15–18.

Cook, R.D. and Weisberg, S. 1982. Residuals and Influence in Regression. Chapman and Hall, London.

Dobson, A.J. and Barnett, A.G. 2018. An Introduction to Generalized Linear Models, 4th Edn. Chapman and Hall/CRC Press, New York.

Hadi, A.S. 1992. A new measure of overall potential influence in linear regression. Computational Statistics and Data Analysis. 14: 1–27.

Hadi, A.S. and Simonoff, J.S. 1993. Procedure for the identification of outliers in linear models. Journal of the American Statistical Association. 88: 1264–1272.

Hoaglin, D.C. and Welsch, R.E. 1978. The hat matrix in regression and ANOVA. The American Statistician. 32: 17–22.

Hosmer, D.W. and Lemeshow, S. 2000. Applied Logistic Regression. 2nd Edn. Wiley, New York.

Huber, P.J. 1981. Robust Statistics. Wiley, New York.

Imon, A.H.M.R. 2005. Identifying multiple influential observations in linear regression. Journal of Applied Statistics. 32: 73–90.

Imon, A.H.M.R. 2006. Identification of high leverage points in logistic regression. Pakistan Journal of Statistics. 22: 147–156.

Imon, A.H.M.R. and Hadi, A.S. 2008. Identification of multiple outliers in logistic regression. Communications in Statistics—Theory and Methods. 37: 1697 1709.

Imon, A.H.M.R. and Hadi, A.S. 2013. Identification of multiple high leverage points in logistic regression. Journal of Applied Statistics. 40: 2601–2616.

Jennings, D.E. 1986. Outliers and residual distributions in logistic regression. Journal of the American Statistical Association. 81: 987–990.

Johnson, W. 1985. Influence measures for logistic regression: another point of view. Biometrika. 72: 59–65.

Madsen, H. and Thyregod, P. 2011. Introduction to General and Generalized Linear Models. CRC Press, New York.

Maronna, R.A., Martin, D.R. and Yohai, V.J. 2006. Robust Statistics: Theory and Methods. Wiley, New York.

McCullagh, P. and Nelder, J.A. 1989. Generalized Linear Models, 2nd Edn. Chapman and Hall, London.

Nelder, N.A. and Wedderburn, R.W.M. 1972. Generalized linear models. Journal of the Royal Statistical Society, Series A. 135: 370–384.

Nurunabi, A.A.M., Imon, A.H.M.R. and Nasser, M. 2010. Identification of multiple influential observations in logistic regression. Journal of Applied Statistics. 37: 1605–1624.

Pierce, D.A. and Schafer, D.W. 1986. Residual in generalized linear model. Journal of the American Statistical Association. 81: 977–990.

Pregibon, D. 1981. Logistic regression diagnostics. Annals of Statistics. 9: 977–986.

Rousseeuw, P.J. and Leroy, A. 1987. Robust Regression and Outlier Detection. Wiley, New York.

Ryan, T.P. 1997. Modern Regression Methods, Wiley, New York.

Serigne N.L. and Ronchetti, E. 2009. Robust and accurate inference for generalized linear models. Journal of Multivariate Analysis. 100: 2126–2136.

Thomas, W. and Cook, R.D. 1990. Assessing influence on predictions from generalized linear models. Technometrics. 32: 59–65.

Williams, D.A. 1987. Generalized linear model diagnostics using the deviance and single case deletions. Applied Statistics. 36: 181–191.

Vellman, P.F. and Welsch, R.E. 1981. Efficient computing of regression diagnostics. The American Statistician. 35: 234–242.

Chapter **4**

The Conditional Breakdown Properties of LAD-LASSO Regression

Boning (Bernice) Feng[1], Avi Giloni*[2] and Jeffrey S. Simonoff[3]

[1]Yale University, School of Public Health, Department of Biostatistics, New Haven, CT, USA, Email: bernice.feng@yale.edu

[2]Yeshiva University, Sy Syms School of Business, Department of Information and Decision Sciences, New York, NY, USA, Email: agiloni@yu.edu

[3]New York University, Leonard N. Stern School of Business, Department of Technology, Operations, and Statistics, New York, NY, USA, Email: jss2@stern.nyu.edu

4.1 INTRODUCTION

It is well-known that many estimators, particularly ones based on least squares, are not robust, being sensitive to the effects of unusual observations (outliers and leverage points). One measure of robustness is the breakdown value, the smallest number of observations that can be sent to arbitrary values that will result in a parameter estimate becoming infinitely large in absolute value (a closely-related measure is the breakdown point, which is the breakdown value divided by the sample size, and thus refers to the smallest proportion of observations

*For Correspondence: Avi Giloni (agiloni@yu.edu)

that can be sent to arbitrary values that will result in a parameter estimate becoming infinitely large in absolute value). This is a worst-case performance measure, as it is based on observations being set to the most disadvantageous values possible.

Consider a standard linear regression situation, in which the data constitute a sample of n observations $\{\mathbf{x}^i, y_i\}$, with $\mathbf{x}^i$ being a p-dimensional vector. The model assumes

$$y_i = \beta_0 + \beta_1 x_{1i} + \cdots + \beta_p x_{pi} + \varepsilon_i,$$

where ε is an n-dimensional error vector. The predictors, $\mathbf{x}_i$, are taken as given and fixed, so the appropriate breakdown value is the conditional breakdown value, in which only response values are allowed to become infinite. It is well-known that the least squares estimator $\hat{\beta}^{OLS}$, which minimizes

$$\sum_{i=1}^{n} (y_i - [\beta_0 + \beta_1 x_{1i} + \cdots + \beta_p x_{pi}])^2,$$

has a breakdown value of one observation, the smallest possible value.

Least absolute deviations (LAD) regression is an alternative regression method that is more resistant to outliers. The LAD estimator $\hat{\beta}^{LAD}$ minimizes

$$\sum_{i=1}^{n} | y_i - [\beta_0 + \beta_1 x_{1i} + \cdots + \beta_p x_{pi}] |. \tag{1}$$

Giloni and Padberg (2004) showed that the breakdown value of $\hat{\beta}^{LAD}$ can be determined via mixed-integer programming, and showed that, depending on the configuration of predictor values, it can be considerably larger than 1.

In recent years, the problem of fitting regression models with large numbers of potential predictors has become increasingly important. A popular approach to this problem has been through the use of regularization methods, in which the usual estimation criterion (sum of squared or sum of absolute residuals) is penalized with a term that forces estimated regression slopes to 0. The LASSO estimator, introduced by Tibshirani (1996), minimizes

$$\sum_{i=1}^{n} (y_i - [\beta_0 + \beta_1 x_{1i} + \cdots + \beta_p x_{pi}])^2 + \lambda \sum_{j=1}^{p} | \beta_j |.$$

Depending on the choice of λ, the LASSO slope estimates are either forced to equal 0 (effectively acting as a variable selection method) or shrunk towards zero. Unfortunately, since the LASSO criterion is still based on least squares, the estimator is not robust, with a breakdown value of 1.

To overcome this weakness, Wang et al. (2007) proposed applying regularization to LAD regression, resulting in the LAD-LASSO, which minimizes

$$\sum_{i=1}^{n} |y_i - [\beta_0 + \beta_1 x_{1i} + \cdots + \beta_p x_{pi}]| + \lambda \sum_{j=1}^{p} |\beta_j|.$$

Various authors have examined the asymptotic properties of the LAD-LASSO and generalizations (including allowing for different values of λ for different predictors), including Wang et al. (2007), Gao and Huang (2010), Chen et al. (2010), Xu and Ying (2010), and Wang (2013).

Despite its motivation as a method resistant to unusual observations, there has been relatively little study of the robustness properties of LAD-LASSO (notable exceptions include Rahardiantoro and Kurnia 2015; Ajeel and Haslem 2020; Cahya et al. 2022). In this chapter, we study the breakdown properties of the estimator, demonstrating that it can achieve a better (sometimes considerably better) breakdown value than that of LASSO. In Section 4.2 we describe the estimator in more detail. In Section 4.3 we discuss how determining the breakdown value of the method can be formulated as a mixed-integer programming problem. Section 4.4 addresses different implementations of the LAD-LASSO, including versions based on centering and/or scaling the predictors. In Section 4.5 we illustrate the calculations of breakdown values of LAD-LASSO methods on several well-known datasets, showing how the breakdown value depends on the choice of λ and how and whether variables are centered and scaled. We further show how using LAD-LASSO as a variable selection method to choose the predictors with nonzero slopes, and then fitting LAD on those predictors (a method closely related to the relaxed LAD-LASSO proposed in Li et al. (2022), can improve the breakdown considerably. We conclude the chapter with discussion of potential future work, including the incorporation of weights to improve the breakdown value further.

4.2 LAD REGRESSION, LAD-LASSO, AND LINEAR PROGRAMMING

The LAD regression problem described in Eq. (1) can be formulated and solved as a linear program (c.f. Giloni and Padberg 2002). Specifically, let

$$\mathbf{y} = \begin{pmatrix} y_1 \\ \vdots \\ y_n \end{pmatrix}, \mathbf{X} = \begin{pmatrix} 1 & x_{11} & \cdots & \cdots & x_{p1} \\ \vdots & & & & \vdots \\ 1 & x_{1n} & \cdots & \cdots & x_{pn} \end{pmatrix} = \begin{pmatrix} \mathbf{x}^1 \\ \vdots \\ \mathbf{x}^n \end{pmatrix}.$$

Next, let $r_i = y_i - [\beta_0 + \beta_1 x_{1i} + \cdots + \beta_p x_{pi}]$ and since the sum of the absolute values of each of the residuals is to be minimized, let $r_i^+ - r_i^- = r_i$, where $r_i^+ \geq 0$ and $r_i^- \geq 0$. The LAD regression problem as an optimization problem is to determine the regression coefficients that minimize $\sum|r_i|$, which is equivalent to minimizing $\sum r_i^+ + r_i^-$, since for any i, setting both $r_i^+ > 0$ and $r_i^- > 0$ will necessarily increase the value of the objective function. Thus, the LAD regression problem can be formulated as

$$\begin{aligned} \min \quad & \mathbf{e}_n^T \mathbf{r}^+ + \mathbf{e}_n^T \mathbf{r}^- \\ \text{s.t.} \quad & \mathbf{X}\beta + \mathbf{r}^+ - \mathbf{r}^- = \mathbf{y} \\ & \beta \text{ free}, \mathbf{r}^+ \geq \mathbf{0}, \mathbf{r}^- \geq \mathbf{0} \end{aligned} \tag{2}$$

where the vector of residuals is $\mathbf{r} = \mathbf{r}^+ - \mathbf{r}^-$, and $\mathbf{e}^T{}_n$ is a vector with all n components equal to 1. The LAD-LASSO problem can be formulated as a nonlinear optimization problem by changing the objective function in the linear program (2) from $\mathbf{e}_n^T\mathbf{r}^+ + \mathbf{e}_n{}^T\mathbf{r}^-$ to $\mathbf{e}_n^T\mathbf{r}^+ + \mathbf{e}_n{}^T\mathbf{r}^- + \lambda\mathbf{e}_p^T|\beta|$. However, by setting $\beta = \beta^+ - \beta^-$, the objective function of the LAD-LASSO problem can be made into the linear function $\mathbf{e}_n^T\mathbf{r}^+ + \mathbf{e}_n^T\mathbf{r}^- + \lambda\mathbf{e}_p^T\beta^+ + \lambda\mathbf{e}_p^T\beta^-$, since if for some j $\beta_j^+ > 0$ and $\beta_j^- > 0$, the LAD-LASSO objective function would be increased. Thus, the LAD-LASSO problem can be formulated as the following linear program:

$$\begin{aligned} \min \quad & \mathbf{e}_n^T \mathbf{r}^+ + \mathbf{e}_n^T \mathbf{r}^- + \lambda \mathbf{e}_p^T \beta^+ + \lambda \mathbf{e}_p^T \beta^- \\ \text{s.t.} \quad & \mathbf{X}\beta^+ - \mathbf{X}\beta^- + \mathbf{r}^+ - \mathbf{r}^- = \mathbf{y} \\ & \beta^+ \geq \mathbf{0}, \beta^- \geq \mathbf{0}, \mathbf{r}^+ \geq \mathbf{0}, \mathbf{r}^- \geq \mathbf{0}. \end{aligned} \tag{3}$$

However, as will be discussed further below, we would like to take a different approach, by formulating the LAD-LASSO problem as an LAD regression problem with an augmented design matrix and response vector. To do so we first reformulate the linear program in Eq. (3) as

$$\begin{aligned} \min \quad & \mathbf{e}_n^T \mathbf{r}^+ + \mathbf{e}_n^T \mathbf{r}^- + \mathbf{e}_p^T \mathbf{c}^+ + \mathbf{e}_p^T \mathbf{c}^- \\ \text{s.t.} \quad & \mathbf{X}\beta + \mathbf{r}^+ - \mathbf{r}^- = \mathbf{y} \\ & -\lambda\beta + \mathbf{c}^+ - \mathbf{c}^- = 0 \\ & \beta \text{ free}, \mathbf{c}^+ \geq \mathbf{0}, \mathbf{c}^- \geq \mathbf{0}, \mathbf{r}^+ \geq \mathbf{0}, \mathbf{r}^- \geq \mathbf{0}, \end{aligned} \tag{4}$$

where $\lambda\mathbf{e}_p^T\beta^+ + \lambda\mathbf{e}_p^T\beta^- = \mathbf{e}_p^T\mathbf{c}^+ + \mathbf{e}_p^T\mathbf{c}^-$ since $\lambda\beta = \mathbf{c}^+ - \mathbf{c}^-$. Next, we note that by defining an augmented design matrix $\mathbf{X}^*$ and augmented response vector $\mathbf{y}^*$, the LAD-LASSO linear program in Eq. (4) can be formulated as an LAD regression problem with design matrix $\mathbf{X}^*$ and response vector $\mathbf{y}^*$. Specifically,

$$\mathbf{y}^* = \begin{pmatrix} y_1 \\ \vdots \\ y_n \\ 0 \\ 0 \\ \vdots \\ \\ 0 \end{pmatrix}, \quad \mathbf{X}^* = \begin{pmatrix} 1 & x_{11} & \dots & \dots & x_{p1} \\ \vdots & & & & \vdots \\ 1 & x_{1n} & \dots & \dots & x_{pn} \\ 0 & -\lambda & 0 & \dots & 0 \\ 0 & 0 & -\lambda & & 0 \\ \vdots & \vdots & & \ddots & \vdots \\ 0 & 0 & \dots & 0 & -\lambda \end{pmatrix} = \begin{pmatrix} \mathbf{x}^{*1} \\ \vdots \\ \mathbf{x}^{*n+p} \end{pmatrix}, \tag{5}$$

with the LAD-LASSO problem formulated as the LAD regression problem in linear programming form as

$$\begin{aligned} \min \quad & \mathbf{e}_n^T \mathbf{r}^+ + \mathbf{e}_n^T \mathbf{r}^- \\ \text{s.t.} \quad & \mathbf{X}^* \beta + \mathbf{r}^+ - \mathbf{r}^- = \mathbf{y}^* \\ & \beta \text{ free}, \mathbf{r}^+ \geq \mathbf{0}, \mathbf{r}^- \geq \mathbf{0}. \end{aligned} \tag{6}$$

The fact that the LAD-LASSO problem can be formulated as a linear program solving an LAD regression problem with an augmented design matrix and augmented response vector is important for two reasons. First, as we demonstrate in the next section, the computational methods of the breakdown value of LAD regression are heavily based upon the LAD regression problem being a linear program. Hence, the LAD-LASSO breakdown value can be computed in a similar manner to that of computing the breakdown value of LAD regression, since it is LAD regression with different data. Second, there is an important property of LAD-LASSO that can be easily seen from its linear programming formulation and solution. Recognizing that the design matrix and y vector for LAD-LASSO in Eq. (5) are $(n + p) \times (p + 1)$ and $(n + p) \times 1$ respectively, Proposition 2 in Giloni and Padberg (2002) describes that an optimal solution to the LAD-LASSO regression problem is of the form

$$\hat{\beta}^{LAD} = \mathbf{X}^{*-1}_B \mathbf{y}^*_B,$$

where $\mathbf{X}^*_B$ is some non-singular $(p + 1) \times (p + 1)$ submatrix of $\mathbf{X}^*$ and $\mathbf{y}^*_B$ the associated rows of the response vector. In the event that the rows in $\mathbf{X}^*_B$ are only among the first n rows, then $\hat{\beta}^{LAD}$ will have $p + 1$ nonzero values. On the other hand, any row of $\mathbf{X}^*_B$ that comes from the last p rows of $\mathbf{X}^*$ will have associated $\hat{\beta}^{LAD}$ values set to 0.

4.3 DETERMINING THE BREAKDOWN VALUE OF LAD-LASSO

Consider the LAD estimated regression parameters $\widehat{\beta}^{LAD}$ based on data $(\mathbf{X}, \mathbf{y})$. If we contaminate m $(1 \le m < n)$ values of the response vector $\mathbf{y}$ in a way so that row i is replaced by $(\mathbf{x}^i, \tilde{\mathbf{y}}_i)$, we obtain some new data $(\mathbf{X}, \tilde{\mathbf{y}})$. The LAD estimated regression parameters applied to $(\mathbf{X}, \tilde{\mathbf{y}})$ are different from the original ones. We can use any norm $\|\cdot\|$ on $\mathbb{R}^p$ to measure the distance $\| \widehat{\beta}^{LAD}(\mathbf{X}, \tilde{\mathbf{y}}) - \widehat{\beta}^{LAD} \|$ of the respective estimates. If we vary over all possible choices, then this distance remains either bounded or not bounded. Let

$$b(m, \mathbf{y} \mid \mathbf{X}) = \sup_{\tilde{y}} \| \widehat{\beta}^{LAD}(\mathbf{X}, \tilde{\mathbf{y}}) - \widehat{\beta}^{LAD}(\mathbf{X}, \mathbf{y}) \|$$

be the maximum bias that results when we replace at most m of the original values of the dependent variable y_i with arbitrary new data. The conditional breakdown value of LAD regression is

$$a(\mathbf{y} \mid \mathbf{X}) = \min_{1 \le m < n} \{m : b(m, \mathbf{y} \mid \mathbf{X}) \text{ is infinite}\};$$

i.e., it is the minimum number of values of $\mathbf{y}$ that, if replaced with arbitrary new data, make the LAD regression technique break down.

Giloni and Padberg (2004) show that the conditional breakdown value of LAD regression can be computed by solving the following mixed-integer program when the design matrix $\mathbf{X}$ is in general position:

$$\min \quad \sum_{i=1}^{n} u_i + l_i = a(\mathbf{y}|\mathbf{X}) \tag{7a}$$

$$\text{s.t.} \quad \mathbf{x}^i \xi + \eta^+ - \eta^- + s_i - t_i = 0 \quad \text{for } i = 1, \ldots, n, \tag{7b}$$

$$s_i - Mu_i \le 0, \quad t_i - Ml_i \le 0 \quad \text{for } i = 1, \ldots, n, \tag{7c}$$

$$\eta_i^+ + \eta_i^- + Mu_i + Ml_i \le M \quad \text{for } i = 1, \ldots, n, \tag{7d}$$

$$u_i + l_i \le 1 \quad \text{for } i = 1, \ldots, n, \tag{7e}$$

$$\sum_{i=1}^{n} \eta_i^+ + \eta_i^- - s_i - t_i \le 0, \quad \sum_{i=1}^{n} s_i + t_i \ge \varepsilon, \tag{7f}$$

$$\xi \text{ free}, \ \eta^+ \ge \mathbf{0}, \ \eta^- \ge \mathbf{0}, \ \mathbf{s} \ge \mathbf{0}, \ \mathbf{t} \ge \mathbf{0}, \tag{7g}$$

$$u_i, l_i \in \{0, 1\} \quad \text{for } i = 1, \ldots, n \tag{7h}$$

where we assume that M is a suitably chosen large number and ε a small number so that constraints in Eqs. (7c) and (7d) are non binding for the solution that results if we set u_i or l_i equal to 1 or $u_i = l_i = 0$.

Mizera and Müller (1999, 2001) provide an enumerative approach to computing the conditional breakdown of LAD regression. Giloni and Padberg (2004) show that this approach provides the same value as the mixed-integer program (when the design matrix is in general position), and Giloni, Sengupta, and Simonoff (2006) describe the enumerative approach through a linear programming framework. Let $N = \{1, \ldots, n\}$ and $E \subseteq N$. Then $a(\mathbf{y}|\mathbf{X}) = |E|$ where $|E|$ is the smallest integer such that

$$\max \frac{\sum_{i \in E} |\mathbf{x}^i \xi|}{\sum_{i \in \mathbb{N}} |\mathbf{x}^i \xi|} \geq \frac{1}{2},$$

where

$$\xi = \begin{pmatrix} 1 & \mathbf{0} \\ & \mathbf{X}_B \end{pmatrix}^{-1} \begin{pmatrix} \gamma \\ \mathbf{0} \end{pmatrix},$$

γ is a positive constant and the algorithm enumerates through all $(p) \times (p+1)$ submatrices $\mathbf{X}_B$ of $\mathbf{X}$.

To compute the breakdown value of the LAD-LASSO regression problem described in Eq. (6), we use the enumerative approach described above with $\mathbf{X}$ substituted with $\mathbf{X}^*$ and $\mathbf{y}$ substituted with $\mathbf{y}^*$. The reason for this is the mixed-integer programming approach is based upon a design matrix that is in general position and the design matrix $\mathbf{X}^*$ is not necessarily in general position, resulting in incorrect breakdown values.

In the next section, we describe the numerical experiments that we conducted to explore the breakdown values of different versions of the LAD-LASSO algorithm. In particular, one of the approaches considered is where the original design matrix and response vectors are centered. This has been a popular approach to constructing LASSO-type estimators, because it avoids the need to differentiate in the objective function the intercept coefficient (which is not shrunk to zero) and the slope coefficients (which are). Instead, the intercept term is removed through centering the response and predictor variables (since the OLS fit based on predictors at their sample mean values equals the sample mean of the response), the LASSO is fit based on a model without an intercept, and the intercept is then added back in at the end. In such a case, the LAD-LASSO regression estimates are determined by first solving an LAD regression problem where there is no intercept or constant term. In such a case, the augmented design matrix used will be $p \times (n + p)$ as opposed to $(p + 1) \times (n + p)$ and will be of the form (with centering and possibly scaling)

$$\mathbf{y}' = \begin{pmatrix} y_1 \\ \vdots \\ y_n \\ 0 \\ 0 \\ \vdots \\ \\ 0 \end{pmatrix}, \mathbf{X}' = \begin{pmatrix} x_{11} & \dots & \dots & x_{p1} \\ \vdots & & & \vdots \\ x_{1n} & \dots & \dots & x_{pn} \\ -\lambda & 0 & \dots & 0 \\ 0 & -\lambda & & 0 \\ \vdots & \vdots & \ddots & \vdots \\ 0 & \dots & 0 & -\lambda \end{pmatrix} = \begin{pmatrix} x'^{1} \\ \vdots \\ x'^{n+p} \end{pmatrix} \tag{8}$$

We note that the conditional breakdown value of the formulation of the LAD-LASSO regression problem is a lower bound on the LAD-LASSO method. This is because the algorithm used to compute the breakdown value permits potential contamination in the response vector described in Eqs. (5) and (8). In other words, the breakdown calculation assumes that contamination can take place in the entire response vector, including the 0 values corresponding to the LASSO constraints, where in reality such contamination would not take place (since those entries do not correspond to real data values). Hence, we also compute the breakdown value of the LAD regression problem on the columns that have nonzero estimated LAD-LASSO parameters. In such a case, if any centering had been performed on the design matrix (as described in the next section), when computing the breakdown value of the LAD regression problem on the columns that have nonzero estimated LAD-LASSO parameters, the original un-centered columns were used including a constant/intercept. Since this breakdown computation is for a traditional LAD regression, either the enumerative approach or the mixed-integer programming approach can be used, and the MIP optimizer Gurobi was used to solve any mixed-integer programs.

4.4 NUMERICAL EXPERIMENTS

Besides the two aforementioned approaches for computing the breakdown value (LAD-LASSO problem vs. remaining columns approach), we now describe the seven approaches considered for computing the breakdown values. The reason for these different approaches is due to the discussions in the literature regarding: (i) whether a constant term should be explicitly estimated by the LAD-LASSO or if the design matrix and response vector should be centered, and (ii) whether the data should be scaled before estimating LAD-LASSO parameters, in order to make all of the slopes on the same scale, thereby avoiding the problem of the penalty term in the LASSO objective function potentially being dominated by individual predictor(s) simply because of their scale. In each of the 7 approaches below, we compute the breakdown value of both the LAD-LASSO problem and the resulting remaining columns after LAD-LASSO is performed (a method closely related to the relaxed LAD-LASSO proposed in Li et al. 2022). Specifically, the 7 approaches in terms of the design matrix $\mathbf{X}^*$ or $\mathbf{X}'$ are as follows:

1. Estimate the breakdown value of the direct LAD-LASSO regression problem. In other words, compute the breakdown value of the LAD regression with augmented design matrix and response vector described in Eq. (5),

$$\mathbf{y}^* = \begin{pmatrix} y_1 \\ \vdots \\ y_n \\ 0 \\ 0 \\ \vdots \\ 0 \end{pmatrix}, \quad \mathbf{X}^* = \begin{pmatrix} 1 & x_{11} & \cdots & \cdots & x_{p1} \\ \vdots & & & & \vdots \\ 1 & x_{1n} & \cdots & \cdots & x_{pn} \\ 0 & -\lambda & 0 & \cdots & 0 \\ 0 & 0 & -\lambda & & 0 \\ \vdots & \vdots & & \ddots & \vdots \\ 0 & 0 & \cdots & 0 & -\lambda \end{pmatrix}.$$

2. Center the response vector and columns of the design matrix by using their respective sample means. We note that the constant term or intercept will no longer be directly computed by the LAD-LASSO linear program. This results in an augmented design matrix and response vector as described in Eq. (8),

$$\mathbf{y}' = \begin{pmatrix} y_1' \\ \vdots \\ y_n' \\ 0 \\ 0 \\ \vdots \\ 0 \end{pmatrix}, \mathbf{X}' = \begin{pmatrix} x_{11}' & \cdots & \cdots & x_{p1}' \\ \vdots & & & \vdots \\ x_{1n}' & \cdots & \cdots & x_{pn}' \\ -\lambda & 0 & \cdots & 0 \\ 0 & -\lambda & & 0 \\ \vdots & \vdots & \ddots & \vdots \\ 0 & \cdots & 0 & -\lambda \end{pmatrix}$$

where

$$y_i' = y_i - \widehat{\mu}(\mathbf{y}), \text{ where } \widehat{\mu}(\mathbf{y}) = \frac{1}{n}\sum_{i=1}^{n} y_i$$

$$x'_{ji} = x_{ji} - \hat{\mu}(\mathbf{x}_j), \text{ where } \hat{\mu}(\mathbf{x}_j) = \frac{1}{n}\sum_{i=1}^{n} x_{ji}$$

3. Center the response vector and columns of the design matrix by using their respective sample medians. We note that the constant term or intercept will no longer be directly computed by the LAD-LASSO linear program. This results in an augmented design matrix and response vector as described in Eq. (8),

$$\mathbf{y}' = \begin{pmatrix} y'_1 \\ \vdots \\ y'_n \\ 0 \\ 0 \\ \vdots \\ 0 \end{pmatrix}, \mathbf{X}' = \begin{pmatrix} x'_{11} & \cdots & \cdots & x'_{p1} \\ \vdots & & & \vdots \\ x'_{1n} & \cdots & \cdots & x'_{pn} \\ -\lambda & 0 & \cdots & 0 \\ 0 & -\lambda & & 0 \\ \vdots & \vdots & \ddots & \vdots \\ 0 & \cdots & 0 & -\lambda \end{pmatrix},$$

where

$y'_i = y_i - m(\mathbf{y})$, where $m(\mathbf{y})$ is the median of $\mathbf{y}$

$x'_{ji} = x_{ji} - m(\mathbf{x}_j)$, where $m(\mathbf{x}_j)$ is the median of $\mathbf{x}_j$.

4. Scale the columns of the design matrix by their respective standard deviations and estimate the breakdown value of the direct LAD-LASSO regression problem. In other words, compute the breakdown value of the LAD regression with augmented design matrix and response vector described in Eq. (5),

$$\mathbf{y}^* = \begin{pmatrix} y_1 \\ \vdots \\ y_n \\ 0 \\ 0 \\ \vdots \\ 0 \end{pmatrix}, \mathbf{X}^* = \begin{pmatrix} 1 & x'_{11} & \cdots & \cdots & x'_{p1} \\ \vdots & & & & \vdots \\ 1 & x'_{1n} & \cdots & \cdots & x'_{pn} \\ 0 & -\lambda & 0 & \cdots & 0 \\ 0 & 0 & -\lambda & & 0 \\ \vdots & \vdots & & \ddots & \vdots \\ 0 & 0 & \cdots & 0 & -\lambda \end{pmatrix},$$

where the scaling of the predictor variables is computed as

$$x'_{ji} = \frac{x_{ji}}{\hat{\sigma}(\mathbf{x}_j)}, \quad \text{where } \hat{\sigma}(\mathbf{x}_j) = \sqrt{\frac{1}{n}\sum_{i=1}^{n}\left(x_{ji} - \hat{\mu}(\mathbf{x}_j)\right)^2}$$

5. Scale the columns of the design matrix by their respective mean absolute deviation from the sample median and estimate the breakdown value of the direct LAD-LASSO regression problem. In other words, compute the breakdown value of the LAD regression with augmented design matrix and response vector described in Eq. (5).

$$\mathbf{y}^* = \begin{pmatrix} y_1 \\ \vdots \\ y_n \\ 0 \\ 0 \\ \vdots \\ 0 \end{pmatrix}, \mathbf{X}^* = \begin{pmatrix} 1 & x'_{11} & \cdots & \cdots & x'_{p1} \\ \vdots & & & & \vdots \\ 1 & x'_{1n} & \cdots & \cdots & x'_{pn} \\ 0 & -\lambda & 0 & \cdots & 0 \\ 0 & 0 & -\lambda & & 0 \\ \vdots & \vdots & & \ddots & \vdots \\ 0 & 0 & \cdots & 0 & -\lambda \end{pmatrix},$$

where the scaling of the predictor variables is computed as

$$x'_{ji} = \frac{x_{ji}}{\widehat{\sigma}(\mathbf{x}_j)}, \text{ where } \widehat{\sigma}(\mathbf{x}_j) = \frac{1}{n}\sum_{i=1}^{n} |x_{ji} - m(\mathbf{x}_j)|$$

6. Center the response vector and columns of the design matrix by using their respective sample means, and scale the columns of the design matrix by their respective standard deviations. We note that the constant term or intercept will no longer be directly computed by the LAD-LASSO linear program. This results in an augmented design matrix and response vector as described in Eq. (8),

$$\mathbf{y}' = \begin{pmatrix} y'_1 \\ \vdots \\ y'_n \\ 0 \\ 0 \\ \vdots \\ 0 \end{pmatrix}, \mathbf{X}' = \begin{pmatrix} x'_{11} & \cdots & \cdots & x'_{p1} \\ \vdots & & & \vdots \\ x'_{1n} & \cdots & \cdots & x'_{pn} \\ -\lambda & 0 & \cdots & 0 \\ 0 & -\lambda & & 0 \\ \vdots & \vdots & \ddots & \vdots \\ 0 & \cdots & 0 & -\lambda \end{pmatrix},$$

where

$$y'_i = y_i - \widehat{\mu}(\mathbf{y}), \text{ where } \widehat{\mu}(\mathbf{y}) = \frac{1}{n}\sum_{i=1}^{n} y_i$$

$$x'_{ji} = \frac{x_{ji} - \widehat{\mu}(\mathbf{x}_j)}{\widehat{\sigma}(\mathbf{x}_j)}, \text{ where } \widehat{\mu}(\mathbf{x}_j) = \frac{1}{n}\sum_{i=1}^{n} x_{ji}$$

and

$$\hat{\sigma}(\mathbf{x}_j) = \sqrt{\frac{1}{n}\sum_{i=1}^{n}(x_{ji} - \hat{\mu}(\mathbf{x}_j))^2}.$$

7. Center the response vector and columns of the design matrix by using their respective sample medians, and scale the columns of the design matrix by their respective mean absolute deviations. We note that the constant term or intercept will no longer be directly computed by the LAD-LASSO linear program. This results in an augmented design matrix and response vector as described in Eq. (8),

$$\mathbf{y}' = \begin{pmatrix} y'_1 \\ \vdots \\ y'_n \\ 0 \\ 0 \\ \vdots \\ 0 \end{pmatrix}, \mathbf{X}' = \begin{pmatrix} x'_{11} & \cdots & \cdots & x'_{p1} \\ \vdots & & & \vdots \\ x'_{1n} & \cdots & \cdots & x'_{pn} \\ -\lambda & 0 & \cdots & 0 \\ 0 & -\lambda & & 0 \\ \vdots & \vdots & \ddots & \vdots \\ 0 & \cdots & 0 & -\lambda \end{pmatrix},$$

where

$y'_i = y_i - m(\mathbf{y})$, where $m(\mathbf{y})$ is the median of $\mathbf{y}$

$$x'_{ji} = \frac{x_{ji} - m(\mathbf{x}_j)}{\hat{\sigma}(\mathbf{x}_j)}, \text{ where } m(\mathbf{x}_j) \text{ is the median of } \mathbf{x}_j$$

and

$$\hat{\sigma}(\mathbf{x}_j) = \frac{1}{n}\sum_{i=1}^{n}|x_{ji} - m(\mathbf{x}_j)|.$$

4.5 EXAMPLES OF BREAKDOWN VALUES OF LAD-LASSO-BASED METHODS

In this section, we illustrate the calculation and interpretation of breakdown values for various datasets that have appeared in the robustness literature. The datasets used are the aircraft data, Coleman data, Hawkins-Bradu-Kass (HBK) data, salinity data, and the modified wood gravity data. Each of these datasets can be found and is discussed in Rousseeuw and Leroy (1987).

For each dataset, we produce a three-panel figure of breakdown values (Figures 4.1–4.5). The top panel plots breakdown values versus the number of active (nonzero slope) predictors over a range of λ values, for the direct LAD-LASSO (L-L), the relaxed LAD-LASSO (RL-L),

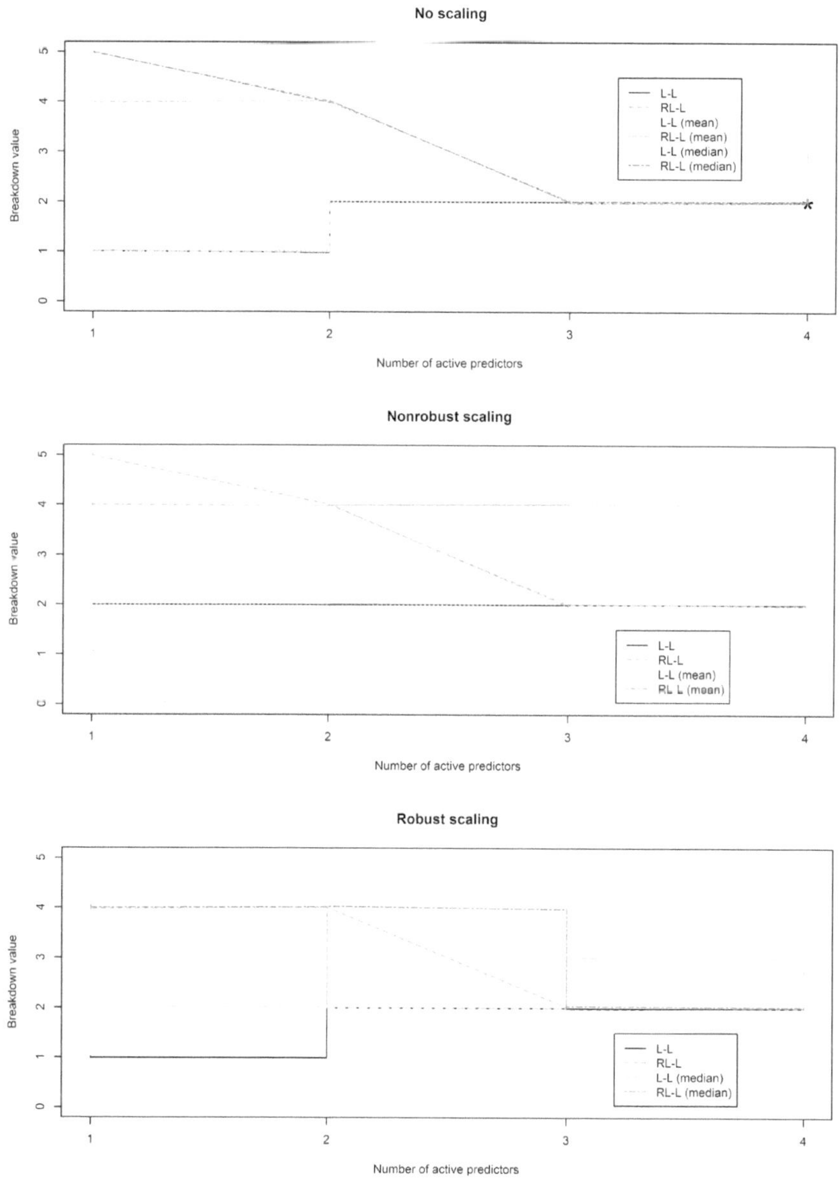

Figure 4.1 Breakdown values for aircraft data.

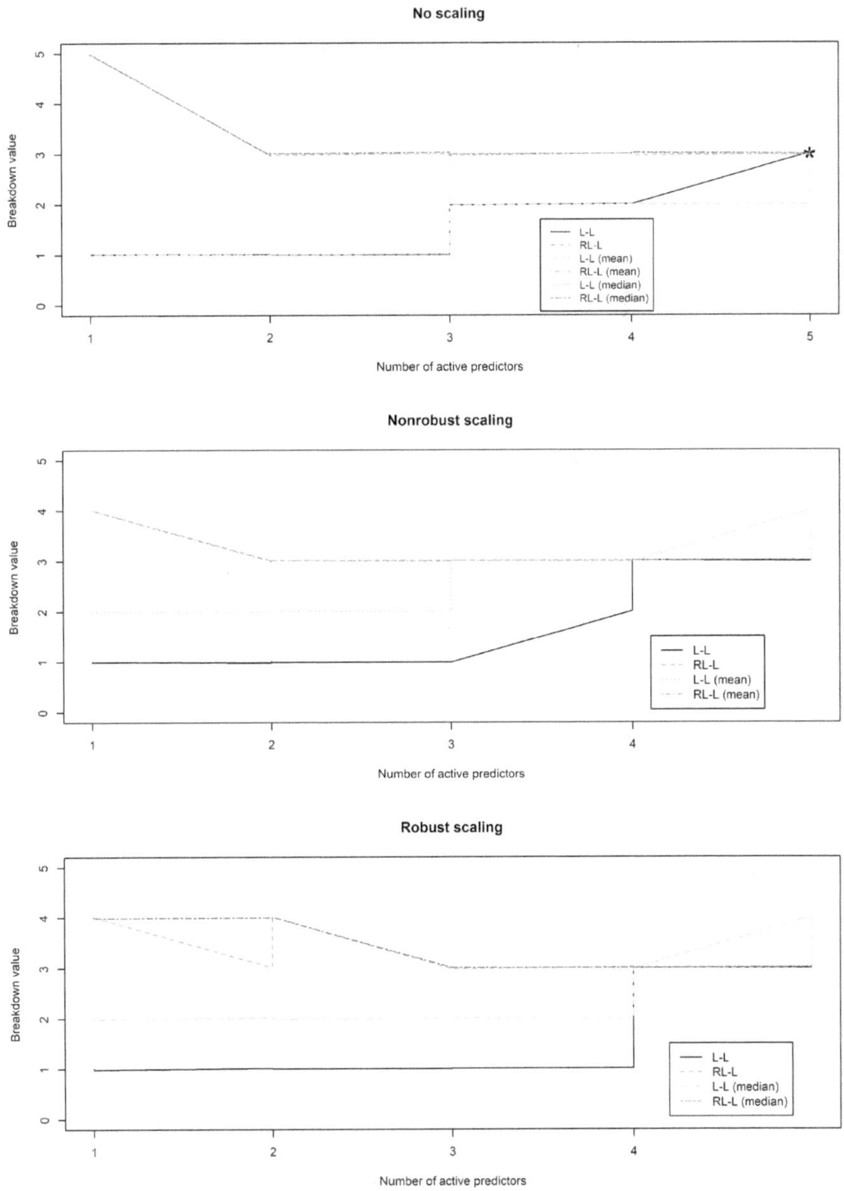

Figure 4.2 Breakdown values for Coleman data.

the versions obtained by estimating the intercept indirectly based on centering variables with sample means (L-L (mean) and RL-L (mean), respectively), and the versions obtained by estimating the intercept indirectly based on centering variables with sample medians (L-L

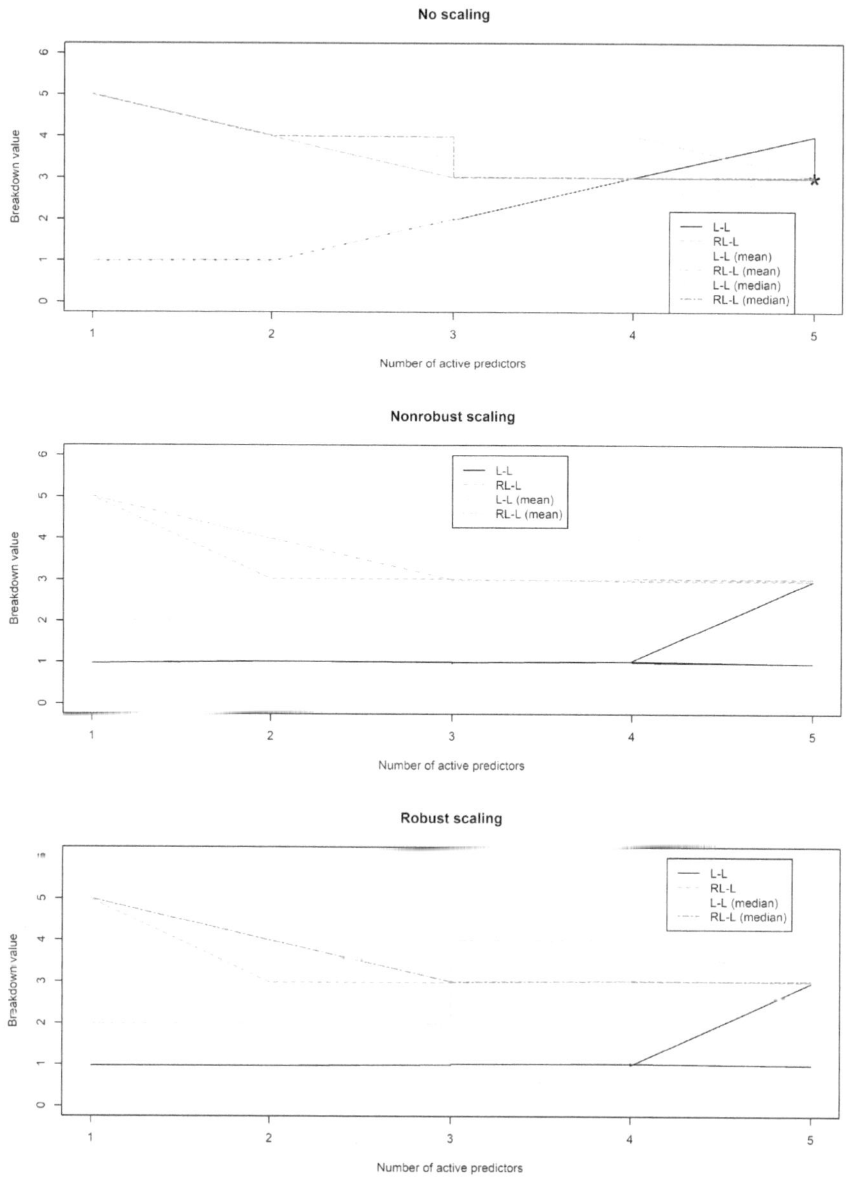

Figure 4.3 Breakdown values for wood data.

(median) and RL-L (median), respectively). The breakdown value of LAD (which corresponds to that of L-L with $\lambda = 0$) is given by an asterisk. The middle panel presents breakdown values when scaling predictors non-robustly using sample standard deviations (and centering using means

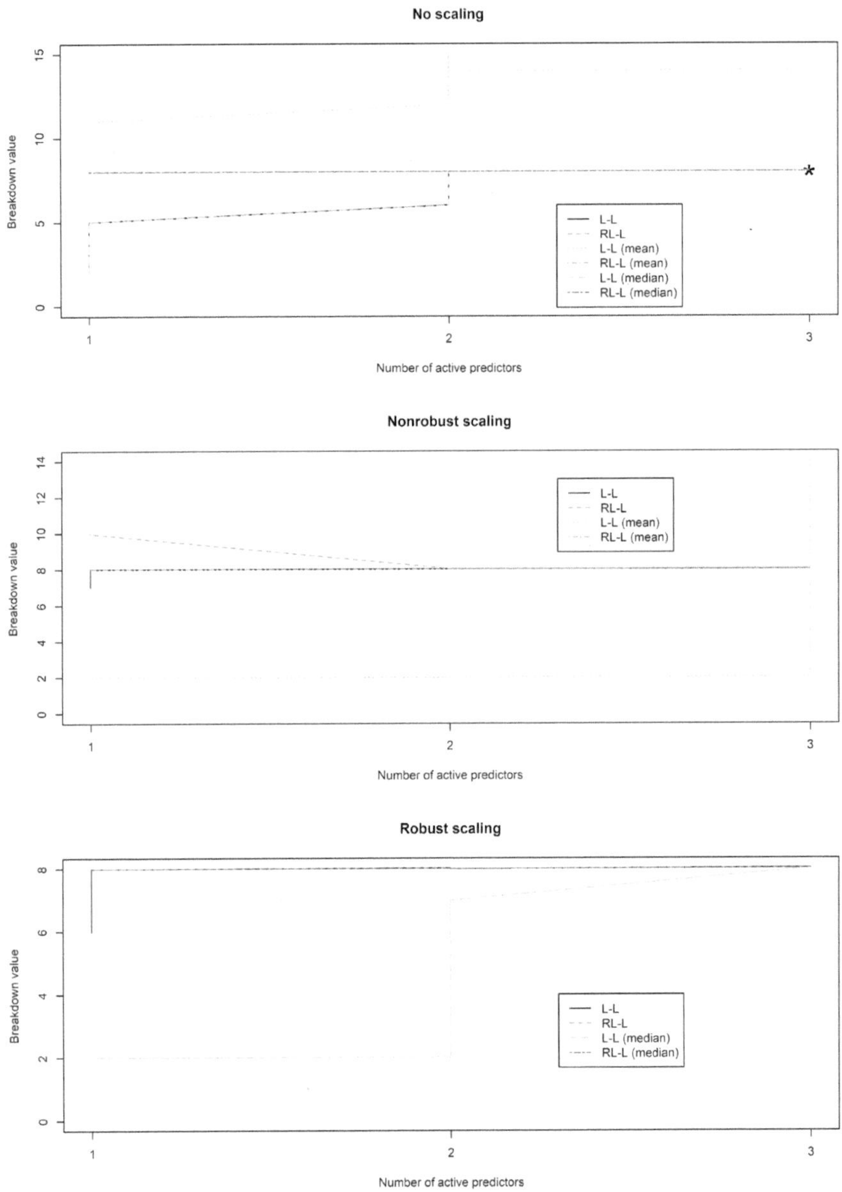

Figure 4.4 Breakdown values for HBK data.

when estimating the intercept indirectly), and the bottom panel presents breakdown values when scaling predictors robustly using sample mean absolute deviations (and centering using medians when estimating the intercept indirectly).

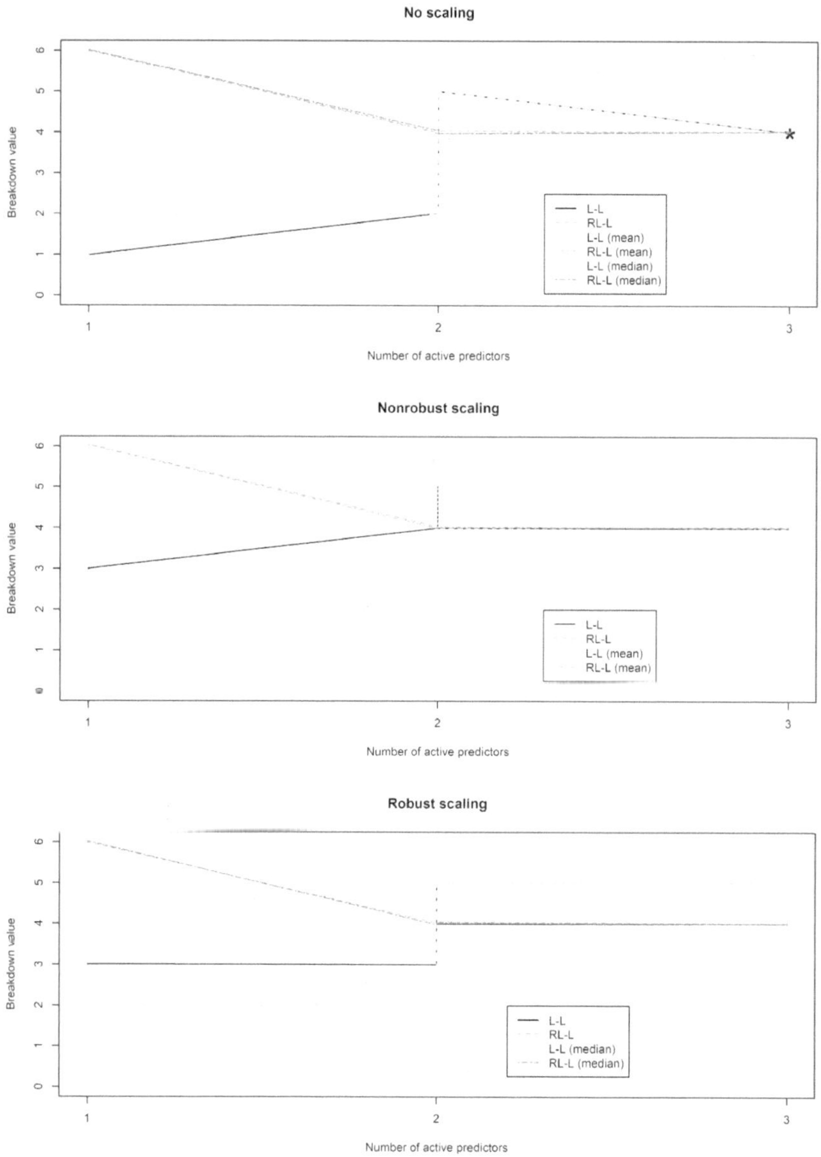

Figure 4.5 Breakdown values for salinity data.

The aircraft data (Figure 4.1) illustrate the general patterns. From the top panel we see that LAD-LASSO tends to have similar breakdown value to LAD itself when the chosen λ leads to all predictors being active, but as more slopes are driven to zero, its breakdown value stays the same or decreases. Distinctions between direct determination of L-L and indirectly fitting the intercept are minor, and perhaps surprisingly, centering robustly using medians isn't any more effective than doing so using means.

On the other hand, using the LAD-LASSO to determine the number of active predictors, but then using LAD on those predictors to estimate the coefficients (the relaxed LAD-LASSO) can make a difference, and the fewer the number of active predictors, the higher the breakdown value can go. Presumably this is reflecting the fact that if a variable is inactive, changes in its values can no longer affect the estimated coefficients, thereby making it easier to avoid breakdown.

The breakdown properties related to scaling are less clear, but once again we see that relaxed LAD-LASSO outperforms LAD-LASSO. When scaling non-robustly using standard deviations, the RL-L methods again tend to gain effectiveness as the number of active predictors decreases. Perhaps more surprisingly, the RL-L results when scaling robustly using mean absolute deviations are much more unstable, with breakdown values sometimes higher and sometimes lower than those using non-robust scaling, and breakdown values counterintuitively sometimes being lower when there are fewer active predictors.

The patterns are similar for the Coleman data (Figure 4.2) and the wood data (Figure 4.3). The other two datasets exhibit somewhat different behavior. When applied to the HBK data, the relaxed LASSOs don't improve on the breakdown from their original L-L method. Further, indirectly estimating the intercept by centering using the mean results in far higher breakdown values for L-L than any other method. We speculate that this could be because 14 observations in the dataset were artificially created to be leverage points that mask each other, with very unusual predictor values relative to the others. When centering using the median, these points are still recognized as unusual, and LAD is affected by them, but when centering by the mean, they are no longer unusual, and their effects on LAD are accordingly less. This notion is only of limited applicability, however, because scaling using the non-robust standard deviation (while still centering using the mean) results in the lowest breakdown values of any method.

The salinity data occupies a somewhat middle position compared to the others. The RL-L methods still generally improve on breakdown as the number of active predictors becomes fewer, but centering can result in higher breakdown values for L-L methods for more active predictors. Further, while RL-L methods can also improve breakdown when scaling variables, scaling L-L methods (either robust or non-robust) sometimes have higher breakdown and sometimes have lower breakdown.

It is important to remember what these results mean. The calculated breakdown values are based on an already-specified value of λ. In practice, this value would be chosen in a data-dependent way, and unusual observations could affect that choice in unanticipated ways. Thus, we cannot say that a particular method, applied to datasets with specific values in a data-dependent way, would lead to better performance. What we can say, however, is that apparently the breakdown properties of LAD-LASSO tend to mirror those of LAD itself, and become worse for choices of λ that lead to fewer active predictors. This tendency, however, can be mitigated by applying LAD to those datasets with fewer predictors (i.e., constructed RL-L estimates), which can result in often noticeable improvements in breakdown.

4.6 CONCLUSION AND FUTURE WORK

In this chapter, we have illustrated how the worst-case robustness of the LAD-LASSO method can be examined through determination of the breakdown value for a given dataset and specified choice of the regularization parameter λ, and how using a version of the relaxed LAD-LASSO can sometimes improve the breakdown considerably. A natural next question is to wonder if the (relaxed) LAD-LASSO method can be modified to improve the breakdown further. Giloni, Sengupta, and Simonoff (2006) showed how weights can be chosen to produce a weighted LAD (WLAD) estimator that can increase the breakdown of LAD regression; it would be interesting to see if a similar improvement can be made in constructing WLAD-LASSO methods. Several authors have proposed versions of WLAD-LASSO (for example, Arslan 2012; Gao and Feng 2018; and Jiang et al. 2021), and it would be interesting to explore the breakdown properties of those methods as well.

REFERENCES

Ajeel, S.M. and Hashem, H.A. 2020. Comparison some robust regularization methods in linear regression via simulation study. Academic Journal of Nawroz University. 9(2): 244–252. https: / /doi.org/10.25007/ajnu.v9n2a818.

Arslan, O. 2012. Weighted LAD-LASSO method for robust parameter estimation and variable selection in regression. Computational Statistics and Data Analysis. 56: 1952–1965.

Cahya, S.D., Sartono, B., Indahwati and Purnaningrum, E. 2022. Performance of LAD-LASSO and WLAD-LASSO on high dimensional regression in handling data containing outliers. Jurnal Teori dan Aplikasi Matematika. 6: 844–856. https://journal.ummat. ac.id/index.php/jtam/article/view/8968.

Chen, X., Wang, Z. and McKeown, M. 2010. Asymptotic analysis of robust LASSOs in the presence of noise with large variance. IEEE Transactions on Information Theory. 56: 5131–5149.

Gao, X. and Feng, Y. 2018. Penalized weighted least absolute deviation regression. Statistics and Its Interface. 11: 79–89.

Gao, X. and Huang, J. 2010. Asymptotic analysis of high-dimensional LAD regression with LASSO. Statistica Sinica. 20: 1485–1506.

Giloni, A. and Padberg, M. 2002. Alternative methods of linear regression. Mathematical and Computer Modelling. 35: 361–374.

Giloni, A. and Padberg, M. 2004. The finite sample breakdown point of l_1-regression. SIAM Journal of Optimization. 14: 1028–1042.

Giloni, A., Sengupta, B. and Simonoff, J.S. 2006. A mathematical programming approach for improving the robustness of least sum of absolute deviations regression. Naval Research Logistics. 53: 261–271.

Giloni, A., Simonoff, J.S. and Sengupta, B. 2006. Robust weighted LAD regression. Computational Statistics and Data Analysis. 50: 3124–3140.

Jiang, Y., Wang, Y., Zhang, J., Xie, B., Liao, J. and Liao, W. 2021. Outlier detection and robust variable selection via the penalized weighted LAD-LASSO method. Journal of Applied Statistics. 48: 234–246.

Li, H., Xu, X., Lu, Y., Yu, X., Zhao, T. and Zhang, R. 2022. Robust variable selection based on relaxed LAD-LASSO. Symmetry. 14: 2161. https://doi.org/10.3390/sym14102161.

Mizera, I. and Müller, C.H. 1999. Breakdown points and variation exponents of robust M-estimators in linear models. Annals of Statistics. 27: 1164–1177.

Mizera, I. and Müller, C.H. 2001. The influence of the design on the breakdown points of l_1-type M-estimators. In: A. Atkinson and W. Müller (Eds.), MODA6—Advances in Model-oriented Design and Analysis, 193–200. Physica-Verlag, Heidelberg.

Rahardiantoro, S. and Kurnia A. 2015. LAD-LASSO: Simulation study of robust regression in high dimensional data. Indonesian Journal of Statistics. 18: 105–107. https://journal.ipb.ac.id/index.php/statistika/article/view/16775.

Rousseeuw, P.J. and Leroy, A.M. 1987. Robust Regression and Outlier Detection. John Wiley & Sons, Inc., New York, NY.

Tibshirani. R. 1996. Regression shrinkage and selection via the LASSO. Journal of the Royal Statistical Society, Series B. 58: 267–288.

Wang, H., Li, G. and Jiang, G. 2007. Robust regression shrinkage and consistent variable selection through the LAD-LASSO. Journal of Business and Economic Statistics. 25: 347–355.

Wang. L. 2013. The L_1 penalized LAD estimator for high dimensional linear regression. Journal of Multivariate Analysis. 120: 135–151.

Xu, J. and Ying, Z. 2010. Simultaneous estimation and variable selection in median regression using Lasso-type penalty. Annals of the Institute of Statistical Mathematics. 62: 487–514.

Chapter 5

Statistical Methods for Identifying Outliers in Univariate and Multivariate Data

K. Manoj* and K. Senthamarai Kannan
[1]Department of Statistics, Manonmaniam Sundaranar University, Tirunelveli

5.1 INTRODUCTION

An outlier in a dataset is an observation that seems inconsistent with the rest of that dataset. The correct definition of outliers often depends on the implicit assumption regarding the data structure and the detection method used. A few definitions are considered adequate to deal with different types of data and methods.

> *Outliers are those that have been given opportunities—and who have had the strength and presence of mind to seize them.*
>
> —Malcolm Gladwell (Author of five *New York Times* bestsellers)

Definition 5.1 In a sample of moderate size taken from a certain population it appears that one or two values are surprisingly far away from the main group (Barnett and Lewis 1978).

*For Correspondence: K. Manoj (manojstatms@gmail.com)

Definition 5.2 An outlier is an observation which deviates so much from other observations as to arouse suspicions that it was generated by a different mechanism (Hawkins 1980).

Outlier detection plays a key role in modeling, inference, and data processing. Outliers can lead to misspecification of the model, biased parameter estimates and poor prediction. Many statistical techniques are sensitive to the presence of outliers. For instance, simple calculations of mean and standard deviation can be distorted by a single grossly inaccurate data point. Outlier validation should be a routine part of any data analysis. Possible outliers should be examined to see if they are erroneous. If a data point is in error, it should be corrected if possible and deleted if not. If there is no reason to believe that the outlier point is in error, it should not be removed without careful consideration. However, the use of more robust techniques may be warranted. Robust strategies often reduce the effect of outliers without eliminating them.

5.2 BACKGROUND OF OUTLIERS

Detection of outliers in the datasets which pointed out the practise of deleting the outliers in the eighteenth century was by Bernoulli (1777). Deletion of outliers is not a proper solution to handle the outliers but this remained a common practise in the past. To address the problem of outliers in the data, the first statistical technique was developed in 1850 Beckman and Cook (1983). Some of the researchers argued that extreme observations should be kept as a part of data as these observations provide very useful information about the data. For example, Bessel and Baeuer (1838) claimed that one should not delete extreme observations just due to their gap from the remaining data Barnett and Lewis (1978). The recommendation of Legendre (1805) is not to rub out the extreme observations "adjusted too large to be admissible". Some of the researchers favored cleaning the data from extreme observations as they distort the estimates. Outliers are always the result of some spurious activity and should be deleted Cousineau and Chartier (2010). Deleting or retaining data is as hot a debated issue today as it was 200 years ago. A number of studies handle the issues of outliers in different aspects Bendre and Kale (1985), Davies and Gather (1993), Iglewicz and Hoaglin (1993), and Barnett and Lewis (1984). Defining outliers by their distance to neighboring examples is a popular approach to finding unusual examples in a dataset known to be distance based outlier detection technique. Verma (1997) opined that detection of outliers in univariate data instead of accommodating the outliers is important because it provides a better estimate of mean and other statistical parameters in an international geochemical reference material.

5.3 MOTIVATION AND IMPORTANCE OF THE STUDY

Most of the existing research on the detection of outlying observations mainly focused on the problem of masking and swamping effects. This chapter concentrates on masking, swamping problems, and treatments of the outliers based on Data mining methods. The discordancy tests (Dixon 1951 and Grubbs 1950, 1969) have been very popular and are still widely used in different fields (e.g., Barnett and Lewis 1978, Li et al. 2003, Verma et al. 2009, Verma and Quiroz-Ruiz 2006, Hayes et al. 2007). Several other discordancy tests are also available for this purpose, most notable among which are the skewness and kurtosis tests (Barnett and Lewis 1994, Verma 1997, Velasco and Verma 1998). The problem of detection of multivariate outliers is a fundamental and important problem in applied statistics. The unreliability of multivariate outlier detection techniques such as Mahalanobis distance and hat matrix leverage has led to development of techniques which have been known in the statistical community for well over a decade. The literature on this subject is vast and growing (e.g., Srivastava von Rosen 1998, Pell 2000, Nag et al. 2005, Nedret and Gulsen 2008, Dang and Serflling 2009).

5.4 IDENTIFICATION OF OUTLIERS

When an outlier is a value that differs significantly from the rest of our data, carefully examining a set of data to find outliers poses some difficulties. By using a stemplot, it is easy to see that some values differ from the rest of the data, but to determine how different the values should be, we need a specific measure that gives us an objective standard.

5.4.1 Determining Outliers

The interquartile range (*IQR*) is used to determine if an extreme value is indeed an outlier. The *IQR* is based upon part of the five number summary of a dataset, namely the first quartile and the third quartile. The calculation of *IQR* involves a single arithmetic operation. Multiplying the *IQR* by 1.5 will give us a way to determine whether a certain value is an outlier. If ($1.5 \times IQR$) is subtracted from the first quartile, any data values that are less than this number are considered outliers. Similarly, if ($1.5 \times IQR$) is added to the third quartile, any data values that are greater than this number are considered outliers.

5.4.1.1 *Strong Outliers*

Some outliers show extreme deviation from the rest of a dataset. In these cases, we can take the steps as given above, changing only the number that we multiply by using the *IQR* and define a certain type of outlier. If we subtract ($3.0 \times IQR$) from the first quartile, any point that is below this number is called a strong outlier. In the same way, the addition of ($3.0 \times IQR$) to the third quartile allows us to define strong outliers by looking at points which are greater than this number.

5.4.1.2 *Weak Outliers*

Besides strong outliers, there is another category for outliers. If a data value is an outlier, but not a strong outlier, then that value is a weak outlier.

Example 5.4.1 First, suppose that we have the dataset 1, 2, 2, 3, 3, 4, 5, 5, 9. The number 9 certainly looks like an outlier. It is much greater than any other value from the rest of the dataset. To objectively determine if 9 is an outlier, we use the above methods. The first quartile is 2 and the third quartile is 5, which means that the interquartile range is 3. We multiply the interquartile range by 1.5, obtaining 4.5, and then add this number to the third quartile. The result, 9.5, is greater than any of our data values. Therefore, there are no outliers.

Example 5.4.2 Now using the same dataset mentioned as before, with the exception that the largest value is 10 rather than 9: 1, 2, 2, 3, 3, 4, 5, 5, 10. The first quartile, third quartile, and interquartile range are identical to Example 5.4.1. When we add $1.5 \times IQR = 4.5$ to the third quartile, the sum is 9.5. Since 10 is greater than 9.5 it is considered to be an outlier. Check whether 10 is a strong or weak outlier. For this we need to look at $3 \times IQR = 9$. When we add 9 to the third quartile and it ends up with a sum of 14. Since 10 is not greater than 14, it is not a strong outlier. Thus, we conclude that 10 is a weak outlier.

5.5 EFFECTS OF OUTLIERS

Outliers may have good or bad effects on the data. If these are the real observations, they point to interesting dimensions of the data. The famous case of Hadlum vs. Hadlum, held in 1949 (Barnett and Lewis 1978) is of statistical interest because of an outlier.

5.5.1 Contamination

Outliers of the third category originate from mixing of two populations in an unbalanced way. The outlier notion quantified the amount of contamination that needed to produce a pre-specified expected number of 'genuine' outliers (Davies and Gather 1993). For example, mixing two populations of 97% and 3% from two different samples may result in 3% of the population from one sample being outliers in the resulting pooled sample.

5.5.2 Masking and Swamping Effects

If the dataset has more than one outlier, the cases may mask each other, thus making and finding the outliers difficult. Sometimes one outlier has a capability to hide the other outliers and sometimes one outlier has the capability to expose an observation as outlier while it is inlier in real terms. Masking effect in cases of some tests that are commonly used as an outlier test for exponential samples are namely, Dixon-type test and the Cochran test (Bendre and Kale 1985).

5.5.2.1 *Masking Effect*

An observation is detected as an inlier in the presence of the extreme observation and by deleting this extreme observation, the observations nearer to it that are also found to be outliers. This phenomenon is considered as the masking effect. When masking occurs to be the mean, and the covariance estimates are skewed towards the group of outliers, the resulting gap of the outliers from the mean is small.

Example 5.5.1. Let x be a univariate vector, as x = [3 4 5 7 9 12 15 22 38]. The use of Grubbs test of outlier detection will just detect one outlier, that is 38. But after deleting this outlier and again applying Grubbs test, 22 will be detected as outlier. So it can be said that 22 is masked by 38. As the mask (38) is removed, 22 appears to be the outlier.

5.5.2.2 *Swamping Effect*

It is said that one outlier swamps the second observation if the latter can be considered as an outlier only under the presence of the first one. In other words, after the deletion of the first outlier the second observation becomes a non-outlying observation. Swamping occurs when a group of outlying instances skewed to be the mean and the covariance estimates toward it and away from the other non-outlying instances, and the resulting distance from these instances to the mean is large, making them look like outliers.

5.5.3 Damaging Effects of Outliers

Estimation of parameters was greatly influenced when outliers were present in the data (Zimmerman 1994), because they may result in an increase in the error variance and decrease the power of the test. If the errors contain outliers, these outliers decrease their normality in a univariate case, sphericity and multivariate normality in case of multivariate altering the odds of making both Type I and Type II errors. In this way the outliers are responsible for creating Type I and Type II errors. Finally, the regression estimates that might be of substantive interest are distorted by the outliers.

5.6 CAUSES OF OUTLIERS

To categorize the different ways in which outliers may arise was attempted by Anscombe and Guttman (1960). It was relevant to consider them in rather more detail. In taking observations, different sources of variability can be encountered. We can distinguish three of these (Barnett and Lewis 1978).

5.6.1 Natural Variability

This is the expression of the way in which observations intrinsically vary over the population; such variation is a natural feature of the population and uncontrollable. Thus, for example, the measurements of heights of men will reflect the amount of variability indigenous to that population.

5.6.2 Measurement Error

Often we must take measurements on members of a population under study. Inadequacies in the measuring instrument superimpose the further degree of variability on the inherent factor. The rounding of obtaining values or mistakes in recording the compound measurement errors are part of it. Controlling this type of variability is possible.

5.6.3 Execution Error

A further source of variability arises in the imperfect by collected data. We may inadvertently choose a biased sample or include individuals who are not the representative of the population we aimed to sample. Again, sensible precautions may reduce such variability.

5.7 UNIVARIATE OUTLIER DETECTION METHODS

Univariate data have an unusual value for a single variable. The results will be concerned with univariate outliers for the dependent variable in the data analysis. There are two kinds of outlier methods: tests of discordancy, and labeling methods. It is also called Formal and Informal methods, respectively.

1. Discordancy tests are designed to test any statistical hypothesis.
2. Labeling methods are to construct a box plot that visually depicts the location, spread, and skewness of the data.

Generally null hypothesis is assumed for a particular distribution and then this hypothesis is checked to observe whether the extreme values belong to the distribution or not at a given level of significance. In this section we have discussed different types of univariate outliers' detection methods with numerical examples.

5.7.1 Discordancy Tests

We consider statistical discordancy tests for deciding whether an outlier score is significant and the corresponding univariate observation is an outlier. On the basis of a test of discordance, we adjudge an outlier to be discordant. We implicitly rejected the working hypothesis in favor of some alternative hypothesis.

Definition 5.7.1 A formal statistical test that is used to decide whether a suspect observation should be considered contaminant or regular is called a discordancy test. Observations tested positive are discordant, and those tested negative are not discordant. The specificity is defined as the probability that a regular observation is correctly classified as not discordant. For example, $P(D_n|R)$ and the sensitivity as the probability that a contaminant observation is correctly classified as discordant $P(D|C)$, where, D_n is not discordant, R is regular as well as D is discordant and C is contaminant, respectively.

5.7.1.1 Properties

Some of the usual properties of datasets that one may encounter which render many statistical discordancy tests unsuitable for efficiently detecting outliers are:

- the datasets are multidimensional
- the datasets can be very large
- the distributions are unknown

- one or more parameters of the distributions are unknown
- the number of outliers is unknown
- the types of outliers are unknown
- the outliers are not necessarily extreme or relatively extreme values.

5.7.1.2 *Models for Discordancy*

Any statistical test inevitably examines two hypotheses: a null hypothesis, or a working hypothesis, which is preserved until significant evidence is found to support its rejection, and the working hypothesis, an alternative hypothesis that must be rejected. For a test of the discordancy of outliers, the working hypothesis would be to express some basic probability model for generating all the data without considering outliers; an alternative hypothesis model reveals a way in which externalities can be combined or modified to explain. Many of the existing works on the test of discordancy is highly intuitive in form and has little regard for the nature of the working and alternative hypothesis.

Example 5.7.1 When concerned with a single upper outlier in a set of independent observations $x_1, x_2, \ldots, x_n$ it is natural, and appealing, to contemplate the statistics

$$\frac{[x_n - x_{n-1}]}{D} \quad \text{or} \quad \frac{[x_n - \overline{x}']}{D} \tag{1}$$

where $x_{(i)}$ is the ith ordered value in the sample (when all observations are placed in ascending order of magnitude), $\overline{x}'$ is the sample mean excluding the outlier $x_{(n)}$, and D is some measure of the spread of the sample (again excluding consideration of $x_{(n)}$).

5.8 OUTLIER DETECTION METHODS

A number of formal outlier tests have been grouped by the following characteristics:

- What is the distributional model for the data? We restrict our discussion to test that the data follow an approximately normal distribution.
- Is the test designed for a single outlier or is it designed for multiple outliers?
- Does the number of outliers need to be specified exactly or can we specify an upper bound for the number of outliers if the test is designed for multiple outliers?

The following are the most commonly used outlier detection tests for normally distributed data. This list is not exhaustive (a large number of outlier tests have been proposed in the literature). The tests given here are essentially based on the criterion of "distance from the mean". This is not the only criterion that could be used.

5.8.1 Grubbs' Test

Grubbs' test (Grubbs 1969 and Stefansky 1972) detects a single outlier in a univariate dataset that follows the normal distribution approximately. Grubbs' test is also called the maximum normed residual test.

Grubbs' test is defined by the following two hypotheses:

H_0: There is no outlier in the dataset

H_1: There is one outlier in the dataset

There are several statistics for the Grubbs' test considering an ordered data sample:

1. Two-sided Test: If the minimum or maximum values are outliers

$$G = \frac{\overline{x} - x_i}{s} \quad \text{or} \quad G = \frac{x_n - \overline{x}}{s} \tag{2}$$

where x_i is the element of the dataset, $\overline{x}$, and s denoting the sample mean and standard deviation and the test statistic is the largest absolute deviation from the sample mean in units of the sample standard deviation. When the calculated value of parameter G is higher or lower than the critical value of choosing statistical significance (α), then the calculated value can be accepted as an outlier.

For the two-sided test, the hypothesis of no outliers is rejected if:

$$G > \frac{n-1}{\sqrt{n}} \sqrt{\frac{t^2_{\left(\frac{\alpha}{2n}, n-2\right)}}{n-2+t^2_{\left(\frac{\alpha}{2n}, n-2\right)}}} \tag{3}$$

with $t_{\left(\frac{\alpha}{2n}, n-2\right)}$ denoting the $\frac{\alpha}{2n}$ percentile of a t-distribution with $(n - 2)$ degrees of freedom. For one-side tests, we use the $\frac{\alpha}{n}$ percentile.

2. The Grubbs' test can also be defined as one of the following one-sided tests:

(a) test whether the minimum value is an outlier

$$G = \frac{\bar{Y} - Y_{min}}{s} \tag{4}$$

with Y_{min} denoting the minimum value.

(b) test whether the maximum value is an outlier

$$G = \frac{Y_{max} - \bar{Y}}{s} \tag{5}$$

with Y_{max} denoting the maximum value.

In the above formulae for the critical regions, the convention used is that t_α is the upper critical value from the t-distribution and $t_{1-\alpha}$ is the lower critical value from the t-distribution.

Example 5.8.1 For instance, we have a dataset used from Example 8.5.1, then perform Grubb's test in *R*. As the *p* value is significant ($G = 2.24747$, p-value $= 0.0195$), the highest value 38 is an outlier.

Check if the minimum value is an outlier. As the *p* value (>0.05) is not significant ($G = 0.87127$, $p = 1$), the minimum value 3 is not an outlier. Normal Probability Plot of data is given in Figure 5.1.

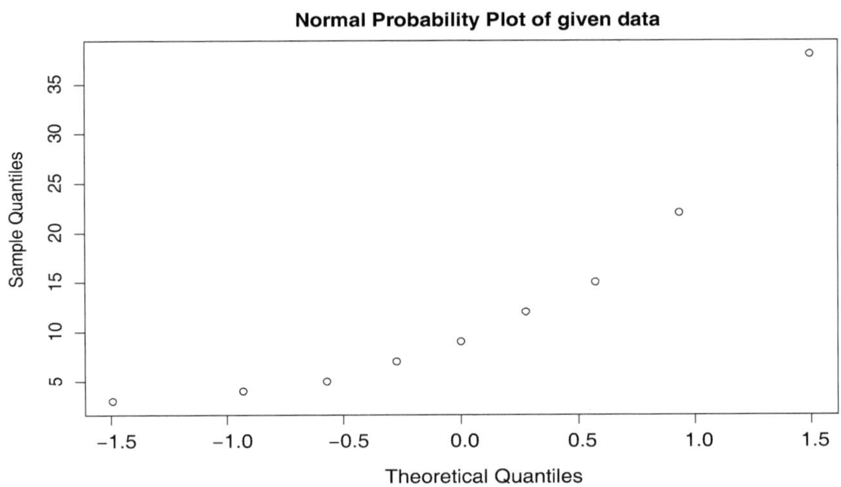

Figure 5.1 Normal probability plot for given data.

5.8.2 Quartile Method

There is no need to use quartile method in statistical tables. To find the outlier using the quartile method, it is necessary to carry out the following steps:

- Calculate the upper quartile: $Q3$ – 75% of the data in the dataset are lower than this.
- Calculate the lower quartile: $Q1$ – 25% of the data in the dataset are higher than this.
- Calculate the gap between the quartiles: $H = Q3\text{˘}Q1$

A value lower than $Q1\text{˘}1.5.H$ and higher than $Q3 + 1.5.H$ is considered to be a mild outlier. A value lower than $Q1 - 3.H$ and higher than $Q3 + 3.H$ is considered to be an extreme outlier.

Example 5.8.2 A survey was conducted to the public, a random sample of 20. They were asked, "How many times do you use Gpay per day?" Their responses are 0 0 0 0 1 1 1 2 2 2 3 3 3 4 4 4 5 5 6 6 7 10 and 15. The observations are in order from smallest to largest, so we can now compute the IQR by finding the median followed by $Q1$ and $Q3$ is 1 and 5, respectively.

The IQR is $Q3 - Q1 = 4$ and $1.5 \times \text{IQR} = 1.5(4) = 6$, 1.5 times the IQR is 6. Our fences will be 6 points below $Q1$ and 6 points above $Q3$. Lower fence: $Q1 - 6 = -5$ and Upper fence: $Q3 + 6 = 11$.

Any observations less than –5 times or greater than 11 times are outliers. Therefore, a single outlier 15 is identified from the outside of fences.

5.8.3 Dixon's Test

The test developed by Dixon and used is appropriate for a small sample size. The test has some limitations that $n \le 30$, were later on extended to $n > 40$. The test first step is organizing the data in an ascending order, and then the next step is to count parameter R.

The test has various test statistics. Suppose, for testing a large set of elements to be an outlier, the sample is arranged in ascending order $X_1 \le X_2 \le \dots \le X_n$, implying that the large sample element is given by X_n. The test statistics proposed by Dixon are:

1. For single outlier x_1 or x_n, respectively:

$$R_{10} = \frac{x_2 - x_1}{x_n - x_1} \quad \text{or} \quad R_{10} = \frac{x_n - x_{n-1}}{x_n - x_1} \tag{6}$$

2. For single outlier x_1 avoiding x_n, or x_n avoiding x_1, respectively:

$$R_{11} = \frac{x_2 - x_1}{x_{n-1} - x_1} \quad \text{or} \quad R_{11} = \frac{x_n - x_{n-1}}{x_n - x_2}, \tag{7}$$

3. For outlier x_1 avoiding x_2 and x_n, or x_n avoiding x_{n-1}, and x_1, respectively:

$$R_{21} = \frac{x_3 - x_1}{x_{n-1} - x_1} \quad \text{or} \quad R_{21} = \frac{x_n - x_{n-2}}{x_n - x_2}, \tag{8}$$

4. For outlier x_1 avoiding x_2, x_n and x_{n-1}, or x_n avoiding x_{n-1}, x_1 and x_2, respectively:

$$R_{22} = \frac{x_3 - x_1}{x_{n-2} - x_1} \quad \text{or} \quad R_{22} = \frac{x_n - x_{n-2}}{x_n - x_3} \tag{9}$$

For testing the smallest sample element to be an outlier, the sample is ordered in descending order implying that the smallest sample element is labeled X_n. All the selection of the test statistics depends on Dixon's criteria. Dixon published several critical values (λ_{ij}) tables for the different statistics r_{ij} for $n > 30$ with the criteria for declaring the appropriate x being an outlier if $r_{ij} > \lambda_{ij}$.

Example 5.8.3 In Example 5.8.2 dataset was used for Dixons outlier test and the results we got are:

1. For greater sample estimates ($r22$ = 0.53333, n = 23, p-value = 0.005267) then the 23rd observation 15 is identified as an outlier.
2. For two-sided sample estimates (Q = 0.33333, n = 23, p-value = 0.04391) then the 23rd observation 15 is identified as an outlier.

5.8.4 Generalized ESD Test for Outliers

The generalized extreme Studentized deviate (ESD) test detects one or more outliers in a univariate dataset that follows an approximately normal distribution (Rosner 1983). The generalized ESD test only requires that an upper bound for the suspected number of outliers be specified. The generalized ESD test is defined for the hypothesis:

H_0: There is no outlier found in the dataset

H_a: There are up to r outliers in the dataset

5.8.4.1 *Test Procedure*

1. Compute the test statistic R_1 from:

$$R_i = \max_i \left\{ \frac{|x_i - \bar{x}|}{s} \right\} \tag{10}$$

   Then find and remove the observation that maximizes $|x_i - \bar{x}|$
2. Compute R_2 in the same way but with the reduced sample of $n - 1$ observations

3. Continue with the process until $R_1, R_2, \ldots, R_r$ have been computed
4. Using the critical value λ_i at the chosen confidence level α, find l, the maximum i such that $R_4 > \lambda_i$. The extreme observations removed at the first l steps are declared as outliers.

For a two-sided outlier problem, the value of λ_i is defined as:

$$\lambda_i = \frac{t_{(p,n-i-1)}(n-i)}{\sqrt{(n-i-1+t^2_{(p,n-i-1)})(n-i+1)}}; \quad i = 1, \ldots, r$$

$$p = 1 - \frac{\alpha/2}{n-i+1} \tag{11}$$

where $t_{(p,d)}$ is the pth percentile of a t-distribution with d degrees of freedom. For the one-sided outlier problem, we substitute $\alpha/2$ by α in the value of p. The tabulated values for several α, $n \leq 500$ and $r \leq 10$, and concludes that this approximation is very accurate when $n > 25$ (Rosner 1983).

It is recommended to use this test with a higher number of outliers than expected and when testing for outliers among data coming from a normal distribution.

Example 5.8.4 Performs Rosner's (1983) generalized extreme Studentized deviate procedure to detect outliers in a univariate sample that follows an approximately normal distribution. The dataset is: 0.25, 0.68, 0.94, 1.15, 1.20, 1.26, 1.26, 1.34, 1.38, 1.43, 1.49, 1.49, 1.55, 1.56, 1.58, 1.65, 1.69, 1.70, 1.76, 1.77, 1.81, 1.91, 1.94, 1.96, 1.99, 2.06, 2.09, 2.10, 2.14, 2.15, 2.23, 2.24, 2.26, 2.35, 2.37, 2.40, 2.47, 2.54, 2.62, 2.64, 2.90, 2.92, 2.92, 2.93, 3.21, 3.26, 3.30, 3.59, 3.68, 4.30, 4.64, 5.34, 5.42 and 6.01 (Figure 5.2).

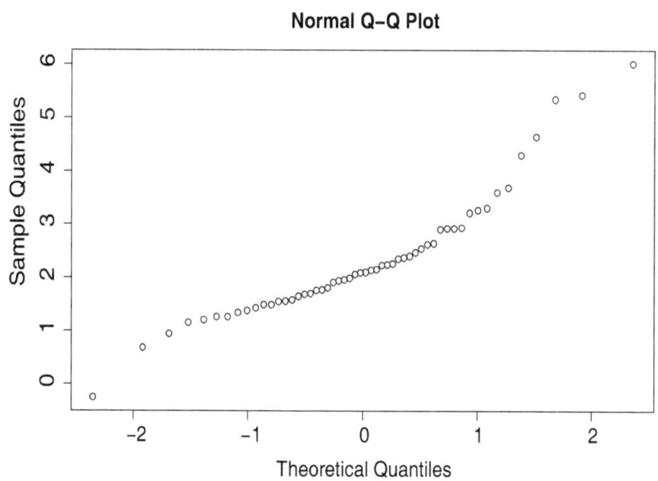

Figure 5.2 Normal probability plot for the given data.

The analysis of R commands is used in the example from NIST (Heckert and Filliben 2003) (see Table 5.1).

Table 5.1 Number of outliers tested with critical value

No. Outliers	Test Stat.	Critical Val.
1	3.118906	3.158794
2	2.942973	3.15143
3	3.179424	3.14389
4	2.810181	3.136165
5	2.81558	3.128247
6	2.848172	3.120128
7	2.279327	3.111796
8	2.310366	3.103243
9	2.101581	3.094456
10	2.067178	3.085425

5.9 LABELING METHODS

Most outlier informal tests generate an interval or criterion for outlier detection instead of hypothesis testing; and any observations beyond the interval or criterion is considered as an outlier. Various location and scale parameters are mostly employed in each labeling method to define a reasonable interval or criterion for outlier detection. There are two reasons for using an outlier labeling method.

1. To find the possible outliers as a screening device before conducting a formal test.
2. To find the extreme values away from the majority of the data regardless of the distribution.

While the formal tests usually require test statistics based on the distribution assumptions and a hypothesis to determine if the largest extreme value is a true outlier of the distribution, most outlier labeling methods present the interval using the location and scale parameters of the data. Although the labeling method is usually simple to use, some observations outside the interval may turn out to be falsely identified outliers after a formal test when the outliers are defined as only observations that deviate from the assumed distribution.

However, if the purpose of the outlier detection is not a preliminary step to find the extreme values violating the distribution assumptions of the main statistical analyses such as the t-test, ANOVA, and regression but mainly to find the extreme values away from the majority of the data regardless of the distribution, the outlier labeling methods may be applicable. In addition, for a large dataset that is statistically problematic, e.g., when it is difficult to identify the distribution of the data or transform it into a proper distribution such as the normal distribution, labeling methods can be used to detect outliers.

5.9.1 Z-Scores

Z-score is a statistical measurement of a score's relationship to the mean in a group of scores. Z-score of 0 means the score is the same as the mean. It can also be positive or negative, indicating whether it is above or below the mean and by how many standard deviations.

This method that can be used to identify outliers in the dataset is the Z-score, using the mean and standard deviation.

$$Z_{score}(i) = \frac{x_i - \overline{x}}{s}, \tag{12}$$

where

$$s = \sqrt{\frac{1}{n-1}\sum_{i=1}^{n}(x_i - \overline{x})^2}$$

The Z-scores based on the property is that if X follows a normal distribution, $N(\mu, \sigma^2)$, then Z follows a standard normal distribution, $z = \frac{x - \mu}{\sigma} \sim N(0,1)$, and Z-scores that exceed 3 in absolute value are generally considered as outliers. This method is simple and it is the same formula as the 3 SD method when the criterion of an outlier is an absolute value of a Z-score of at least 3. A possible maximum Z-score is dependent on the sample size and it computed as $(n-1)/\sqrt{n}$ (Shiffler 1988).

5.9.1.1 *Limitations*

- Since there is no Z-score that exceeds 3 in a sample size less than or equal to 10, then the Z-score is not very good for outlier labeling, particularly in small datasets.
- The standard deviation can be inflated by a few or a single observation having an extreme value. Thus, it can cause a masking problem, i.e., the less extreme outliers go undetected because of the most extreme outlier(s).
- To use Z-scores to identify possible outliers, this can be misleading (particularly for small sample sizes) due to the fact that the maximum Z-score is at most $(n-1)/\sqrt{n}$.

5.9.1.2 *Interpretation of Z-Scores*

We can interpret z-scores in the following ways:

1. If Z-score < 0 then the element is less than the mean.
2. If Z-score > 0 then the element is greater than the mean.
3. If Z-score = 0 then the element is equal to the mean.
4. If Z-score = 1 or 2 then the element is 1 standard deviation greater than the mean; 2 standard deviations greater than the mean, respectively.
5. If Z-score = –1 or –2 then the element is 1 standard deviation less than the mean; 2 standard deviations less than the mean, respectively.
6. If the number of elements in the set is large, about 68% of the elements have a Z-score between –1 and 1; about 95% have a Z-score between –2 and 2; and about 99% have a Z-score between –3 and 3 (Figure 5.3).

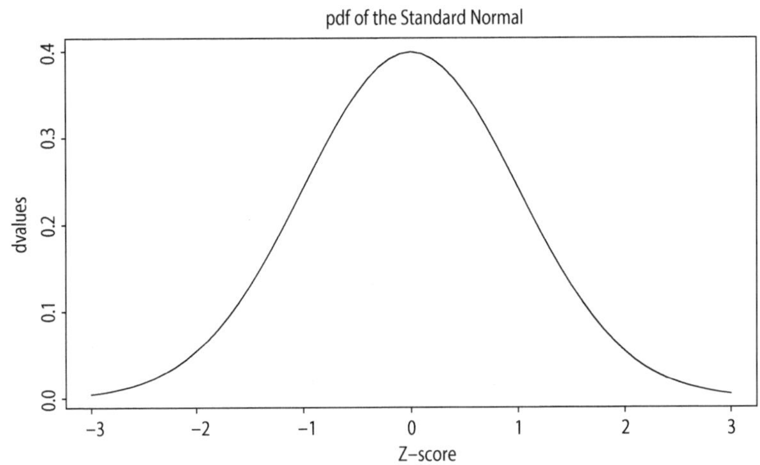

Figure 5.3 Standard Scores of normal distribution.

As a general rule, Z-scores less than –1.96 or greater than 1.96 are considered unusual and generally very interesting. They are also synonymous with being statistically significant and outliers.

Example 5.9.1 In this study, the diabetes data was obtained from the primary health center. This data has 50 observations for the patient's diabetes levels (Kannan et al. 2015).

Table 5.2 clearly represents the observation that 525 is an outlier. Now, we can see the 48th and 49th observations also may be considered as outliers, but due to masking problems the two observations are not under consideration for outliers.

Table 5.2 Computation of the Z-scores

Obs. No.	x_i	Z-score	Obs. No.	x_i	Z-score
1	70	−0.83	26	110	−0.24
2	75	−0.76	27	113	−0.20
3	81	−0.67	28	114	−0.18
4	84	−0.63	29	117	−0.14
5	84	−0.63	30	117	−0.14
6	84	−0.63	31	119	−0.11
7	85	−0.61	32	121	−0.08
8	85	−0.61	33	121	−0.08
9	86	−0.60	34	127	0.01
10	92	−0.51	35	130	0.05
11	93	−0.49	36	131	0.07
12	95	−0.46	37	132	0.08
13	95	−0.46	38	134	0.11
14	96	−0.45	39	135	0.13
15	96	−0.45	40	136	0.14
16	96	−0.45	41	139	0.19
17	98	−0.42	42	153	0.39
18	99	−0.40	43	155	0.42
19	101	−0.38	44	166	0.59
20	101	−0.38	45	169	0.63
21	105	−0.32	46	172	0.68
22	106	−0.30	47	175	0.72
23	108	−0.27	48	236	1.62
24	109	−0.26	49	236	1.62
25	110	0.24	50	**525**	**5.90**

5.9.2 Modified Z-Scores

Based on the Z-scores that were used, two estimators, i.e., the sample mean ($\overline{x}$) and sample standard deviation(s) can be affected by a few extreme values or by even a single extreme value. To resolve this problem, the median and the median of the absolute deviation (MAD) are employed in the modified Z-scores instead of the mean and standard deviation of the sample, respectively (Iglewicz and Hoaglin 1993).

$$MAD = median\{|x_i - \tilde{x}|\} \tag{13}$$

where $\tilde{x}$ is the sample median.

$$M_i = \frac{0.6745(x_i - \tilde{x})}{MAD} \quad (14)$$

where $E(MAD) = 0.675\sigma$ for large normal data.

The observations are labeled outliers when $|M_i| > 3.5$ through the simulation based on pseudo-normal observations for sample sizes of 10, 20, and 40. The M_i score is effective for normal data in the same way as the Z-score (Iglewicz and Hoaglin 1993).

5.9.2.1 *Numerical Example*

Table 5.3 shows that in the computed data values of the modified Z-scores in absolute value, out of these 3 observations (236, 236, and 525), may well be outliers.

Table 5.3 Computation of Z-Scores compared with the modified Z-Scores

i	x_i	Case-1	Case-2	mz-score	i	x_i	Case-1	Case-2	mz-score
1	70	−0.83	−1.35	−1.42	26	110	−0.24	−0.23	0.0000
2	75	−0.76	−1.21	−1.2425	27	113	−0.2	−0.15	0.1065
3	81	−0.67	−1.04	−1.0295	28	114	−0.18	−0.12	0.1420
6	84	−0.63	−0.96	−0.923	29	117	−0.14	−0.03	0.2485
5	84	−0.63	−0.96	−0.923	30	117	−0.14	−0.03	0.2485
6	84	−0.63	−0.96	−0.923	31	119	−0.11	0.02	0.3195
7	85	−0.61	−0.93	−0.8875	32	121	−0.08	0.08	0.3905
8	85	−0.61	−0.93	−0.8875	33	121	−0.08	0.08	0.3905
9	86	−0.6	−0.9	−0.852	34	127	0.01	0.25	0.6035
10	92	−0.51	−0.73	−0.639	35	130	0.05	0.33	0.7100
11	93	−0.49	−0.7	−0.6035	36	131	0.07	0.36	0.7455
12	95	−0.46	−0.65	−0.5325	37	132	0.08	0.39	0.7810
13	95	−0.46	−0.65	−0.5325	38	134	0.11	0.44	0.8520
14	96	−0.45	−0.62	−0.497	39	135	0.13	0.47	0.8875
15	96	−0.45	−0.62	−0.497	40	136	0.14	0.5	0.9230
16	96	−0.45	−0.62	−0.497	41	139	0.19	0.58	1.0295
17	98	−0.42	−0.56	−0.426	42	153	0.39	0.97	1.5265
18	99	−0.4	−0.54	−0.3905	43	155	0.42	1.03	1.5975
19	101	−0.38	−0.48	−0.3195	44	166	0.59	1.34	1.9880
20	101	−0.38	−0.48	−0.3195	45	169	0.63	1.42	2.0945
21	105	−0.32	−0.37	−0.1775	46	172	0.68	1.5	2.2010
22	106	−0.3	−0.34	−0.142	47	175	0.72	1.59	2.3075
23	108	−0.27	−0.29	−0.071	49	**236**	1.62	3.29	4.4730
24	109	−0.26	−0.26	−0.0355	48	**236**	1.62	3.29	4.4730
25	110	−0.24	−0.23	0.0000	50	**525**	5.9	–	14.7325

5.9.3 Median Absolute Deviation (MADe)

This method is one of the basic robust methods which are largely unaffected by the presence of extreme values of the dataset. It is similar to the SD method. However, the median and MADe are employed in this method instead of the mean and standard deviation. It can be defined as follows:

$$2MAD_e : Median \pm 2MAD_e \tag{15}$$

$$3MAD_e : Median \pm 3MAD_e \tag{16}$$

where $MAD_e = 1.483 \times MAD$ for large normal data and is an estimator of the spread in a data similar to the standard deviation.

$$MAD = median\{|x_i - median(x)|\},\ i = 1, 2, \ldots, n \tag{17}$$

The MAD is scaled by a factor of 1.483; it is similar to the standard deviation in normal distribution or absolute deviation around the median as stated in the title and is a robust measure of central tendency.

Example 5.9.2 This method was computed from the dataset results as follows: from the equations MAD_e = 28.1694, *Median* = 110, MAD = 19. Here the 2 MAD_e method has identified 6 outliers which are 172, 525, 175, 236, 169 and 236. Also, the 3 MAD_e method has identified 3 outliers which are: 525, 236, and 236.

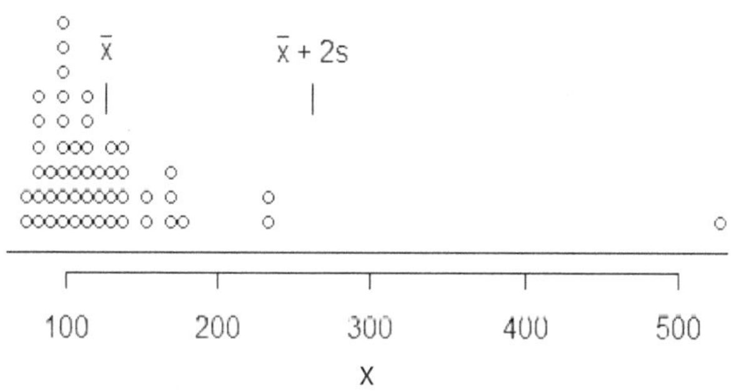

Figure 5.4 Dotplot for visualize the data with outliers.

In Figure 5.4, the extreme value at x = 525 has dragged $\bar{x} + 2s$ which is the outlier cutoff, above the same two points at x = 236, 236. Only the point at x = 525 is therefore caught as an outlier, even though in the point at x = 236, 236 is clearly also an outlier.

5.9.4 Tukey's Method (Box Plot)

Box plot, is a simple graphical tool to visualize information about continuous univariate data, such as the median, lower quartile, upper quartile, lower extreme and upper extreme of a dataset. This method uses *IQR* to find outliers and filter out very large or very small numbers.

$$\text{Low outliers} = Q_1 - 1.5(Q_3 - Q_1) = Q_1 - 1.5\,(IQR) \quad (18)$$

$$\text{High outliers} = Q_1 + 1.5(Q_3 - Q_1) = Q_1 + 1.5\,(IQR) \quad (19)$$

where, Q_1 is the first quartile, Q_3 is the third quartile, *IQR* is the interquartile range. These equations give two values, or 'fences'. A fence cordons off the outliers from all of the values that are contained in the bulk of the data.

5.9.4.1 Procedures

The following steps are given for finding outliers using *IQR*:

- Find the *IQR* and Median.
- Find Q_1 and Q_3. Q_1 can be thought of as a median in the lower half of the data. Q_3 can be thought of as a median for the upper half of data. Subtract Q_1 from Q_3.
- Compute 1.5 * *IQR* and subtract from Q_1 to get lower fence.
- Add to Q_3 to get upper fences.
- Add fences to the data to identify outliers.

In this method the result obtained from the dataset is: the *IQR* (Inter Quartile Range) is the distance $Q_1 = 95.25$, $Q_3 = 133.5$ and *IQR* = 38.25. Thus, the inner fences are 37.875, 190.875 and outer fences are –19.5, 248.25. (Table 5.4.) The two extreme values are 236 and 525 and are identified as probable outliers in this method. Figure 5.5 is a box plot for the dataset.

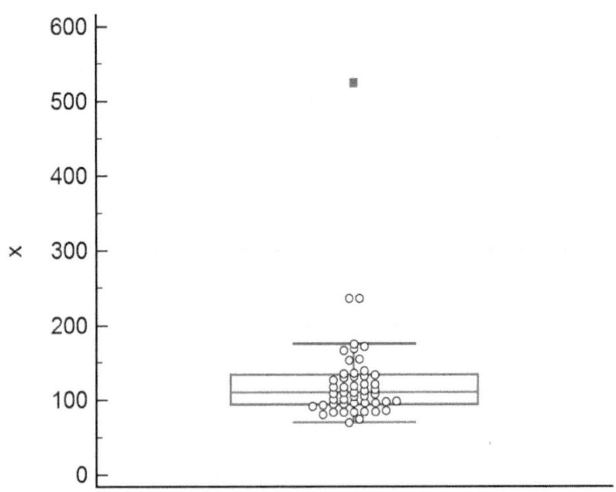

Figure 5.5 Box and Whisker plot for visualizing outliers.

Table 5.4 Tukey's method outlier detection using *IQR*

Sample Size	**50**
Lowest value	70.0000
Highest value	525.0000
Arithmetic mean	126.3400
Median	110.0000
Standard deviation	67.5533
Coefficient of Skewness	4.4967 ($P < 0.0001$)
Coefficient of Kurtosis	25.1797 ($P < 0.0001$)
Suspected outliers (Tukey 1977)	
Outside values	236 236
Far-out values	525

5.10 MULTIVARIATE OUTLIER DETECTION METHODS

In statistical analysis, it can be challenging to identify outliers in multivariate data. The multivariate outlier is the combination of unusual scores on at least two variables. Both types of outliers can influence the outcome of statistical analyses. A multivariate outlier need not be an extreme in any of its components. The idea of extremeness inevitably arises from some form of 'ordering' of the data. They are based on (robust) estimation of location and scatter, or on quantiles of the data. The basis for multivariate outlier detection is the Mahalanobis distance. The standard method for multivariate outlier detection is robust estimation of the parameters in the Mahalanobis distance and the comparison with a critical value of the χ^2 distribution (Rousseeuw and Zomeren 1990). However, also values larger than this critical value are not necessarily outliers, they could still belong to the data distribution.

5.10.1 Multivariate Normality Assumption

Multivariate normality is defined when each variable under consideration is normally distributed with respect to each other variable. Multivariate normal distributions take the form of symmetric three-dimensional bells when the x axis is the values of a given variable, the y axis is the count for each value of the x variable, and the z axis is the values of any other variable under consideration.

The variable $d^2 = (x - \mu)'\Sigma^{-1}(x - \mu)$ has a chi-square distribution with p degrees of freedom, and for 'large' samples the observed Mahalanobis distances have an approximate chi-square distribution. This result can be used to evaluate (subjectively) whether a data point may be an outlier and whether the observed data may have a multivariate normal distribution (Figure 5.6).

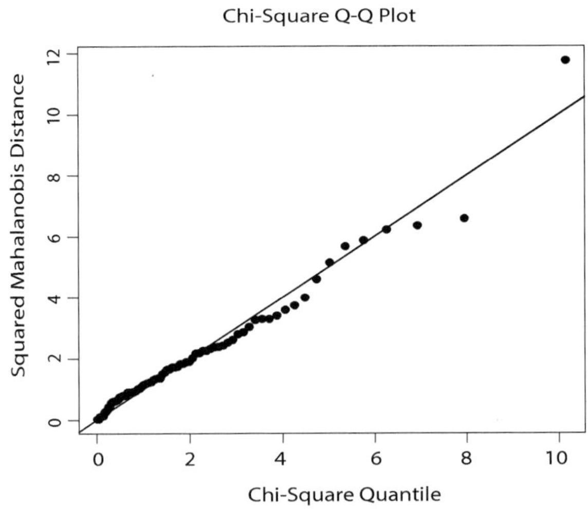

Figure 5.6 Q-Q Plot for evaluating multivariate normality and outliers.

5.10.2 Mahalanobis Distance (MD_i)

Mahalanobis distance (*MD*) is a statistical measure of the extent to which cases are multivariate outliers, based on a χ^2 distribution, assessed using $p < .001$. The shape and size of multivariate data are measured by the covariance matrix.

A familiar distance measure which takes into account the covariance matrix is the Mahalanobis distance. A classical approach for detecting outliers is to compute the Mahalanobis distance for each observation $x_i (i = 1, \ldots, n)$:

$$MD_i = \sqrt{(x_i - \bar{x})^T \Sigma^{-1} (x_i - \bar{x})} \quad (20)$$

where,

$$\Sigma = \frac{1}{n-1} \sum_{i=1}^{n} (x_i - \bar{x})(x_i - \bar{x})^T \quad (21)$$

The distance explained how far it is from the center of the cloud, taking into account the shape of the cloud as well. It is well known that this approach suffers from the masking effect, by which multiple outliers do not necessarily have a large MD_i.

An iterative Mahalanobis distance type of method for the detection of multiple outliers in multivariate data that tackles problems which can arise with the method,

- It will be referred to as method
- Outlying observations may not necessarily have large values for D_i because masking may occur, i.e., one outlier masks the appearance of another outlier.

The critical chi-square values for 2 to 10 degrees of freedom at a critical α of .001 are shown below. For $df > 10.$, a maximum MD_i that is larger than the critical chi-square value for $df = k$ at a critical α value of .001 indicates the presence of one or more multivariate outliers.

5.10.2.1 *Numerical Illustration*

We have used in this example dataset (Chicco and Jurman 2020) retrieved from UCI Repository which contains the medical records of 299 patients who had heart failure. It was collected during their follow-up period, where each patient profile had 13 clinical features.

The large blue point on the plot is the center of the data. The black points are the measured observations for creatinine phosphokinase and platelets. The 17 points outside the ellipse are identified as outliers at a significance level of 0.95 based on Mahalanobis distance. These points are listed below (Figure 5.7, Tables 5.5 and 5.6).

Table 5.5 Survival of patients with heart failure dataset information (Chicco and Jurman 2020).

Feature	Explanation	Measurement	Range
Age	Age of the patient	Years	[40, …, 95]
Anaemia	Decrease of red blood cells or hemoglobin	Boolean	0, 1
High blood pressure	If a patient has hypertension	Boolean	0, 1
Creatinine phosphokinase (CPK)	Level of the CPK enzyme in the blood	mcg/L	[23, …, 7861]
Diabetes	If the patient has diabetes	Boolean	0, 1
Ejection fraction	Percentage of blood leaving the heart at each contraction	Percentage	[14, …, 80]
Sex	Woman or man	Binary	0, 1
Platelets	Platelets in the blood	kiloplatelets/mL	[25.01, …, 850.00]
Serum creatinine	Level of creatinine in the blood	mg/dL	[0.50, …, 9.40]
Serum sodium	Level of sodium in the blood	mEq/L	[114, …, 148]
Smoking	If the patient smokes	Boolean	0, 1
Time	Follow-up period	Days	[4, …, 285]
(target) death event	If the patient died during the follow-up period	Boolean	0, 1

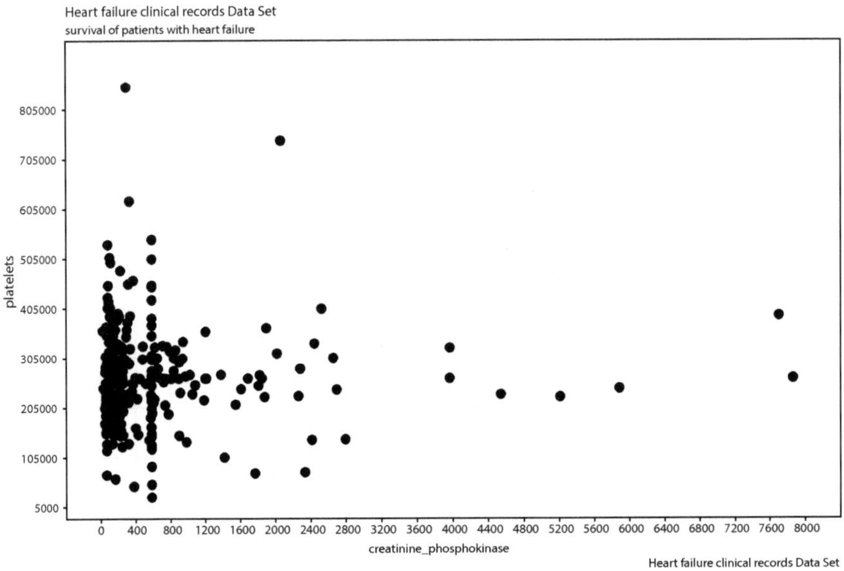

(a)

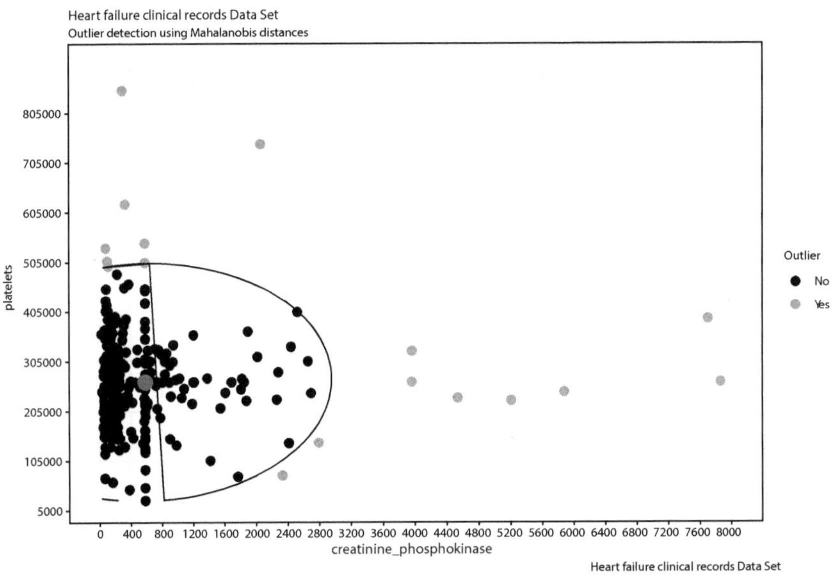

(b)

Figure 5.7 The plots depict for identifying multivariate outliers using Mahalanobis distance.

Table 5.6 Outliers points detected by Mahalanobis distance.

Observation	Creatinine_ phosphokinase	Platelets	mdist	p	Outlier
2	7861	263358	56.314647	0.0000000000005907855	Yes
53	3964	263358	12.157582	0.0022909447786599074	Yes
61	7702	390000	55.093805	0.0000000000010877581	Yes
70	113	497000	6.000263	0.0497805200093326822	Yes
73	5882	243000	29.955374	0.0000003128046725073	Yes
104	5209	226000	22.990742	0.0000101770945106459	Yes
106	328	621000	13.494869	0.0011738876224143557	Yes
110	292	850000	36.176039	0.0000000139467463279	Yes
118	102	507000	6.514409	0.0384958588453466824	Yes
135	4540	231000	16.826798	0.0002218743726946240	Yes
164	2334	75000	7.144366	0.0280944528062658933	Yes
172	3966	325000	12.461813	0.0019676675532171459	Yes
225	582	504000	6.057396	0.0483785844347474139	Yes
228	2794	141000	6.906763	0.0316384766020072344	Yes
241	81	533000	7.941623	0.0188581234894770917	Yes
288	582	543000	8.179902	0.0167400496335352519	Yes
297	2060	742000	25.921603	0.0000023506902237951	Yes

5.11 OUTLIERS IN REGRESSION

An observation that is substantially different from all other ones can make a large difference in the results of regression analysis. Outliers occur very frequently in real data, and they often go unnoticed because nowadays much data is processed by computers, without careful inspection or screening. Outliers may be a result of keypunch errors, misplaced decimal points, recording or transmission errors, exceptional phenomena such as earthquakes or strikes, or members of a different population slipping into the sample. Robust regression is an alternative to least squares regression when data is contaminated with outliers or influential observations and it can also be used for the purpose of detecting influential observations.

The Regression analysis is to fit the equations to the observed variables. The usual methods of linear model are defined as:

$$y_i = \beta_0 + \beta_1 x_{i1} + \cdots + \beta_p x_{ip} + e_i \quad \text{for} \quad i = 1, \ldots, n \tag{22}$$

where n is the sample size, y_i is the response variable, and $x_{i1}, \ldots, x_{ip}$ are called explanatory variables. e_i is the error term that is assumed as normally distributed by mean zero and standard deviation σ.

$$\beta = \begin{bmatrix} \beta_1 \\ \vdots \\ \beta_p \end{bmatrix} \tag{23}$$

from the data

$$\textit{Cases}\overset{\textit{Variables}}{\begin{bmatrix} x_{11} & \cdots & x_{1p} & y_1 \\ \vdots & & \vdots & \vdots \\ x_{i1} & \cdots & x_{ip} & y_i \\ \vdots & & \vdots & \vdots \\ x_{n1} & \cdots & x_{np} & y_n \end{bmatrix}} \tag{24}$$

Applying a regression estimator to such a dataset yields

$$\hat{\beta} = \begin{bmatrix} \hat{\beta}_1 \\ \vdots \\ \hat{\beta}_p \end{bmatrix} \tag{25}$$

where the estimates $\hat{\beta}_j$ are called the regression coefficients. Vectors and matrices will be denoted by boldface throughout. Although the actual β_j are unknown, one can multiply the explanatory variables with these $\hat{\beta}_j$ and obtain

$$\hat{y}_i = x_{i1}\hat{\beta}_1 + \cdots + x_{ip}\hat{\beta}_p. \tag{26}$$

where $\hat{y}_i$ is called the predicted or estimated value of y_i. The residual r_i of the ith case is the difference between what is actually observed and what is estimated:

$$r_i = y_i - \hat{y}_i. \tag{27}$$

As an illustration, at the effect of outliers in the simple regression model

$$y_i = \beta_1 x_i + \beta_2 + e_i \tag{28}$$

in which the slope β_1 and the intercept β_2 are to be estimated. In the simple regression model, one can make a plot of the (x_i, y_i), which is sometimes called a scatter plot, in order to visualize the data structure.

5.11.1 Outliers and Leverage

Outliers play an important role in regression. It is common practice to distinguish between two types of outliers. Outliers in the response variable represent model failure. Such observations are called outliers. Outliers with respect to the predictors are called leverage points. They can affect the regression model, too. Their response variables need not be outliers. In regression it helps to make a distinction between the two types of leverage points:

1. Good leverage point; 2. Bad leverage point.

5.11.1.1 Good Leverage Point

A good leverage point has limited effect on giving a distorted view of the how majority of points are associated. Good leverage points improve the precision of the regression coefficients.

5.11.1.2 Bad Leverage Point

A bad leverage point has a grossly effect estimate of the slope of the regression line if an estimator with a small breakdown point is used. Bad leverage points reduce the precision of the regression coefficients.

5.11.2 Leverage Point (h_i)

An observation with an extreme value on a predictor variable is called a point with high leverage. In linear regression, identification of leverage points may be quite easy to detect. In the linear regression model, the leverage score for ith data unit is defined as

$$h_{ii} = (H)_{ii} \tag{29}$$

The ith diagonal of the hat matrix $H = X(X'X)^{-1}X'$. Leverage values fall between 0 and 1. Investigate observations with leverage values greater than $3p/n$, where p is the number of model terms (with constant) and n is the number of observations (Figure 5.8).

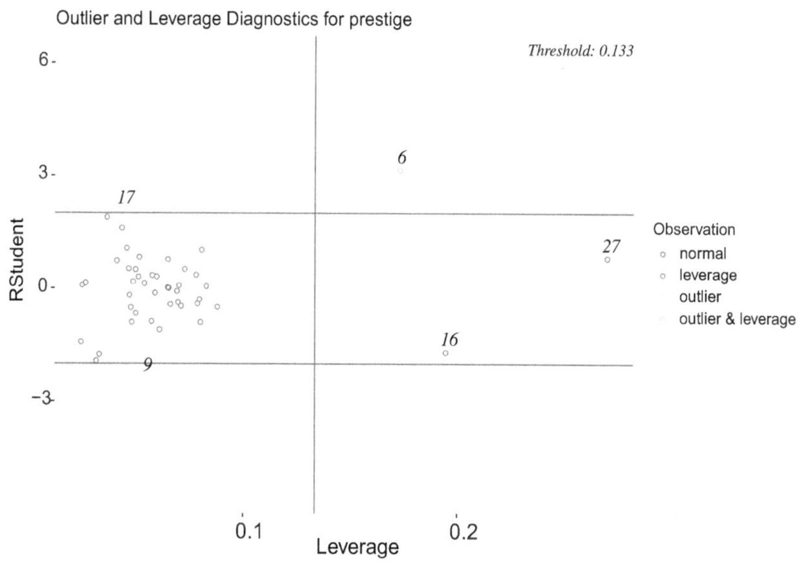

Figure 5.8 Graph for detecting influential observations [Duncan Data (Table 5.7)].

Table 5.7 Duncan's occupational prestige data.

	Type	Income	Education	Prestige		Type	Income	Education	Prestige
accountant	prof	62	86	82	store.clerk	wc	29	50	16
pilot	prof	72	76	83	carpenter	bc	21	23	33
architect	prof	75	92	90	electrician	bc	47	39	53
author	prof	55	90	76	RR.engineer	bc	81	28	67
chemist	prof	64	86	90	machinist	bc	36	32	57
minister	prof	21	84	87	auto.repairman	bc	22	22	26
professor	prof	64	93	93	plumber	bc	44	25	29
dentist	prof	80	100	90	gas.stn.attendant	bc	15	29	10
reporter	wc	67	87	52	coal.miner	bc	7	7	15
engineer	prof	72	86	88	streetcar.motorman	bc	42	26	19
undertaker	prof	42	74	57	taxi.driver	bc	9	19	10
lawyer	prof	76	98	89	truck.driver	bc	21	15	13
physician	prof	76	97	97	machine.operator	bc	21	20	24
welfare.worker	prof	41	84	59	barber	bc	16	26	20
teacher	prof	48	91	73	bartender	bc	16	28	7
conductor	wc	76	34	38	shoe.shiner	bc	9	17	3
contractor	prof	53	45	76	cook	bc	14	22	16
factory.owner	prof	60	56	81	soda.clerk	bc	12	30	6
store.manager	prof	42	44	45	watchman	bc	17	25	11
banker	prof	78	82	92	janitor	bc	7	20	8
bookkeeper	wc	29	72	39	policeman	bc	34	47	41
mail.carrier	wc	48	55	34	waiter	bc	8	32	10
insurance.agent	wc	55	71	41					

5.11.3 Cook's Distance (D_i)

Dennis Cook (1977) introduced a distance measure in 1977 for commonly used estimates to study the influence of a data point when performing least squares regression analysis.

$$D_i = \frac{\sum_{j=1}^{n} (\widehat{Y}_i - \widehat{Y}_j(i))^2}{pMSE} \tag{30}$$

The following equation equally expressed as

$$D_i = \frac{e_i^2}{pMSE}\left[\frac{h_{ii}}{(1-h_{ii})^2}\right] \tag{31}$$

$$D_i = \frac{(\widehat{\beta} - \widehat{\beta}^{-i})^T (X^T X)(\widehat{\beta} - \widehat{\beta}^{-i})}{(1+p)s^2} \tag{32}$$

where, $\widehat{\beta}$ is the LS estimate of β and $\widehat{\beta}^{-i}$ is the LS estimate of β on the dataset without case i. $\widehat{Y}_j$ is the prediction from the full regression model for observation j; $\widehat{Y}_{j(i)}$ is the prediction for observation j from a refitted regression model in which observation i has been omitted. h_{ii} is the ith diagonal elements of the hat matrix $X(X^T X)^{-1}X^T$. MSE is the Mean square error p is the number of fitted parameters (see Figures 5.9 and 5.10).

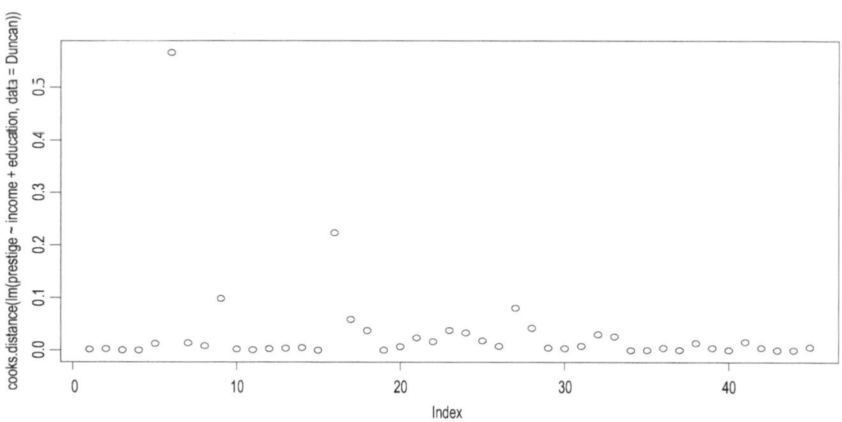

Figure 5.9 Cook's distance scatter plot.

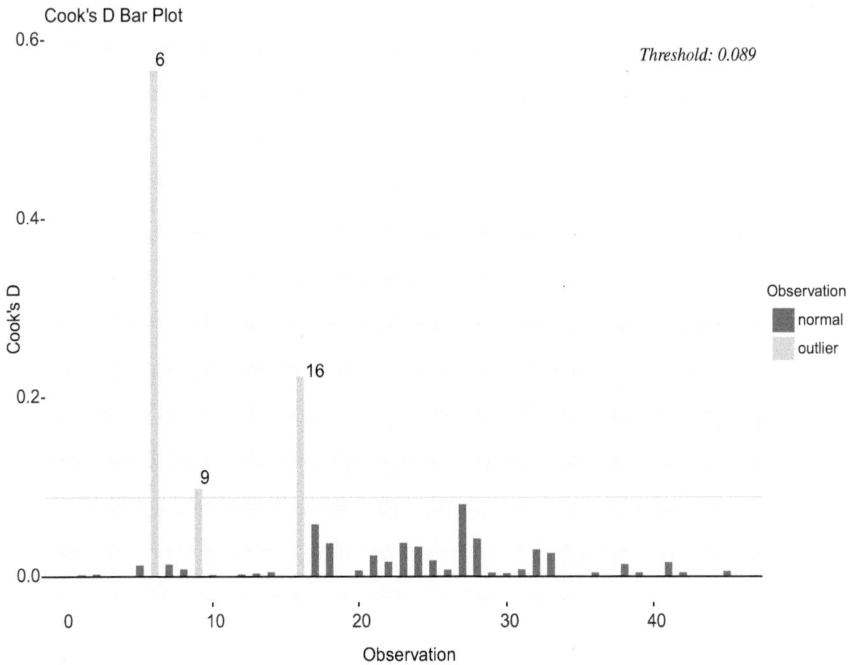

Figure 5.10 A data point having a large Cook's *D* indicates that the data point strongly influences the fitted values.

5.11.3.1 Numerical Illustration

The Duncan data frame in Table 5.7 has 45 rows and 4 columns. Data on the prestige and other characteristics of 45 U.S. occupations in 1950 (Duncan and Reiss 1961).

5.11.4 Difference in Fits (DFFITS)

DFFITS is the diagnostic tool for statistical regression models and it shows the influence point. It is a standardized function of the difference between the predicted value for the observation when it is included in the dataset and when (only) it is excluded in the dataset. This quantity measures how much the regression function changes at the *i*th observation when the *i*th variable is deleted.

$$DFFITS_i = \frac{\widehat{Y}_i - \widehat{Y}_{i(i)}}{s_{(i)}\sqrt{v_{ii}}} \tag{33}$$

$$= \frac{v_{ii}}{1-v_{ii}}\left(\frac{e_i}{s_{(i)}\sqrt{1-v_{ii}}}\right) \tag{34}$$

$$(n-p'-1)s_{(i)}^2 = (n-p')s^2 - \frac{e_i^2}{1-v_{ii}}$$

For small sample datasets, the values of 1 or greater values are considered as 'suspicious'. In large samples of datasets, values of $2\sqrt{p/n}$ influence diagnostics for prestige (Figure 5.11).

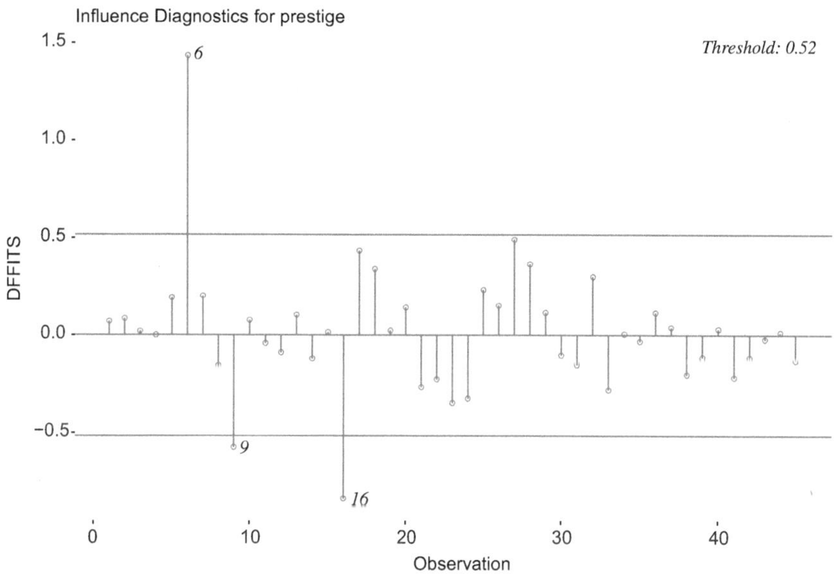

Figure 5.11 DFFITS: difference in fits, is used to identify influential data points.

5.12 APPLICATIONS OF OUTLIER DETECTION

Outlier's detection can be applied on lot of datasets for various purpose, some of which are discussed below:

- Outlier detection techniques are applied to detect Credit Card Fraud, Fraudulent Applications, and Fraudulent Usage of Credit Card Associated with credit card thefts.
- Loan application processing to detect fraudulent applications or potentially problematic customers.
- Intrusion detection is the way of detecting unauthorized access in computer networks.

- A very small percentage of false alarms in millions of data items can have a huge impact on an analyst.
- Labeled data not usually available for intrusions: This gives preference to semi-supervised and unsupervised outlier detection techniques. Intrusion detection systems have been classified into host based and network based intrusion detection systems Denning (1987).
- Detecting mobile phone fraud by monitoring phone activity or suspicious trades in the equity markets.
- Monitoring the performance of computer networks such as to detect network bottlenecks.
- Monitoring processes to detect faults in motors, generators, pipelines or space instruments on shuttles.
- To detect faulty production runs for example cracked beams.
- Identifying novel features or misclassified features in satellite images.
- Detecting novelties in images—for robot neotaxis or surveillance systems.
- Detection of background-independent moving image features by motion segmentation.
- Time-series monitoring safety critical applications such as drilling or high-speed milling.
- Medical condition monitoring, such as heart-rate monitors and identifying novel molecular structures.
- Detecting unexpected entries in databases for data mining to detect errors, frauds, or valid but unexpected entries.
- Detecting mislabeled data in a training dataset.

Overall, outlier detection plays a crucial role in various domains and applications for identifying unusual patterns or behaviors in data.

5.13 SUMMARY AND DISCUSSION

In this chapter we discussed several outliers' identification methods for both univariate and multivariate observations. In addition, several problems are addressed while detecting outliers using different methods. Especially, dealing with outliers when swamping and masking problems occurred. Also we suggested various treatments, if the observations were considered as outliers. We should carefully handle outlying observations, otherwise it may lead to biased results. However, the main benefit of the outliers in cross sectional data is that they reveal interesting facts.

Outliers have importance as they appear different from the remaining data and having some genuine causes.

REFERENCES

Anscombe, F.J. and Guttman, I. 1960. Rejection of outliers. Technometrics. 2(2): 123–147.

Barnett, V. and Lewis, T. 1978. Outliers in Statistical Data. Wiley.

Barnett, V. and Lewis, T. 1984. Outliers in Statistical Data, 2nd Edn. Wiley.

Barnett, V. and Lewis, T. 1994. Outliers in Statistical Data, 3rd Edn. Wiley.

Beckman, R.J. and Cook, R.D. 1983. Outlier..........s. Technometrics. 25(2): 119–149.

Bendre, S.M. and Kale, B.K. 1985. Masking effect on tests for outliers in exponential models. Journal of the Royal Statistical Society. 80(392): 1020–1025.

Bernoulli, D. 1777. The most probable choice between several discrepant observations and the formation therefrom of the most likely induction, in C.G. Allen (1961). Biometrika. 48: 3–13.

Bessel, F.W and Baeuer, J.J. 1838. Gradmessung in Ostpreussen und ihre Verbindung mit Preussischen und Russischen Dreiecksketten. Berlin: Dümmler.

Chicco, D. and Jurman, G. 2020. Machine learning can predict survival of patients with heart failure from serum creatinine and ejection fraction alone. BMC Medical Informatics and Decision Making. 20(1): 16. Retrieved from https://doi.org/10.1186/s12911-020-1023-5 doi: 10.1186/s12911-020-1023-5.

Cook, D.R. 1977. Detection of influential observation in linear regression. Technometrics. 15–18.

Cousineau, D. and Chartier, S. 2010. Outliers detection and treatment: A review. International Journal of Psychological Research. 3: 58–67.

Dang, X. and Serflling, R. 2009. Nonparametric depth-based multivariate outliers' identifiers and masking robustness properties. Journal of Statistical Planning and Inference.

Davies, L. and Gather, U. 1993. The identification of multiple outliers. Journal of the American Statistical Association. 88(423): 782–792.

Denning, D. 1987, Feb. An intrusion-detection model. Software Engineering, IEEE Transactions on, SE-13. (2): 222–232.

Dixon, W.J. 1951. Ratios involving extreme values. Annals Mathematics Statistical. 22(1): 68–78.

Duncan, O.D. and Reiss, A.J. 1961. A socioeconomic index for all occupations. Class: Critical Concepts. 1(1): 388–426.

Grubbs, F.E. 1950. Sample criteria for testing outlying observations. Annals Mathematics Statistical. 21(1): 27–58.

Grubbs, F.E. 1969. Procedures for detecting outlying observations in samples. American Statistical Association and American Society for Quality. 62(11): 1- 21.

Hawkins, D.M. 1980. Identification of Outliers. Chapman and Hall.

Hayes, K., Kinsella, A. and Coffey, N. 2007. A note on the use of outlier criteria in Ontario laboratory quality control schemes. Clinical Biochemistry. 40(3–4): 147–152.

Heckert, N. and Filliben, J. 2003. 2003-06-01 00:06:00. NIST/SEMATECH e-Handbook of Statistical Methods; Chapter 1: Exploratory Data Analysis.

Iglewicz, B. and Hoaglin, D.C. 1993. How to detect and handle outliers. Wisconsin: ASQC Quality Press. 16.

Kannan, K.S., Manoj, K. and Arumugam, S. 2015. Labeling methods for identifying outliers. International Journal of Statistics and Systems (IJSS). 10(2): 231–238.

Legendre, A.M. 1805. Nouvelles méthodes pour la détermination des orbites des comètes. Appendice sur la méthodes des moindres carrés. Didot, Paris. 80.

Li, X., Zhang, H., Ranish, J. and Aebersold, R. 2003. Automated statistical analysis of protein abundance ratios from data generated by stable-isotope dilution and tandem mass spectrometry. Analytical Chemistry. 75(23): 6648–6657.

Nag, A.K., Mitra, A. and Mitra, S. 2005. Multiple outlier detection in multivariate data using self-organizing maps title. Computational Statistics. 20(2): 245–264.

Nedret, B. and Gulsen, K. 2008. A comparison of multiple outlier detection methods for regression data. Communications in Statistics—Simulation and Computation. 37(3): 521–545.

Pell, R.J. 2000. Multiple outlier detection for multivariate calibration using robust statistical techniques. Chemometrics and Intelligent Laboratory Systems. 52(1): 87–104.

Rosner, B. 1983. Percentage points for a generalized esd many-outlier procedure. Technometrics. 25: 165.

Rousseeuw, P.J. and Zomeren, B.C.v. 1990. Unmasking multivariate outliers and leverage points. Journal of the American Statistical Association. 85(411): 633–639.

Shiffler, R.E. 1988. Maximum z scores and outliers. The American Statistician. 42(1): 79–80. Retrieved from https://www.tandfonline.com/doi/abs/10.1080/00031305.1988.10475530 doi: 10.1080/00031305.1988.10475530.

Srivastava, M.S. and von Rosen, D. 1998. Outliers in multivariate regression models. Journal of Multivariate Analysis. 65(2): 195–208.

Stefansky, W. 1972. Rejecting outliers in factorial designs. Technometrics. 14(2): 469–479. Retrieved 2022-12-29, from http://www.jstor.org/stable/1267436.

Tukey, J.W. 1977. Exploratory data analysis, Vol. 2: 131–160.

Velasco, F. and Verma, S. 1998. Importance of skewness and kurtosis statistical tests for outlier detection and elimination in evaluation of geochemical reference materials. Mathematical Geology. 30(1): 109–128.

Verma, S.P. 1997. Sixteen statistical tests for outlier detection and rejection in evaluation of international geochemical reference materials: Example of microgabbro pm-s. Geostandards Newsletter. 21(1): 59–75.

Verma, S.P., Díaz-González, L. and González-Ramírez, R. 2009. Relative efficiency of single-outlier discordancy tests for processing geochemical data on reference materials and application to instrumental calibrations by a weighted least-squares linear regression model. Geostandards and Geoanalytical Research. 33(1): 29–49.

Verma, S.P. and Quiroz-Ruiz, A. 2006. Critical values for 22 discordancy test variants for outliers in normal samples up to sizes 100, and applications in science and engineering. Revista mexicana de ciencias geológicas. 23(3): 302–319.

Zimmerman, D. 1994. A note on the influence of outliers on parametric and nonparametric tests. Journal of General Psychology. 4(121): 391–401.

Chapter 6

Statistical Modeling of Maximum Rainfall in Guinea

Malick Kebe[1], Saralees Nadarajah*[2], M. Masoom Ali[3]
[1]Department of Mathematics, Howard University, Washington DC, USA, malick.kebe@biston.howard.edu
[2]Department of Mathematics, University of Manchester, Manchester M13 9PL, UK, mbbsssn2@manchester.ac.uk
[3]Department of Mathematical Sciences, Ball State University, Muncie, IN, USA, mali@bsu.edu

6.1 INTRODUCTION

The Republic of Guinea is a West African nation located between the latitudes 7–13 degrees north. It borders the Atlantic Ocean, stretching along it for 300 kilometres. Guinea has an area of 245,857 square kilometres (94,926 square miles). It is bordered by Mali to the northeast, Senegal to the north, Guinea-Bissau to the northwest, Liberia and Sierra Leone to the south, Ivory Coast to the southeast, and the Atlantic Ocean to the west. Guinea is a member of the Sahel nations and a coastal west African country.

Known for its fertile land, abundance of natural resources, biodiversity, and number of rivers either passing through it or originating from it which earns it the name "Water tower of west Africa", Guinea is a tropical country with just two seasons: A dry season and a rainy season. The dry season starts in November and ends in March while the wet

*For Correspondence: Saralees Nadarajah (mbbsssn2@manchester.ac.uk)

season covers the rest of the year. Guinea's capital Conakry is the world's most humid capital with an annual rainfall exceeding 4,000 millimeters.

Politically, Guinea is subdivided into eight administrative regions (including the special region of Conakry, Guinea's capital), it has four natural geographic subdivisions characterized by distinct topographies and climates. In order of percentage of the country covered, they are: upper Guinea (La Haute-Guinee) covering 38% of the country, forested Guinea (Guinee Forestiere) covering 23% of the country, middle Guinea (La Moyenne-Guinee) covering 20% of the country and maritime/lower Guinea (La Guinee Maritime) covering 18% of the country. Both upper Guinea and forested Guinea are forested and mountainous.

Figure 6.1 shows a map delineating the natural geographic partitions of Guinea. Maritime/lower Guinea is a coastal zone. It includes the coast and the coastal plain. Middle Guinea is an area of highlands with an altitude of hundreds of meters above sea level. It is a mountainous area with a mountain microclimate. Many rivers have their source in this region. Upper Guinea is a region of plateaus and savannas in the north-east of the country. Forested Guinea is located in the south-east, and owes its name to the humid forest that covers most of its territory.

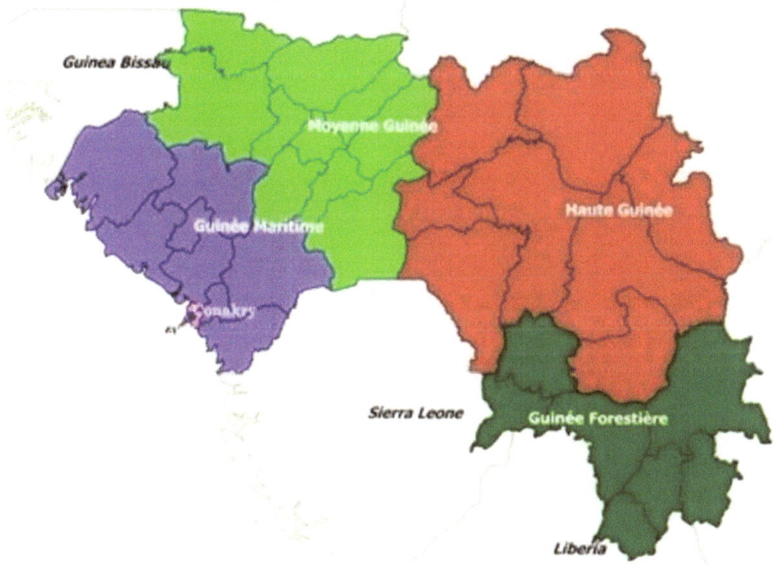

Figure 6.1 Guinea's four natural regions.

In this paper, we focus on rain. Rain is extremely important to the earth as it is a crucial element to the Earth's water cycle. Its existence is essential to the survival of animals, including human beings. It has the power to adorn crops and acts as a carrier for indispensable nutrients.

These crops grow to be healthy and strong and as a consequence produce food for the populace and a means of living to farmers all over the world. On the flip side, rainfall can be destructive. In fact, it can be among the most destructive forces on earth when it takes the form of tsunamis or hurricanes. Extreme rainfall—a rare event—can cause considerable property losses and loss of life.

Flash floods can be a consequence of too much rain. According to Creutin et al. (2013), flash floods are defined as those flood events where the rise in water is either during or within a few hours of the rainfall. In other words, an atypical downpour of rain that cannot be redirected fast enough. These floods generally follow an atypical downpour of rain. Extreme value theory comes in handy because it helps to model these extreme events which while rare can be catastrophic. Knowing more about such events may help in the protection of property and life.

Recently, heavy rains have brought about flooding in varying areas of Guinea. As a result, parts of the country have suffered much in terms of economic losses, deaths, and forced human displacement (Floodlist.com 2022).

From 25th to 26th of August 2022, sustained heavy rains fell across the eastern Kankan region according to the Guinea National Red Cross and National Disaster Management Agency. The International Federation of Red Cross (IFRC) reported over 29,000 people affected, 93 houses along with 320 water points and 375 latrines destroyed and 335 people displaced. Of the 24,135 people (3,448 households) that have been assessed and classified as most affected, there were 10,839 males, and 13,296 females; 3,928 of the most affected population were children under the ages of 5 years, 873 were pregnant women, 1,732 were breastfeeding women, 87 were persons with disabilities and 649 were elderly persons. Cases of diarrhea had been reported. Moreover, 2,460 hectares of agricultural land were flooded and 873 livestock destroyed or missing.

Less than a month later, severe floods gutted Conakry killing three people—two of them died by electrocution—and affecting 2,576 people, including 137 children. The Ministry of Territorial Administration believed that the "anarchic and cumbersome constructions which were obstructing water passages" were to blame and promised to take measures aimed at addressing this problem. Three years prior, a flood in the same city resulted in the death of five people and locals pointed to the improper use of necessary infrastructure as a reason for the extensive damage that resulted. To many, blocked drainage channels exacerbated the problem and made the damage much more deleterious than it could have been. Indeed, images on social media appeared to confirm those remarks.

The Kankan 2020 floods affected 19,814 people or 3,963 households (Floodlist.com 2020). In total, 27 neighborhoods of the urban commune and 12 sub-prefectures were impacted. In 2021, heavy rain affected 69,671 people, or 9,953 households with 1,972 made homeless, 21 injured and 5 deaths recorded in prefectures of Siguiri (43,815 people affected, 6,280 households), Gueckedou (9,305 people affected, 1,364 households), and Conakry (16,551 people affected, or 2,309 households) (Floodlist.com 2021).

Between 30th of August 2021 and 31st of August 2021 heavy rain fell over much of Guinea. As a result, 5 people died, 57 were injured, and thousands were negatively affected from the damage caused by the torrential rain and heavy flooding in the capital city Conakry, as well as the far away prefectures of Gueckedou and Siguiri. Reports showed that 43,815 individuals (6,280 households) were affected in Siguiri, 9,305 individuals (1,364 households) in Gueckedou, and 16,551 people (2,309 households) in Conakry. The extreme rain resulted in the destruction of 867 houses, 763 water points, and the displacement of over 2,500 people. There were also severe impacts on agriculture and livestock (Floodlist.com 2021).

Between 26th to 28th of July 2015, local media said that 450 millimeters of rain fell in Conakry (Floodlist.com 2015). Houses and roads suffered damage and a minimum of 4 people lost their life as a direct consequence of the flooding. This flood was different in the sense that around that time, neighboring countries also saw non-negligible rain. In their Climate Prediction Center's Africa Hazards Outlook for 30th July 2015–5th August 2015, the National Oceanic and Atmospheric Administration reported that localized areas in Senegal, Guinea, Sierra Leone, and Liberia received greater than 200 millimeters of rainfall.

Because of all these effects, which are a consequence of heavy rainfall, it is all the more important to model extreme values of rainfall. Too much rainfall causes much damage—especially for developing nations like Guinea. Floods cause displacements and change lives for the worst. Along with physical destruction and loss of life, there is also plenty of economic impact both for individuals as well as for the country. Statistically speaking, these flood inducing bouts of rain are extreme and relatively rare. Typically, rare situations do not worry people but in certain settings like climate and finance, for example, they are extremely important since the consequences which follow from such events are notable and far reaching. This is why Extreme Value Theory is so important—it allows us to model and quantify extreme values as well as draw inferences about them. Playing a part in helping to mitigate these conditions for the benefit of the population and their goods is one motivation behind this paper. Modeling these extremes could be one tool authorities have in their planning and management toolbox to address the disasters.

There has been much research looking at rainfall in or around Guinea. Rowell et al. (1992) attributed variation in rainfall to changes in sea surface temperature in the Gulf of Guinea. Aguilar et al. (2009) considered changes in extreme rainfall as well as temperatures in Western Central Africa, the Republic of Guinea, and Zimbabwe using 27 indices primarily based on station level thresholds calculated over a base period. Ngugen et al. (2011) found that sea-surface temperature anomalies in the southeast Atlantic in winter could potentially predict Guinean coastal rainfall variability in the spring. Sall et al. (2007) discussed the asymmetry of the rainfall distribution over the mountain range of the Fouta Djallon-Guinea—in relation with orography and local atmospheric circulation. Sambou et al. (2018) studied spatial and temporal rainfall variability of the transboundary catchment area of Kayanga/Geba River between the Republic of Guinea, Senegal and Guinea-Bissau. Cook and Vizy (2019) showed that there is a positive linear relationship between sea surface temperature and rainfall in the Guinean coast. Kante et al. (2019) considered a 19-year period of time (1981–2010) and examined the relationship between thunderstorms and categorized rainfalls in the context of climate in Guinea over that period. Loua et al. (2019) considered changes in temperature and rainfall over a 57-year period (1960–2016) over the Conakry airport. As an extension of previous studies that had limited themselves to the monsoon seasons, using National Meteorological Service of Guinea data from 1981–2010, Kante et al. (2020) showed that between July and August (rainfall season peak), the coastline, particularly Conakry, recorded the largest amount of rainfall. Wourou et al. (2020) probed both the interannual variability of rainfall in the Guinean Coast and its links with sea surface temperatures throughout the years from 1901 to 2016. Son and Seo (2020) presented climatological characteristics and physical mechanisms of Guinea Coast precipitation in June.

But none of these papers model extreme rainfall in Guinea. This paper provides the first statistical modeling of extreme rainfall in Guinea. We fit extreme value distributions to rainfall data from 12 stations throughout the Republic of Guinea. We discover the wettest and driest areas with respect to maximum rainfall. Knowing these areas could possibly lead to policy changes which might benefit the country and its citizens. We also like to get a sense of the return level curves which will allow planners to take necessary precautions and bring about necessary changes.

The contents of the paper are organized as follows. Section 6.2 describes the data from 12 locations in Guinea: Boke, Conakry, Faranah Kankan, Kindia, Kissidougou, Koundara, Labe, Mamou, Macenta, Siguiri and Zerekore. Section 6.3 describes the method used to analyze the data. Section 6.4 presents the results of the method and their discussion. The paper is concluded in Section 6.5.

6.2 DATA

The data consists of rainfall in millimeters for 12 stations aptly named after the cities in which they are located: Boke, Conakry, Faranah Kankan, Kindia, Kissidougou, Koundara, Labe, Mamou, Macenta, Siguiri, and Zerekore. The years of data are different for different stations. For Kindia, Boke, Kankan, Labe and Zerekore, the data are from 1991 to 2020. For Conakry, Koundara, Faranah, and Mamou, the data covers a 30-year period from 1990 to 2020. For Macenta and Siguiri, the data are from 1982 to 2020. For Kissidougou, the data are from 1981 to 2009. For Mamou, the data are from 1990 to 2017. The data was obtained from the National Meteorology Service of Guinea and was originally arranged by months out of the year.

The distribution of the data for each station was observed and errors resulting from data processing were not found. There were no anomalies or impossible values. Missing values were disregarded from the analysis.

Each of the four aforementioned natural geographic areas contain three stations for representativeness: the upper Guinea region includes Faranah, Kankan and Siguiri stations; the forest Guinea region includes Kissidougou, Macenta, and Nzerekore stations; the middle Guone region includes Koundara, Labe and Mamou stations; the coastal Guinea region is home for Boke, Conakry, and Kindia stations. The 12 stations chosen correspond to stations where data are regularly collected and reliable (Kante et al. 2020).

Figure 6.2 displays a map showing the positions of the weather stations in Guinea. They are spread out over the nation and thus give a good representation of the geography and meteorology. All the different zones are accounted for. For example, each of the four geographic regions of Guinea is represented by three weather stations.

Typically, an event is called extreme if it lies above or below some pre-defined threshold. These thresholds are determined differently leading to inconsistent meanings. In our case, the extreme is defined as the maximum numerical value of rainfall collected by year and indicated by the weather stations. In the literature, some researchers investigate daily and even sub-daily extremes while others study extremes over a duration of days, weeks, or months. We picked a 1-year interval due to interest in the maximal rainfall per year of the data for each of the 12 weather stations.

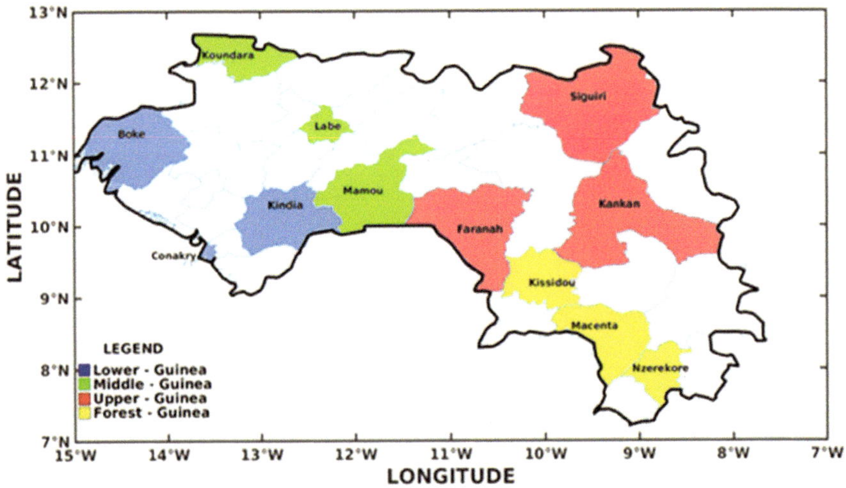

Figure 6.2 Geographic location of weather stations.

We aggregated the data to get one single value for each year. Table 6.1 gives the number of years of record, mean, minimum, maximum, median, standard deviation, coefficient of variation, skewness, and kurtosis.

6.3 METHOD

Let X denote a random variable representing the monthly maximum rainfall. According to extreme value theory (Leadbetter et al. 1983, Resnick 1987, and Embrechts et al. 1997), the cumulative distribution function of X can be approximated by

$$F_X(x) = \exp\left[-\left(1+\xi\frac{x-\mu}{\sigma}\right)^{-1/\xi}\right] \tag{1}$$

for $\mu - \sigma/\xi \leq x < \infty$ if $\xi > 0$, $-\infty < x < \infty$ if $\xi = 0$ and $-\infty < x \leq \mu - \sigma/\xi$ if $\xi < 0$, where $-\infty < \mu < \infty$ denotes a location parameter, $\sigma > 0$ denotes a scale parameter and $-\infty < \xi < \infty$ denotes a shape parameter. Note that if $\xi > 0$ then X has a heavy tail bounded below by $\mu - \sigma/\xi$. If $\xi < 0$ then X has a short tail bounded above by $\mu - \sigma/\xi$.

The distribution in (1) is known as the Generalized Extreme Value (GEV) distribution. The GEV distribution was fitted to the data in Section 6.2 by the method of maximum likelihood. Suppose $x_1, x_2, \ldots, x_n$ is an enumeration of the data in Section 6.2. The maximum likelihood estimates of μ, σ and ξ were obtained by maximizing

Table 6.1 The mean, minimum, maximum, median, standard deviation, coefficient of variation, skewness, and the kurtosis of the maximum rainfall values

Station	Count	Mean	Minimum	Maximum	Median	Standard deviation	Coefficient of variation	Skewness	Kurtosis
Kankan	30	389.993	247	612	355	103.099	0.264	0.690	2.251
Kindia	30	513.567	387	636	513	76.960	0.150	0.075	1.734
Faranah	29	415.621	282	1528	371.0	220.692	0.531	4.626	23.896
Boke	30	671.267	157	2031	587.0	314.547	0.469	2.770	12.993
Conakry	30	1253.733	833	1677	1230.5	237.386	0.189	0.157	2.113
Koundara	29	392.552	209	673	380.0	104.573	0.266	0.565	3.169
Kissidougou	29	444.276	308	572	435.0	75.979	0.171	0.052	1.992
Labe	30	423.667	303	635	419.0	80.480	0.190	0.530	2.935
Macenta	30	578.733	399	934	560.0	133.854	0.231	0.780	3.039
Mamou	28	447.500	323	729	432.5	89.126	0.199	1.043	4.662
Siguiri	39	351.821	219	634	344.0	80.720	0.229	0.902	5.127
Zerekore	31	371.839	277	539	359.0	68.001	0.183	0.684	2.875

Note: The mean maximum rainfall is smallest for Siguiri and largest for Conakry. The minimum of the maximum rainfall is smallest for Boke and largest for Conakry. The maximum of the maximum rainfall is smallest for Zerekore and largest for Boke. The median maximum rainfall is smallest for Siguiri and largest for Conakry. The standard deviation of the maximum rainfall is smallest for Zerekore and largest for Boke. The coefficient of variation of the maximum rainfall is smallest for Kindia and largest for Faranah.

The mean values in all of the stations are greater than the median, suggesting right skewed distributions. This is confirmed by the positive values of skewness. The skewness is smallest for Kindia and largest for Faranah.

The kurtosis values for Faranah, Boke, Koundara, Macenta, Mamou, Siguiri are greater than 3. Graphically, their distributions are more peaked and have fatter tails than the normal distribution. They are leptokurtic. The stations at Kindia, Conakry, Kissidougou, Labe, and Zerekore have kurtosis values less than 3. Graphically, they are less peaked and more pinched tails than the normal distribution. They are platykurtic. The kurtosis is smallest for Kindia and largest for Faranah.

$$L(\mu, \sigma, \xi) = \frac{1}{\sigma^n} \prod_{i=1}^{n} \left\{ \left(1 + \xi \frac{x_i - \mu}{\sigma}\right)^{-\frac{1}{\xi}-1} \exp\left[-\left(1 + \xi \frac{x_i - \mu}{\sigma}\right)^{-\frac{1}{\xi}} \right] \right\}$$

over all possible values of μ, σ and ξ. The maximum likelihood estimates are the values of μ, σ and ξ corresponding to the maximum of $L(\mu, \sigma, \xi)$. The maximization was performed using the command fgev in the *R* package evd (Stephenson 2018, R Core Team 2022). Other distributions (for example, the normal distribution) may provide better fits to the monthly maximum rainfall. But the GEV distribution is theoretically justified.

Let $\widehat{\mu}$, $\widehat{\sigma}$ and $\widehat{\xi}$ denote the maximum likelihood estimates of μ, σ and ξ, respectively. A quantity of interest based on (1) is the *T*-year return level loosely interpreted as the monthly maximum rainfall expected on average once in every *T* years. Let x_{T} denote the *T*-year return level corresponding to (1). It must satisfy

$$F_X(x_T) = 1 - \frac{1}{T}. \tag{2}$$

Inverting (2), we obtain

$$x_T = \widehat{\mu} + \frac{\widehat{\sigma}}{\widehat{\xi}} \left\{ \left[-\log\left(1 - \frac{1}{T}\right) \right]^{-\widehat{\xi}} - 1 \right\}. \tag{3}$$

See equation (3.4) in Coles (2001).

6.4 RESULTS AND DISCUSSION

The GEV distribution was fitted to monthly maximum rainfall data from all 12 weather stations. Table 6.2 shows the estimates, standard errors, and 95% confidence intervals for the GEV parameters.

All stations except for Faranah have shape parameter estimates not significantly different from zero since zero is included in the 95% confidence intervals. Faranah has a significantly positive shape parameter estimate. The standard error of the shape parameter estimates is smallest for Boke and largest for Kindia.

The location parameter estimate is smallest for Siguiri and largest for Conakry. Its standard error is smallest for Zerekore and largest for Conakry. The scale parameter estimate is smallest for Zerekore and largest for Conakry. Its standard error is also smallest for Zerekore and largest for Conakry.

Table 6.2 Parameter estimates, standard errors, and confidence intervals for the fitted GEV distribution

		μ	σ	ξ
3*Kankan	Parameter estimates	336.559	70.102	0.176
	Standard errors	15.303	12.323	0.198
	Confidence interval (95%)	(306.565, 366.553)	(45.949, 94.256)	(−0.212, 0.565)
3*Kindia	Parameter estimates	492.374	79.361	−0.435
	Standard errors	18.047	15.890	0.259
	Confidence interval (95%)	(457.001, 527.747)	(48.216,110.505)	(−.942,0.073)
3*Faranah	Parameter estimates	349.616	59.989	0.279
	Standard errors	11.422	9.284	0.121
	Confidence interval (95%)	(327.230, 372.003)	(37.793, 74.185)	(0.042, 0.515)
3*Boke	Parameter estimates	548.739	205.671	0.029
	Standard errors	40.365	27.472	0.075
	Confidence interval (95%)	(469.625, 627.854)	(151.827, 259.516)	(−0.119, 0.177)
3*Conakry	Parameter estimates	1165.954	225.104	−0.260
	Standard errors	47.822	36.096	0.176
	Confidence interval (95%)	(1072.222, 1259.686)	(154.357, 295.851)	(−.605, 0.085)
3*Koundara	Parameter estimates	348.578	89.718	−0.104
	Standard errors	18.762	13.355	0.135
	Confidence interval (95%)	(311.804, 385.352)	(63.542, 115.893)	(−0.369, 0.162)
3*Kissidougou	Parameter estimates	420.617	77.677	−0.357
	Standard errors	16.478	12.933	0.195
	Confidence interval (95%)	(388.320, 452.914)	(50.329, 101.026)	(−0.740, 0.025)
3*Labe	Parameter estimates	389.652	68.778	−0.099
	Standard errors	14.329	10.396	0.148
	Confidence interval (95%)	(361.567, 417.737)	(48.402, 89.155)	(−0.388, 0.191)
3*Macenta	Parameter estimates	513.183	96.821	0.102
	Standard errors	21.833	17.356	0.219
	Confidence interval (95%)	(470.390,555.976)	(62.803, 130.839)	(−0.327, 0.532)
3*Mamou	Parameter estimates	407.685	68.548	0.003
	Standard errors	14.730	10.800	0.147
	Confidence interval (95%)	(378.813, 436.556)	(47.381, 89.715)	(−0.285, 0.291)
3*Siguiri	Parameter estimates	317.601	68.672	−0.078
	Standard errors	14.114	8.450	0.091
	Confidence interval (95%)	(293.858, 341.344)	(52.109, 85.234)	(−0.258, 0.101)
3*Zerekore	Parameter estimates	340.846	53.438	−0.004
	Standard errors	11.260	8.449	0.172
	Confidence interval (95%)	(318.776,362.916)	(36.877, 69.999)	(−0.341, 0,334)

The fit of the GEV distribution for each station was checked by probability and quantile plots. The plots are shown in Figures 6.3 and 6.4 for the 12 stations.

Probability plots are plots of $\widehat{F}(x_{(i)})$, the observed probabilities, versus $i/(n + 1)$, the expected probabilities, where $x_{(1)} \le x_{(2)} \le \cdots \le x_{(n)}$ are the data arranged in increasing order and

$$\widehat{F}(x) = \exp\left[-\left(1+\widehat{\xi}\frac{x-\widehat{\mu}}{\widehat{\sigma}}\right)^{-\frac{1}{\widehat{\xi}}}\right].$$

Quantile plots are plots of $x_{(i)}$, the observed quantiles, versus $\widehat{F}^{-1}(i/(n+1))$, the expected quantiles, where $\widehat{F}^{-1}(\cdot)$ denotes the inverse function of $\widehat{F}(\cdot)$. The diagonal straight lines in Figure 6.3 correspond

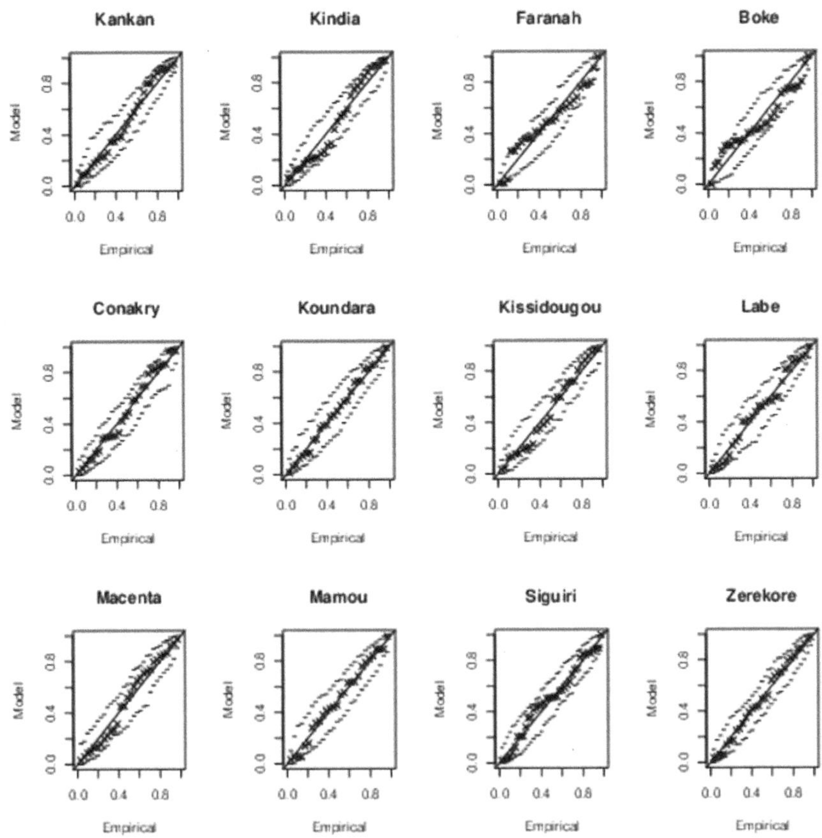

Figure 6.3 Observed probabilities against expected probabilities for the fit of the GEV distribution.

to the observed and expected probabilities being equal. The diagonal straight lines in Figure 6.4 correspond to the observed and expected quantiles being equal. The dashed lines in Figures 6.3 and 6.4 are the 95% simulated confidence intervals. The closer the plotted points (in Figures 6.3 and 6.4) are to the diagonal lines the better the fit. The plotted points must lie within the simulated confidence intervals for the fit to be considered adequate. Hence, the fit of the GEV distribution for monthly maximum rainfall from the 12 locations is adequate.

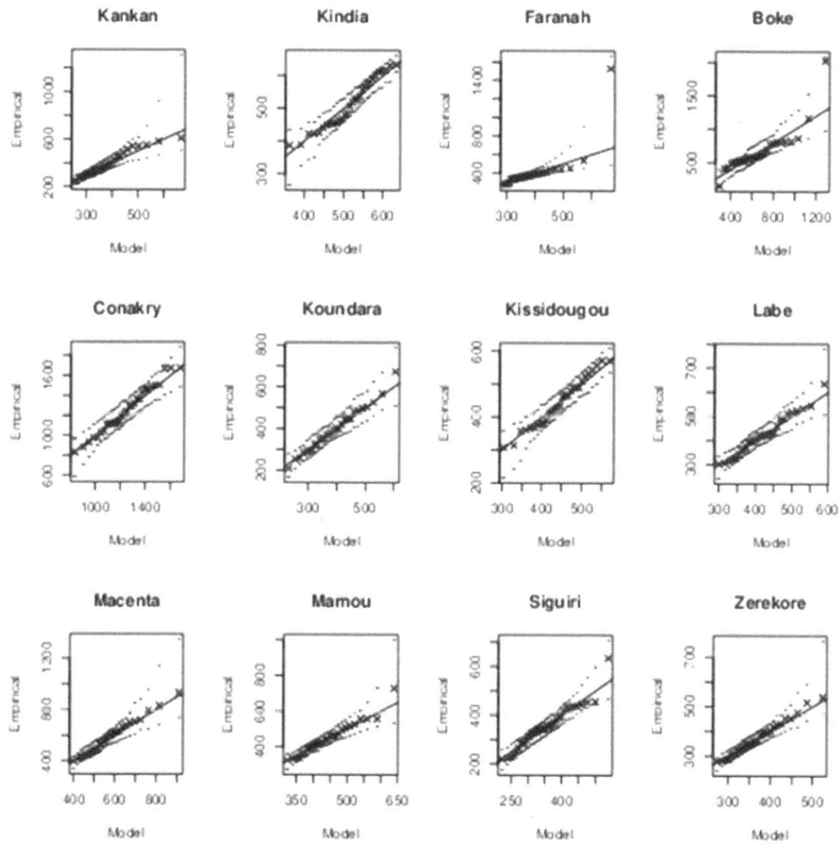

Figure 6.4 Observed quantiles against expected probabilities for the fit of the GEV distribution.

In addition to using probability and quantile plots, we tested goodness of fit using the Kolmogorov-Smirnov, Anderson-Darling and Cramer-von Mises tests. The Kolmogorov-Smirnov test gave the *p*-values of 0.060, 0.097, 0.066, 0.087, 0.068, 0.078, 0.079, 0.068, 0.059, 0.056, 0.091, and 0.098 for the 12 locations. The Anderson-Darling test gave the *p*-values of

0.099, 0.099, 0.087, 0.064, 0.062, 0.090, 0.078, 0.065, 0.055, 0.080, 0.063, and 0.068. The Cramer-von Mises test gave the p-values of 0.089, 0.061, 0.095, 0.064, 0.087, 0.057, 0.096, 0.051, 0.058, 0.056, 0.070, and 0.051.

Return levels can be used to help us predict the occurrence likelihood of catastrophic events of interest such as rainfall levels which can lead to flooding. We can interpret the return level x_T as the value expected to be surpassed on an average once every T years. x_{100} can be interpreted as the amount of rain we might expect that will be exceeded once in a hundred years (this could be used to inform flood mitigation, for example). It is important to emphasize that a 100-year rain storm could materialize once, twice or more, or none at all in a 100-year interval. A 100-year flood is a flood with a 1% change of occurring in any given year. Plots of x_T for T = 2, …, 100 are shown in Figures 6.5–6.8 for the four natural geographical areas.

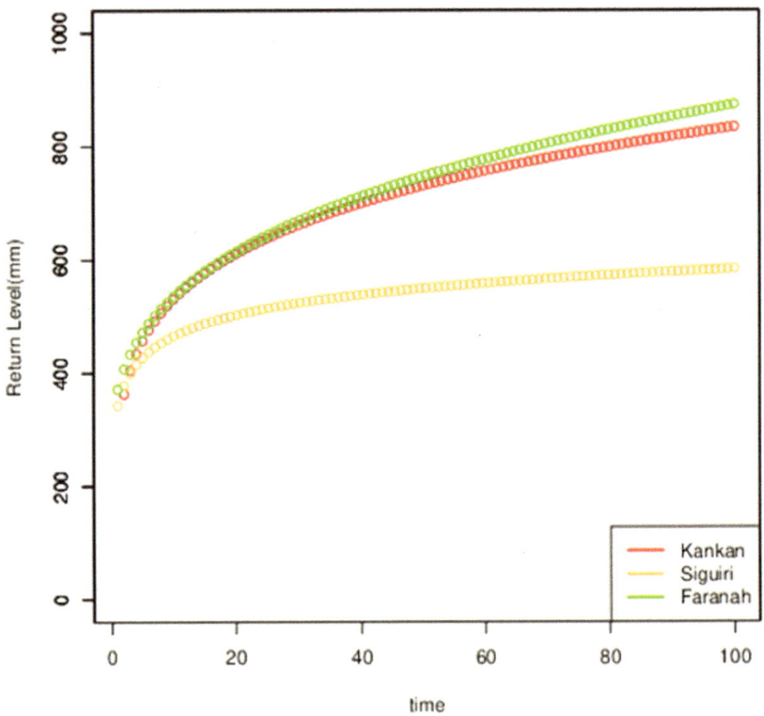

Figure 6.5 Estimates of x_T versus T = 2, 3, …, 100 for upper Guinea.

In upper Guinea, Kankan and Faranah have similar return curves. The return curve for Siguiri is notably lower. Yet many disasters have taken place in Siguiri. As a whole, upper Guinea is the driest natural region.

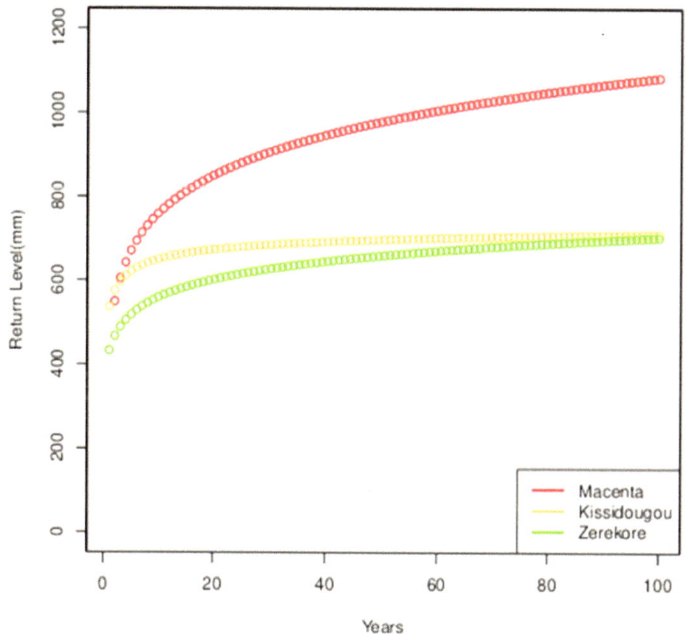

Figure 6.6 Estimates of x_T versus $T = 2, 3, \ldots, 100$ for forested Guinea.

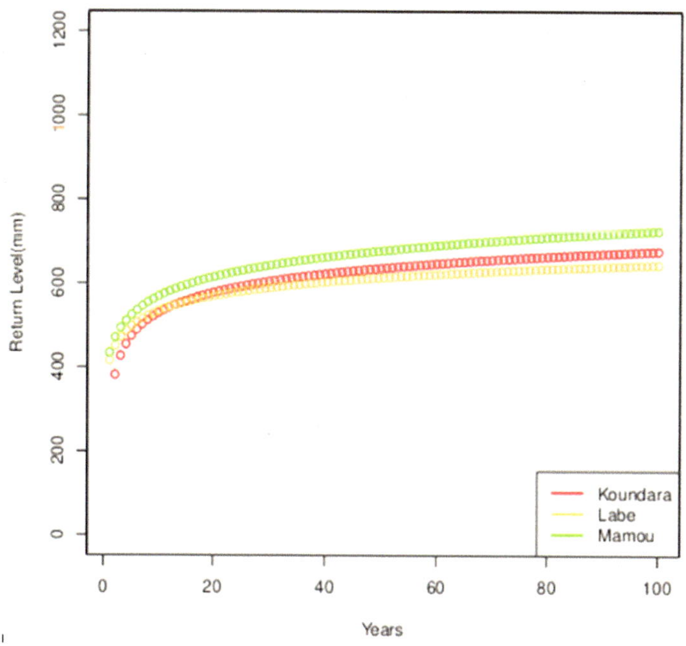

Figure 6.7 Estimates of x_T versus $T = 2, 3, \ldots, 100$ for middle Guinea.

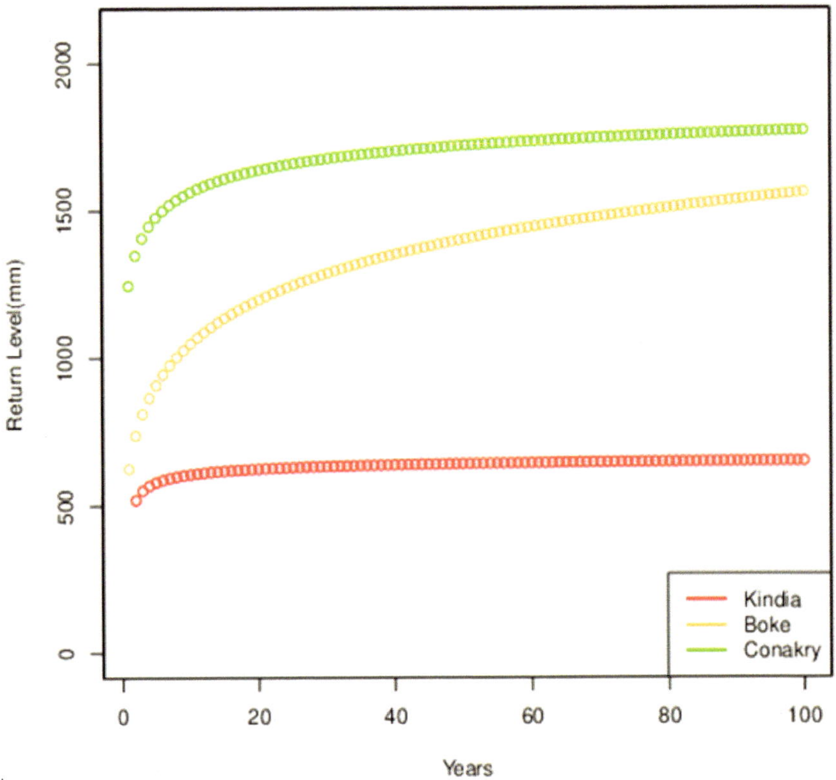

Figure 6.8 Estimates of x_T versus $T = 2, 3, \ldots, 100$ for coastal/maritime Guinea.

Forested Guinea owes its name to the humid forest that covered most of its territory. While this has a longer rainy season compared to the other regions, Kissidougou and Zerekore return curves are relatively low and do not top 680 millimeters. Macenta's return curve exceeds 1,000 millimeters and highlights the fact that Macenta's station has a higher contribution rainfall than other stations in its region. Macenta's station has a higher 100-year return curve than the other two stations in its area.

We must remember that we are looking at extreme rain and not say cumulative rainfall. For example, in the middle Guinea region, Labe in general receives more rain than Mamou, but the return curve shows that Mamou is more likely to get extreme rain. Koundara receives less rain and less extreme rain than Mamou and Labe and this could be due to the effects of mountainous ranges around those two locations which are not as pronounced in Koundara.

The coastal area is the rainiest and has the most potential for rain. Kindia has a lower return curve than both Conakry and Boke, which come in first and second respectively in terms of return levels. Indeed, Conakry is the rainiest city in Guinea as well as the station with the highest return curve. This confirms observations from Kante et al. (2020) and Loua et al. (2019) and reminds us that Conakry is among the wettest capitals in the world. The average rainfall at Conakry is about 4,300 millimeters a year, making it the wettest capital city in the world.

According to Kante et al. (2020), Koundara's contribution is lower than contributions from other stations in middle Guinea, Faranah's monthly rainfall contributions are slightly higher than other stations during certain months, Macenta's station has a higher rainfall than other stations in its region, and Kissidou station comes in second position whose rainfall exceeds that of Nzerekore. Figures 6.5–6.8 corroborate these remarks even though we are using return level curves for comparison.

Figure 6.9 shows all 12 return plots juxtaposed against each other. The return level estimates are largest for Conakry, second largest for Boke, and smallest for Siguiri.

Finally, we investigated to see if there are significant trends in the maximum rainfall for each station. We fitted (1) with the location parameter $\mu = a + b \times (Year - StartYear)$, where b is the trend parameter. By comparing the fit of this model with the earlier fit of the GEV distribution, we can see if the trend is significant or not. The presence of a linear trend in the dataset was tested under the null hypothesis that assumes no trend. We failed to reject the null with a 5% significance level for any of the stations.

But Aguilar et al. (2009) and Loua et al. (2019) reported a decreasing trend over the years; however, both worked with datasets of measurements of over 50 years. With our limited data, no trend was detected. A bigger dataset could have returned some significant trends in a subset of the weather stations. There is also the fact that we aggregate data at the yearly level which could veil some of the granularity and subtleties in the datasets.

Another possible reason for the lack of a trend could stem from the land area. Guinea's area is 245,857 square kilometres (94,926 square miles, or roughly the size of the United Kingdom) and as such we are less likely to see as much variation in maximal rainfall as say for a country like India or even the Democratic Republic of Congo. Additionally, the cycles of rain and drought lead one to believe that there is at least no positive trend. The changes in sea surface temperatures have been linked to rainfall levels in west Africa and this cyclical nature may also explain why we saw no trends at the yearly level.

The lack of trend does not however imply that damages caused by extreme rainfall should be minimized although this news could be

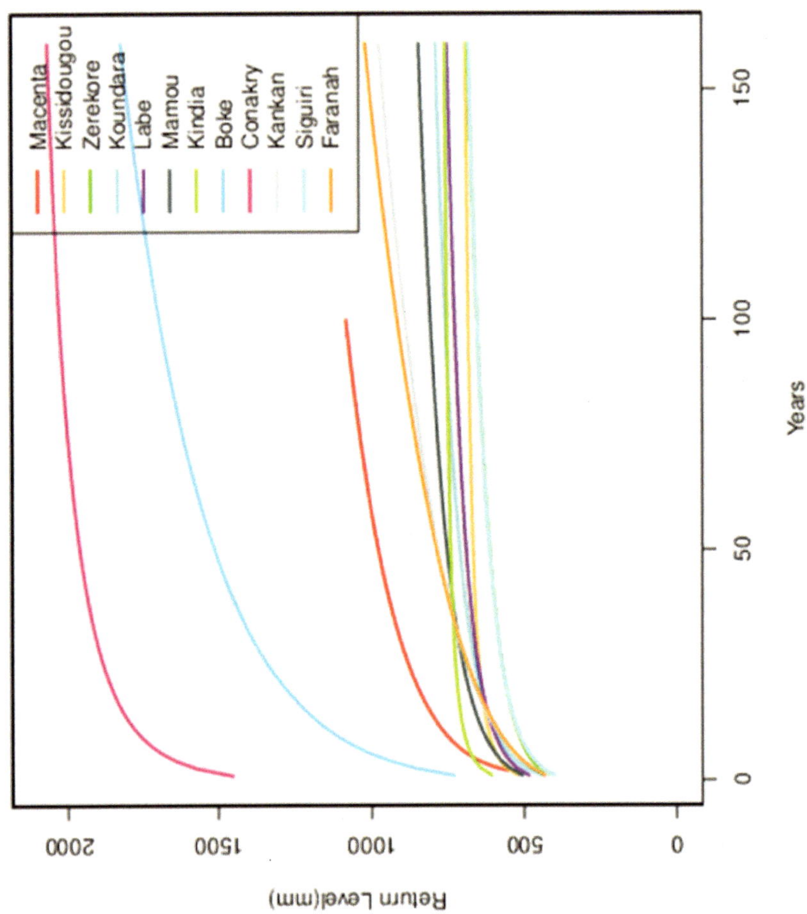

Figure 6.9 Estimates of x_T versus $T = 2, 3, \ldots, 100$ for all 12 stations.

considered celebratory in the sense that statistically increasing maximal rainfall over the years would imply an even bigger load on authorities charged with mitigating damage caused by such events. It is important to note that the occasional heavy rainfall can still be quite destructive as we have seen in several recent events in Guinea mentioned earlier.

While our analysis is showing that none of the 12 stations are showing any trend, Senegal, a country bordering Guinea, is showing trends at four of its 12 synoptic stations (N'dri and Nadarajah 2022). Other countries like Zimbabwe, Madagascar, Malawi have seen significant trends over the years too (Otto et al. 2022).

6.5 CONCLUSIONS

This paper has provided the first extreme value statistical analysis of maximum rainfall using data from 12 weather stations peppered across Guinea. As assessed by probability plots, quantile plots Kolmogorov-Smirnov tests, Anderson-Darling tests, and Cramer-von Mises tests, the generalized extreme value distribution allowed for an adequate fit to the data from all stations. Using historical data, we generated return curves. The wettest areas with respect to return levels are Conakry, Boke, and Macenta. On the other hand, the driest areas with respect to return levels are Koundara, Kissidougou, and Labe.

None of the stations exhibited significant trends at the 5% significance level. While this does not hold true for all of Guinea, it is likely that this statement holds true for the overwhelming majority of the Guinean territory based on representativeness and zone coverage of the 12 weather stations examined.

It is recommended that authorities consider making decisions based on return curves, invest in developing more sophisticated forecasting systems and transparently partnering up with organizations able to transfer these technical competencies, bolster warning communication systems, invest in better infrastructure, and take advantage of new and existing technologies to harness the power of nature in beneficial ways.

Authorities may want to focus on areas that are more likely to get higher quantities of extreme rainfall. Thresholds could be set so that say thresholds above 100-year return levels could be considered risky localities and given more attention. Other areas could be given attention and resources according to the slope of their return curves: areas with steeper return curves might need more focused attention and larger resources.

Among the recommendations are partnering up with organizations and researchers to build post-modern, robust forecasting systems capable of simulating extreme rainfall all around the country while taking into account its great variability from season to season and from year to year. It is critical to take into account climate change as well as monitor sea surface temperatures since these two factors would allow for more accurate and reliable models. Climate monitoring has benefits besides helping to predict rainfall. For example, there is evidence that early climate monitoring can be helpful for governments and international agencies since El Nino and La Nina events can be reliable predictors for disease outbreaks (Anyamba et al. 2019).

Climate Risk and Early Warning System (CREWS) is an organization that realizes the importance and cost-saving measure of forecasting and whose mission includes ensuring warning capacity gets strengthened in line with international standards. Funded by Australia, Germany, France, Luxembourg, and the Netherlands, CREWS provides countries like Guinea with technical assistance. While the idea of CREWS is noble, there is little known about the efficacy of the models and the level to which they have helped the nations. Therefore closer, more transparent work with such organizations as well as independently strengthening in-house forecasting systems will be beneficial. There is more knowledge to discover so funding of researchers, high level training and more demand for research by policymakers (Kumwenda et al. 2017) will result in fresh and novel solutions from researchers in Guinea. Water related professional courses (Odume and Slaughter 2018) as well as continued development for meteorologists will aid in preparation and damage containment as well as for a fast recovery.

From an agricultural standpoint, one thing to do is plant crops on floating rafts, a practice known as aquaponics (Surnar et al. 2015) to minimize damage resulting from flooding. Furthermore, a rain collection system could then be used to offer water to the populace to satisfy certain domestic needs. In high rain areas, long structures which support solar panels on the top and a water harvesting system on the bottom could be used as mixed production systems. Similar systems have been successfully debuted in Kenya in 2022 (Randall-Hoggis et al. 2021). In addition to harnessing solar energy, these structures have the added benefit of providing a nourishing environment for plants who would otherwise die from harsh, exposed environments. Researchers in China are currently working on solar panels which can use the friction of rain droplets to generate electricity irrespective of the time of the day. If testing goes well, and those materials show promise, they could be a good investment for Guinea to use in its tool box against extreme rain. These panels may be used to meet some unmet low level electricity demands.

Although some researchers (Aguilar et al. 2009, Loua et al. 2019) have come to the conclusion that rainfall has been declining over the years, rare floods can and do occur and are often accompanied by catastrophic results. These resulting floods tend to have a significant effect on the economic and medical well being of people. The need to do a satisfactory job and invest in better forecasting is underscored by the potential havoc that these floods bring about. Investment in more flood resilient infrastructure as well as adoption of novel ideas could help cut down on calamities caused by these seemingly unpredictable deluges which bring with them years of pain and forever uprooted lives. These suggestions may not only save lives but also harness the power of floods and redirect them to other, more productive uses. Flood-inducing, extreme quantities of rain cannot be fully stopped, but a timely warning system, better, more robust infrastructure and creative tools to harness some of the power of these rainstorms will behoove habitants of rain prone parts of Guinea.

Some of the positive actions to be taken by the government of Guinea could include: further vegetables and other commodities less reliable on rain planted on the driest areas; increased agricultural and electricity production based on water taking place in the wettest areas; increased electricity production based on solar energy taking place in the driest areas; and so on.

Future work are to use the following techniques for modeling extreme rainfall in Guinea: generalized Pareto distribution, gamma generalized extreme values, spatial temporal modeling via multivariate time series, Bayesian modeling of extreme values, regional analysis based on L moments, bootstrapping methods, nonparametric copulas, time varying copulas, quantile regression methods, neural networks, wavelet transformation analysis and other spectral methods.

Compliance with Ethical Standards

We have complied with all ethical standards. Research did not involve human participants and/or animals.

Funding

No funds received for this research.

Data Availability

The datasets used and analyzed are available from the corresponding author on reasonable request.

Conflict of Interest

The authors have no conflicts of interest.

ACKNOWLEDGEMENTS

The authors are grateful to the Guinean Meteorological Service (GMS) for providing the rainfall data.

REFERENCES

Aguilar, E., Barry, A.A., Brunet, M., Ekang, L., Fernandes, A., Massoukina, M., et al. 2009. Changes in temperature and precipitation extremes in western central Africa, Guinea Conakry, and Zimbabwe, 1955–2006. Journal of Geophysical Research. 114: D02115.

Anyamba, A., Chretien, J.P., Britch, S.C., Soebiyanto, R.P., Small, J.L., Jepsen, R., et al. 2019. Global disease outbreaks associated with the 2015–2016 El Nino event. Scientific Reports. 9: 1930.

Coles, S.G. 2001. An Introduction to Statistical Modeling of Extreme Values. Springer Verlag, New York.

Cook, K.H. and Vizy, E.K. 2019. Contemporary climate change of the African monsoon systems. Current Climate Change Reports. http://link.springer.com/article/10.1007/s40641-019-00130-1.

Creutin, J.D., Borga, M., Gruntfest, E., Lutoff, C., Zoccatelli, D. and Ruin, I. 2013. A space and time framework for analyzing human anticipation of flash floods. Journal of Hydrology. 482: 14–24.

Embrechts, P., Klüppelberg, C. and Mikosch, T. 1997. Modelling Extremal Events for Insurance and Finance. Springer Verlag, Berlin.

Floodlist.com 2015. Guinea—450mm of rain in 3 days—Floods in Conakry leave 4 dead. https://floodlist.com/africa/guinea-450mm-rain-floods-conakry.

Floodlist.com 2020. Guinea—50,000 affected by floods in Kankan. https://floodlist.com/africa/guinea-floods-kankan-september-2020.

Floodlist.com 2021. Guinea—floods leave 5 dead, 800 homes destroyed, and 2,500 people displaced. https://floodlist.com/africa/guinea-floods-september-2021.

Floodlist.com 2022. Guinea—deadly floods strike in Conakry, thousands affected in Kankan region. https://floodlist.com/africa/guinea-floods-conakry-september-2022.

Kante, I.K., Sall, S.M., Badiane, D. and Diouf, I. 2019. Seasonal variability of rainfall and thunderstorms in Guinea over the period 1981 to 2010. African Journal of Environment Science and Technology. 13: 324–341.

Kante, I.K., Sall, S.M., Badiane, D., Diouf, I., Dieng, A.L., Diaby, I., et al. 2020. Analysis of rainfall dynamics in Conakry, Republic of Guinea. Atmospheric and Climate Sciences. 10: article ID 96548.

Kumwenda, S., Niang, E.H.A., Orondo, P.W., William, P., Oyinlola, L., Bongo, G.N., et al. 2017. Challenges facing young African scientists in their research careers: A qualitative exploratory study. Malawi Medical Journal. 29: 1–4.

Leadbetter, M.R., Lindgren, G. and Rootzén, H. 1983. Extremes and Related Properties of Random Sequences and Processes. Springer Verlag, New York.

Loua, R.T., Bencherif, H., Mbatha, N., Begue, N., Hauchecorne, A., Bamba, Z., et al. 2019. Study on temporal variations of surface temperature and rainfall at Conakry airport, Guinea: 1960–2016. Climate. 7: 93.

N'dri, K.I. and Nadarajah, S. 2022. Statistical modeling of monthly maximum temperature in Senegal. Environmental Research Communications. 4: Article number 075002.

Nguyen, H., Thorncroft, C.D. and Zhang, C. 2011. Guinean coastal rainfall of the west African Monsoon. Quaterly Journal of the Royal Meteorological Society. 137: 1828–1840.

Odume, N. and Slaughter, A. 2018. How we manage extreme water conditions is the next big challenge for African countries. Quartz Africa, https://qz.com/africa/1205679/climate-change-africas-challenge-with-both-floodsand-droughts/

Otto, F.E.L., Zachariah, M., Wolski, P., Pinto, I., Barimalala, R., Nhamtumbo, B., et al. 2022. Climate change increased rainfall associated with tropical cyclones hitting highly vulnerable communities in Madagascar, Mozambique and Malawi. https://www.worldweatherattribution.org/climate-change-increased-rainfall-associated-with-tropical-cyclones-hitting-highly-vulnerable-communities-in-madagascar-mozambique-malawi/

R Core Team. 2022. R: A Language and Environment for Statistical Computing. R Foundation for Statistical Computing. Vienna, Austria.

Randle-Boggis, R.J., Lara, E., Onyango, J., Temu, E.J. and Hartley, S.E. 2021. Agrivoltaics in east Africa: Opportunities and challenges. In: AIP Conference Proceedings. 2361: Article number 090001.

Resnick, S.I. 1987. Extreme Values, Regular Variation, and Point Processes. Springer Verlag, New York.

Rowell, D.P., Folland, C.K., Maskell, K., Owen, J.A. and Ward, M.N. 1992. Modelling the influence of global sea surface temperatures on the variability and predictability of seasonal Sahel rainfall. Geophysical Research Letters. 19(9): 905–908.

Sall, S.M., Viltard, A. and Sauvageot, H. 2007. Rainfall distribution over the Fouta Djallon–Guinea. Atmospheric Research. 86: 149–161.

Sambou, S., Dacosta, H. and Paturel, J.E. 2018. Spatio-temporal variability of rainfall from 1932 till 2014 in the Kayanga/Geba River watershed (Republic of Guinea, Senegal, Guinea-Bissau). Physio-Geo. 12: 61–78.

Son, J.H. and Seo, K.H. 2020. Mechanisms for the climatological characteristics and interannual variations of the Guinea coast precipitation: Early summer west African monsoon. Atmosphere. 11: Article number 396.

Stephenson, A. 2018. evd: Functions for extreme value distributions. R package version 2.3-3.

Surnar, S., Sharma, O.P. and Saini, V.P. 2015. Aquaponics: Innovative farming. International Journal of Fisheries and Aquatic Studies. 2: 261–263.

Worou, K., Goosse, H., Fichefet, T., Guichard, F. and Diakhate, M. 2020. Interannual variability of rainfall in the Guinean Coast region and its links with sea surface temperature changes over the twentieth century for the different seasons. Climate Dynamics. 55: 449–470.

Chapter **7**

Statistical Outliers and a New Generalized Probability Distribution with Applications

Aafaq A. Rather*[1], Rushika Kinjawadekar[2], Aijaz Ahmad[3], Ajjaz Maqbool Dar[4], Rifat Nisa[5] and Berihan R. Elemary[6]

[1]Symbiosis Statistical Institute, Symbiosis International (Deemed University), Pune-411004, India

[2]Marathwada Mitra Mandal, Pune, India

[3]Department of Mathematics, Bhagwant University, Ajmer, Rajasthan, India

[4]Department of Mathematics, Govt. College for Womens Nawakadal, Srinagar-190002, India

[5]Department of Statistics, Cluster University, Srinagar, Kashmir, India

[6]Department of Statistics and Insurance, Faculty of Commerce, Damietta University, Egypt

7.1 INTRODUCTION

Statistics has the ability of summarizing big datasets. Some statistical tools provide a researcher with the ability of describing a whole dataset by using one single value. The mean is a good example of such tools. However, as good as these features can be, there is an anomaly in

*For Correspondence: Aafaq A. Rather (aafaq7741@gmail.com)

statistics that can distort the accuracy of such analytical tools: the outlier. In statistics, an outlier is a value that differs significantly from the rest of the data in a dataset. Outliers can occur due to measurement errors, data entry errors, or even genuine extreme values that are part of the data distribution. Outliers are observations that deviate from the normal pattern or distribution of the data. They can have a significant impact on statistical analyses and may indicate errors in data collection or measurement. Table 7.1 shows the dataset that represents test results from a group of 10 students, and student 4 has a grade of 20, which is significantly lower than the grades of the other students. This value is identified as an outlier since it deviates significantly from the main pattern of the dataset. It's important to identify and handle outliers appropriately since they can impact the accuracy of statistical analyses and conclusions. Depending on the nature and cause of the outlier, one may choose to remove it from the dataset, transform the data, or use robust statistical methods that are less sensitive to outliers.

Usually, there is a logical explanation for the existence of an outlier. Table 7.1 shows that all the 10 students of the class were exposed to the same lesson, same teacher, and tested with the same methods. It was expected that all of them would perform at the same level. However, student 4 may claim that he was sick during the test. This would be a satisfying explanation for the poor grade.

Table 7.1 Outlier of test results of students out of 100 marks

Students	1	2	3	4	5	6	7	8	9	10
Grade	88	88	86	20	90	91	96	93	87	92

Example: Let us find the outliers for the given data 2, 4, 6, 8, 10, 12, 14, 16, 18, 20, and 22.

First, the dataset is divided into two halves: 2, 4, 6, 8, 10, 12 and 14, 16, 18, 20, 22. The first quartile (mid-value of first half) is denoted by Q_1 and the third quartile (mid-value of second half) is denoted by Q_3.

Therefore,
$$Q_1 = \frac{6+8}{2} = \frac{14}{2} = 7$$

$$Q_3 = \frac{16+18}{2} = \frac{34}{2} = 17$$

$$\text{Interquartile Range } (IQR) = Q_3 - Q_1 = 17 - 7 = 10$$

$$1.5 \times IQR = 1.5 \times 10 = 15$$

$$\text{Upper boundary} = Q_3 + 1.5 \times IQR = 17 + 15 = 32$$

$$\text{Lower boundary} = Q_1 - 1.5 \times IQR = 7 - 15 = -8$$

–8 and 32 are the outlier boundaries and there are no data beyond these points. Hence, there are no outliers.

7.2 LENGTH BIASED WEIGHTED SUSHILA (LBWS) DISTRIBUTION

Weighted distributions are an important tool in statistics and probability theory that allow for the modeling of data that is not equally likely to be observed. The idea was first introduced by Fisher in (1934), and later developed by Rao (1965) in a more unified manner. Weighted models are particularly useful when standard distributions are not appropriate for recording observations with equal probabilities. In such situations, weighted distributions can be formulated to record observations according to some weighted function, which takes into account the relative importance of each observation. One important type of weighted distribution is the length-biased distribution, which arises when the weight function considers only the length of the units being observed. This concept was first introduced by Cox in (1969) and Zelen in (1974), and has been studied extensively in the literature. Size-biased distributions are another type of weighted distribution where the sampling mechanism selects units with probability proportional to some measure of the unit size. The characterization and estimation of length biased weighted generalized uniform distribution have been discussed by Rather and Subramanian in (2018a). Subramanian and Rather obtained the length biased version of Shanker distribution with applications in (2020). Recently, Rather and Ozel have obtained a new length-biased power Lindley distribution with properties and applications in (2021).

Sushila distribution was introduced by Shanker et al. (2013) which is a mixed distribution between exponential distribution and gamma distribution. Sushila distribution is a two-parameter continuous distribution. Elgarhy and Shawki (2017) discussed exponentiated Sushila distribution. Rather and Subramanian (2018b), obtained the length-biased version of Sushila distribution and its statistical properties. Recently, Rather and Subramanian (2019), reviewed the weighted Sushila distribution with properties and its applications.

The probability density function (pdf) of Sushila distribution is given by

$$f(x;\alpha,\theta)=\frac{\theta^2}{\alpha(\theta+1)}\left(1+\frac{x}{\alpha}\right)e^{-\frac{\theta}{\alpha}x};\alpha,\theta,x>0$$

and its cumulative distribution function (cdf) is given by

$$F(x;\alpha,\theta)=1-\frac{\alpha(\theta+1)+\theta x}{\alpha(\theta+1)}e^{-\frac{\theta}{\alpha}x}$$

Rather and Subramanian (2019) obtained the probability density function of weighted Sushila distribution and it is given by

$$f(x)=\frac{\theta^{c+2}}{\alpha^{c+1}(\theta+c+1)\Gamma(c+1)}x^{c}\left(1+\frac{x}{\alpha}\right)e^{-\frac{\theta}{\alpha}x};\ x>0,\alpha,\theta,c>0 \tag{1}$$

and its mean is given by

$$E(X)=\mu_1'=\frac{\alpha(\theta\Gamma(c+2)+\Gamma(c+3))}{\theta(\theta+c+1)\Gamma(c+1)} \tag{2}$$

Suppose X is a non-negative random variable with probability density function $f(x)$. Let $w(x)$ be the non-negative weight function, then the probability density function of the weighted random variable X_w is given by

$$f_w(x)=\frac{w(x)f(x)}{E(w(x))},\quad x>0,$$

where $w(x)$ be a non-negative weight function and $E(w(x))=\int w(x) f(x)dx<\infty$.

For different weighted models, we have different choice of the weight function $w(x)$. When $w(x) = x^a$, then the resulting distribution is termed as weighted distribution. In this paper, we have to find the length-biased version of weighted Sushila distribution, so we will take $a = 1$ in weights x^a in order to get the length biased weighted Sushila distribution and its pdf is given by (see Figures 7.1 and 7.2).

$$f_l(x)=\frac{x\,f(x)}{E(x)},\quad x>0 \tag{3}$$

Substitute Eqs. (1) and (2) in Eq. (3), we will get the required pdf of length biased weighted Sushila distribution as

$$f_l(x)=\frac{\theta^{c+3}}{\alpha^{c+2}(\theta\Gamma(c+2)+\Gamma(c+3))}x^{c+1}\left(1+\frac{x}{\alpha}\right)e^{-\frac{\theta}{\alpha}x} \tag{4}$$

and the cumulative density function of LBWS distribution is obtained as

$$F_l(x)=\int_0^x f_l(x)dx$$

$$F_l(x) = \int_0^x \frac{\theta^{c+3}}{\alpha^{c+2}(\Gamma(c+2)+\Gamma(c+3))} x^{c+1}\left(1+\frac{x}{\alpha}\right)e^{-\frac{\theta}{\alpha}x}dx$$

After simplification, we will get the cumulative distribution function of LBWS distribution

$$F_l(x) = \frac{\theta}{(\theta\Gamma(c+2)\Gamma(c+3))}\left(\gamma\left((c+2),\frac{0}{\alpha}x\right)+\frac{1}{\theta}\gamma\left((c+3),\frac{\theta}{\alpha}x\right)\right) \tag{5}$$

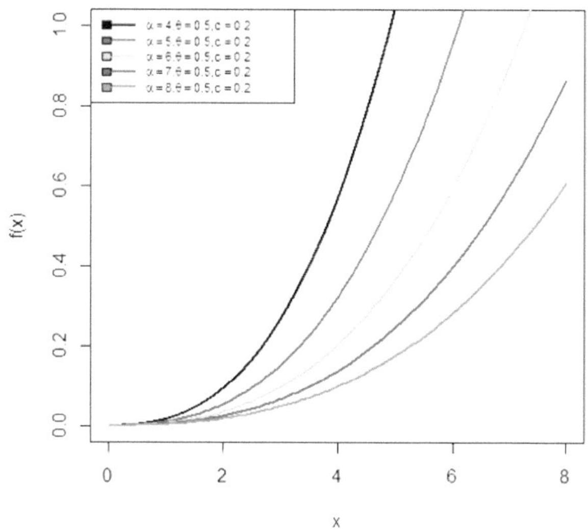

Figure 7.1 Pdf plot of length biased weighted Sushila distribution.

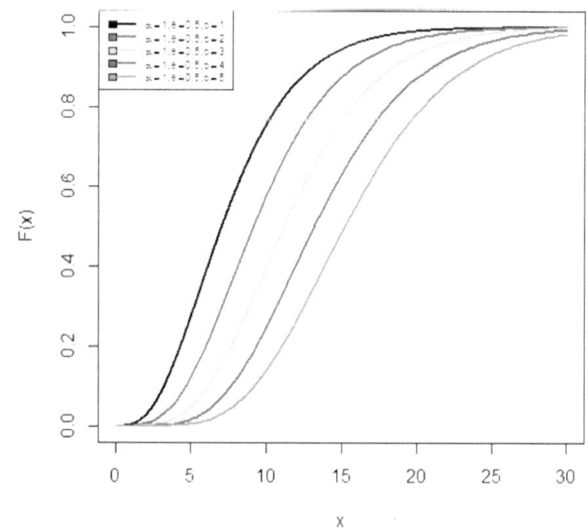

Figure 7.2 Cdf plot of length biased weighted Sushila distribution.

7.3 RELIABILITY ANALYSIS

In this section, we will study the survival function, failure rate, reverse hazard rate, and Mills ratio of the LBWS distribution.

The survival function or the reliability function of the length biased weighted Sushila distribution is given by

$$S(x) = 1 - \frac{\theta}{\theta\Gamma(c+2)+\Gamma(c+3)}\left(\gamma\left((c+2),\frac{\theta}{\alpha}x\right)+\frac{1}{\theta}\gamma\left((c+3),\frac{\theta}{\alpha}x\right)\right)$$

The hazard function is also known as the hazard rate, instantaneous failure rate, or force of mortality and is given by

$$h(x) = \frac{\theta^{c+3}x^{c+1}\left(1+\frac{x}{\alpha}\right)e^{-\frac{\theta}{\alpha}x}}{\alpha^{c+2}\left((\theta\Gamma(c+2)+\Gamma(c+3))-\left(\theta\gamma\left((c+2),\frac{\theta}{\alpha}x\right)+\gamma\left((c+3),\frac{\theta}{\alpha}x\right)\right)\right)}$$

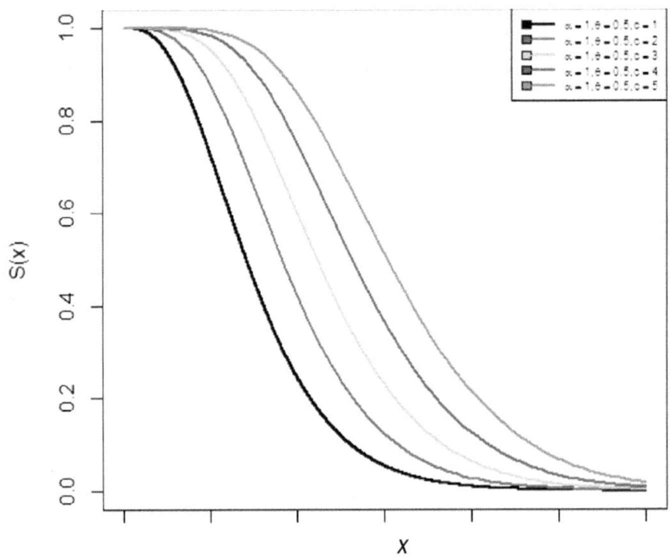

Figure 7.3 Showing the survival function curves of LBWS distribution.

The reverse hazard rate is given by

$$h_r(x) = \frac{\theta^{c+2}x^{c+1}\left(1+\frac{x}{\alpha}\right)e^{-\frac{\theta}{\alpha}x}}{\alpha^{c+2}\left(\gamma\left((c+2),\frac{\theta}{\alpha}x\right)+\frac{1}{\theta}\gamma\left((c+3),\frac{\theta}{\alpha}x\right)\right)}$$

and the Mills ratio of the LBWS distribution is

$$\text{Mills Ratio} = \frac{1}{h_r(x)} = \frac{\alpha^{c+2}\left(\gamma\left((c+2),\frac{\theta}{\alpha}x\right)+\frac{1}{\theta}\gamma\left((c+3),\frac{\theta}{\alpha}x\right)\right)}{\theta^{c+2}x^{c+1}\left(1+\frac{x}{\alpha}\right)e^{-\frac{\theta}{\alpha}x}}$$

7.4 MOMENTS AND ASSOCIATED MEASURES

In this section, we will discuss the structural properties of LBWS distribution.

Theorem: If a random variable X follows LBWS distribution with parameters α, θ and c, then its rth order moment $E(X^r)$ about origin is given by

$$E(X^r) = \frac{\alpha^r(\theta\Gamma(c+r+2)+\Gamma(c+r+3))}{\theta^r(\theta\Gamma(c+2)+\Gamma(c+3))}$$

Proof: The raw moments can be written as

$$E(X^r) = \mu_r' = \int_0^\infty x^r f_l(x)dx$$

$$E(X^r) = \int_0^\infty \frac{\theta^{c+3}}{\alpha^{c+2}(\theta\Gamma(c+2)+\Gamma(c+3))} x^{c+r+1}\left(1+\frac{x}{\alpha}\right)e^{-\frac{\theta}{\alpha}x}dx$$

$$E(X^r) = \frac{\theta^{c+3}}{\alpha^{c+2}(\theta\Gamma(c+2)+\Gamma(c+3))}\left(\int_0^\infty x^{(c+r+2)-1}e^{-\frac{\theta}{\alpha}x}dx + \frac{1}{\alpha}\int_0^\infty x^{(c+r+3)-1}e^{-\frac{\theta}{\alpha}x}dx\right)$$

After simplification, we obtain

$$E(X^r) = \frac{\alpha^r(\theta\Gamma(c+r+2)+\Gamma(c+r+3))}{\theta^r(\theta\Gamma(c+2)+\Gamma(c+3))} \tag{6}$$

Putting $r = 1$ in Eq. (6), we will get the mean of LBWS distribution which is given by

$$E(X) = \mu_1' = \frac{\alpha(\theta\Gamma(c+3)+\Gamma(c+4))}{\theta(\theta\Gamma(c+2)+\Gamma(c+3))}$$

and putting $r = 2$, we obtain the second moment as

$$E(X^2) = \frac{\alpha^2(\theta\Gamma(c+4)+\Gamma(c+5))}{\theta^2(\theta\Gamma(c+2)+\Gamma(c+3))}$$

Therefore,

$$\text{Variance} = \left(\frac{\alpha}{\theta}\right)^2 \left(\frac{\theta\Gamma(c+4)+\Gamma(c+5)}{\theta\Gamma(c+2)+\Gamma(c+3)} - \left(\frac{\theta\Gamma(c+3)+\Gamma(c+4)}{\theta\Gamma(c+2)+\Gamma(c+3)}\right)^2 \right)$$

$$\text{S.D.}\,(\sigma) = \frac{\alpha}{\theta}\sqrt{\left(\frac{\theta\Gamma(c+4)+\Gamma(c+5)}{\theta\Gamma(c+2)+\Gamma(c+3)} - \left(\frac{\theta\Gamma(c+3)+\Gamma(c+4)}{\theta\Gamma(c+2)+\Gamma(c+3)}\right)^2 \right)}$$

$$\text{Coefficient of Variation (C.V.)} = \left(\frac{\theta\Gamma(c+2)+\Gamma(c+3)}{\theta\Gamma(c+3)+\Gamma(c+4)}\right) \times \sqrt{\left(\frac{\theta\Gamma(c+4)+\Gamma(c+5)}{\theta\Gamma(c+2)+\Gamma(c+3)} - \left(\frac{\theta\Gamma(c+3)+\Gamma(c+4)}{\theta\Gamma(c+2)+\Gamma(c+3)}\right)^2 \right)}$$

$$\text{Coefficient of Dispersion } (\gamma) = \frac{\alpha}{\theta}\left(\frac{\theta\Gamma(c+4)+\Gamma(c+5)}{\theta\Gamma(c+2)+\Gamma(c+3)} - \left(\frac{\theta\Gamma(c+3)+\Gamma(c+4)}{\theta\Gamma(c+2)+\Gamma(c+3)}\right)^2 \right) \times \left(\frac{\theta\Gamma(c+2)+\Gamma(c+3)}{\theta\Gamma(c+3)+\Gamma(c+4)}\right)$$

7.4.1 Harmonic Mean

Theorem: If a random variable X follows LBWS distribution, then its harmonic mean is given by

$$H.M. = \frac{\theta(\theta\Gamma(c+1)+\Gamma(c+2))}{\alpha(\theta\Gamma(c+2)+\Gamma(c+3))}$$

Proof: The Harmonic mean of the LBWS model can be obtained as

$$H.M. = E\left(\frac{1}{x}\right) = \int_0^\infty \frac{1}{x} f_l(x)\,dx$$

$$H.M. = \int_0^\infty \frac{\theta^{c+3}}{\alpha^{c+2}(\theta\Gamma(c+2)+\Gamma(c+3))} x^c \left(1+\frac{x}{\alpha}\right) e^{-\frac{\theta}{\alpha}x}\,dx$$

$$H.M. = \frac{\theta^{c+3}}{\alpha^{c+2}(\theta\Gamma(c+2)+\Gamma(c+3))}\left(\int_0^\infty x^{(c+1)-1} e^{-\frac{\theta}{\alpha}x}\,dx + \frac{1}{\alpha}\int_0^\infty x^{(c+2)-1} e^{-\frac{\theta}{\alpha}x}\,dx \right)$$

$$H.M. = \frac{\theta^{c+3}}{\alpha^{c+2}(\theta\Gamma(c+2)+\Gamma(c+3))}\left(\frac{\alpha^{c+1}\Gamma(c+1)}{\theta^{c+1}}+\frac{\alpha^{c+1}\Gamma(c+2)}{\theta^{c+2}}\right)$$

$$H.M. = \frac{\theta(\theta\Gamma(c+1)+\Gamma(c+2))}{\alpha(\theta\Gamma(c+2)+\Gamma(c+3))}$$

7.4.2 Moment Generating Function (MGF) and Characteristic Function

Theorem: The MGF and characteristic function of a random variable X following LBWS distribution are as

$$M_x(t) = \left(\frac{1}{\theta\Gamma(c+2)+\Gamma(c+3)}\right)\sum_{j=0}^{\infty}\frac{t^j\alpha^j}{j!\theta^j}(\theta\Gamma(c+j+2)+\Gamma(c+j+3))$$

and

$$M_x(it) = \left(\frac{1}{\theta\Gamma(c+2)+\Gamma(c+3)}\right)\sum_{j=0}^{\infty}\frac{(it)^j\alpha^j}{j!\theta^j}(\theta\Gamma(c+j+2)+\Gamma(c+j+3))$$

Proof: Let X have a length biased weighted Sushila distribution, then the MGF of X is obtained as

$$M_x(t) = E(e^{tx}) = \int_0^{\infty} e^{tx} f_l(x)dx$$

Using Taylor's series

$$M_x(t) = E(e^{tx}) = \int_0^{\infty}\left(1+tx+\frac{(tx)^2}{2!}+\cdots\right)f_l(x)dx$$

$$M_x(t) = \int_0^{\infty}\sum_{j=0}^{\infty}\frac{t^j}{j!}x^j f_l(x)dx$$

$$M_x(t) = \sum_{j=0}^{\infty}\frac{t^j}{j!}\mu_j'$$

$$M_x(t) = \sum_{j=0}^{\infty}\frac{t^j}{j!}\frac{\alpha^j(\theta\Gamma(c+j+2)+\Gamma(c+j+3))}{\theta^j(\theta\Gamma(c+2)+\Gamma(c+3))}$$

$$M_x(t) = \left(\frac{1}{\theta\Gamma(c+2)+\Gamma(c+3)}\right)\sum_{j=0}^{\infty}\frac{t^j\alpha^j}{j!\theta^j}(\theta\Gamma(c+j+2)+\Gamma(c+j+3))$$

Similarly, the characteristic function of LBWS distribution can be obtained as

$$\varphi_x(t) = M_x(it)$$

$$M_x(it) = \left(\frac{1}{\theta\Gamma(c+2)+\Gamma(c+3)}\right)\sum_{j=0}^{\infty}\frac{(it)^j\alpha^j}{j!\theta^j}(\theta\Gamma(c+j+2)+\Gamma(c+j+3))$$

7.5 LIKELIHOOD RATIO TEST

Let $X_1, X_2, \ldots, X_n$ be a random sample from the weighted Sushila distribution. To test the hypothesis

$$H_0 : f(x) = f(x;\, \alpha,\, c,\, \theta) \text{ against } H_1 : f(x) = f_l(x;\, \alpha,\, c,\, \theta)$$

in order to test whether the random sample of size n comes from the weighted Sushila distribution or length biased weighted Sushila distribution, the following test statistic is used

$$\Delta = \frac{L_1}{L_0} = \prod_{i=1}^{n}\frac{f_l(x;\alpha,c,\theta)}{f(x;\alpha,c,\theta)}$$

$$\Delta = \prod_{i=1}^{n}\frac{\theta(\theta+c+1)+\Gamma(c+1)x_i}{\alpha(\theta\Gamma(c+2)+\Gamma(c+3))}$$

$$\Delta = \left(\frac{\theta(\theta+c+1)+\Gamma(c+1)}{\alpha(\theta\Gamma(c+2)+\Gamma(c+3))}\right)^n\prod_{i=1}^{n}x_i$$

We reject the null hypothesis, if

$$\Delta = \left(\frac{\theta(\theta+c+1)+\Gamma(c+1)}{\alpha(\theta\Gamma(c+2)+\Gamma(c+3))}\right)^n\prod_{i=1}^{n}x_i > k$$

or, $$\Delta^* = \prod_{i=1}^{n}x_i > k\left(\frac{\alpha(\theta\Gamma(c+2)+\Gamma(c+3))}{\theta(\theta+c+1)\Gamma(c+1)}\right)^n$$

$$\Delta^* = \prod_{i=1}^{n}x_i > k^*, \quad \text{where } k^* = k\left(\frac{\alpha(\theta\Gamma(c+2)+\Gamma(c+3))}{\theta(\theta+c+1)\Gamma(c+1)}\right)^n$$

For large sample size n, $2\ log\Delta$ is distributed as chi-square distribution with one degree of freedom and also p-value is obtained from the chi-square distribution. Thus, we reject the null hypothesis, when the probability value is given by

$p(\Delta^* > \beta^*)$, where $\beta^* = \prod_{i=1}^{n} x_i$ is less than a specified level of significance and $\prod_{i=1}^{n} x_i$ is the observed value of the statistic Δ^*.

7.6 ENTROPIES

The concept of entropy is important in different areas such as probability and statistics, physics, communication theory, and economics. Entropies quantify the diversity, uncertainty, or randomness of a system. Entropy of a random variable X is a measure of variation of the uncertainty.

7.6.1 Renyi Entropy

Theorem: If a random variable X follows LBWS distribution, then the Renyi entropy of X is given by

$$e(\beta) = \frac{1}{1-\beta} \log\left(\left(\frac{\theta^{c+3}}{\alpha^{c+2}(\theta\Gamma(c+2)+\Gamma(c+3))}\right)^{\beta} \sum_{i-0}^{\infty} \binom{\beta}{i} \frac{\alpha^{\beta(c+1)+1}\Gamma(\beta(c+1)+i+1)}{(\beta\theta)^{\beta(c+1)+i+1}}\right)$$

Proof: For a given probability distribution, Renyi entropy is given by

$$e(\beta) = \frac{1}{1-\beta}\log\left(\int f^{\beta}(x)dx\right)$$

where, $\beta > 0$ and $\beta \neq 1$

$$e(\beta) = \frac{1}{1-\beta}\log\int_{0}^{\infty}\left(\frac{\theta^{c+3}}{\alpha^{c+2}(\theta\Gamma(c+2)+\Gamma(c+3))}x^{c+1}\left(1+\frac{x}{\alpha}\right)e^{-\frac{\theta}{\alpha}x}\right)^{\beta}dx$$

$$e(\beta) = \frac{1}{1-\beta}\log\left(\left(\frac{\theta^{c+3}}{\alpha^{c+2}(\theta\Gamma(c+2)+\Gamma(c+3))}\right)^{\beta}\int_{0}^{\infty}x^{\beta(c+1)}e^{-\frac{\beta\theta}{\alpha}x}\left(1+\frac{x}{\alpha}\right)^{\beta}dx\right) \quad (7)$$

Using binomial expansion in Eq. (7), we get

$$e(\beta) = \frac{1}{1-\beta} \log\left(\left(\frac{\theta^{c+3}}{\alpha^{c+2}(\theta\Gamma(c+2)+\Gamma(c+3))}\right)^{\beta}\sum_{i=0}^{\infty}\binom{\beta}{i}\left(\frac{1}{\alpha}\right)^{i}\int_{0}^{\infty}x_i^{(\beta(c+1)+i+1)-1}e^{-\frac{\beta\theta}{\alpha}x_i}dx\right)$$

$$e(\beta)=\frac{1}{1-\beta}\log\left(\left(\frac{\theta^{c+3}}{\alpha^{c+2}(\theta\Gamma(c+2)+\Gamma(c+3))}\right)^{\beta}\sum_{i=0}^{\infty}\binom{\beta}{i}\frac{\alpha^{\beta(c+1)+1}\Gamma(\beta(c+1)+i+1)}{(\beta\theta)^{\beta(c+1)+i+1}}\right)$$

7.6.2 Tsallis Entropy

A generalization of Boltzmann-Gibbs (B-G) statistical mechanics initiated by Tsallis has focussed a great deal to attention. This generalization of B-G statistics was proposed firstly by introducing the mathematical expression of Tsallis entropy (Tsallis 1988) for a continuous random variable is defined as follows:

Theorem: Show that the Tsallis entropy of order λ of the LBWS distribution is

$$S_\lambda=\frac{1}{\lambda-1}\left(\left(\frac{\theta^{c+3}}{\alpha^{c+2}(\theta\Gamma(c+2)+\Gamma(c+3))}\right)^{\lambda}\sum_{i=0}^{\infty}\binom{\lambda}{i}\frac{\alpha^{\lambda(c+1)+1}\Gamma(\lambda(c+1)+i+1)}{(\lambda\theta)^{\lambda(c+1)+i+1}}\right)$$

Proof: Tsallis entropy is defined as

$$S_\lambda=\frac{1}{\lambda-1}\left(1-\int_0^{\infty}f^{\lambda}(x)dx\right)$$

$$S_\lambda=\frac{1}{\lambda-1}\int_0^{\infty}\left(\frac{\theta^{c+3}}{\alpha^{c+2}(\theta\Gamma(c+2)+\Gamma(c+3))}x^{c+1}\left(1+\frac{x}{\alpha}\right)e^{-\frac{\theta}{\alpha}x}\right)^{\lambda}dx$$

$$S_\lambda=\frac{1}{\lambda-1}\left(\left(\frac{\theta^{c+3}}{\alpha^{c+2}(\theta\Gamma(c+2)+\Gamma(c+3))}\right)^{\lambda}\int_0^{\infty}x^{\lambda(c+1)}e^{-\frac{\lambda\theta}{\alpha}x}\left(1+\frac{x}{\alpha}\right)^{\lambda}dx\right) \tag{8}$$

Using binomial expansion in Eq. (8), we get

$$S_\lambda=\frac{1}{\lambda-1}\left(\left(\frac{\theta^{c+3}}{\alpha^{c+2}(\theta\Gamma(c+2)+\Gamma(c+3))}\right)^{\lambda}\sum_{i=0}^{\infty}\binom{\lambda}{i}\left(\frac{1}{\alpha}\right)^{i}\int_0^{\infty}x_i^{(\lambda(c+1)+i+1)-1}e^{-\frac{\lambda\theta}{\alpha}x_i}dx\right)$$

$$S_\lambda=\frac{1}{\lambda-1}\left(\left(\frac{\theta^{c+3}}{\alpha^{c+2}(\theta\Gamma(c+2)+\Gamma(c+3))}\right)^{\lambda}\sum_{i=0}^{\infty}\binom{\lambda}{i}\frac{\alpha^{\lambda(c+1)+1}\Gamma(\lambda(c+1)+i+1)}{(\lambda\theta)^{\lambda(c+1)+i+1}}\right)$$

7.7 ORDER STATISTICS

Let $X_{(1)}, X_{(2)}, \ldots, X_{(n)}$ be the order statistics of a random sample $X_1, X_2, \ldots, X_n$ drawn from the continuous population with probability density function $f_x(x)$ and cumulative density function with $F_x(x)$, then the pdf of rth order statistics $X_{(r)}$ is given by

$$f_{X(r)}(x) = \frac{n!}{(r-1)!(n-r)!} f_X(x)[F_X(x)]^{r-1}[1-F_X(x)]^{n-r} \tag{9}$$

Using Eqs. (4) and (5) in Eq. (9), the probability density function of rth order statistics $X_{(r)}$ of LBWS distribution is given by

$$f_{X(r)}(x) = \frac{n!}{(r-1)!(n-1)!}\left(\frac{\theta^{c+3}}{\alpha^{c+2}(\theta\Gamma(c+2)+\Gamma(c+3))} x^{c+1}\left(1+\frac{x}{\alpha}\right)e^{-\frac{\theta}{\alpha}x}\right)$$

$$\times\left(\frac{\theta}{\theta\Gamma(c+2)+\Gamma(c+3)}\left(\gamma\left((c+2),\frac{\theta}{\alpha}x\right)+\frac{1}{\theta}\gamma\left((c+3),\frac{\theta}{\alpha}x\right)\right)\right)^{r-1}$$

$$\times\left(1-\frac{\theta}{\theta\Gamma(c+2)+\Gamma(c+3)}\left(\gamma\left((c+2),\frac{\theta}{\alpha}x\right)+\frac{1}{\theta}\gamma\left((c+3),\frac{\theta}{\alpha}x\right)\right)\right)^{n-r}$$

Therefore, the probability density function of higher order statistics $X(n)$ can be obtained as

$$f_{X(n)}(x) = \left(\frac{n\theta^{c+3}}{\alpha^{c+2}(\theta\Gamma(c+2)+\Gamma(c+3))} x^{c+1}\left(1+\frac{x}{\alpha}\right)e^{-\frac{\theta}{\alpha}x}\right)$$

$$\times\left(\frac{\theta}{\theta\Gamma(c+2)+\Gamma(c+3)}\left(\gamma\left((c+2),\frac{\theta}{\alpha}x\right)+\frac{1}{\theta}\gamma\left((c+3),\frac{\theta}{\alpha}x\right)\right)\right)^{n-1}$$

and the pdf of 1st order statistic $X(1)$ can be obtained as

$$f_{X(1)}(x) = \left(\frac{n\theta^{c+3}}{\alpha^{c+2}(\theta\Gamma(c+2)+\Gamma(c+3))} x^{c+1}\left(1+\frac{x}{\alpha}\right)e^{-\frac{\theta}{\alpha}x}\right)$$

$$\times\left(1-\frac{\theta}{\theta\Gamma(c+2)+\Gamma(c+3)}\left(\gamma\left((c+2),\frac{\theta}{\alpha}x\right)+\frac{1}{\theta}\gamma\left((c+3),\frac{\theta}{\alpha}x\right)\right)\right)^{n-1}$$

7.8 BONFERRONI AND LORENZ CURVES

The Bonferroni and the Lorenz curves are not only used in economics to study income and poverty, but it is also being used in other fields like reliability, medicine, insurance, and demography. The Bonferroni and Lorenz curves are given by:

Theorem: The Bonferroni and the Lorenz Curves for a random variable X following the LBWS distribution are respectively given by

$$B(p)=\left(\frac{\theta}{p(\theta\Gamma(c+3)+\Gamma(c+4))}\left(\gamma\left((c+3),\frac{\theta}{\alpha}q\right)+\frac{1}{\theta}\gamma\left((c+4),\frac{\theta}{\alpha}q\right)\right)\right)$$

and

$$L(p)=pB(p)=\frac{\theta}{(\theta\Gamma(c+3)+\Gamma(c+4))}\left(\gamma\left((c+3),\frac{\theta}{\alpha}q\right)+\frac{1}{\theta}\gamma\left((c+4),\frac{\theta}{\alpha}q\right)\right)$$

Proof: We know the Bonferroni and Lorenz curve of a random X having density function $f(x)$ is given by

$$B(p)=\frac{1}{p\mu_1'}\int_0^q x f_l(x)dx$$

and

$$L(p)=\frac{1}{\mu_1'}\int_0^q x f_l(x)dx$$

where

$$\mu_1'=E(X)=\left(\frac{\alpha}{\theta}\right)\left(\frac{\theta\Gamma(c+3)+\Gamma(c+4)}{\theta\Gamma(c+2)+\Gamma(c+3)}\right) \quad \text{and} \quad q=F^{-1}(p)$$

$$B(p)=\left(\frac{\alpha}{p\theta}\right)\left(\frac{\theta\Gamma(c+3)+\Gamma(c+4)}{\theta\Gamma(c+2)+\Gamma(c+3)}\right)$$
$$\int_0^q \frac{\theta^{c+3}}{\alpha^{c+2}(\theta\Gamma(c+2)+\Gamma(c+3))}x^{c+2}\left(1+\frac{x}{\alpha}\right)e^{-\frac{\theta}{\alpha}x}dx$$

After simplification, we get

$$B(p)=\frac{\theta}{p(\theta\Gamma(c+3)+\Gamma(c+4))}\left(\gamma\left((c+3),\frac{\theta}{\alpha}q\right)+\frac{1}{\theta}\gamma\left((c+4),\frac{\theta}{\alpha}q\right)\right)$$

and

$$L(p) = pB(p) = \frac{\theta}{(\theta\Gamma(c+3)+\Gamma(c+4))}\left(\gamma\left((c+3), \frac{\theta}{\alpha}q\right)+\frac{1}{\theta}\gamma\left((c+4), \frac{\theta}{\alpha}q\right)\right)$$

7.9 MAXIMUM LIKELIHOOD ESTIMATORS

In this section, we will discuss the maximum likelihood estimators of the parameters of length biased weighted Sushila distribution. Consider X_1, X_2, ..., X_n to be the random sample of size n from the LBWS distribution, then the likelihood function is given by

$$L(x;\alpha,c,\theta) = \frac{\theta^{n(c+3)}}{\alpha^{n(c+2)}(\theta\Gamma(c+2)+\Gamma(c+3))^n}\prod_{i=1}^{n}\left(x_i^{c+1}\left(1+\frac{x_i}{\alpha}\right)e^{-\frac{\theta}{\alpha}x_i}\right)$$

The log likelihood function is

$$\log L = n(c+3)\log\theta - n(c+2)\log\alpha - n\log(\theta\Gamma(c+2)+\Gamma(c+3)) + (c+1)$$

$$\sum_{i=1}^{n}\log x_i + \sum_{i=1}^{n}\log\left(1+\frac{x_i}{\alpha}\right) - \frac{\theta}{\alpha}\sum_{i=1}^{n}x_i \tag{10}$$

The maximum likelihood estimates of α, c, θ can be obtained by differentiating Eq. (10) with respect to α, c, θ and must satisfy the normal equation

$$\frac{\partial\log L}{\partial\alpha} = -\frac{n(c+2)}{\alpha} - \sum_{i=1}^{n}\frac{x_i}{\alpha(\alpha+x_i)} + \frac{\theta}{\alpha^2}\sum_{i=1}^{n}x_i = 0$$

$$\frac{\partial\log L}{\partial\theta} = \frac{n(c+3)}{\theta} - \frac{n\Gamma(c+2)}{(\theta\Gamma(c+2)+\Gamma(c+3))} - \frac{1}{\alpha}\sum_{i=1}^{n}x_i = 0$$

$$\frac{\partial\log L}{\partial c} = n\log\theta - n\log\alpha - n\psi(c+2) + \sum_{i=1}^{n}\log x_i = 0$$

where $\psi(.)$ is the digamma function.

Because of the complicated form of likelihood equations, algebraically it is very difficult to solve the system of nonlinear equations. Therefore, we use R and Wolfram mathematics for estimating the required parameters.

7.10 DATA ANALYSIS

In this section, we have used a real lifetime dataset in length biased weighted Sushila distribution and the model has been compared with Sushila distribution.

The below data is related to the strength of 31 windows of aircraft glass reported by Fuller et al. (1994).

18.83	20.80	21.657	23.03	23.23	24.05	24.321	25.50	25.52	25.80	26.69
26.77	26.78	27.05	27.67	29.90	31.11	33.20	33.73	33.76	33.89	34.76
35.75	35.91	36.98	37.08	37.09	39.58	44.045	45.29	45.431		

To evaluate the criteria for a given model, like "Bayesian Information Criterion (*BIC*), Akaike Information Criterion (*AIC*), Akaike Information Criterion Corrected (*AICC*) and $-2\log L$. The better distribution is which corresponds to lesser values of *AIC, BIC, AICC* and $-2\log L$. For calculating *AIC, BIC, AICC* and $-2\log L$ can be evaluated by using the formulas as follows:

$$AIC = 2K - 2\log L,\ BIC = k \log n - 2\log L,\ AICC = AIC + \frac{2k(k+1)}{(n-k-1)}$$

where k is the number of parameters, n is the sample size, and $-2\log L$ is the maximized value of log likelihood function and are shown in Table 7.2.

Table 7.2 Performance of distributions

Distribution	Parameter Estimates			–2 logL	AIC	AICC	BIC
	$\hat{\alpha}$	$\hat{\theta}$	$\hat{c}$				
Sushila	0.137249849 (0.108963418)	0.008869415 (0.006921270)	—	252.493	256.493	257.3818	259.3609
LBW Sushila	0.2567681 (0.8926121)	0.1574252 (0.547785)	16.9002969 (4.7500810)	204.2438	208.2438	217.4127	219.5658

From Table 7.2, we can see that the length biased weighted Sushila distribution has the lesser values of *AIC, BIC, AICC* and $-2\log L$ as compared to Sushila distribution. Hence, we can conclude that the length biased weighted Sushila distribution shows better performance than the Sushila distribution.

7.11 CONCLUSION

In this chapter, the concept of outliers has been discussed and a new generalization of the Sushila distribution, known as the length biased weighted Sushila distribution, has been introduced. This new distribution has three parameters and is generated using the length-biased technique. The parameters are estimated using the maximum likelihood estimator. Mathematical properties of the new distribution, along with reliability measures, have been discussed. The distribution has also been applied to real-life datasets, and the results have been compared to those obtained using the Sushila distribution. The length biased weighted Sushila distribution has been found to provide a better fit than the Sushila distribution.

REFERENCES

Cox, D.R. 1969. Some sampling problems in technology. In: New Development in Survey Sampling, Johnson, N.L. and Smith, H., Jr. (Eds.). New York Wiley-Interscience. 506–527.

Elgarhy, E. and Shawki, A.W. 2017. Exponentiated Sushila distribution, International Journal of Scientific Engineering and Science. 1(7): 9–12.

Fisher, R.A. 1934. The effects of methods of ascertainment upon the estimation of frequencies. Annals of Eugenics. 6: 13–25.

Fuller, E.J., Frieman, S., Quinn, J., Quinn, G. and Carter, W. 1994. Fracture mechanics approach to the design of glass aircraft windows: A case study. SPIE Proceedings. 2286: 419–430.

Rao, C.R. 1965. On discrete distributions arising out of method of ascertainment In: Classical and Contagious Discrete, G.P. Patil (Ed.). Pergamum Press and Statistical Publishing Society. Calcutta. 320–332.

Rather, A.A. and Subramanian, C. 2018a. Characterization and estimation of length biased weighted generalized uniform distribution. International Journal of Scientific Research in Mathematical and Statistical Sciences. 5(5). 72–76.

Rather, A.A. and Subramanian, C. 2018b. Length biased Sushila distribution. Universal Review. 7(XII): 1010–1023.

Rather, A.A. and Subramanian C. 2019. On weighted Sushila distribution with properties and its applications. International Journal of Scientific Research in Mathematical and Statistical Science. 6(1): 105–117.

Rather, A.A. and Ozel Gamze. 2021. A new length-biased power Lindley distribution with properties and its applications. Journal of Statistics and Management Systems. DOI: 10.1080/09720510.2021.1920665.

Shanker, R., Shambhu S., Uma S. and Ravi S. 2013. Sushila distribution and its application to waiting times data. International Journal of Business Management. 3(2): 1–11.

Subramanian, C. and Rather, A.A., 2020. A new extension of Shanker distribution with real life data. Journal of Xi'an University of Architecture and Technology. XII(III): 2001–2009.

Tsallis, C. 1988. Possible generalization of Boltzmann-Gibbs statistics. Journal of Statistical Physics. 52: 479–487.

Zelen, M. 1974. Problems in cell kinetic and the early detection of disease. In: Reliability and Biometry, F. Proschan and R.J. Sering (Eds.), SIAM. Philadelphia. 701–706.

Chapter 8

Outlier Detection Based on Range Statistic
Empirical Evaluation and Comparisons

Dania Dallah[1], Hana Sulieman[2] and Ayman Alzaatreh[3]

[1,2,3]Department of Mathematics and Statistics,
American University of Sharjah, UAE

8.1 INTRODUCTION

In real datasets, it often happens that some observations are different from the majority. Such observations are called outliers. When they exist, outliers can have a significant influence on the data. Generally, they tend to disrupt many statistical characteristics of the data and hence their analyses. There exist numerous reasons for their presence, such as, human error (such as data entry error), sampling error (including items or people that are outside the targeted population), and natural variation in the data. For these reasons, statisticians have created numerous different methods to help detect those anomalous observations. Actively monitoring outliers and finding the suitable outlier detection technique can help flag errors in the dataset. This cleans the dataset from any

*For Correspondence: Ayman Alzaatreh (email: aalzaatreh@aus.edu)

insignificant outcome and fits the data for better results. However, outliers can be informative for the data. Sometimes, outliers are an integral part of the data structure and can provide valuable information for the study. As a result, it is important to understand these data values and how they function as separate variables.

There are a wide variety of applications for outlier detection. Some include credit cards, tax fraud detection, cybersecurity, insurance, healthcare, and many more areas. For example, one can identify any suspicious transaction behavior that is different from normal ones in credit cards. Hence, this helps secure any financial transaction and prevent intrusions and identity theft. In the medical field, the medical records contain electronic entries that describe a patient's condition, treatment, and their laboratory results. Thus, many hospitals require techniques to detect anomalous data that indicate the existence of a disease (Smiti 2020).

In the field of statistics there exist numerous outlier detection approaches, each comes with its own characteristics and practical benefits. Below is a brief description of the most widely used methods.

8.1.1 IQR and Boxplot

The most common approach in detecting outliers using descriptive statistics is using the interquartile range (IQR). The concept of IQR is used in creating boxplots, which is a method to present outliers graphically using numerical data based on their quartiles. Boxplots were proposed by Tukey (1977) who defined outliers as the observations that are below the boxplot lower whisker defined by Q_3 – 1.5IQR or above the boxplot upper whisker defined by Q_3 + 1.5IQR.

Schwertman et al. (2007) enhanced and modified the boxplot concept by introducing a more general fencing technique to provide greater flexibility in determining the likelihood that a data value from a normal population falls beyond a certain limit or boundary. The generalized fences are defined as

$$F_n = q_2 \pm \frac{IQR}{k_n} Z_\alpha$$

where q_i is the ith quartile, $IQR = q_3 - q_1$, k_n is the appropriate adjustment relating the expected value of the IQR to the standard deviation for various sample sizes, and Z_α is chosen such that $P(Z \geq Z_\alpha) = \alpha$ for the standard normal variable Z and the total outside rate is 2α.

Moreover, Dovoedo and Chakraborti (2015) proposed a new method for detecting outliers in a univariate dataset based on the boxplot. The method was applied to data that follow the location-scale family distribution including commonly used distributions such as the normal, *t*, and Cauchy distributions. The authors use a modified version of the Tukey boxplot that adjusts the whisker length based on the interquartile range and the sample size, and defines outliers as observations that fall outside of the adjusted whiskers. The whiskers are determined using a robust estimate of the scale of the data, such as the IQR or the median absolute deviation (MAD), rather than the sample standard deviation. They demonstrate the effectiveness of the method through simulations and real data examples. The proposed method can be more effective in detecting outliers in univariate data especially for datasets with heavy-tailed or skewed distributions.

Because of its simplicity and easy interpretation, the boxplot method has been widely used in various application areas. For example, Li et al. (2016) presented a novel approach to identifying insider threats using outlier detection based on the boxplot method. Insider threats are a growing concern in cybersecurity, and detecting them early is crucial to prevent data breaches and other security incidents. The authors proposed using outlier mining techniques to identify suspicious activities by insiders, who may be employees, contractors, or other authorized users with access to sensitive information. Then, they introduce the boxplot method as a tool for identifying outliers in a dataset, reflecting possible anomalous behavior among employees. The proposed method involves constructing a boxplot of the user's activity data, such as login times, file accesses, and network traffic, and identifying outliers as activities that deviate significantly from the typical pattern of the user's behavior. Outliers may indicate unusual or suspicious behavior that warrants further investigation, such as an employee accessing sensitive information outside of their normal work hours or sending large amounts of data to an external recipient. The authors demonstrate the effectiveness of their method through a case study involving data collected from a financial institution. In short, they show that the boxplot method was able to identify a small number of potential insider threats, which were subsequently investigated and found to be valid.

8.1.2 Grubbs' Test

Grubbs (1949) proposed two tests for identifying outliers in a dataset: Grubbs' test for one outlier and the Grubbs' test for multiple outliers. The tests assume that the data follow a normal distribution. The Grubbs' test for one outlier involves calculating the maximum deviation of a data point from the mean of the dataset, and comparing it to a critical value calculated from the t-distribution. If the maximum deviation is greater than the critical value, then the data point is an outlier. The Grubbs' test for multiple outliers involves iteratively identifying and removing outliers from the dataset until no more outliers are found. This process is repeated until all outliers have been identified and removed from the dataset.

8.1.3 Extreme Studentized Deviate

Rosner (1983) proposed the extreme Studentized deviate (ESD), a sequential version of the Grubbs' test, which is a statistical test for identifying outliers based on the assumption of normality procedure. The modified test is designed to detect multiple outliers by iteratively removing the most extreme value from the dataset and recalculating the test statistic until no outliers remain. All is needed is an upper bound r where the technique performs r different tests for each outlier. The effectiveness of the proposed method is demonstrated through simulations and an application to real-world data. The results show that the modified Grubbs' test performs well in detecting multiple outliers, even in datasets with a high proportion of potential outliers. The generalized ESD test is used in the following hypotheses:

H_0 : There are no outliers in the dataset.

H_a : There are up to r outliers in the dataset.

The procedure starts with ordering the values from smallest to largest. Then, a maximum number of the suspected outliers needs to be specified to know how many tests are needed to fully detect all the outliers. A series of test statistics are calculated by removing the data point that is furthest away from the mean, and then recomputing the test statistic. The test statistic is

$$R_i = \frac{\max_i |x_i - \bar{x}|}{s}$$

with $\bar{x}$ and s denoting the sample mean and sample standard deviation, respectively.

For a significance level α, the critical regions corresponding to the r test statistics are constructed by computing the r critical values. The critical region is

$$\lambda_i = \frac{(n-i)t_{p,n-i-1}}{\sqrt{\left(n-i-1+t_{p,n-i-1}^2\right)(n-i+1)}}, i = 1, 2, \ldots, r$$

Schwertman et al. (2007) conducted a comparison of their sequential fences approach with the generalized ESD test and Tukey's box-plot method. The authors show that the sequential fences method performs well in identifying outliers in both real-world and simulated datasets.

For the real-world dataset, the sequential fences procedure successfully identified all four actual outliers, the ESD method was able to identify only three of the four outliers at the same level of confidence, and the "box-plot rule" fails to identify any of the outliers. For a simulated dataset, it appears that in general, the sequential fences method is superior to the "box-plot rule" in identifying outliers, particularly when dealing with smaller sample sizes and multiple outliers for any sample size. The "box-plot rule" may be slightly better at identifying outliers for larger sample sizes or extreme outliers only when there are only two outliers present. On the other hand, they concluded that the ESD method is way superior in detecting the first outliers than any of the other two methods, but needs far more intense computations.

Moreover, Cohn et al. (2013) proposed a modified version of the extension of Grubbs' Test for multiple outliers, known as Grubbs-Beck test statistic. The left-sided test statistic differs from the Rosner (1975) ESD statistic in that it only considers low outliers and does not include the suspected outlier in the computation of the mean and standard deviation of the trimmed data. The results show that the modified test statistic performs well in detecting multiple low outliers, and can be used to improve the accuracy of flood frequency analysis.

8.1.4 Range Statistic

In statistics, the range indicates roughly how diverse the data values are. It is a useful measure of the spread out of the dataset and can also be an effective tool to gesture the existence of outliers. Existence of one extreme value in either end of the dataset will result in an entirely different and inflated range.

Harter (1961) was the first to introduce the range statistic and use it in an attempt to detect outliers. He defined the test statistic as $W = R/\sigma$ from a sample of normal distribution, where R is the range of a sample and σ is the population standard deviation. In a random sample of n observations drawn from a normal population with σ standard deviation, Harter (1961) defined W such that:

1. R is non-negative
2. R is such that the distribution function of $W = R/\sigma$ is independent of σ.

While W can be a useful tool for detecting outliers, it has a primary shortcoming. When an independent estimate of σ is used, the statistic W becomes highly influenced by the existence of unusual values in the data because the standard deviation itself is inflated in value by the unusual or outlier values. Consequently, using W as a tool to identify outliers can be misleading.

In this chapter, we propose to standardize the range R by a more robust statistic defined by the IQR. We define the statistic $K = R/\text{IQR}$ and explore the distributional behavior of K and its performance in detecting anomalous observations in simulated data.

This chapter is organized as follows. Section 8.2, introduces the methodology and the proposed standardization of the range statistics, called K. The probability distribution of K is also explored through simulation studies based on univariate normal data. Section 8.3 presents an experimental study to assess the performance of K in detecting outliers and compares it to the performance of W. Finally, conclusions and some insight for future work are presented in Section 8.4.

8.2 METHODOLOGY

8.2.1 Probability Density Function of the Range Statistic

The probability density function of the Range Statistic can be constructed using the notion of order statistic.

In an ordered random sample, $X_1, X_2, \ldots, X_n$, the **probability density function** of the ith order statistics, X_1 is given by Arnold et al. (2008)

$$f_{x_{(i)}}(x) = \frac{n!}{(i-1)!(n-i)!} f_x(x)[F_x(x)]^{i-1}[1-F_x(x)]^{n-i} \tag{1}$$

whereas the **cumulative probability distribution** function is given by

$$F_{x_{(i)}}(x) = \sum_{i=r}^{n} \binom{n}{i} [F_x(x)]^{i}[1-F_x(x)]^{n-i} \tag{2}$$

It can also be shown that the **joint distribution of two order** X_i, and X_j can be equal to

$$f_{i,j}(x_i, x_j) = \frac{n!}{(i-1)!(j-i-1)!(n-j)!} \cdot [F_x(x_i)^{i-1}][F_x(x_j) - F_x(x_i)]^{j-i-1}$$

$$[1-F_x(x_j)]^{n-j} f_x(x_i) f_x(x_j) \tag{3}$$

$$-\infty < x_i < x_j < \infty, 1 \le i \le j \le n$$

Based on Eq. (3), and taking $i = 1$, $X_1 = \min \{x_1, \ldots, X_n\}$ and $j = n$, $X_n = \max \{X_1, \ldots, X_n\}$, one can deduce that the pdf of the range is as follows:

PDF of R

$$f_R(r) = \int_{-\infty}^{\infty} n(n-1)\{F(x_1+r) - F(x_1)\}^{n-2} f(x_1) f(x_1+r)\, dx_1 \tag{4}$$

where $0 < r < \infty$

Furthermore, when $i = \frac{n+1}{4}$ and $j = \frac{3(n+1)}{4}$, we obtain the joint distribution of the two-order statistic $X_i = Q_1$, and $X_j = Q_3$ following Eq. (3). This results in the following pdf of IQR statistic:

$$f_Q(q) = \frac{n!}{\left(\frac{n}{4}-\frac{3}{4}\right)!\left(\frac{n}{2}-\frac{1}{2}\right)!\left(\frac{n}{4}-\frac{1}{4}\right)!}$$

$$\int_{-\infty}^{\infty} [F_x(x_i)]^{\frac{n}{4}-\frac{3}{4}} [F_x(q+x_i)-F_x(x_i)]^{\frac{n}{2}-\frac{1}{2}} \cdot [1-F_x(q+x_i)]^{\frac{n}{4}-\frac{1}{4}} \cdot f_x(x_i) f_x(q+x_i)\, dx_i, \tag{5}$$

where $Q = IQR$ and $0 < q < \infty$. In examining Eqs. (4) and (5), it became evident that finding an analytical closed form for the probability density function of K is nearly impossible. Alternatively, the probability behavior of K can be explored empirically through simulation experiments in order to shed some light on its distribution characteristics and establish some understanding of performance. The empirical work is presented in the next subsection.

8.2.2 Empirical Exploration

We utilize the program RStudio as a programming tool for the simulation experiment that helps in exploring the probability distribution of W and K. The way the simulation works is by randomly sampling from a normal distribution with mean equal zero and standard deviation of 1. Multiple random samples of specified sample size, n, are generated for from $N(0, 1)$ and the statistics W and K are calculated for each sample. 100,000 samples are simulated for each of the following selected sample sizes: n = 20, 30, 50, 100, 500, 1,000. To understand their distributions, we visualize them by empirically graphing their distributions using the kernel density function. We summarize our results in Figure 8.1.

Summary statistics of the distributions of W and K are included in Table 8.1.

It can be seen that the distribution of W has a higher mean than that of distribution of K for all sample sizes. On the other hand, K has larger standard deviation than W as n increases, but becomes smaller for n = 500 and 1,000. Similarly, K shows to have higher skewness and kurtosis, but decreases for n = 500 and 1,000.

The distribution of W seems to be a bit symmetrical but tends to shift to the right as the sample size increases making the graphs appear to get less symmetric with slightly extended right tail. In general, the

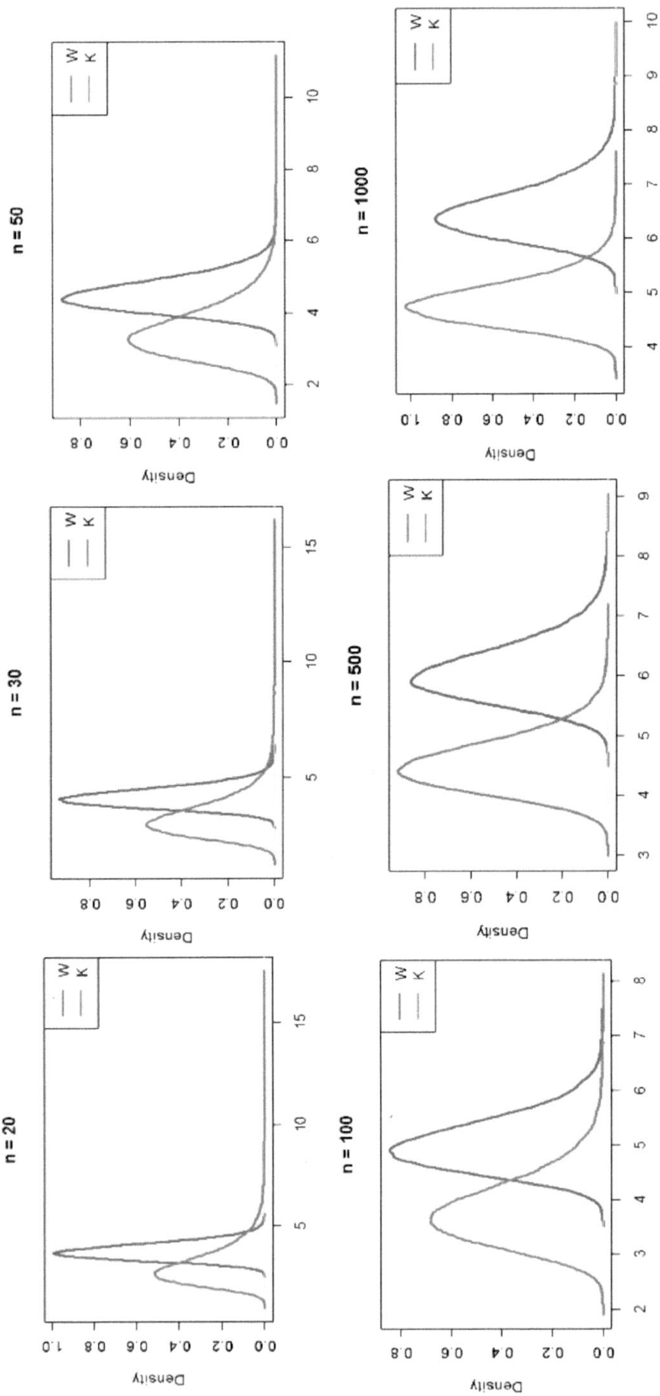

Figure 8.1 Empirical exploration of the distributions of W and K for $n = 20$, 30, 50, 100, 500, and 1,000.

Table 8.1 Summary statistics for the distributions of W and K

		Mean	Standard Deviation	Skewness	Kurtosis
$n = 30$	W	4.122268	0.4356989	0.4435724	3.153828
	K	3.300196	0.8529568	1.260766	6.5794
$n = 50$	W	4.522148	0.4652672	0.5004537	3.259883
	K	3.511475	0.7275207	0.9012815	4.616724
$n = 100$	W	5.029702	0.4871028	0.5366889	3.384648
	K	3.817455	0.6094634	0.6532409	3.726032
$n = 500$	W	6.080118	0.486926	0.5747558	3.486537
	K	4.529131	0.4468052	0.505479	3.43373
$n = 1{,}000$	W	6.485454	0.474598	0.5855118	3.58377
	K	4.819501	0.4044197	0.5119087	3.486777

distribution of the range statistic tends to be skewed to the right for data coming from a normal distribution. On the other hand, the distribution of K is clearly more skewed to the right. Unlike the W distribution, K tends to get more symmetrical as the sample size increases. In addition, the distribution of W and K tend to get further apart from each other as n gets larger. In conclusion, the PDFs for W and K agree in behavior for each sample size n with K showing slightly more skewness to the right than W.

In the next section, we present a comparison of the performance of W and K in detecting outliers based on the upper 5% tail threshold for the outlier region.

8.3 OUTLIER DETECTION

With the construction of the empirical distributions of W and K in the preceding section, it becomes clear that large values of these non-negative two statistics can flag the existence of outliers. Hence, an outlier region can be defined in the upper tail of their distributions. Table 8.2 gives the simulated 95th percentiles (upper 5% of the distribution tail) for various sample sizes n = 20, 30, 50, 100, 500, and 1,000.

Table 8.2 95th percentiles of W and K

	95th percentile W	95th percentile of K
$n = 20$	4.486456	4.934946
$n = 30$	4.898379	4.872126
$n = 50$	5.355476	4.842119
$n = 100$	5.900913	4.914124
$n = 500$	6.953673	5.318908
$n = 1{,}000$	7.337845	5.54065

To compare the performance of W and K methods, a simulation study is conducted to estimate empirically the probability that at least one outlier is detected using each method. In the simulation study, a random sample of size n-c generated from $N(0, 1)$ is contaminated with n random sample generated from $N(\mu, 1)$. The resulting data of size n is then standardized. Based on the standardized data, the observed values of the range statistics W and K, are computed and compared with their respective 95th percentiles (see Table 8.2). A data point is flagged as an outlier with respect to the method W or K if the observed value exceeds the 95th percentile.

Table 8.3 Summary of the percent of outliers detected from simulated data for different sample size and mean

Contamination = 5 data points			
Sample Size	**Mean Contamination**	**Method W**	**Method K**
	mean = 2	3.00%	4.38%
	mean = 3	0.71%	3.01%
n = 20	mean = 4	0.09%	2.09%
	mean = 5	0.01%	1.57%
	mean = 10	0.00%	0.29%
	mean = 2	4.35%	7.02%
	mean = 3	1.99%	12.84%
n = 30	mean = 4	0.42%	26.95%
	mean = 5	0.08%	47.34%
	mean = 10	0.00%	99.30%
	mean = 2	6.47%	8.82%
	mean = 3	4.64%	20.07%
n = 40	mean = 4	1.99%	42.22%
	mean = 5	0.52%	67.14%
	mean = 10	0.00%	99.99%
	mean = 2	8.29%	10.68%
	mean = 3	8.39%	25.98%
n = 50	mean = 4	5.23%	52.07%
	mean = 5	2.22%	77.15%
	mean = 10	0.00%	100.0%
	mean = 2	12.60%	13.77%
	mean = 3	25.09%	39.08%
n = 100	mean = 4	35.80%	72.57%
	mean = 5	41.48%	93.84%
	mean = 10	17.30%	100.0%
	mean = 2	13.07%	13.17%
	mean = 3	45.27%	47.29%
n = 500	mean = 4	85.59%	87.73%
	mean = 5	99.12%	99.35%
	mean = 10	100.0%	100.0%
	mean = 2	11.71%	11.90%
	mean = 3	43.86%	44.03%
n = 1,000	mean = 4	86.52%	86.97%
	mean = 5	99.48%	99.55%
	mean = 10	100.0%	100.0%

The simulation study is conducted for various values of n = 20, 30, 50, 100, 500, 1,000 and μ = 2, 3, 4, 5, 10. The number of contaminated data points is chosen to be 5. The simulation study is repeated 10,000 times and the percentage of which each method can flag at least one outlier for each sample size is reported in Table 8.3.

It is expected that when the mean contamination increases, the likelihood of existing outlier in the dataset increases making the identification of such outlier values easier. Except for n = 20, the performance of K conforms to this postulation. The larger the mean value used for contaminating the generated dataset, the higher the percentage of times at least one outlier is deducted by K, for n = 30, 40, 50, 100, 500, and 1,000. For n = 20, the five contaminated observations represent the top 25% points of the data. It is very likely that the third quartile, Q_3, be one of those contaminated points resulting in a reduced value of K which reduces the likelihood of identifying the induced outliers in the dataset. On the other hand, Table 8.3 demonstrates that W fails to detect outliers when data contaminated with high values of mean contamination for n less than or equal to 100. For n = 500 and above, W and K begin having similar percentages. They become equally capable in detecting an outlier in a dataset.

8.4 CONCLUSION

Numerous outlier detection techniques are found in the field of statistics that help give crucial information about the dataset. Outlier detection is a commonly applied research field that is used in a variety of domains. In this chapter, we explore the use of the range statistic standardized by the standard deviation and by IQR, W and K statistics, respectively, to identify outliers in univariate data. By randomly simulating repeated samples from normal distribution and mean-contaminating the data, we examined the performance of W and K statistics in detecting induced outliers for different sample sizes. The numerical experiments have shown that standardizing the range statistic by the IQR provides a more capable measure of correctly identifying outliers in data with sample sizes less than 100. For sample size greater than 100, the statistics W and K demonstrated comparable accuracy in detecting the anomalous observations in the dataset.

This chapter introduces a range statistic-based measure to identify outliers in univariate data. The numerical simulations showed that the measure has a high degree of accuracy in identifying anomalous observations in small and large sample sizes. Future enhancement of the work will include creating a sequential approach using the K statistic to identify outliers and assess its performance on real datasets.

REFERENCES

Arnold, B.C., Balakrishnan, N. and Nagaraja, H.N. (2008). A first course in order statistics. 2008. Classic Ed. SIAM, Philadelphia, Pennsylvania.

Cohn, T.A., England, J.F., Berenbrock, C.E., Mason, R.R., Stedinger, J.R. and Lamontagne, J.R. 2013. A generalized Grubbs-Beck test statistic for detecting multiple potentially influential low outliers in flood series. Water Resources Research. 49(8): 5047–5058.

Dovoedo, Y.H. and Chakraborti, S. 2015. Boxplot-based outlier detection for the location-scale family. Communications in Statistics-simulation and Computation. 44(6): 1492–1513.

Grubbs, F.E. 1949. Sample criteria for testing outlying observations. University of Michigan.

Harter, H.L. 1961. Use of tables of percentage points of range and Studentized range. Technometrics. 3(3): 407–411.

Li, A., Feng, M., Li, Y. and Liu, Z. 2016. Application of outlier mining in insider identification based on boxplot method. Procedia Computer Science. 91: 245–251.

Rosner, B. 1975. On the detection of many outliers. Technometrics. 17(2): 221–227.

Rosner, B. 1983. Percentage points for a generalized ESD many-outlier procedure. Technometrics. 25(2): 165–172.

RStudio Team. 2020. RStudio: Integrated Development for R. RStudio, PBC, Boston, MA. URL http://www.rstudio.com/

Schwertman, N.C. and de Silva, R. 2007. Identifying outliers with sequential fences. Computational Statistics and Data Analysis. 51(8): 3800–3810.

Smiti, A. 2020. A critical overview of outlier detection methods. Computer Science Review. 38: 100306.

Tukey, J.W. 1977. Exploratory Data Analysis, Vol. 2, pp. 131–160. Reading, Massachussets: Addison-Wesley.

Chapter **9**

Exponentiated Reduced Modified Weibull Distribution

Saad J. Almalki[1] and Saralees Nadarajah[2*]

[1]Department of Mathematics and Statistics, Taif University, P.O. Box 888, Taif, KSA, saad8582@hotmail.com

[2]Department of Mathematics, University of Manchester, Manchester M13 9PL, UK, mbbsssn2@manchester.ac.uk

9.1 INTRODUCTION

The Weibull distribution has been extensively generalized using different techniques by several authors, see Almalki and Nadarajah (2014) for a comprehensive review of known generalizations. If $G(x)$ is a cumulative distribution function (CDF), then the distribution with CDF, $F(x) = G(x)^{\theta}$, is called exponentiated distribution. Using this exponentiation several lifetime distributions have been generalized. Kundu and Gupta (1999) generalized the standard exponential distribution to the generalized exponential (GE) distribution with CDF

$$F(x; \alpha, \theta) = (1 - e^{-\alpha x})^{\theta}$$

for $x > 0$, where $\alpha > 0$ and $\theta > 0$. This distribution has been extensively studied by several authors (for example, Gupta and Kundu 2001a, 2001b, Raqab 2004, Gupta and Kundu 2007, and Kundu and Gupta 2008).

*For Correspondence: Saralees Nadarajah (mbbsssn2@manchester.ac.uk)

Mudholkar and Srivastava (1993) generalized the Weibull distribution to the exponentiated Weibull (EW) distribution. The CDF of the EW distribution is

$$F(x;\alpha,\gamma,\theta)=(1-e^{-\alpha x^{\gamma}})^{\theta} \tag{1}$$

for $x > 0$, where $\alpha > 0$, $\gamma > 0$ and $\theta > 0$. The exponentiated Pareto (EP) distribution was proposed by Gupta et al. (1998), while the generalized Rayleigh (GR) distribution was studied by Surles and Padgett (2001, 2005) and Kundu and Raqab (2005). The CDF of the GR distribution is

$$F(x;\alpha,\theta)=(1-e^{-\alpha x^{2}})^{\theta}$$

for $x > 0$, where $\alpha > 0$ and $\theta > 0$. Using the same technique, Nadarajah and Kotz (2006) introduced the exponentiated gamma (EGam) distribution, the exponentiated Gumbel (EGum) distribution and the exponentiated Fréchet (EF) distribution. The generalized modified Weibull (GMW) distribution that generalizes the modified Weibull (MW) distribution of Lai et al. (2003) was proposed by Carrasco et al. (2008). The CDF of the GMW distribution is

$$F(x)=(1-e^{-\alpha x^{\gamma}e^{\lambda x}})^{\theta} \tag{2}$$

for $x > 0$, where $\alpha > 0$, $\gamma > 0$, $\lambda > 0$ and $\theta > 0$. Sarhan and Kundu (2009) introduced a three-parameter distribution called the generalized linear failure rate (GLFR) distribution with CDF

$$F(x)=(1-e^{-\alpha x-\beta x^{2}})^{\theta}$$

for $x > 0$, where $\alpha > 0$, $\beta > 0$ and $\theta > 0$.

Recently, different versions of exponentiated distributions have been introduced by several authors: Sarhan and Apaloo (2013), El-Gohary et al. (2013), and Sarhan et al. (2013) introduced the exponentiated modified Weibull extension (EMWEx) distribution, the generalized Gompertz (GG) distribution, the exponentiated generalized linear exponential (EGLE) distribution and the beta exponentiated Nadarajah-Haghighi distribution, respectively.

The unimodal hazard rate function (HRF) has many applications in reliability and survival analysis. The HRFs of the GMW and EW distributions can exhibit unimodal shapes. There are other distributions exhibiting unimodal HRFs: the inverse Weibull (IW) distribution, Kumaraswamy inverse Weibull (KumIW) distribution due to Shahbaz et al. (2012), the beta inverse Weibull (BIW) distribution due to Hanook et al. (2013), the generalized Weibull (GW) distribution due to Mudholkar and Kollia (1994), the extended Weibull distribution due to Marshall and Olkin (1997), and the beta Weibull distribution due to Famoye et al. (2005) among others.

Almalki (2018) introduced a three-parameter modified Weibull distribution called the reduced new modified Weibull (RNMW) distribution. Its CDF is

$$F(x) = 1 - e^{-\alpha\sqrt{x} - \beta\sqrt{x}e^{\lambda x}}$$

for $x > 0$, where $\alpha > 0$, $\beta > 0$ and $\lambda > 0$. Here, α, β are scale parameters and λ is an acceleration parameter. The corresponding probability density function (PDF) and HRF are

$$f(x) = \frac{1}{2\sqrt{x}}\left[\alpha + \beta(1 + 2\lambda x)e^{\lambda x}\right]e^{-\alpha\sqrt{x} - \beta\sqrt{x}e^{\lambda x}} \tag{3}$$

and

$$h(x) = \frac{\alpha + \beta(1 + 2\lambda x)e^{\lambda x}}{2\sqrt{x}}, \tag{4}$$

respectively, for $x > 0$, where $\alpha > 0$, $\beta > 0$ and $\lambda > 0$. A particular case of the RNMW distribution is studied in Almalki et al. (2021).

Unfortunately, the RNMW distribution can exhibit only bathtub-shaped HRFs. The aim of this paper is to introduce a four-parameter generalization that can accommodate monotonically increasing, unimodal, and bathtub-shaped HRFs. The new distribution will be called the exponentiated reduced modified Weibull (ERMW) distribution.

The contents of this paper are organized as follows. The new distribution is introduced in Section 9.2. Some special cases of the ERMW distribution are presented in Section 9.3. The shapes of the PDF and the HRF of the ERMW distribution are studied in Section 9.4. The moments of the ERMW distribution are given in Section 9.5. Section 9.6 discusses maximum likelihood estimates (MLEs) of the unknown parameters of the ERMW distribution. It also presents a simulation study to assess the performance of the MLEs. Two real data applications of the ERMW distribution are described in Sections 9.7. One of the datasets has a bathtub-shaped HRF and the other has a unimodal HRF. Some conclusions and future work are presented in Section 9.8.

9.2 THE ERMW DISTRIBUTION

Let X denote a random variable having the ERMW distribution. Its CDF is defined by exponentiating the CDF of the RNMW distribution. That is, we take the CDF of X as

$$F(x) = \left[1 - e^{-\alpha\sqrt{x} - \beta\sqrt{x}e^{\lambda x}}\right]^{\theta} \tag{5}$$

for $x > 0$, where $\alpha > 0$, $\beta > 0$, $\lambda > 0$ and $\theta > 0$. Here, α and β are scale parameters, λ is an acceleration parameter and θ is a shape parameter. The corresponding PDF and HRF are

$$f(x) = \frac{\theta}{2\sqrt{x}}\left[\alpha + \beta(1+2\lambda x)e^{\lambda x}\right]\left[1 - e^{-\alpha\sqrt{x}-\beta\sqrt{x}e^{\lambda x}}\right]^{\theta-1} e^{-\alpha\sqrt{x}-\beta\sqrt{x}e^{\lambda x}} \tag{6}$$

and

$$h(x) = \frac{\theta\left[\alpha + \beta(1+2\lambda x)e^{\lambda x}\right]\left[1 - e^{-\alpha\sqrt{x}-\beta\sqrt{x}e^{\lambda x}}\right]^{\theta-1} e^{-\alpha\sqrt{x}-\beta\sqrt{x}e^{\lambda x}}}{2\sqrt{x}\left\{1 - \left[1 - e^{-\alpha\sqrt{x}-\beta\sqrt{x}e^{\lambda x}}\right]^{\theta}\right\}}, \tag{7}$$

respectively, for $x > 0$, where $\alpha > 0$, $\beta > 0$, $\lambda > 0$ and $\theta > 0$.

If the parameter θ is a positive integer, the ERMW distribution can be interpreted as the lifetime distribution of a parallel system of θ independent and identical components, where each component is a serial system with two independent components: one of which follows the Weibull distribution with parameters α and $1/2$, and the other follows the modified Weibull distribution of Lai et al. (2003) with parameters β, $1/2$ and λ.

Using the series expansion

$$\left(1 - e^{-\alpha\sqrt{x}-\beta\sqrt{x}e^{\lambda x}}\right)^{\theta-1} = \sum_{j=0}^{\infty} \frac{(-1)^j \Gamma(\theta)}{\Gamma(\theta - j) j!} e^{-j\alpha\sqrt{x}-j\beta\sqrt{x}e^{\lambda x}},$$

the PDF in Eq. (6) can be expressed as

$$f(x) = \sum_{j=0}^{\infty} \frac{(-1)^j \Gamma(\theta)}{\Gamma(\theta - j)(j+1)!} f_{RNMW}(x; \alpha_{j+1}, \beta_{j+1}, \lambda)$$

for $x > 0$, where $\phi_{j+1} = (j + 1)\phi$ and f_{RNMW} $(x; \alpha, \beta, \lambda)$ is the PDF in Eq. (3). So, the PDF of the ERMW distribution is a linear combination of the PDFs of the RNMW distribution.

The PDF in Eq. (6) can also be expressed as

$$f(x) = h_{RNMW}(x) e^{-\alpha\sqrt{x}-\beta\sqrt{x}e^{\lambda x}} \left(1 - e^{-\alpha\sqrt{x}-\beta\sqrt{x}e^{\lambda x}}\right)^{\theta-1} \tag{8}$$

for $x > 0$, where $h_{RNMW}(x)$ is the HRF of the RNMW distribution in Eq. (4).

Figure 9.1 shows possible shapes of the PDF in (6). We can see that monotonically decreasing, unimodal and monotonically decreasing followed by unimodal shapes are possible for the PDF. Monotonically decreasing shapes correspond to small values of θ. Unimodal shapes correspond to large values of θ. The location and the magnitude of the mode become larger as θ becomes larger.

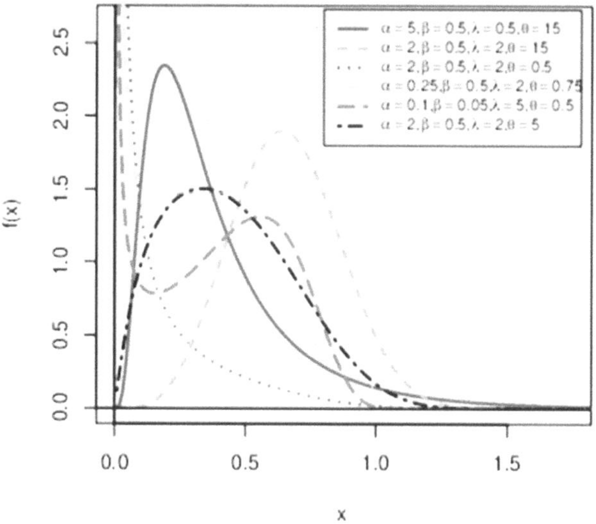

Figure 9.1 PDF of the ERMW distribution for selected α, β, λ and θ.

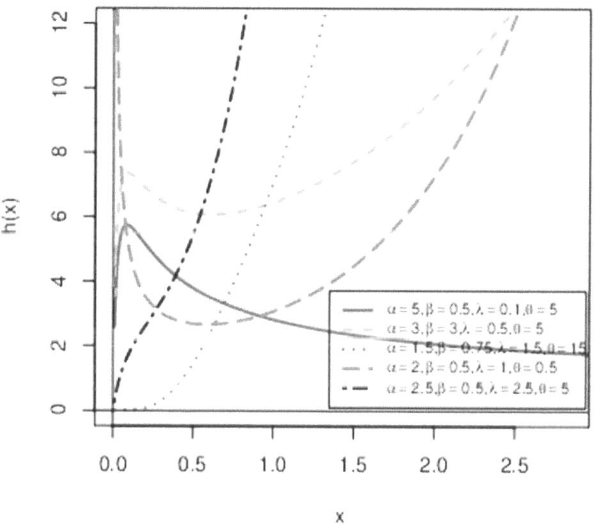

Figure 9.2 The HRF of the ERMW distribution.

Figure 9.2 shows possible shapes of the HRF in Eq. (7). We can see that monotonically increasing, bathtub and unimodal shapes are possible for the HRF.

9.3 SPECIAL CASES

The ERMW distribution includes as special cases distributions which are special cases of several distributions widely used in survival analysis:

1. **The reduced generalized modified Weibull (RGMW) distribution:** If $\alpha = 0$, the ERMW distribution reduces to the RGMW distribution specified by the CDF

$$F(x) = \left(1 - e^{-\beta\sqrt{x}e^{\lambda x}}\right)^{\theta},$$

a special case of the GMW distribution of Carrasco et al. (2008) when its first shape parameter is equal 1/2. The RGMW distribution has just three parameters. Its HRF can be monotonically increasing, bathtub-shaped, unimodal, or unimodal followed by increasing. Not many distributions have this property, especially with three parameters. Figure 9.3 shows possible shapes of the PDF and the HRF of the RGMW distribution.

2. **The reduced exponentiated Weibull (REW) distribution:** When $\alpha = \lambda = 0$ or $\beta = \lambda = 0$, the ERMW distribution reduces to the REW distribution specified by the CDF

$$F(x) = \left(1 - e^{-\alpha\sqrt{x}}\right)^{\theta},$$

a special case of the EW distribution due to Mudholkar and Srivastava (1993). The PDFs and the HRFs of the REW distribution are shown in Figure 9.4.

3. **The reduced Lai et al.'s modified Weibull (RLMW) distribution**: For $\alpha = 0$ and $\theta = 1$, the ERMW distribution reduces to the RLMW distribution specified by the CDF

$$F(x) = 1 - e^{-\beta\sqrt{x}e^{\lambda x}},$$

a special case of the MW distribution due to Lai et al. (2003).

4. **The reduced Weibull (RW) distribution:** By setting $\beta = \lambda = 0$ and $\theta = 1$, we can reduce the ERMW distribution to the RW distribution specified by the CDF

$$F(x) = 1 - e^{-\alpha\sqrt{x}}.$$

We can also get a RW distribution by setting $\lambda = 0$ and $\theta = 1$.

9.4 SHAPE

In this section, we discuss shapes of the PDF in Eq. (6) and the HRF in Eq. (7).

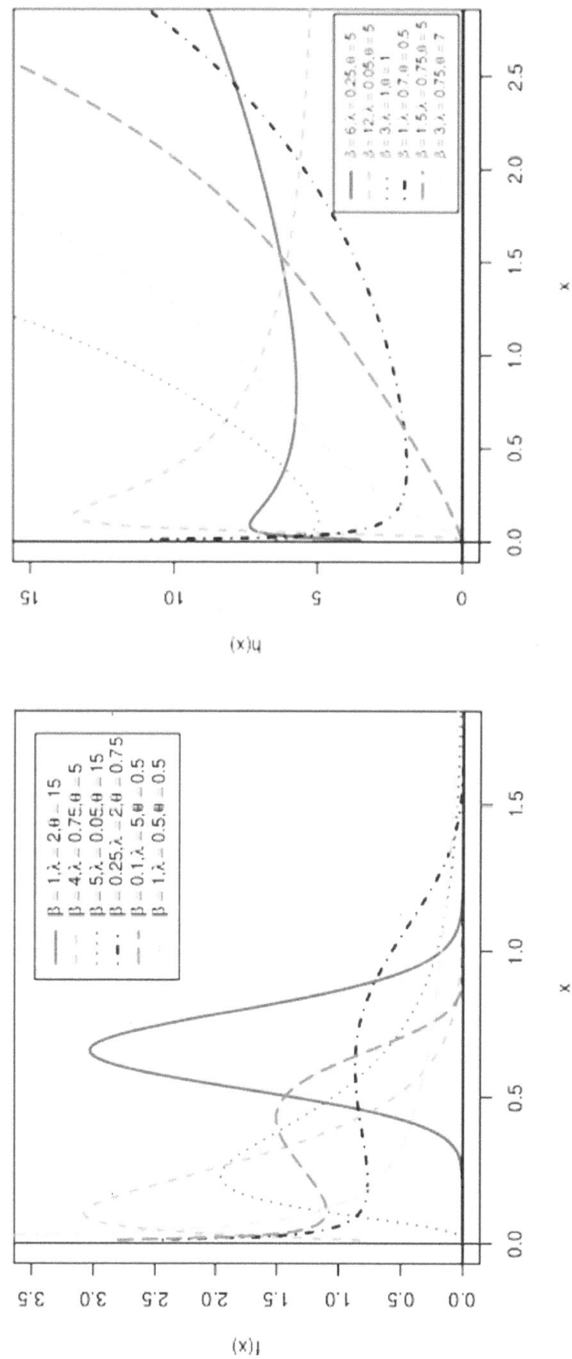

(a) (b)

Figure 9.3 (a) The PDF of the RGMW distribution, (b) The HRF of the RGMW distribution.

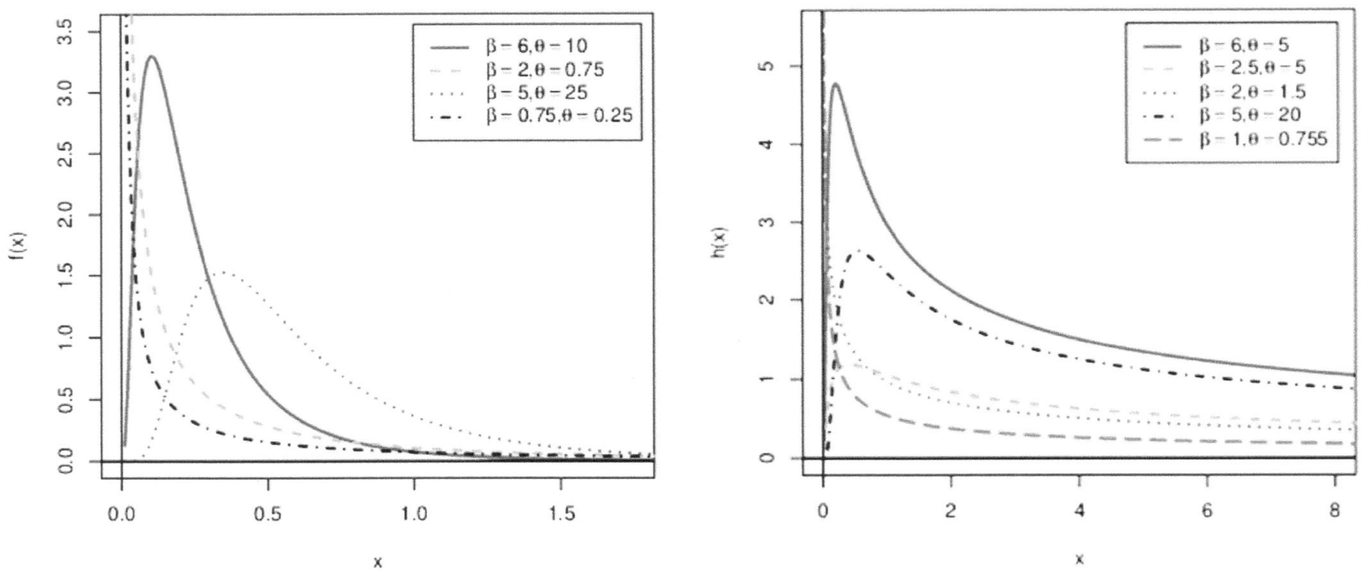

(a) (b)

Figure 9.4 (a) The PDF of the REW distribution, (b) The HRF of the REW distribution.

The derivative of log $f(x)$ with respect to x is

$$\frac{d\log f(x)}{dx}=\frac{\beta\lambda(3+2\lambda x)e^{\lambda x}}{\alpha+\beta(1+2\lambda x)e^{\lambda x}}+\frac{(\theta-1)[\alpha+\beta(1+2\lambda x)e^{\lambda x}]}{2\sqrt{x}(e^{\alpha\sqrt{x}+\beta\sqrt{x}e^{\lambda x}}-1)}$$

$$-\frac{1}{2x}-\frac{\alpha}{2\sqrt{x}}-\frac{\beta(1+2\lambda x)e^{\lambda x}}{2\sqrt{x}}.$$

So, the modes of $f(x)$ at say $x = x_0$ are the roots of $\frac{d\log f(x)}{dx}=0$. The mode corresponds to a local maximum (see Figure 9.3) if $\frac{d\log f(x)}{dx}>0$ for all $x < x_0$ and $\frac{d\log f(x)}{dx}<0$ for all $x > x_0$. The mode corresponds to a local minimum (see Figure 9.3) if $\frac{d\log f(x)}{dx}<0$ for all $x < x_0$ and $\frac{d\log f(x)}{dx}>0$ for all $x > x_0$.

The derivative of log $h(x)$ with respect to x is

$$\frac{d\log h(x)}{dx}=\frac{d\log f(x)}{dx}+\frac{\theta\left[\alpha+\beta(1+2\lambda x)e^{\lambda x}\right]\left(1-e^{-\alpha\sqrt{x}-\beta\sqrt{x}e^{\lambda x}}\right)^{\theta-1}}{2\sqrt{x}e^{\alpha\sqrt{x}+\beta\sqrt{x}e^{\lambda x}}\left[1-\left(1-e^{-\alpha\sqrt{x}-\beta\sqrt{x}e^{\lambda x}}\right)^{\theta}\right]}.$$

So, the modes of $h(x)$ at say $x = x_0$ are the roots of $\frac{d\log h(x)}{dx}=0$. The mode corresponds to a local maximum (see, for example, the top left plot in Table 9.6) if $\frac{d\log h(x)}{dx}>0$ for all $x < x_0$ and $\frac{d\log h(x)}{dx}<0$ for all $x > x_0$. The mode corresponds to a local minimum (see, for example, Table 9.5) if $\frac{d\log h(x)}{dx}<0$ for all $x < x_0$ and $\frac{d\log h(x)}{dx}>0$ for all $x > x_0$. The mode corresponds to a point of inflexion (see, for example, the bottom right plot in Figure 9.4) if either $\frac{d\log h(x)}{dx}>0$ for all $x \neq x_0$ or $\frac{d\log h(x)}{dx}<0$ for all $x \neq x_0$.

The property of the long useful period of the bathtub shaped HRF seen in the RNMW distribution of Almalki (2018) also holds for the ERMW distribution, see Figure 9.5a.

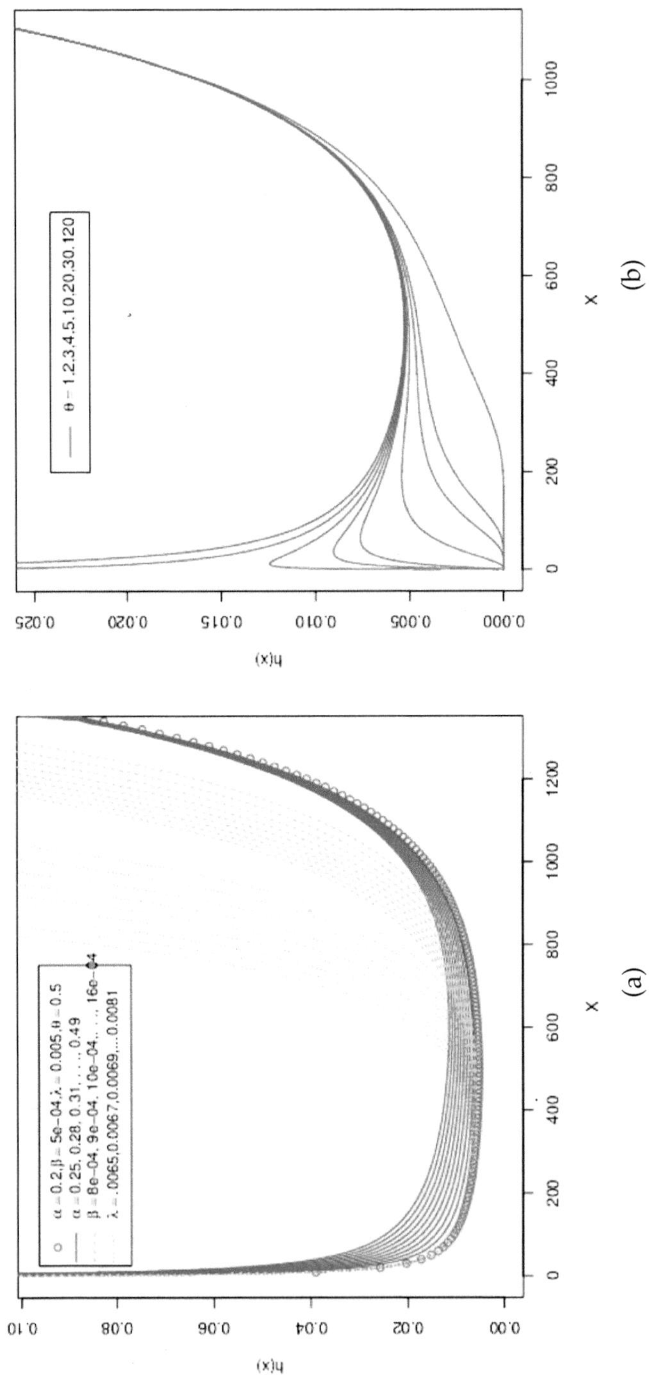

Figure 9.5 The HRF of the ERMW distribution as α changes.

Figure 9.5a shows how the bathtub-shaped HRFs of the ERMW distribution are affected by the parameters β, λ and α. The curve of open red circles is the bathtub shaped HRF when $\alpha = 0.2$, $\beta = 5 \times 10^{-4}$, $\lambda = 0.005$, $\theta = 0.5$. The blue curves in this figure are the HRFs at the same values of α, β and λ while θ increases from 0.25 to 0.49 in steps of 0.03. Increasing values of this parameter have the effect of scaling the bathtub shape slightly up. The green dash curves and the pink dot curves are, respectively, the HRFs when $\beta = 8 \times 10^{-4}, 9 \times 10^{-4}, \ldots, 16 \times 10^{-4}$ and $\lambda = 0.0065, 0.0067, \ldots, 0.0081$. Increasing values of these parameters have the effect of shortening the constant period (useful life period). That is, increasing values of β and λ reduce the survival time of the system.

Figure 9.5b shows the role of the shape parameter θ and how it turns the HRF from being bathtub shaped to unimodal to increasing. The HRF is bathtub shaped when $\theta \leq 2$. It is unimodal when $\theta > 2$. The HRF increases for large values of θ.

Figures 9.6a, b, c, d show how the unimodal HRFs of the ERMW distribution are affected by the parameters β, λ, θ and α (Here, by unimodal HRF we mean an HRF taking a unimodal shape followed by a monotonic increasing shape.). The curve of open red circles in Figure 9.6a is a unimodal HRF of the ERMW distribution. Increasing values of α have the effect of scaling the unimodal shape. A slight increase of β and λ has the effect of giving an earlier start for the monotonic increasing part of the HRF. That is, the survival time of the system decreases as β and λ increase slightly. A sharp increase of β and λ has the effect of turning the unimodal HRF into an increasing HRF, see Figures 9.6c and 9.6d. Increasing values of θ have the effect of changing the HRF from being bathtub shaped to unimodal to increasing (see Figure 9.6b).

9.5 MOMENTS

Let X denote an ERMW random variable. The rth raw moment of X say μ'_r can be expressed as

$$\mu'_r = \sum_{j=0}^{\infty} \frac{(-1)^j \Gamma(\theta)}{(j+1)!\Gamma(\theta - j)} \mu'_{r,RNMW(\alpha_{j+1},\beta_{j+1},\lambda)} \tag{9}$$

for $r \geq 1$, where $\mu'_{r,RNMW(\alpha,\beta,\lambda)}$ denotes the rth raw moment of a random variable having the RNMW distribution with parameters (α, β, λ). Using Eq. (9) and expressions for $\mu'_{r,RNMW(\alpha,\beta,\lambda)}$ given in Almalki (2018), one can obtain expressions for the mean, variance, skewness, and kurtosis of X.

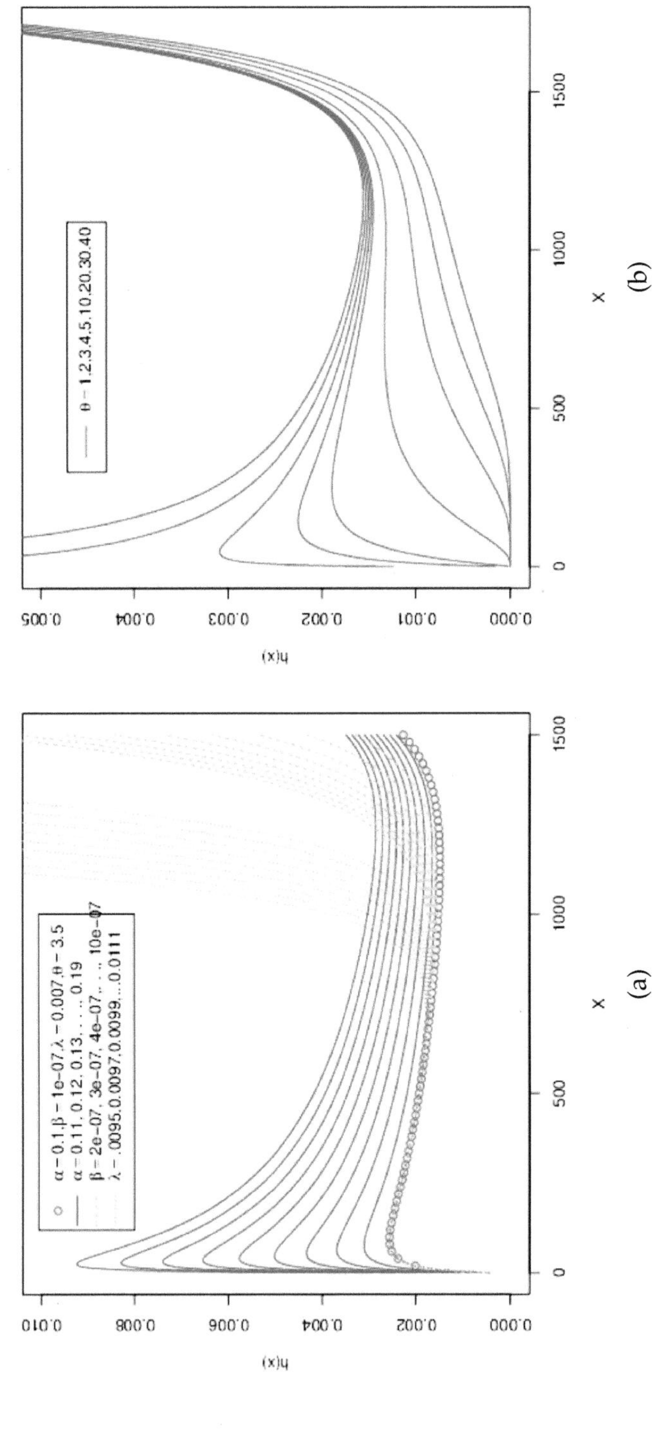

Figure 9.6 *(Contd.)*

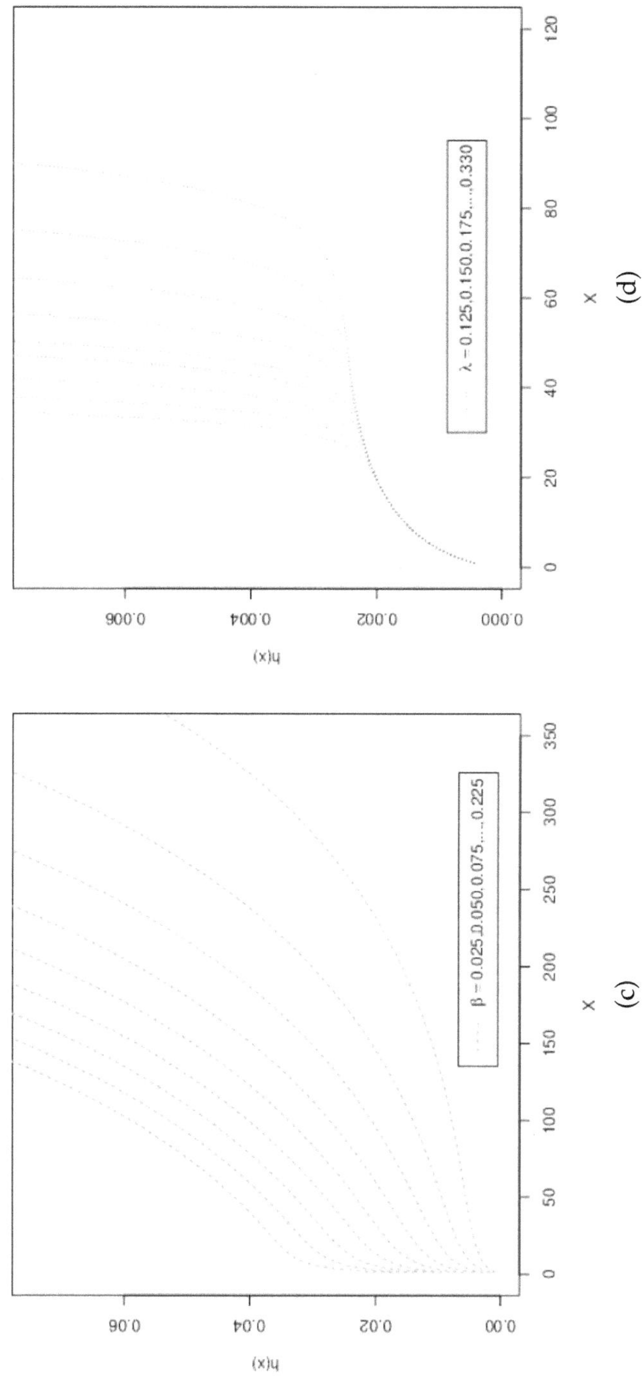

Figure 9.6 The HRF of the ERMW distribution as θ changes.

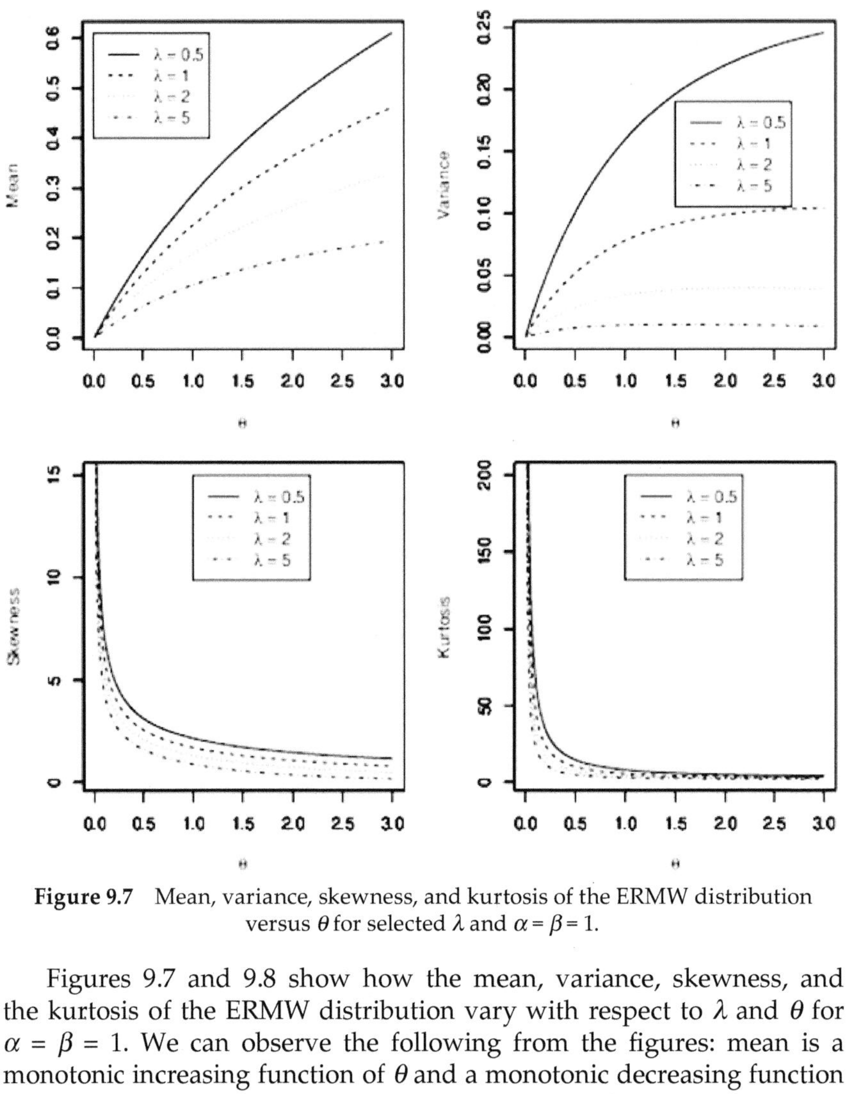

Figure 9.7 Mean, variance, skewness, and kurtosis of the ERMW distribution versus θ for selected λ and $\alpha = \beta = 1$.

Figures 9.7 and 9.8 show how the mean, variance, skewness, and the kurtosis of the ERMW distribution vary with respect to λ and θ for $\alpha = \beta = 1$. We can observe the following from the figures: mean is a monotonic increasing function of θ and a monotonic decreasing function of λ; variance is a monotonic increasing function of θ and a monotonic decreasing function of λ; skewness is a monotonic decreasing function of θ and a monotonic decreasing function of λ; kurtosis is a monotonic decreasing function of θ and a monotonic decreasing function of λ.

9.6 ESTIMATION

Suppose $x_1, x_2, \ldots, x_n$ is a random sample from the ERMW distribution with unknown parameters $(\alpha, \beta, \lambda, \theta)$. Section 9.6.1 estimates these

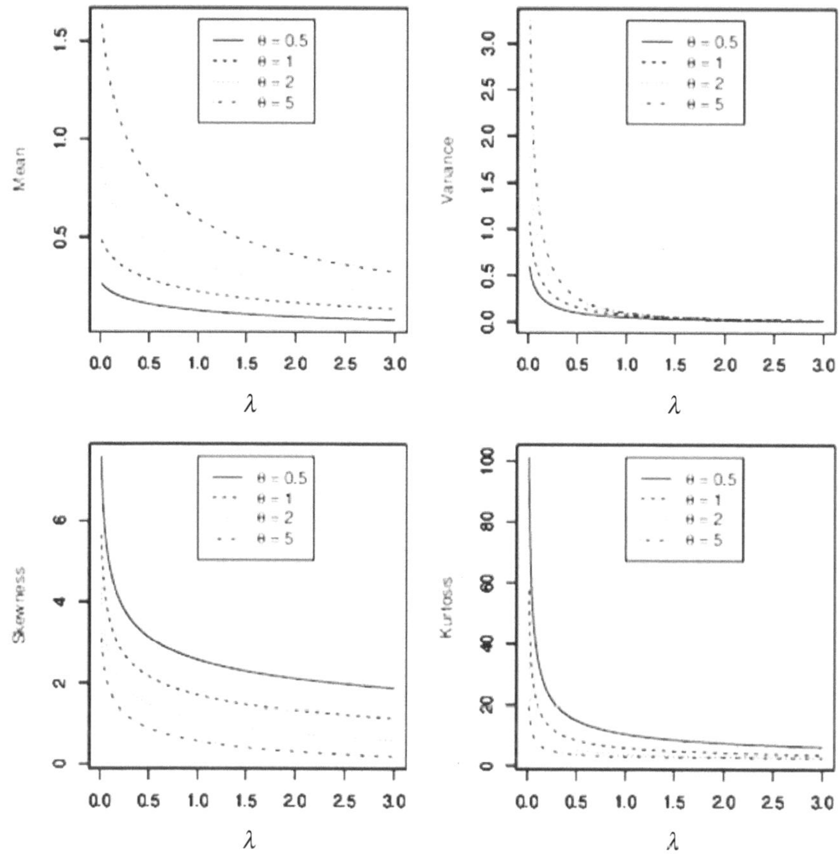

Figure 9.8 Mean, variance, skewness, and kurtosis of the ERMW distribution versus λ for selected θ and $\alpha = \beta = 1$.

parameters by the method of maximum likelihood. Section 9.6.2 assesses the finite sample performance of the MLEs with respect to biases, mean squared errors, coverage probabilities and coverage lengths.

9.6.1 Maximum Likelihood Estimation

Given the data $x_1, \ldots, x_n$, the log-likelihood function $(\alpha, \beta, \lambda, \theta)$ is

$$l(\alpha,\beta,\lambda,\theta) = n\log(\theta) + \sum_{i=1}^{n}\left[\log\left(h_{RNMW}(x_i)\right) - \alpha\sqrt{x_i} - \beta\sqrt{x_i}e^{\lambda x_i}\right]$$

$$+(\theta-1)\sum_{i=1}^{n}\log\left[1-e^{-\alpha\sqrt{x_i}-\beta\sqrt{x_i}e^{\lambda x_i}}\right].$$

By setting the first partial derivatives of l with respect to α, β, λ and θ to zero, we obtain the normal equations

$$\sum_{i=1}^{n} \frac{1}{h_{\mathrm{RNMW}}(x_i)\left(2\sqrt{x_i}\right)} - \sum_{i=1}^{n} \sqrt{x_i} + (\theta - 1)\sum_{i=1}^{n} \frac{\sqrt{x_i}}{e^{\alpha\sqrt{x_i} + \beta\sqrt{x_i}e^{\lambda x_i}} - 1} = 0, \tag{10}$$

$$\sum_{i=1}^{n} \frac{(0.5 + \lambda x_i)e^{\lambda x_i}}{h_{\mathrm{RNMW}}(x_i)\sqrt{x_i}} - \sum_{i=1}^{n} e^{\lambda x_i}\sqrt{x_i} + (\theta - 1)\sum_{i=1}^{n} \frac{\sqrt{x_i}e^{\lambda x_i}}{e^{\alpha\sqrt{x_i} + \beta\sqrt{x_i}e^{\lambda x_i}} - 1} = 0,. \tag{11}$$

$$\beta\sum_{i=1}^{n} \frac{\sqrt{x_i}\left(\frac{3}{2} + \lambda x_i\right)e^{\lambda x_i}}{h_{\mathrm{RNMW}}(x_i)} - \beta\sum_{i=1}^{n} e^{\lambda x_i}\sqrt{x_i^3} + (\theta - 1)\sum_{i=1}^{n} \frac{\sqrt{x_i^3}e^{\lambda x_i}}{e^{\alpha\sqrt{x_i} + \beta\sqrt{x_i}e^{\lambda x_i}} - 1} = 0, \tag{12}$$

$$\frac{n}{\theta} + \sum_{i=1}^{n} \log\left[1 - e^{-\alpha\sqrt{x_i} - \beta\sqrt{x_i}e^{\lambda x_i}}\right] = 0. \tag{13}$$

The MLEs can be obtained by solving the nonlinear Eqs. (10)–(13), numerically for α, β, λ and θ. This can be done using R, Matlab and Mathcad, among other packages.

To obtain confidence intervals for α, β, λ and θ, we use the observed information matrix since the expected information matrix is very complicated and the MLEs of the unknown parameters cannot be obtained analytically.

According to Cox and Hinkley (1974), if $\hat{\alpha}$, $\hat{\beta}$, $\hat{\lambda}$ and $\hat{\theta}$ are the MLEs of α, β, λ and θ, respectively, then $\sqrt{n}(\hat{\alpha} - \alpha, \hat{\beta} - \beta, \hat{\lambda} - \lambda, \hat{\theta} - \theta)$ converges in distribution to a quadrivariate normal random vector with zero means, and covariance matrix $E(I^{-1})$, where

$$I = -\left[l_{\alpha\alpha}\ l_{\alpha\beta}\ l_{\alpha\lambda}\ l_{\alpha\theta}\ \ l_{\beta\beta}\ l_{\beta\lambda}\ l_{\beta\theta}\ \ l_{\lambda\lambda}\ l_{\lambda\theta}\ \ l_{\theta\theta}\right].$$

Explicit expressions for the elements of this matrix are given in the appendix. We estimate the covariance of $(\hat{\alpha}, \hat{\beta}, \hat{\lambda}, \hat{\theta})$ by $n^{-1}\hat{I}^{-1}$, where

$$\hat{I} = -\left[l_{\hat{\alpha}\hat{\alpha}}\ l_{\hat{\alpha}\hat{\beta}}\ l_{\hat{\alpha}\hat{\lambda}}\ l_{\hat{\alpha}\hat{\theta}}\ \ l_{\hat{\beta}\hat{\beta}}\ l_{\hat{\beta}\hat{\lambda}}\ l_{\hat{\beta}\hat{\theta}}\ \ l_{\hat{\lambda}\hat{\lambda}}\ l_{\hat{\lambda}\hat{\theta}}\ \ l_{\hat{\theta}\hat{\theta}}\right],$$

where $\hat{\alpha}$, $\hat{\beta}$, $\hat{\lambda}$ and $\hat{\theta}$ are the MLEs of α, β, λ and θ, respectively.

9.6.2 Simulation Study

In this section, we assess the performance of the MLEs with respect to sample size n. The assessment is based on a simulation study:

1. Generate ten thousand samples of size n from (5). The inversion method was used to generate samples, i.e., variants of the ERMW distribution were generated by solving

$$\left(1-e^{-\alpha\sqrt{X}-\beta\sqrt{X}e^{\lambda X}}\right)^{\theta} = U,$$

where $U \sim U(0, 1)$ is a uniform variate on the unit interval.
2. Compute the MLEs for the ten thousand samples, say $(\widehat{\alpha}_i, \widehat{\beta}_i, \widehat{\lambda}_i, \widehat{\theta}_i)$ for $i = 1, 2, \ldots, 10{,}000$.
3. Compute the standard errors of the MLEs for the ten thousand samples, say $(s_{\widehat{\alpha}_i}, s_{\widehat{\beta}_i}, s_{\widehat{\lambda}_i}, s_{\widehat{\theta}_i})$ for $i = 1, 2, ..., 10{,}000$. The standard errors were computed by inverting the observed information matrices.
4. Compute the biases and mean squared errors given by

$$bias_h(n) = \frac{1}{10{,}000} \sum_{i=1}^{10{,}000} (\widehat{h}_i - h),$$

$$MSE_h(n) = \frac{1}{10{,}000} \sum_{i=1}^{10000} (\widehat{h}_i - h)^2$$

for $h = \alpha, \beta, \lambda, \theta$.
5. Compute the coverage probabilities and coverage lengths given by

$$CP_h(n) = \frac{1}{10{,}000} \sum_{i=1}^{10{,}000} I\left\{\widehat{h}_i - 1.959964 s_{\widehat{h}_i} < h < \widehat{h}_i + 1.959964\, s_{\widehat{h}_i}\right\},$$

$$CL_h(n) = \frac{3.919928}{10{,}000} \sum_{i=1}^{10{,}000} s_{\widehat{h}_i}$$

for $h = \alpha, \beta, \lambda, \theta$, where $I\{\cdot\}$ denotes the indicator function.

We repeated these steps for $n = 10, 11, \ldots, 100$ with $\alpha = 1$, $\beta = 1$, $\lambda = 1$ and $\theta = 1$, so computing $bias_h(n)$, $MSE_h(n)$, $CP_h(n)$ and $CL_h(n)$ for $h = \alpha, \beta, \lambda, \theta$ and $n = 10, 11, \ldots, 100$.

Figure 9.9 shows how the four biases vary with respect to n. Figure 9.10 shows how the four mean squared errors vary with respect to n. Figure 9.11 shows how the four coverage probabilities vary with respect to n. Figure 9.12 shows how the four coverage lengths vary with respect to n. The broken line in Figure 9.9 corresponds to the biases being zero. The broken line in Figure 9.10 corresponds to the mean squared errors being zero. The broken line in Figure 9.11 corresponds to the nominal coverage probability of 0.95.

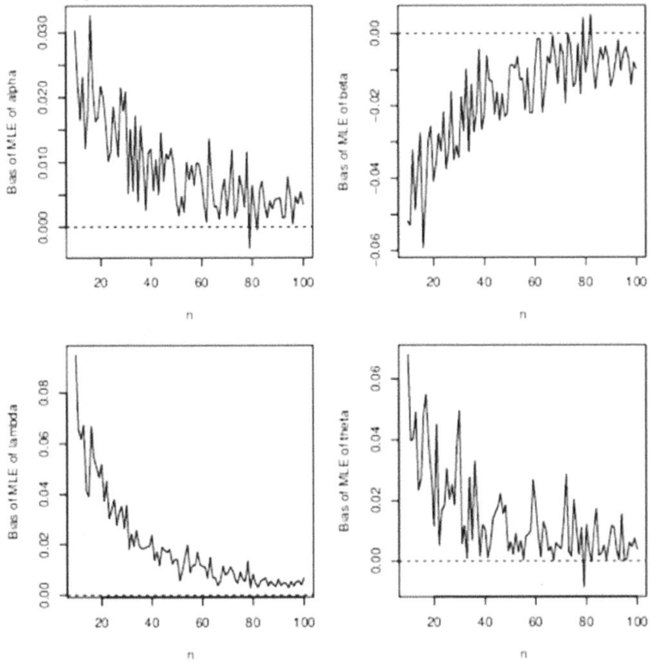

Figure 9.9 Biases of the MLEs of $(\alpha, \beta, \lambda, \theta)$ versus $n = 10, 11, \ldots, 100$.

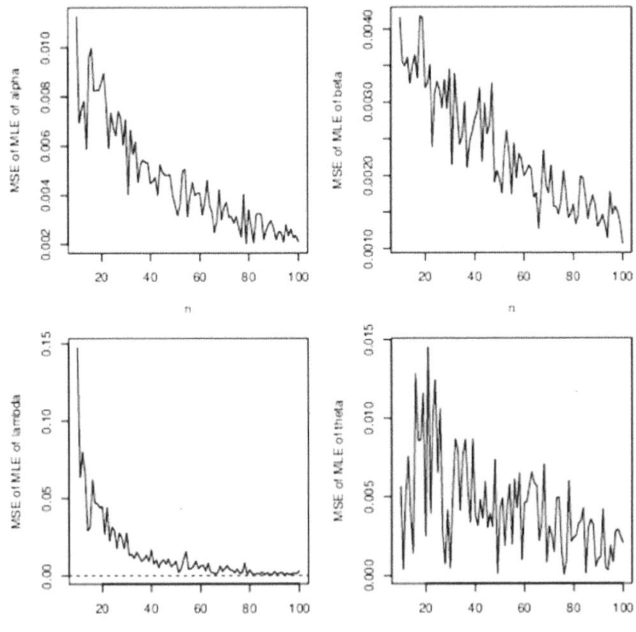

Figure 9.10 Mean squared errors of the MLEs of $(\alpha, \beta, \lambda, \theta)$ versus $n = 10, 11, \ldots, 100$.

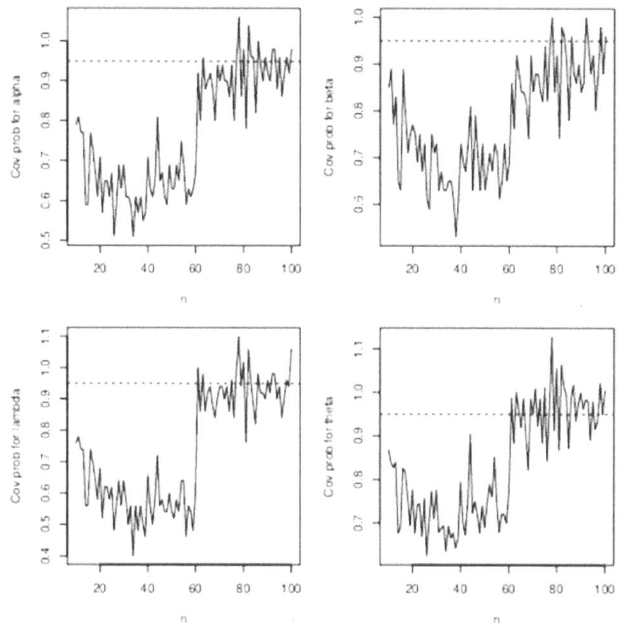

Figure 9.11 Coverage probabilities of the MLEs of $(\alpha, \beta, \lambda, \theta)$ versus $n = 10, 11, \ldots, 100$.

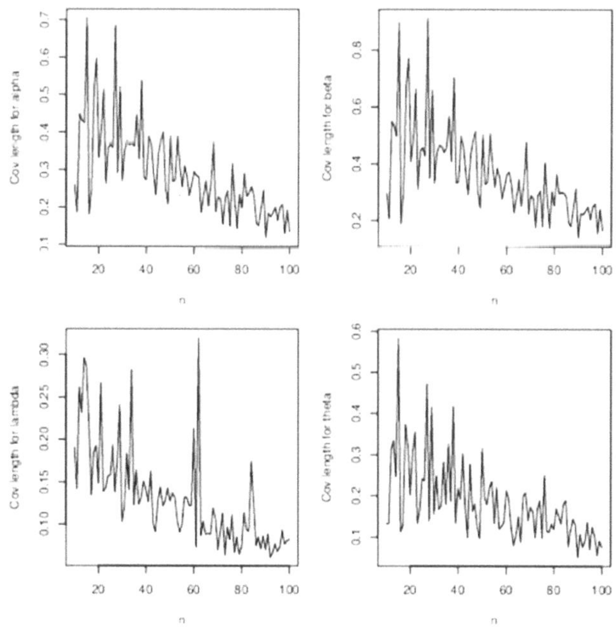

Figure 9.12 Coverage lengths of the MLEs of $(\alpha, \beta, \lambda, \theta)$ versus $n = 10, 11, \ldots, 100$.

The following observations can be drawn from the figures: the biases for α, λ and θ are generally positive; the biases for β are generally negative; the biases appear smallest for the parameter, α; the biases appear largest for the parameter, λ; the biases for each parameter either decrease or increase to zero as $n \to \infty$; the mean squared errors appear smallest for the parameter, β; the mean squared errors appear largest for the parameter, λ; the mean squared errors for each parameter decrease to zero as $n \to \infty$; the coverage probabilities for each parameter are reasonably close enough to the nominal level for all n greater than or equal to sixty; the coverage lengths appear smallest for the parameter, λ; the coverage lengths appear largest for the parameter, β; the coverage lengths for each parameter decrease to zero as $n \to \infty$. These observations are for only one choice for $(\alpha, \beta, \lambda, \theta)$, namely that $(\alpha, \beta, \lambda, \theta) = (1, 1, 1, 1)$. But the results were similar for other choices.

Section 9.7 presents two real data applications. The sample size for the first dataset is 72. The sample size for the second dataset is 44.

Hence, the biases for $\widehat{\alpha}$, $\widehat{\beta}$, $\widehat{\lambda}$ and $\widehat{\theta}$ can be expected to be less than 0.01, 0.02, 0.02, and 0.02, respectively, for both datasets. The mean squared errors for $\widehat{\alpha}$, $\widehat{\beta}$, $\widehat{\lambda}$ and $\widehat{\theta}$ can be expected to be less than 0.005, 0.003, 0.02, and 0.005, respectively, for both data-sets. The coverage probabilities can be expected to be accurate for $\widehat{\alpha}$, $\widehat{\beta}$, $\widehat{\lambda}$ and $\widehat{\theta}$ for the second dataset. The coverage lengths for $\widehat{\alpha}$, $\widehat{\beta}$, $\widehat{\lambda}$ and $\widehat{\theta}$ can be expected to be less than 0.35, 0.4, 0.15, and 0.2, respectively, for both datasets. Hence, the point estimates given in Section 9.7 can be considered accurate enough for both data sets. The confidence intervals given for the second data set can also be considered accurate enough. However, the confidence intervals given for the first data set must be interpreted conservatively.

9.7 APPLICATIONS

This section uses two well-known datasets to show how the proposed distribution can be applied in practice to provide good fits. The HRF of the first dataset is unimodal followed by an increasing shape. That of the second dataset is bathtub-shaped. The proposed distribution will be compared to the four-parameter GMW distribution, the three-parameter RGMW distribution, the two-parameter REW distribution, and the three-parameter EW distribution. The Kolmogorov-Smirnov (K-S) statistic (the distance between the empirical and fitted CDFs), the Akaike information criterion (AIC), the Bayesian information criterion (BIC), and the corrected Akaike information criterion (CAIC) will be used to compare the fits of the distributions.

The likelihood ratio test will be used to compare the GMW distribution against the RGMW distribution by testing the hypotheses $H_0: \gamma = 1/2$ versus $H_1: H_0$ is false. Under the null hypothesis, the likelihood ratio test statistic $\omega = 2(L_{GMW} - L_{RGMW})$ has an asymptotic χ^2 distribution with one degree of freedom. The EW distribution will be compared against the REW distribution by testing the hypotheses $H_0: \gamma = 1/2$ versus $H_1: H_0$ is false. Under the null hypothesis, the likelihood ratio test statistic $\omega = 2(L_{EW} - L_{REW})$ has an asymptotic χ^2 distribution with one degree of freedom.

9.7.1 Infected Pigs Data

The data in Table 9.1 are survival times of 72 pigs infected by virulent tubercle bacilli (Greenwich 1992). The TTT plot in Figure 9.13 shows a concave shape and then a convex shape followed by a concave shape. This corresponds to the HRF being unimodal followed by an increasing shape. Hence, the ERMW distribution is suitable for this data, see Section 9.2.

Table 9.1 Infected pigs data

43	45	53	56	56	57	58	66	67	73	74	79	80	80	81	81
81	82	83	83	84	88	89	91	91	92	92	97	99	99	100	100
101	102	102	102	103	104	107	108	109	113	114	118	121	123	126	128
137	138	139	144	145	147	156	162	174	178	179	184	191	198	211	214
243	249	329	380	403	511	522	598								

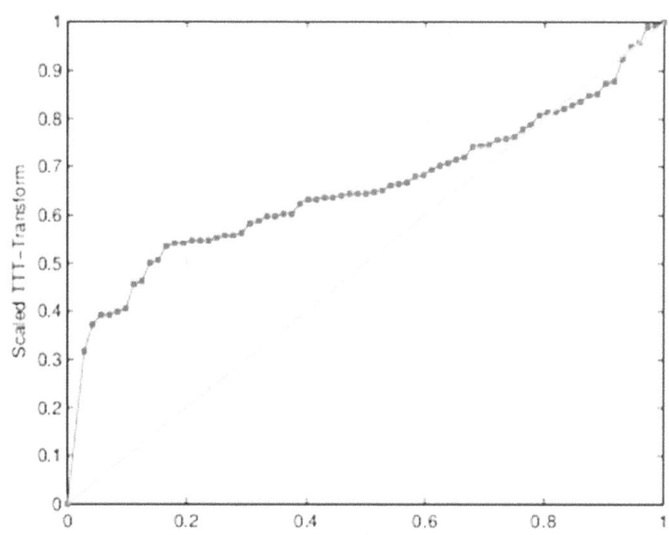

Figure 9.13 TTT-transform plot for the infected pigs data.

Table 9.2 shows the MLEs of the parameters, their standard errors, AIC values, BIC values, and the CAIC values for the fitted ERMW, GMW, RGMW, REW, and EW distributions.

Table 9.2 MLEs of parameters, standard errors, AIC, BIC, and CAIC for the distributions fitted to the infected pigs dataset.

Model	$\hat{\alpha}$	$\hat{\beta}$	$\hat{\gamma}$	$\hat{\lambda}$	$\hat{\theta}$	AIC	BIC	CAIC
ERMW	0.412	6.396×10^{-8}	–	0.022	57.562	807.8	816.9	808.4
	(0.042)	(7.109×10^{-7})	–	(0.019)	(22.924)			
RGMW	0	0.422	–	1.108×10^{-8}	63.403	807.2	814.0	807.5
	–	(0.108)	–	4.885×10^{-4}	(50.557)			
REW	0.423	0	–	0	36.517	805.2	809.7	805.3
	0.042	–	–	–	(25.683)			
GMW	–	0.496	0.470	4.639×10^{-9}	69.877	808.7	817.8	809.3
	–	(0.247)	(0.008)	4.573×10^{-4}	(50.012)			
EW	–	0.471	0.478	0	65.135	806.9	813.7	807.2
	–	(0.254)	0.081	–	(49.625)			

The ERMW distribution has the smallest negative log-likelihood of 399.898. The negative log-likelihood of the GMW distribution is 400.350. Among the remaining distributions, the REW distribution has the smallest negative log-likelihood of 400.572. The REW distribution has the smallest values for the AIC, the BIC, and the CAIC. Given these observations, one could be tempted to say that the REW distribution provides the best fit.

Table 9.3 shows the K-S test statistics for the fitted distributions. The values of K-S statistics are about the same for all fitted distributions. All of them are about 0.11.

Table 9.3 K-S statistics for the distributions fitted to the infected pigs dataset

Model								K-S								
ERMW								0.112								
RGMW								0.110								
REW								0.111								
GMW								0.114								
EW								0.110								

Figure 9.14a shows the histogram of the data and the estimated PDFs. Figure 9.14b shows the empirical survival function of the data and the estimated ones. Figure 9.14d shows the nonparametric HRF of the data and the estimated HRFs.

Figures 9.14a and 9.14b show that all of the distributions provide good fits. Figure 9.14d shows that all of the distributions provide good

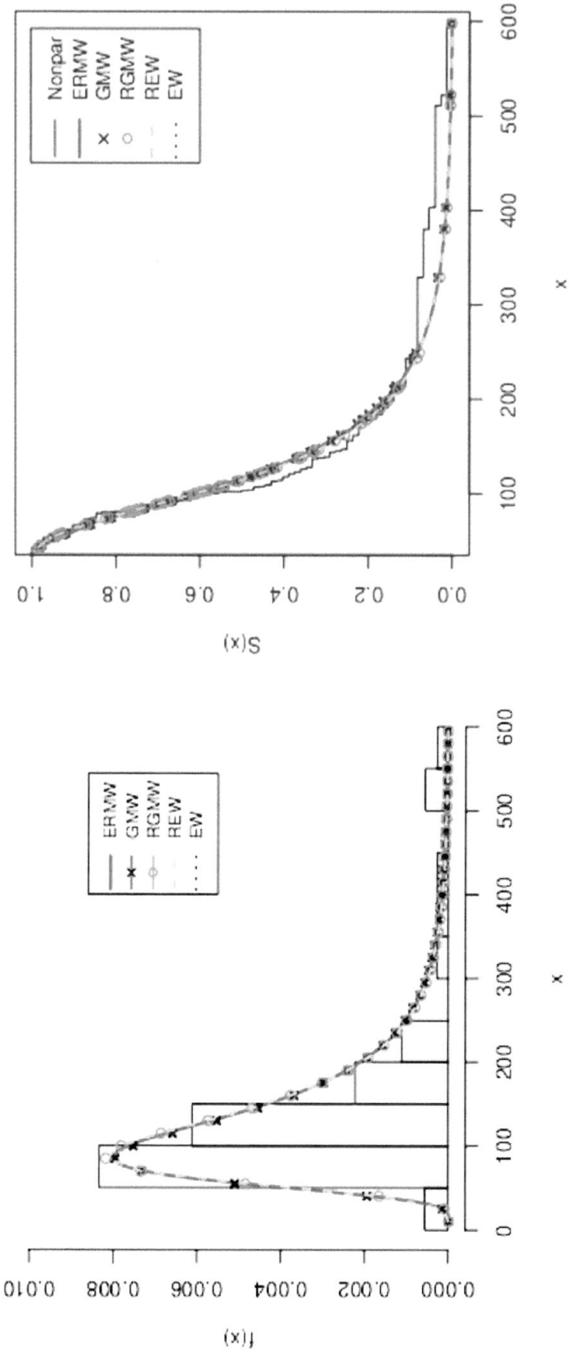

(a)

(b)

Figure 9.14 *(Contd.)*

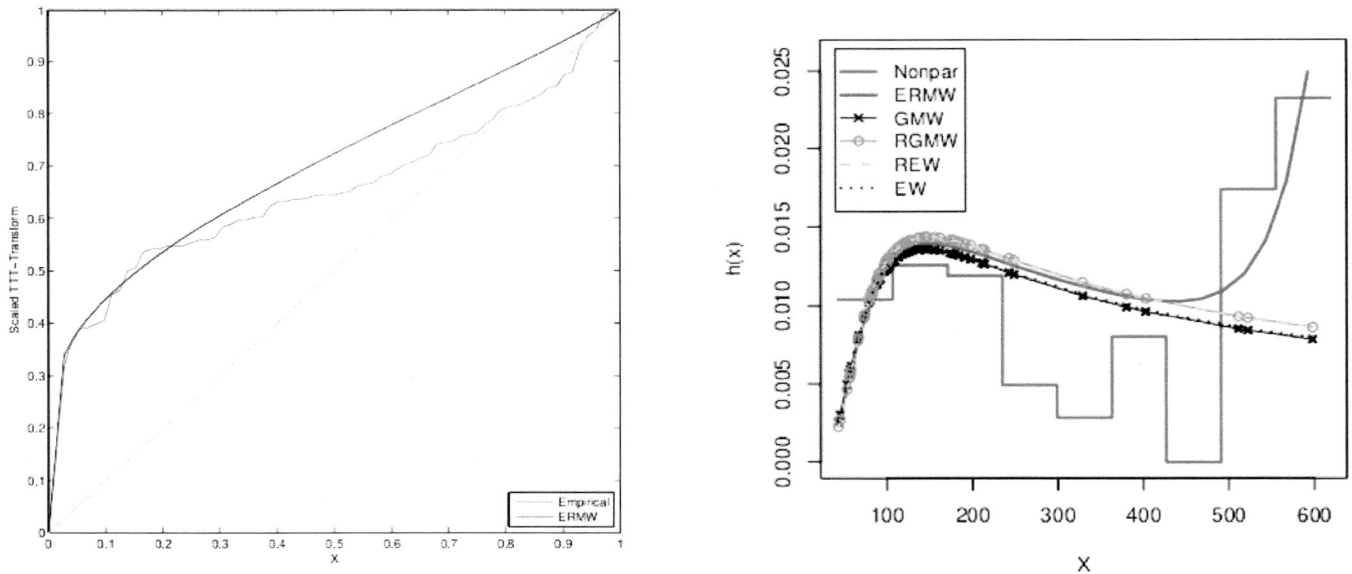

(c) (d)

Figure 9.14 For infected pigs data: (a) Histogram and fitted PDFs; (b) Empirical and fitted survival functions; (c) Empirical and fitted TTT-transforms; (d) Nonparametric and fitted HRFs.

fits to the first and middle parts of the nonparametric HRF. But the GMW, RGMW, REW and the EW distributions do not appear to capture the last part of the nonparametric HRF well. Only the ERMW distribution appears to provide a good fit to the last part of the nonparametric HRF. Since the differences in the likelihood and the differences in the AIC, BIC, CAIC measures are small among the fitted distributions, we conclude that the ERMW distribution gives the best fit. The ERMW distribution is the only one that captures the nonparametric HRF well in all three parts.

The variance-covariance matrix for the fitted ERMW distribution is I^{-1} = [1.739×10^{-3} – 4.108×10^{-10} 4.366×10^{-6} 0.913 – 4.108×10^{-10} 5.054×10^{-13} – 1.351×10^{-8} – 2.119×10^{-7} 4.366×10^{-6} – 1.351×10^{-8} 3.655×10^{-4} 2.195×10^{-3} 0.913 – 2.119×10^{-7} 2.195×10^{-3} 525.529].

So, approximate 95% confidence intervals for the parameters α, β, λ and θ are [0.331, 0.494], [0, 1.457×10^{-6}], [0, 0.06] and [12, 102], respectively.

Finally, we give results of likelihood ratio tests. The log-likelihood ratio statistic for testing $H_0 : \gamma = 1/2$ in Eq. (2) versus $H_1 : H_0$ is false is $\omega = 0.445$ and the corresponding p-value is 0.505. So, there is no evidence to reject the null hypothesis at any level of significance. Therefore, the RGMW distribution can be chosen as an alternative to the GMW distribution for modeling the dataset.

The log-likelihood ratio statistic for testing $H_0 : \gamma = 1/2$ in Eq. (1) versus $H_1 : H_0$ is false is $\omega = 0.294$ and the corresponding p-value is 0.588. So, the null hypothesis could not be rejected. Therefore, the REW distribution can be chosen as an alternative to the EW distribution for modeling the dataset.

9.7.2 Kumar Data

Kumar et al. (1989) presented data consisting of times between failures (TBF) in days of load-haul-dump machines (LHD) used to pick up rock or waste. The TTT plot in Figure 9.15 shows a convex shape followed by a concave shape. This corresponds to a bathtub-shaped HRF. Hence, the ERMW, GMW, RGMW, and the EW distributions are appropriate for modeling this data. The REW distribution is not appropriate as it does not exhibit a bathtub-shaped HRF.

Tables 9.4 and 9.5 show the MLEs of the parameters, their standard errors, AIC values, BIC values, CAIC values, and K-S test statistics for the fitted ERMW, GMW, RGMW, REW, and EW distributions.

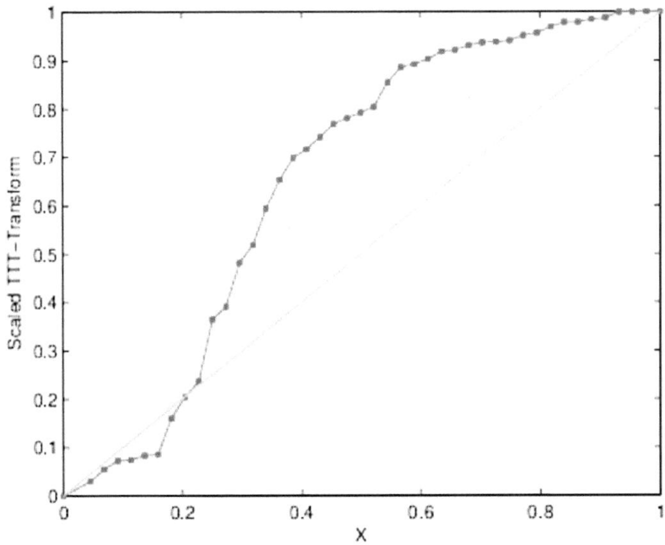

Figure 9.15 TTT-transform plot for Kumar data.

Table 9.4 MLEs of parameters, standard errors, AIC, BIC, and CAIC for the distributions fitted to Kumar data.

Model	$\hat{\alpha}$	$\hat{\beta}$	$\hat{\gamma}$	$\hat{\lambda}$	$\hat{\theta}$	AIC	BIC	CAIC
ERMW	0.099	4.86×10^{-4}	–	0.067	1.476	407.8	414.9	408.8
	(0.049)	(7.593×10^{-4})	–	(0.017)	(0.034)			
RGMW	0	2.919×10^{-3}	–	0.046	0.521	414.9	420.3	415.5
	–	(6.530×10^{-3})	–	0.023	(0.270)			
REW	0.292	0	–	0	3.686	450.6	454.1	450.9
	0.041	(0)	–	–	(0.880)			
GMW	–	4.590×10^{-6}	1.578	0.060	0.266	412.7	419.8	413.7
	–	(7.693×10^{-6})	0.463	0.019	(0.061)			
EW	–	1.984×10^{-7}	3.460	0	0.310	423.9	429.3	424.5
	–	(3.158×10^{-7})	0.347	–	(0.061)			

Table 9.5 K-S statistics for the distributions fitted to kumar data

Model	K-S
ERMW	0.069
RGMW	0.094
REW	0.234
GMW	0.090
EW	0.175

The ERMW distribution has the smallest negative log-likelihood of –199.882, the smallest AIC value, the smallest BIC value, the smallest

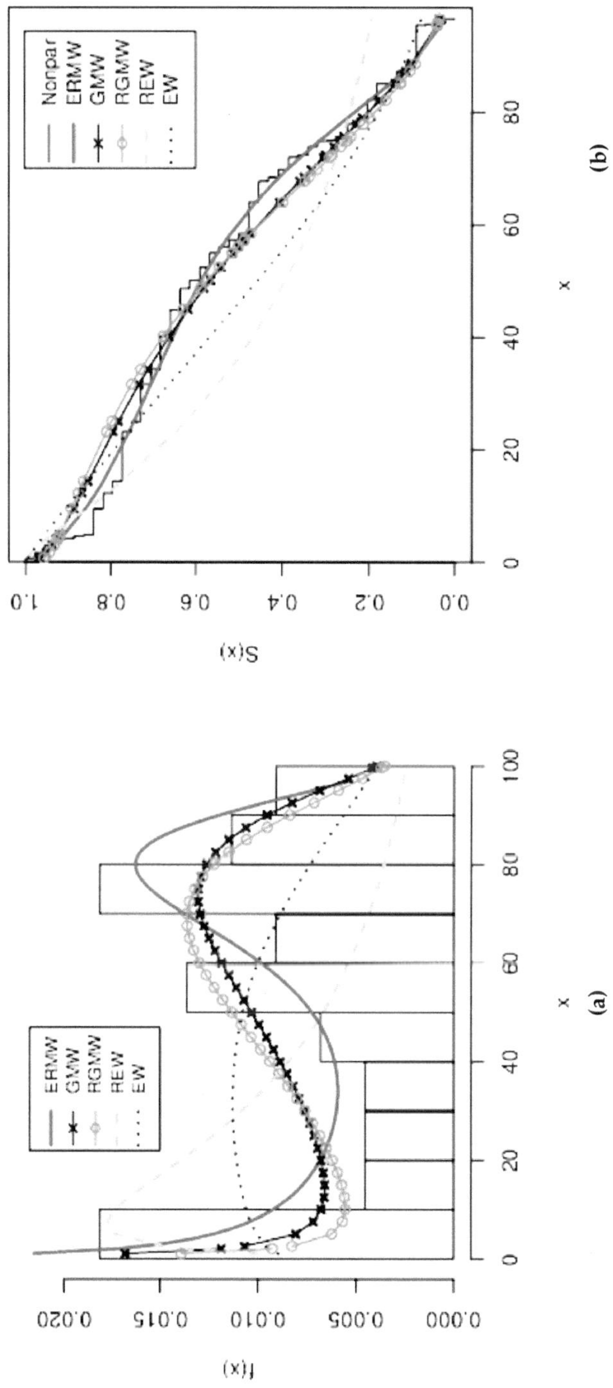

Figure 9.16 *(Contd.)*

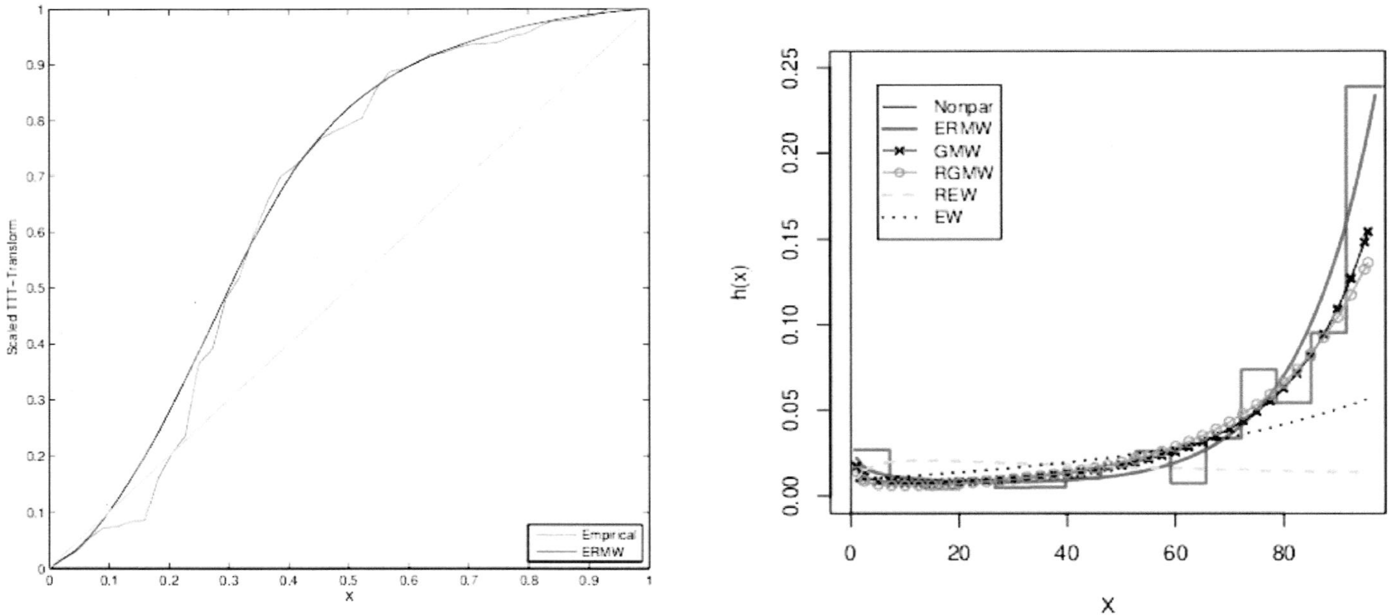

(c) (d)

Figure 9.16 For Kumar data: (a) Histogram and fitted PDFs; (b) Empirical and fitted survival functions; (c) Empirical and fitted TTT-transforms; (d) Nonparametric and fitted HRFs.

CAIC value, and the smallest K-S test statistic. So, the ERMW distribution provides the best fit with respect to these criteria. This observation is confirmed by the diagnostic plots in Figure 9.16.

The variance-covariance matrix for the fitted ERMW distribution is I^{-1} = [2.43×10^{-3} – 1.575×10^{-5} 3.267×10^{-4} 0.023 – 1.575×10^{-5} 5.766×10^{-7} – 1.264×10^{-5} – 1.043×10^{-4} 3.267×10^{-4} – 1.264×10^{-5} 2.83×10^{-4} 2.176×10^{-3} 0.023 – 1.043×10^{-4} 2.176×10^{-3} 0.282].

So, approximate 95% confidence intervals for the parameters α, β, λ and θ are [2.075×10^{-3}, 0.195], [0, 1.974×10^{-3}], [0.034, 0.100] and [0.436, 2.516], respectively.

Finally, we give results of likelihood ratio tests. The log-likelihood ratio statistic for testing $H_0 : \gamma = 1/2$ in (2) versus $H_1 : H_0$ is false is $\omega = 4.227$ and the corresponding p-value is 0.04. So, there is evidence that the null hypothesis can be rejected at the 5% significance level. Therefore, the RGMW distribution cannot be chosen as an alternative to the GMW distribution for modeling the dataset.

As reported before, the REW distribution does not exhibit bathtub shaped HRFs. The following result of the likelihood ratio test is an outcome of this inability. The log-likelihood ratio statistic for testing $H_0 : \gamma = 1/2$ in (1) versus $H_1 : H_0$ is false is $\omega = 28.631$ and the corresponding p-value is 8.759×10^{-8}. So, the null hypothesis can be rejected at any level of significance greater than or equal 8.759×10^{-8}.

9.8 CONCLUSIONS AND FUTURE WORK

We have generalized the RNMW distribution of Alamlki (2018) to accommodate unimodal HRFs. The generalized distribution is referred to as the ERMW distribution. It includes as special cases several distributions which are special cases of well known lifetime distributions like the RGMW, REW, RW, RLMW, and RW distributions. We have discussed maximum likelihood estimation of the ERMW distribution and presented two real data applications. One of the data sets has a unimodal HRF and the other has a bathtub-shaped HRF. The ERMW distribution was shown to give the best fit for both datasets.

The distribution introduced in Almalki and Yuan (2013), a precursor to the RNMW distribution, is a competing risk model based on two different modified Weibull distributions. This can be generalized to a competing risk model composed of n modified Weibull distributions. The model can be defined formally as follows.

Definition 9.1 Modified Weibull competing risk model: Suppose X_1, X_2, ..., X_n are independent random variables following a Weibull distribution or some of its generalizations. Let $F_i(x)$ denote the *CDF* of X_i, $i = 1, 2, \ldots, n$. Let $Z = \min(X_1, \ldots, X_n)$. Then, the *CDF*, the *PDF* and the *HRF* of Z are

$$F(z) = 1 - \prod_{i=1}^{n} [1 - F_i(z)],$$

$$f(z) = \sum_{i=1}^{n} [h_i(z)] \prod_{j=1}^{n} S_j(z)$$

and

$$h(z) = \sum_{i=1}^{n} [h_i(z)],$$

respectively, where $S_i(z) = 1 - F_i(z)$.

If $n = 2$ the above definition includes the distribution of Almalki and Yuan (2013) and several other known distributions:

1. **The linear failure rate (LFR) distribution:** given by the CDF
$$F(x) = 1 - e^{-\alpha x - \beta x^2} \tag{14}$$
for $x > 0$. This distribution is based on exponential and Rayleigh distributions.
2. **The additive Weibull (AddW) distribution of Xie and Lai (1996):** given by the CDF
$$F(x) = 1 - e^{-\alpha x^{\theta} - \beta x^{\gamma}} \tag{15}$$
for $x > 0$. This distribution is based on two different Weibull distributions.
3. **The β distribution of Bousquet et al. (2006):** given by the CDF
$$F(x) = 1 - e^{-\alpha x - \beta x^{\gamma}} \tag{16}$$
for $x > 0$. This distribution is based on exponential and Weibull distributions.
4. **Sarhan and Zaindin's (2009) modified Weibull distribution:** Sarhan and Zaindin (2009) introduced a three-parameter modified Weibull (SZMW) distribution. It is the same as the β distribution of Bousquet et al. (2006).

A future work is to study the modified Weibull competing risk model further and consider parameter estimation based on outlier detection.

REFERENCES

Almalki, S.J. and Yuan, J. 2013. The new modified Weibull distribution. Reliability Engineering and System Safety. 111: 164–170.

Almalki, S.J. and Nadarajah, S. 2014. Modifications of the Weibull distribution: A review. Reliability Engineering and System Safety. 124: 32–55.

Almalki, S. 2018. Reduced new modified Weibull distribution. Communications in Statistics—Theory and Methods. 47: 2297–2313.

Almalki, S.J., Abushal, T.A., Alsulami, M.D. and Abd-Elmougod, G.A. 2021. Analysis of type-II censored competing risks' data under reduced new modified Weibull distribution. Complexity. 2021: Article number 9932840.

Bousquet, N., Bertholon, H. and Celeux, G. 2006. An alternative competing risk model to the Weibull distribution in lifetime data analysis. Lifetime Data Analysis. 12: 481–504.

Carrasco, J.M., Ortega, E.M. and Cordeiro, G.M. 2008. A generalized modified Weibull distribution for lifetime modeling. Computational Statistics and Data Analysis. 53: 450–462.

Cox, D.R. and Hinkley, D.V. 1974. Theoretical Statistics. Chapman and Hall, London.

El-Gohary, A., Alshamrani, A. and Al-Otaibi, A.N. 2013. The generalized Gompertz distribution. Applied Mathematical Modeling. 37: 13–24.

Famoye, F., Lee, C. and Olumolade, O. 2005. The beta-Weibull distribution. Journal of Statistical Theory and Applications. 4: 121–136.

Greenwich, M. 1992. A unimodal hazard rate function and its failure distribution. Statistical Papers. 33: 187–202.

Gupta, R.C., Gupta, P.L. and Gupta, R.D. 1998. Modeling failure time data by Lehman alternatives. Communications in Statistics: Theory and Methods. 27: 887–904.

Gupta, R.D. and Kundu, D. 2001a. Exponentiated exponential family: An alternative to gamma and Weibull distributions. Biometrical Journal. 43: 117–130.

Gupta, R.D. and Kundu, D. 2001b. Generalized exponential distribution: Different method of estimations. Journal of Statistical Computation and Simulation. 69: 315–337.

Gupta, R.D. and Kundu, D. 2007. Generalized exponential distribution: Existing results and some recent developments. Journal of Statistical Planning and Inference. 137: 3537–3547.

Hanook, S., Shahbaz, M.Q., Mohsin, M. and Golam Kibria, B.M. 2013. A note on beta inverse Weibull distribution. Communications in Statistics—Theory and Methods. 42: 320–335.

Kumar, U., Klefsjö, B. and Granholm, S. 1989. Reliability investigation for a fleet of load haul dump machines in a Swedish mine. Reliability Engineering and System Safety. 26: 341–361.

Kundu, D. and Gupta, R.D. 1999. Generalized exponential distribution. Australian and New Zealand Journal of Statistics. 41: 173–188.

Kundu, D. and Raqab, M.Z. 2005. Generalized Rayleigh distribution: Different methods of estimations. Computational Statistics and Data Analysis. 49: 187–200.

Kundu, D. and Gupta, R.D. 2008. Generalized exponential distribution: Bayesian estimations. Computational Statistics and Data Analysis. 52: 1873–1883.

Lai, C.D., Xie, M. and Murthy, D.N.P. 2003. A modified Weibull distribution. IEEE Transactions on Reliability. 52: 33–37.

Marshall, A.W. and Olkin, I. 1997. A new method for adding a parameter to a family of distributions with application to the exponential and Weibull families. Biometrika. 84: 641–652.

Mudholkar, G.S. and Kollia, G.D. 1994. Generalized Weibull family: A structural analysis. Communications in Statistics: Theory and Methods. 23: 1149–1171.

Mudholkar, G.S. and Srivastava, D.K. 1993. Exponentiated Weibull family for analyzing bathtub failure-rate data. IEEE Transactions on Reliability. 42: 299–302.

Nadarajah, S. and Kotz, S. 2006. The exponentiated type distributions. Acta Applicandae Mathematica. 92: 97–111.

Raqab, M.Z. 2004. Generalized exponential distribution: Moments of order statistics. Statistics. 38: 29-41.

Sarhan, A.M. and Kundu, D. 2009. Generalized linear failure rate distribution. Communications in Statistics: Theory and Methods. 38: 642–660.

Sarhan, A.M. and Zaindin, M. 2009. Modified Weibull distribution. Applied Sciences. 11: 123–136.

Sarhan, A.M., Ahmad, A.E.B.A. and Alasbahi, I.A. 2013. Exponentiated generalized linear exponential distribution. Applied Mathematical Modelling. 37: 2838–2849.

Sarhan, A.M. and Apaloo, J. 2013. Exponentiated modified Weibull extension distribution. Reliability Engineering and System Safety. 112: 137–144.

Shahbaz, M.Q., Shahbaz, S. and Butt, N.S. 2012. The Kumaraswamy-inverse Weibull distribution. Pakistan Journal of Statistics and Operation Research. 8: 479–489.

Surles, J.G. and Padgett, W.J. 2001. Inference for reliability and stress-strength for a scaled Burr type X distribution. Lifetime Data Analysis. 7: 187–200.

Surles, J.G. and Padgett, W.J. 2005. Some properties of a scaled Burr type X distribution. Journal of Statistical Planning and Inference. 128: 271–280.

Xie, M. and Lai, C.D. 1996. Reliability analysis using an additive Weibull model with bathtub-shaped failure rate function. Reliability Engineering and System *Safety*. 52: 87–93.

Appendix

The elements of the observed information matrix $I(\vartheta)$ in Section 9.6.1 are:

$$l_{\alpha\alpha} = -\sum_{i=1}^{n}\left[\frac{h_{\alpha}(x_i;\ \underline{\vartheta})}{h(x_i;\ \underline{\vartheta})}\right]^2 + (1-\theta)\sum_{i=1}^{n}\frac{x_i e^{\alpha\sqrt{x_i}+\beta\sqrt{x_i}e^{\lambda x_i}}}{\left(e^{\alpha\sqrt{x_i}+\beta\sqrt{x_i}e^{\lambda x_i}}-1\right)^2},$$

$$l_{\alpha\beta} = -\sum_{i=1}^{n}\frac{h_{\alpha}(x_i;\ \underline{\vartheta})h_{\beta}(x_i;\underline{\vartheta})}{h(x_i;\underline{\vartheta})^2} + (1-\theta)\sum_{i=1}^{n}\frac{x_i e^{\alpha\sqrt{x_i}+\beta\sqrt{x_i}e^{\lambda x_i}+\lambda x_i}}{\left(e^{\alpha\sqrt{x_i}+\beta\sqrt{x_i}e^{\lambda x_i}}-1\right)^2},$$

$$l_{\alpha\lambda} = -\sum_{i=1}^{n}\frac{h_{\alpha}(x_i;\underline{\vartheta})h_{\lambda}(x_i;\underline{\vartheta})}{h(x_i;\underline{\vartheta})^2} + (1-\theta)\sum_{i=1}^{n}\frac{\beta x_i^2 e^{\alpha\sqrt{x_i}+\beta\sqrt{x_i}e^{\lambda x_i}+\lambda x_i}}{\left(e^{\alpha\sqrt{x_i}+\beta\sqrt{x_i}e^{\lambda x_i}}-1\right)^2},$$

$$l_{\alpha\theta} = \sum_{i=1}^{n}\frac{\sqrt{x_i}}{e^{\alpha\sqrt{x_i}+\beta\sqrt{x_i}e^{\lambda x_i}}-1},$$

$$l_{\beta\beta} = -\sum_{i=1}^{n}\left[\frac{h_{\beta}(x_i;\underline{\vartheta})}{h(x_i;\underline{\vartheta})}\right]^2 + (1-\theta)\sum_{i=1}^{n}\frac{x_i e^{\alpha\sqrt{x_i}+\beta\sqrt{x_i}e^{\lambda x_i}+2\lambda x_i}}{\left(e^{\alpha\sqrt{x_i}+\beta\sqrt{x_i}e^{\lambda x_i}}-1\right)^2},$$

$$l_{\beta\lambda} = -\sum_{i=1}^{n}\frac{h(x_i;\underline{\vartheta})h_{\beta\lambda}(x_i;\underline{\vartheta}) - h_{\beta}(x_i;\underline{\vartheta})h_{\lambda}(x_i;\underline{\phi})}{h(x_i;\underline{\vartheta})^2} - \sum_{i=1}^{n}x_i e^{\lambda x_i}$$
$$+(\theta-1)\sum_{i=1}^{n}\frac{\sqrt{x_i^3}e^{\lambda x_i}\left[(1-\beta\sqrt{x_i}e^{\lambda x_i})e^{\alpha\sqrt{x_i}+\beta\sqrt{x_i}e^{\lambda x_i}}-1\right]}{\left(e^{\alpha\sqrt{x_i}+\beta\sqrt{x_i}e^{\lambda x_i}}-1\right)^2},$$

$$l_{\beta\theta} = \sum_{i=1}^{n}\frac{\sqrt{x_i}e^{\lambda x_i}}{e^{\alpha\sqrt{x_i}+\beta\sqrt{x_i}e^{\lambda x_i}}-1},$$

$$l_{\lambda\lambda} = \sum_{i=1}^{n}\left[\frac{h(x_i;\underline{\vartheta})h_{\lambda\lambda}(x_i;\underline{\vartheta}) - h_{\lambda}(x_i;\underline{\vartheta})^2}{h(x_i;\underline{\vartheta})^2} - \beta\sqrt{x_i^5}e^{\lambda x_i}\right]$$

$$+(\theta-1)\sum_{i=1}^{n}\frac{\sqrt{x_i^5}e^{\lambda x_i}\left[\left(1-\beta\sqrt{x_i}e^{\lambda x_i}\right)e^{\alpha\sqrt{x_i}+\beta\sqrt{x_i}e^{\lambda x_i}}-1\right]}{\left(e^{\alpha\sqrt{x_i}+\beta\sqrt{x_i}e^{\lambda x_i}}-1\right)^2},$$

$$l_{\lambda\theta}=\sum_{i=1}^{n}\frac{\sqrt{x_i^3}e^{\lambda x_i}}{e^{\alpha\sqrt{x_i}+\beta\sqrt{x_i}e^{\lambda x_i}}-1},$$

$$l_{\theta\theta}=-\frac{n}{\theta^2},$$

where

$$h_{\alpha}(x_i;\underline{\vartheta})=\frac{1}{2\sqrt{x_i}},$$

$$h_{\beta}(x_i;\underline{\vartheta})=\frac{(0.5+\lambda x_i)e^{\lambda x_i}}{\sqrt{x_i}},$$

$$h_{\lambda}(x_i;\underline{\vartheta})=\beta\sqrt{x_i}\left(\frac{3}{2}+\lambda x_i\right)e^{\lambda x_i},$$

$$h_{\beta\lambda}(x_i;\underline{\vartheta})=\sqrt{x_i}\left(\frac{3}{2}+\lambda x_i\right)e^{\lambda x_i},$$

$$h_{\lambda\lambda}(x_i;\underline{\vartheta})=\beta\sqrt{x_i}\left(\frac{5}{2}+\lambda x_i\right)e^{\lambda x_i},$$

and $h(x_i;\underline{\vartheta})$ is the HRF of the RNMW distribution.

Chapter **10**

Robust Weighted Random Forest Regression with Alternative Bootstrap

Aylin Alin

Faculty of Science, Dokuz Eylul University, Izmir, Turkiye.
Email: aylin.alin@deu.edu.tr

10.1 INTRODUCTION

Random Forest Method (RFM) introduced by Breiman (2001) is a highly data adaptive machine learning tool. It allows to consider both regression and classification problems, and is capable of handling large datasets, missing values in predictors, multi-collinearity problem or datasets where the sample size is much smaller than the number of predictors. It is applied in wide range of areas such as for predicting energy consumption (Ahmad et al. 2017), for automatic selection of molecular descriptors (Cano et al. 2017), for real time radar derived rainfall forecasting (Yu et al. 2017), for image classification (Mahapatra 2014, Xu et al. 2012), for modeling statistical arbitrage (Krauss et al. 2017), for diagnosing aviation turbulence (Williams 2014), for outcome prediction in antibody incompatible kidney transplantation (Shaikhina et al. 2019), for genetic data analysis (Chen and Ishwaran 2012, Díaz-Uriarte and Alvarez de Andres 2006), for the predictive validitý of the items from 2 mindfulness instruments (Sauer et al. 2015), and for ecological prediction (Prasad et al. 2006).

Random Forest is an ensemble learning method that is a committee of decision trees. Final predictions are aggregated from these trees. Even though a brief description will be provided in the following section, more detailed information about Random Forest including its consistency properties, and its superiorities compared to other machine learning tools can be found in Breiman (2001, 2004), Denil et al. (2013), Siroky (2009), Chen and Ishwaran (2012); Verikas et al. (2011); Cutler et al. (2012), and references therein. In this paper, we focus on the case that the response variable y is measured at least in interval scale, and the method is called *Random Forest Regression*. We study its robustness against the outlying data points (outliers) in response. Outliers in explanatory variables are not studied since the values of response variable determine the splits in trees. Outliers are the data points that are located far from all other data points or the ones that do not fit the distributional structure. They can be caused by human error during data collection, recording, by the participants that purposefully report incorrect data, by selecting data points from a different population than the rest of the sample, or by a contaminated error distribution. Whatever the cause, outliers profoundly affect the results of an experiment Jones (2019). They generally increase the error variance, reduce the power of the test, altering the odds of making both Type I and II errors. All these problems can be prevented by using robust methods.

Even though Random Forest is a nonparametric technique and can even be used with problematic predictors with multi-collinearity, as pointed out by Roy and Larocque (2012), it is not clear if, and to what extent, it is sensitive to outliers in the regression context. Robustness of the method is still an active research topic. Li (2018), Sage (2018), and Meng et al. (2019) studied and reviewed the robustness of Random Forest for both classification and regression, and they proposed different approaches to improve its robustness. To robustify the method, one can modify the aggregation statistics used for aggregating predictions from each decision tree or use more robust loss function to perform splits at each node of a tree. In this paper, to amplify the robustness of the original Random Forest Regression algorithm, we introduce the following approaches:

- Robust splitting criterion
- Robust aggregation statistic
- Robust bagging
- Weights.

Herein, we implement robust weights that have never been used with the Random Forest method, and a different bootstrap technique for bagging that is more robust requiring less data than traditional bootstrap yet giving the same and even better prediction performance. This bootstrap technique is called *Sufficient m-out-of-n (Smoon) Bootstrap* introduced by Alin et al. (2017). As will be explained later, it will contribute to the robustness since it does not use all the sample values unlike the original bootstrap. Genuer et al. (2017) proposed using moon bootstrap in Random Forest for classification. However, as far as we know, there is no version of random forest regression based on the Smoon bootstrap method. With different combinations of proposed modifications in original algorithm for regression, we end up with 6 new robust versions that are not available in the literature. We empirically investigate their prediction performances on contaminated and uncontaminated simulated data-sets as well as on real data. We present the theoretical background of why proposed alternative bagging works. As proved by our numerical calculations, proposed weights combined with robust loss function and sufficient moon bootstrap improve the prediction performance and reduce the complexity of the trees. The layout of the paper is as follows. In Section 10.2, we present brief information on the original random forest regression, we introduce the proposed modifications in Section 10.3, and we compare their performances on simulated and real datasets in Section 10.4.

10.2 RANDOM FOREST REGRESSION

We focus on regression problems with k predictors and a continuous response variable. Let

$$y_i = h(x_i, \chi) + \varepsilon_i \tag{1}$$

be a regression function with $\chi = \{(x_i, y_i), i = 1, \ldots, n\}$ being the training sample with $x_i = (x_{i1}, \ldots, x_{ik})$ is the ith observation vector of size k, and ε is the random noise that is assumed to have zero mean and constant variance σ^2. We have a sequence of training sets $\{\chi^l\}$, $l = 1, \ldots, L$ each consisting of n independent observations from the same underlying distribution as χ. Our aim is to use $\{\chi^l\}$, and to get a better predictor than the single training set predictor $h(x_i, \chi)$ by aggregation. The most common way for aggregation is taking the expected value of predictors in sequence of L training sets, that is, $h_A(x_i) = E_{\chi^L} f(x_i, \chi^L, \theta^L)$. Hence, Random Forest is classified as an ensemble method that is the collection of weak learners $f(x, \chi^l, \theta^l)$. Since we do not have the luxury of having the set of L training samples, the common approach is to use nonparametric bootstrap to aggregate the predictions. Let $f^b(x_i, \chi^b, \theta^b)$ be the bth weak learner with b being the bth i.i.d. random vector that selects points to be in the weak learner. The classical ensemble predictor is the mean of B weak learner predictors. In Random Forest Regression, a weak learner is a regression tree and the ensemble predictor is

$$\hat{h}^B(x_i) = E_{\chi^B} f(x_i, \chi^B, \theta^B) = \frac{1}{B}\sum_{b=1}^{B} f^b(x_i, \chi^b, \theta^b) \tag{2}$$

where χ^B is the set of sequences $\{\chi^b\}$, $b = 1,\ldots,B$, and $f^b(x, \chi^b, \theta^b)$ is the prediction from bth regression tree. Each $f^b(x_i, \chi^b, \theta^b)$ is obtained from a tree built on a nonparametric bootstrap sample $\chi^b = \{(x_i^b, y_i^b), i = 1, \ldots, n\}$ taken from the sample χ, and the predictor in Eq. (2) is bootstrap aggregated predictor, and the whole process is called *bootstrap aggregation (bagging)*. The law of large numbers ensures convergence as $b \to \infty$.

The variance of average of trees is

$$\rho\sigma^2 + \frac{1-\rho}{B}\sigma^2 \tag{3}$$

where σ^2 is the variance of a tree, and ρ is the correlation between the bootstrap trees. Bagging is for reducing the variance without affecting the low bias of the weak learner.

The variance of the average of independently built bootstrap trees is σ^2/B, and bagging reduces the variance by $1/B$. However, there may be correlation between the trees. Hence, to reduce the variance of the aggregated results, additional randomization is included by randomly selecting s variables to be used for deciding the best split at each node. This randomization is for decreasing the first component of Eq. (3) in addition to low variance advantage of bagging. The idea of the original Random Forest algorithm can be summarized as below:

Step 1: *A bootstrap sample of size n is selected with replacement from the training sample.*

Step 2: *An unpruned tree is built using randomly selected ω variables out of k variables at each node to determine descending nodes.*

Step 3: *The tree is grown till specific criterion is met. For this study, a node is terminal if the minimum node size n_{min} is reached.*

Step 4: *The steps 1–3 are repeated for B times and the predictions from each of B trees is averaged.*

One of the main choices to be made when constructing a decision tree is to decide for the splitting criterion. To split a node into its two descendant nodes we have to decide for the split variable and for the split cut. In the original Random Forest Regression, the best split is the one minimizing the loss function, the sum of squared residuals

$$SSR = \sum_{i \in t_R}^{n_{tR}} (y_i - \bar{y}_R)^2 + \sum_{i \in t_L}^{n_{tL}} (y_i - \bar{y}_L)^2 \tag{4}$$

where n_{tR} and n_{tL} are the sizes of the right and left splits of node t, and $\bar{y}_R$ and $\bar{y}_L$ are the means of the response values at the left and right splits. In the original Random Forest, the prediction of an observation i in the bth tree is the mean of the terminal node where ith observation falls $(t^b_{ter_i})$ (Eq. 5).

$$f^b\left(x_i, \chi^b, \theta^b\right) = \frac{1}{n_{t^b_{ter_i}}} \sum_{j=1}^{n} y_j I\left(j \in t^b_{ter_i}\right) \tag{5}$$

Both the mean and *SSR* are sensitive to the small perturbations in the data. Even though Random Forest is flexible for violation of ordinary assumptions that most of the original models cannot tolerate, the method will be vulnerable to the noise. Hence, in the following section we introduce modifications on the original algorithm to make it robust to the noise.

10.3 MODIFICATIONS ON RANDOM FOREST REGRESSION

We introduce the following modifications on the original Random Forest algorithm.

1. Robust splitting criterion and prediction
2. Robust bagging
3. Robust aggregation statistic
4. Weights.

Below are the details of those modifications.

10.3.1 Robust Splitting Criterion and Prediction

Instead of nonrobust *SSR*, we propose to minimize the loss function, the sum of absolute deviation from median (Eq. 6) to decide for the best split variable and the split value.

$$SAMD = \sum_{i \in t_R}^{n_{tR}} \left|y_i - \tilde{y}_R\right| + \sum_{i \in t_L}^{n_{tL}} \left|y_i - \tilde{y}_L\right| \tag{6}$$

The statistics $\tilde{y}_R$ and $\tilde{y}_L$ are the medians of the response values at the left and right splits of node t. Here, for the predicted value of the corresponding data point we use the median (Eq. 7) of the terminal node in the bth tree including ith observation.

$$f^{R_b}(x_i, \chi^b, \theta^b) = \underset{j \in t^b_{ter_i}}{median}(\{y_j\}) \tag{7}$$

10.3.2 Robust Bagging

In the original Random forest, B trees are grown on bootstrap samples of the same size with the original training sample randomly selected with replacement from the training sample. In the classical bootstrap, the approximate proportion of bootstrap samples in which any given data point will appear j times is $(j!e)^{-1}$, meaning that a particular point fails to appear in about $(1-1/n)^n \rightarrow e^{-1} \approx 36.79\%$, appears only once in about e^{-1} of resamples, but appears multiple times in the remaining $1-2/e \approx 26.4\%$ of resamples (Martin and Roberts 2014). Thus, if the original training sample contains outliers, these observations will potentially appear many times in the bootstrap samples affecting the trees. Herein, as a solution we propose using sufficient m-out-of-n bootstrap (Smoon bootstrap) introduced by Alin et al. (2017).

The idea of the sufficient bootstrap Singh and Sedory (2011) is to use only the distinct units in a bootstrap sample. Some may confuse it with subsampling bootstrap that is based on resampling without replacement from the original sample at smaller than the original sample size. In sufficient bootstrap, using distinct units will reduce both the chance of an outlier point to be included multiple times in a bootstrap sample and the tree size that is important for big data. The expected length of a sufficient bootstrap resample is $n[1-(1-1/n)^n]$. In m-out-of-n(moon) bootstrap, Swanepoel (1986) uses a smaller sample of size m drawn from the empirical distribution. It has been proposed as a solution to the inconsistency problem of bootstrap. This inconsistency can occur, for example, when the statistic of interest is an insufficiently smooth function of the true underlying distribution function, or when certain moment conditions on the underlying distribution, such as finite variance, fail to hold (Alin et al. 2017). For the studies on optimal choice of m, refer Wei et al. (2016) and references therein. Genuer et al. (2017) used moon bootstrap for classification but unlike the original version of the moon bootstrap they used samples of size m without replacement. One of the recent bootstrap methods is the sufficient moon bootstrap that uses the distinct units in a bootstrap sample of size m. It has been proposed by Alin et al. (2017) to improve the sufficient bootstrap in terms of consistency.

As pointed out by Meng (2013), using all data would not be required to obtain accurate estimations in learning methods, and that sampling approaches is an important and reliable way to deal with big data. Hence, we propose Smoon bootstrap for bagging to reduce the sample size and the chance of having outliers in the bootstrap samples, to build less complex trees with smaller number of leaves and still getting consistent estimates with less error in the predictions. Breiman (1996) proved why ordinary bagging works. Herein, we introduce the theorem given below depicting how bagging with Smoon bootstrap works equally well or even better.

Theorem 10.3.1 *Let* $\chi^b_{m_v} = \{(x^{b_m}_s, y^{b_m}_s), s = 1, \ldots, v\}$ *be a Smoon bootstrap sample set of sequences consisting v distinct units in a bootstrap sample of* $\chi^b_m = \{(x^{b_m}_i, y^{b_m}_i), i = 1, \ldots, m\}$ *selected with simple random sampling with replacement. Then,*

$$\hat{h}^B_{m_v}(x_i) = E_{\chi^B_{m_v}} f^B_{m_v}(x_i, \chi^B_{m_v}, \theta^B_{m_v}) = \frac{1}{B}\sum_{b=1}^{B} f^b_{m_v}(x_i, \chi^b_{m_v}, \theta^b_{m_v})$$

with $f^b_{m_v}(x_i, \chi^b_{m_v}, \theta^b_{m_v}) = \frac{1}{n_{t^b_{m_v ter_i}}}\sum_{j=1}^{n} y_j I(j \in t^b_{m_v ter_i})$ *is an unbiased estimator of* $h(x_i, \chi)$*, and for any convex* ϕ *and* $m = \frac{2}{3}n$*,*

$$E_{\chi^B_{m_v}} \phi(y_i - \hat{h}^B_{m_v}(x_i)) \le E_{\chi^B}\phi(y_i - \hat{h}^B(x_i)).$$

Proof. For the sake of simplicity, we will give the proof for only the root node.

Following Alin et al. (2017) and Pathak (1962), $f^b_{m_v}(x_i, \chi^b_{m_v}, \theta^b_{m_v})$ is an unbiased estimator for $f(x_i, \chi^l, \theta^l)$, and since $\hat{h}^B_{m_v}(x_i)$ is the mean of all $f^b_{m_v}(x_i, \chi^b_{m_v}, \theta^b_{m_v})$, it is also an unbiased estimator of $h(x_i)$.

- For $m = n$;
  Following Sinha and Sen (1989), we have $E_{\chi^b_{m_v}} \phi(f^b_{m_v}(x_i, \chi^b_{m_v}, \theta^b_{m_v}) - f(x_i, \chi^l, \theta^l)) \le E_{\chi^b}\phi(f^b(x_i, \chi^b, \theta^b) - f(x_i, \chi^l, \theta^l))$ that leads to $E_{\chi^B_{m_v}} \phi(y_i - \hat{h}^B_{m_v}(x_i)) \le E_{\chi^B}\phi(y_i - \hat{h}^B(x_i))$.
- For $m = \frac{2}{3}n$;

  Following Alin et al. (2017) and Pathak (1962),

$$V(f^b_{m_v}(x_i, \chi^b_{m_v}, \theta^b_{m_v})) \approx \left(\frac{1}{m} - \frac{1}{2n}\right) s^2_n$$

  where s^2_n is the sample variance. Since

$$V(f^b(x_i, \chi^b, \theta^b)) = \left(\frac{1}{n}\right)s_n^2,$$

$$RE(f_{m_v}^b(x_i, \chi_{m_v}^b, \theta_{m_v}^b), f^b(x_i^b, \chi^b, \theta^b)) \approx \frac{2m}{2n-m},$$

$f_{m_v}^b(x_i, \chi_{m_v}^b, \theta_{m_v}^b)$ will be at least as efficient as $f^b(x_i^b, \chi^b, \theta^b)$ for $m = \frac{2}{3}n$ with $E_{\chi_{m_v}^B}\phi(y_i - \hat{h}_{m_v}^B(x_i)) \leq E_{\chi^B}\phi(y_i - \hat{h}^B(x_i))$

10.3.3 Robust Aggregation

The classical aggregation statistic used in the original random forest is the mean of the predictions from B trees (Eq. 2) where $f^b(x, \chi, v^b)$ is the mean as in Eq. (5). We replace the mean with the median of all predictions from bootstrap trees (Eq. 8) as the robust aggregation method.

$$\hat{h}^R(x_i) = median(f^1(x_i, \chi^1, \theta^1), \ldots, f^B(x_i, \chi^B, \theta^B)) \quad (8)$$

The function $f^b(x, \chi^b, \theta^b)$ can be mean (Eq. 5) or the median (Eq. 7) of the terminal node that the ith observation falls.

10.3.4 Weights

Herein, the weight function in Eq. (9) called the *Fair Function* is introduced into the Random Forest algorithm. This function is one of the weight functions studied by Cummins and Andrews (1995), and has also been used by Serneels et al. (2005) for Partial Robust M-Regression. The reason we choose this particular function is that as pointed out by Serneels et al. (2005), many numerical experiments indicate that the fair function combined with c = 4 gives a good compromise between robustness and statistical efficiency.

$$\omega(c)_i = \frac{1}{\left(1 + \left|\frac{y_{i_{scaled}}}{c}\right|\right)^2} \quad (9)$$

In Eq. (9), $y_{i_{scaled}} = \frac{y_i - \tilde{y}}{MAD}$ with MAD is the median absolute deviations, and $\tilde{y}$ is the median of the response variable. The MAD gives a measure of variation that is more robust than the sample standard deviation Cummins and Andrews (1995). We calculate the weighted response $y_{w_i} = \omega_i y_i$ and split the node with the loss function in Eq. (10). We recalculate the weights and weighted response at each node.

$$SAMD_w = \sum_{i \in t_R}^{n_{t_R}} \left| y_{w_i} - \tilde{y}_{w_R} \right| + \sum_{i \in t_L}^{n_{t_L}} \left| y_{w_i} - \tilde{y}_{w_L} \right| \tag{10}$$

All the 8 variations studied in this paper have been summarized in Table 10.1. The versions RF and RFM have also been studied by Roy and Larocque (2012) who compared 6 different versions including the Quantile Random Forest algorithm proposed by Meinshausen (2006). In this study, we have introduced 6 new robust Random Forest Regression algorithms.

10.4 NUMERICAL RESULTS

We studied the performance of the proposed algorithms on four different simulation designs as well as on five real data sets using Mean Prediction Square Error (Eq. 11) and Mean of Absolute Prediction Error (Eq. 12);

$$MPSE = \frac{1}{n_{test}} \sum_{i=1}^{n_{test}} \left(h_{i(test)} - \hat{h}_{i(test)} \right)^2 \tag{11}$$

Table 10.1 The algorithms

Methods	*Weighted*	*Splitting Rule*	*Prediction in Terminal Node*	*Bootstrap*	*Aggregation*
RF	No	SSR	Mean	Ordinary	Mean
SmoonRF	No	SSR	Mean	Smoon	Mean
RFM	No	SAMD	Median	Ordinary	Median
SmoonRFM	No	SAMD	Median	Smoon	Median
WRF	$c = 4$	$SAMD_w$	Mean	Ordinary	Mean
WSmoonRF	$c = 4$	$SAMD_w$	Mean	Smoon	Mean
WRFM	$c = 4$	$SAMD_w$	Median	Ordinary	Median
WSmoonRFM	$c = 4$	$SAMD_w$	Median	Smoon	Median

$$MAPE = \frac{1}{n_{test}} \sum_{i=1}^{n_{test}} | h_{i(test)} - \hat{h}_{i(test)} || \tag{12}$$

where $h_{i(test)}$ and $\hat{h}_{i(test)}$ are the ith observed value of the test set and its prediction from the Random Forest built on the training sample, respectively. Simulation designs 1 and 2 are from Sage (2018). We have training samples of size 200 and test sets with the size of 800 for both designs. For simulation designs 3 and 4 both are from Li (2018), we have training and test samples each with size 1,000. For real data, 10% of the data have been left as the test set. For simulated datasets we study the contaminated as well as non-contaminated data while no contamination has been introduced on real data. For contamination probability we consider 10% and 20%. Only the training samples have been contaminated as in Roy and Larocque (2012) where they explain the reason as contamination are considered as errors or anomalies, and they really want to predict the clean data that they see as the true population. They claim that if the test data is also contaminated, this becomes a characteristic of the population itself and cannot really be considered a contamination. For Smoon bootstrap, m has been taken as $\frac{2}{3}n$ as proposed in Theorem 10.3.1. For the weights, the value of the tuning constant c has been taken 4 as suggested by Serneels et al. (2005). Simulations have been repeated 100 times. Minimum node size for a terminal node is kept 15 for all simulated data and real data. Twenty-five trees have been created for all forests in all numerical calculations including both the simulated and real data as Breiman (1996) explains that more than 25 bootstrap replicates is love's labor cost. Along with many authors, Cutler et al. (2012) suggest $\frac{k}{3}$ for ω for Regression Forest, and that is the reason that we use the same approach for all calculations.

10.4.1 Simulation Design 1

This simulation design also studied by Sage (2018) is generated as a binary tree with 6 predictors and 7 leaves. It is defined by

$$\begin{aligned} y_i = \delta[&I((x_{1i} \le 0),(x_{2i} \le 0)) \\ &+ 2I((x_{1i} \le 0),(x_{2i} > 0),(x_{4i} \le 0)) \\ &+ 3I((x_{1i} \le 0),(x_{2i} > 0),(x_{4i} > 0),(x_{6i} \le 0)) \\ &+ 4I((x_{1i} \le 0),(x_{2i} > 0),(x_{4i} > 0),(x_{6i} > 0)) \\ &+ 5I((x_{1i} \le 0),(x_{3i} \le 0)) \\ &+ 6I((x_{1i} \le 0),(x_{3i} \le 0),(x_{5i} \le 0)) \\ &+ 7I((x_{1i} \le 0),(x_{3i} \le 0),(x_{5i} > 0))] \\ &+ \varepsilon_i \end{aligned}$$

where $\mathbf{X} \sim N_6(\mathbf{0}, \mathbf{I}_6)$, and $\varepsilon_i \sim (0,1)$. The effect size δ is taken 0.2, 0.4, 0.6 and 0.8 representing the range from moderate to high signal-to-noise ratio. We contaminate the error in training data by $\varepsilon_{cont} = (1 - p)N(0, 1) + pN(0, 5)$ where p is the probability of contamination. The smaller the signal to noise ratio, the more leptokurtic the contaminated data get (Figure 10.1) for which sufficient bootstrap combined with weighted data, robust loss function and mean as the aggregation and prediction statistic (WSmoonRF) gives better prediction performance based on the MPSE values in Table 10.2. However, for heavily contaminated data,

Table 10.2 MPSE and MAPE values, and standard deviations for design 1

δ	Method	MPSE (Standard Deviations)			MAPE (Standard Deviations)		
		Contamination			Contamination		
		0	0.1	0.2	0	0.1	0.2
0.8	RF	40.8376(2.4503)	41.0583(2.4209)	41.4074(2.3942)	8.7596(0.2939)	8.8055(0.3027)	8.8327(0.3262)
	SmoonRF	41.4213(2.2579)	41.7384(2.4444)	42.0407(2.4297)	8.9258(0.3028)	8.9775(0.3315)	8.9932 (0.3390)
	RFM	52.3916(4.2376)	51.1288(4.3129)	50.1467(4.3378)	8.4349(0.3565)	8.4261(0.3631)	8.4336(0.3555)
	SmoonRFM	51.4968(4.5766)	50.5114(4.7986)	49.1527(4.4513)	**8.4209**(0.3619)	**8.4219**(0.3629)	**8.4150**(0.3445)
	WRF	**39.8776**(2.4623)	**40.0156**(2.2491)	40.4356(2.2322)	8.5907(0.3097)	8.6285(0.2817)	8.6867(0.3012)
	WSmoonRF	40.0947(2.4778)	40.3382(2.3087)	**40.3956**(2.1889)	8.7177(0.3118)	8.7623(0.3032)	8.7579(0.3222)
	WRFM	52.4409(4.8665)	51.1171(3.9872)	50.4421(4.2389)	8.4514(0.4025)	8.4260(0.3527)	8.4630(0.3625)
	WSmoonRFM	51.6654(4.9197)	50.3787(4.5934)	49.1281(4.4017)	8.4262(0.3976)	8.4253(0.3515)	8.4177(0.3402)
0.6	RF	24.2829(1.3543)	24.5482(1.4376)	25.0077(1.4063)	6.8152(0.2117)	6.8599(0.2210)	6.9144(0.2434)
	SmoonRF	24.6140(1.3931)	25.0325(1.5160)	25.1167(0.4543)	6.9082(0.2279)	6.9755(0.2429)	6.9762(0.2622)
	RFM	29.6751(2.4671)	29.1527(2.6257)	29.1153(2.4419)	6.6798(0.2711)	6.6724(0.2896)	6.7030(0.2687)
	SmoonRFM	29.1062(2.5058)	28.9906(2.6312)	28.0634(2.4331)	6.6635(0.2493)	6.6611(0.2646)	**6.6396**(0.2518)
	WRF	**23.7082**(1.3054)	**24.0473**(1.3828)	24.4331(1.4019)	6.6936(0.2118)	6.7522(0.2085)	6.8181(0.2382)
	WSmoonRF	23.9385(1.2681)	24.1181(1.3534)	**24.2782**(1.4014)	6.7835(0.2077)	6.8252(0.2213)	6.8416(0.2364)
	WRFM	29.5784(2.2002)	29.3860(2.5331)	28.9651(2.3699)	6.6718(0.2479)	6.6834(0.2661)	6.6931(0.2552)
	WSmoonRFM	29.3057(2.4659)	28.8073(2.5175)	28.3251(2.6477)	**6.6617**(0.2434)	**6.6525**(0.2558)	6.6542(0.2623)
0.4	RF	12.5611(0.6977)	12.7886(0.7439)	13.2048(0.8307)	4.9306(0.1442)	4.9773(0.1513)	5.0502(0.1783)
	SmoonRF	12.7310(0.6825)	12.8799(0.7452)	13.0754(0.7393)	4.9838(0.1462)	5.0148(0.1595)	5.0386(0.1666)
	RFM	14.0762(1.1525)	14.0877(1.2036)	14.1435(1.0344)	4.9204(0.1823)	4.9311(0.1839)	4.9454(0.1651)
	SmoonRFM	13.9029(1.1001)	13.8820(1.2346)	13.6949(0.9979)	4.8950(0.1705)	4.9043(0.1832)	**4.8943**(0.1524)
	WRF	**12.3166**(0.6945)	12.5971(0.7203)	12.9635(0.8123)	**4.8722**(0.1440)	4.9287(0.1417)	4.9951(0.1723)
	WSmoonRF	12.3682(0.7078)	**12.5367**(0.7307)	**12.7372**(0.7371)	4.9050(0.1481)	4.9457(0.1555)	4.9750(0.1642)
	WRFM	14.0836(1.1669)	14.0697(1.1315)	14.1010(1.0231)	4.9139(0.1798)	4.9251(0.1713)	4.9372(0.1638)
	WSmoonRFM	13.8822(1.1348)	13.8129(1.1695)	13.7526(1.0417)	4.8972(0.1692)	**4.8996**(0.1773)	4.9018(0.1606)

Table 10.2 *(COntd.)*

Table 10.2 MPSE and MAPE values, and standard deviations for design 1. (*C0ntd.*)

δ	Method	MPSE (Standard Deviations) Contamination			MAPE (Standard Deviations) Contamination		
		0	0.1	0.2	0	0.1	0.2
0.2	RF	5.4898(0.2757)	5.7402(0.3338)	6.0701(0.4186)	3.2519(0.0859)	3.3207(0.0993)	3.4120(0.1223)
	SmoonRF	5.5015(0.2907)	5.6564(0.3352)	5.8465(0.3735)	3.2589(0.0903)	3.3034(0.0988)	3.3541(0.1114)
	RFM	5.6233(0.3417)	5.6602(0.3398)	5.7786(0.3520)	3.2709(0.0953)	3.2820(0.0949)	3.3114(0.1020)
	SmoonRFM	5.5316(0.3151)	5.5753(0.3492)	**5.5991**(0.3090)	3.2481(0.0905)	3.2612(0.0985)	**3.2662**(0.0895)
	WRF	5.4302(0.2872)	5.7048(0.3649)	6.0934(0.4371)	3.2338(0.0899)	3.3070(0.1017)	3.4151(0.1268)
	WSmoonRF	**5.3997**(0.2789)	**5.5614**(0.3258)	5.7540(0.3610)	**3.2299**(0.0862)	3.2767(0.0978)	3.3268(0.1086)
	WRFM	5.6280(0.3495)	5.6654(0.3652)	5.7751(0.3462)	3.2711(0.0975)	3.2840(0.1020)	3.3105(0.0987)
	WSmoonRFM	5.5349(0.3267)	**5.5614**(0.3377)	5.6032(0.3171)	3.2496(0.0905)	**3.2577**(0.0953)	3.2669(0.0925)

SmoonRFM and WSmoonRFM have smaller MAPE values. Even for non-contaminated data WRF and WSmoonRF both perform better than classical Random Forest.

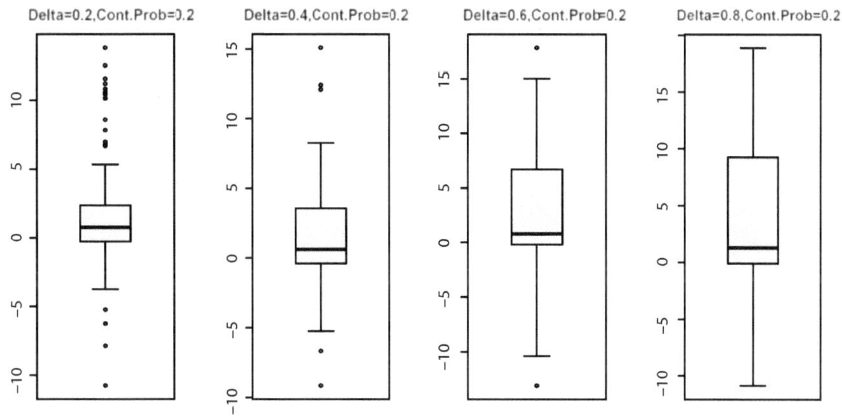

Figure 10.1 Box plots for simulation design 1.

10.4.2 Simulation Design 2

$$
\begin{aligned}
y_i = \delta[&x_{1i} + 0.707x_{2i}^2 + I(x_{3i} > 0) + 0.873\log(|x_{1i}|)x_{3i} \\
&+ 0.894x_{2i}x_{4i} + 2I(x_{5i} > 0) + 0.464\exp(x_{6i})] \\
&+ \varepsilon_i
\end{aligned}
$$

where $\mathbf{X} \sim N_6(\mathbf{0}, \mathbf{I}_6)$, and $\varepsilon_i \sim (0, 1)$. We consider the effect sizes δ determining the signal-to-noise ratio is taken 0.15, 0.3, 0.6 representing the range from moderate to high signal-to-noise ratio. As in simulation design 1, we contaminate the error in training data by $\varepsilon_{cont} = (1 \quad p)$ $N(0, 1) + pN(0, 5)$. Unlike simulation design 1, contaminated data with larger signal-to-noise ratio have a significant amount of outliers. The MPSE and MAPE values for the design are given in Table 10.3. Random Forest with sufficient bootstrap has the smallest MPSE and MAPE for uncontaminated data. In terms of MPSE, it is the best one even for contaminated data for high signal-to-noise ratio. As signal-to-noise ratio gets smaller, the amount of outliers get larger (Figure 10.2), and the random forest algorithm with median used for both prediction and aggregation combined with sufficient bootstrap and robust loss function (SmoonRFM) has the smallest MPSE.

All numerical results SmoonRFM and WSmoonRFM compete with each other for both MPSE and MAPE. As in Simulation 1, by using sufficient bootstrap we decrease the complexity of the trees by 48% less node number yet with better prediction performance.

Figure 10.2 Box plots for simulation design 2.

10.4.3 Simulation Design 3

This design is from Li (2018). We generate data from an underlying model $y = x^2 + \varepsilon$ where $x \sim U[-5, 5]$ and $\varepsilon \sim N(0, 1)$. We contaminate the training data as $\varepsilon_{cont} = (1 - p)N(0, 1) + p2t(2)$ with $p = 0, 0.10, 0.20$ where $t(2)$ is t-distribution with 2 degrees of freedom. Since we only have one independent variable, instead of Random Forest we will build trees with bagging. The results are listed in Table 10.4. Both MPSE and MAPE values point to WSmoonRFM as the best in prediction for contaminated data.

10.4.4 Simulation Design 4

We generate data from an underlying model $y_i = \sum_{k=1}^{10} x_{ki}^2 + \varepsilon_i$ where $x \sim N_{10}(0, 1)$ and $\varepsilon_i \sim N(0, 1)$. We contaminate the training data as $\varepsilon_{cont} = (1 - p)N(0, 1) + p15t(2)$ with $p = 0, 0.10, 0.20$ where $t(2)$ is t-distribution with 2 degrees of freedom. The response in this design is heavily contaminated compared to the Design 3 (Figure 10.3), and SmoonRFM is the one with the smallest error (Table 10.5).

Table 10.3 MPSE and MAPE values, and standard deviations for design 2

δ	Method	MPSE (Standard Deviations) Contamination			MAPE (Standard Deviations) Contamination		
		0	0.1	0.2	0	0.1	0.2
0.6	RF	8.0751(0.4759)	8.3021(0.5477)	8.6538(0.5864)	3.8238(0.0962)	3.8783(0.1096)	3.9659(0.1290)
	SmoonRF	**8.0736**(0.4628)	**8.2422**(0.5473)	**8.4494**(0.4971)	**3.8188**(0.0960)	**3.8601**(0.1130)	3.9099(0.1086)
	RFM	8.4265(0.5099)	8.5192(0.5490)	8.6547(0.5456)	3.8987(0.1096)	3.9235(0.1127)	3.9560(0.1119)
	SmoonRFM	8.3773(0.5135)	8.4364(0.5435)	8.4655(0.5226)	3.8806(0.1021)	3.8948(0.1109)	**3.9028**(0.1025)
	WRF	8.2372(0.4946)	8.5255(0.5582)	8.8901(0.5868)	3.8574(0.1028)	3.9264(0.1161)	4.0168(0.1285)
	WSmoonRF	8.2254(0.4854)	8.3895(0.5456)	8.6150(0.5202)	3.8490(0.0975)	3.8878(0.1135)	3.9447(0.1119)
	WRFM	8.4617(0.5472)	8.5526(0.5429)	8.6730(0.5667)	3.9059(0.1113)	3.9295(0.1129)	3.9590(0.1149)
	WSmoonRFM	8.4011(0.5183)	8.4468(0.5346)	8.5049(0.5190)	3.8850(0.1035)	3.8952(0.1101)	3.9094(0.1021)
0.3	RF	4.3354(0.1972)	4.5933(0.2834)	4.9397(0.3418)	2.8525(0.0664)	2.9337(0.0881)	3.0441(0.1100)
	SmoonRF	**4.3067**(0.1966)	4.4791(0.2622)	4.6602(0.2737)	**2.8422**(0.0655)	2.8976(0.0842)	2.9568(0.0951)
	RFM	4.4778(0.2234)	4.5376(0.2259)	4.5975(0.2478)	2.8974(0.0746)	2.9154(0.0752)	2.9385(0.0802)
	SmoonRFM	4.4061(0.2117)	**4.4470**(0.2282)	**4.4758**(0.2268)	8731(0.0684)	**2.8860**(0.0738)	**2.8964**(0.0749)
	WRF	4.4003(0.1991)	4.6795(0.2823)	5.0215(0.3641)	2.8721 (0.0662)	2.9609(0.0889)	3.0710(0.1162)
	WSmoonRF	4.3539(0.1996)	4.5172(0.2677)	4.7208(0.2612)	2.8552(0.0662)	2.9080(0.0849)	2.9749(0.0877)
	WRFM	4.4953(0.2134)	4.5349(0.2376)	4.6042(0.2509)	2.9028(0.0691)	2.9181(0.0766)	2.9382(0.0792)
	WSmoonRFM	4.4170(0.2092)	4.4472(0.2295)	4.4823(0.2115)	2.8760(0.0679)	2.8875(0.0736)	2.8977(0.0692)
0.15	RF	3.4125(0.1533)	3.7010(0.2471)	3.9993(0.3312)	2.5478(0.0625)	2.6576(0.0888)	2.7522(0.1132)
	SmoonRF	**3.3702**(0.1512)	3.5518(0.2374)	3.7128(0.2343)	**2.5316**(0.0623)	2.6063(0.0890)	2.6560(0.0876)
	RFM	3.5024(0.1606)	5541(0.2337)	6053(0.1915)	2.5835(0.0627)	2.6034(0.0878)	2.6188(0.0727)
	SmoonRFM	3.4219(0.1610)	**3.4656**(0.2218)	**3.4706**(0.1703)	2.5492(0.0655)	2.5719(0.0840)	2.5695(0.0677)
	WRF	3.4401(0.1525)	3.7623 (0.2774)	1024(0.3314)	2.5579(0.0622)	2.6802(0.1007)	2.7901(0.1134)
	WSmoonRF	3.3853(0.1477)	3.5778(0.2501)	3.7615(0.2486)	2.5367(0.0619)	2.6156(0.0922)	2.6743(0.0910)
	WRFM	3.5062(0.1621)	3.5437(0.2253)	3.5994(0.1838)	2.5837(0.0650)	2.5985(0.0838)	2.6158(0.0701)
	WSmoonRFM	3.4293(0.1537)	3.4662(0.2180)	3.4716(0.1704)	2.5535(0.0630)	**2.5711**(0.0818)	**2.5694**(0.0663)

Table 10.4 MPSE and MAPE values, and standard deviations for design 3

Method	MPSE (Standard Deviations) Contamination			MAPE (Standard Deviations) Contamination		
	0	**0.1**	**0.2**	**0**	**0.1**	**0.2**
RF	1.1071(0.0431)	1.2702(0.1053)	1.7260(0.5354)	0.8383(0.0191)	0.8853(0.0286)	0.9777(0.0647)
SmoonRF	1.0764(0.0446)	1.1515(0.0670)	1.3907(0.3120)	0.8268(0.0204)	0.8531(0.0259)	0.9111(0.0497)
RFM	1.1700(0.0562)	1.1887(0.0462)	1.2316(0.0575)	0.8619(0.0239)	0.8691(0.0194)	0.8840(0.0227)
SmoonRFM	1.1258(0.0522)	1.1322(0.0531)	1.1598(0.0468)	0.8454(0.0234)	0.8471(0.0209)	0.8587(0.0196)
WRF	1.0880(0.0407)	1.2077(0.0737)	1.5864(0.4575)	0.8318(0.0199)	0.8710(0.0274)	0.9513(0.0606)
WSmoonRF	**1.0645**(0.0394)	1.1332(0.0565)	1.3293(0.2592)	**0.8227**(0.0191)	0.8490(0.0229)	0.8999(0.0496)
WRFM	1.1553(0.0475)	1.1610(0.0535)	1.2009(0.0513)	0.8573(0.0215)	0.8581(0.0209)	0.8742(0.0235)
WSmoonRFM	1.1113(0.0476)	**1.1204**(0.0493)	**1.1424**(0.0403)	0.8409(0.0209)	**0.8444**(0.0207)	**0.8530**(0.0173)

Table 10.5 MPSE and MAPE values, and standard deviations for design 4

Method	MPSE (Standard Deviations) Contamination			MAPE (Standard Deviations) Contamination		
	0	0.1	0.2	0	0.1	0.2
RF	**16.9083**(0.8925)	23.9822(7.9268)	58.2243(9.0685)	**3.1982**(0.0764)	3.5522(0.3148)	3.9554(1.1302)
SmoonRF	16.9879(0.8923)	20.7231(6.3758)	30.8612(7.2385)	3.2047(0.0757)	3.4424(0.2411)	3.7145(0.7638)
RFM	17.9848(1.0541)	18.0989(1.0888)	18.2741(1.0939)	3.2379(0.0819)	3.2507(0.0854)	3.2739(0.0862)
SmoonRFM	17.9903(1.0987)	**18.0254**(1.0961)	**18.2038**(1.0843)	3.2326(0.0851)	**3.2433**(0.0853)	**3.2645**(0.0831)
WRF	17.6434(0.9505)	24.9857(8.3516)	55.0932(6.5062)	3.2708(0.0786)	3.6408(0.3167)	4.0561(0.9487)
WSmoonRF	17.7616(0.9482)	21.3844(6.3206)	33.7899(3.5069)	3.2834(0.0803)	3.5195(0.2518)	3.8269(0.8687)
WRFM	18.3515(1.0765)	18.4961(1.1238)	18.6204(1.1507)	3.2714(0.0856)	3.2879(0.0876)	3.3030(0.0899)
WSmoonRFM	18.4206(1.0923)	18.4426(1.1427)	18.4863(1.1259)	3.2724(0.0857)	3.2786(0.0905)	3.2896(0.0888)

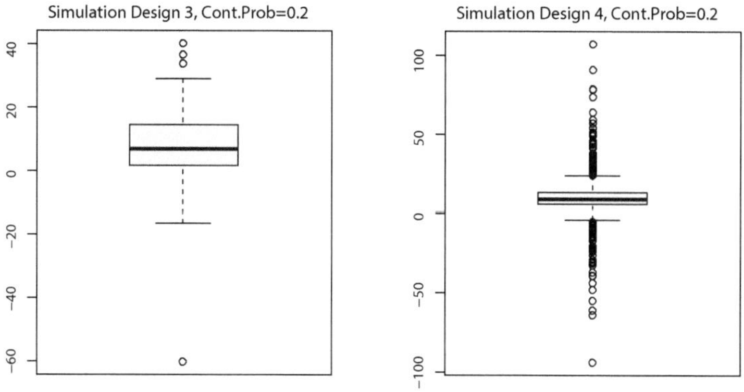

Figure 10.3 Box plots for simulation designs 3 and 4.

10.4.5. Real Data Examples

We studied 5 data sets with different sizes. Four of those data are from UCI Machine Learning Repository (Dua and Graff 2017), Forest fire data (Cortez and Morais 2007), Energy efficiency data (Tsanas and Xifara 2012), Parkinson disease (Little et al. 2007), and Boston housing (Harrison and Rubinfeld 1978) data. Body fat data is from StatLib Library at Carnegie Mellon University (Roger 1995). In all datasets, any row with a missing value at one of the futures has been removed. Only the numerical independent variables have been used to be compatible with simulation design where no categorical variable has been included. Some examples include ID or day and month columns like the Forest Fire data. Those columns have also been excluded. Multi-collinearity is common problem in all datasets. Except Body Fat data, all the responses in the rest of the 4 datasets are non-normally distributed (Figure 10.5 and Table 10.8). The distribution of response in Forest Fire and Housing data are leptokurtic whereas the response in Energy and Parkinson data have a platykurtic distribution (Table 10.8). Forest fire and Housing data have serious outlier problems (Figure 10.4). The MPSE and MAPE values are given in Tables 10.6 and 10.7.

Table 10.6 Prediction mean square error (MPSE) values for real datasets

Methods	Body Fat	Energy	Forest Fire	Housing	Parkinson
RF	1.5809	16.4106	0.1860	6.1258	0.9847
SmoonRF	1.8758	15.5735	0.1866	5.9927	0.9679
RFM	1.7963	10.5266	0.1805	5.7410	0.8642
SmoonRFM	2.3797	**9.7947**	**0.1784**	5.8353	0.8963
WRF	1.7623	13.6450	0.1889	5.5678	0.9984
WSmoonRF	**1.5279**	13.2559	0.1863	5.4584	0.9323
WRFM	2.2780	12.2416	0.1800	6.0166	0.8799
WSmoonRFM	2.2114	10.4772	**0.1785**	**5.4516**	**0.8587**

Table 10.7 Prediction mean square error (MAPE) values for real datasets

Methods	Body Fat	Energy	Forest Fire	Housing	Parkinson
RF	0.3402	1.6655	0.1247	0.6626	0.3156
SmoonRF	0.3757	1.6113	0.1254	0.6425	0.3104
RFM	0.3434	1.1677	0.1189	0.5305	0.2705
SmoonRFM	0.4273	**1.1390**	0.1182	0.5337	0.2794
WRF	0.3572	1.5196	0.1238	0.5909	0.3133
WSmoonRF	**0.3172**	1.4709	0.1228	0.5988	0.3008
WRFM	0.4098	1.3777	0.1194	0.5440	0.2689
WSmoonRFM	0.4016	1.2225	**0.1173**	**0.5271**	**0.2567**

Table 10.8 Skewness and Kurtosis values for real datasets

	Body Fat	Energy	Forest Fire	Housing	Parkinson
Skewness	0.1735	0.3604	0.8906	1.1088	0.4028
Kurtosis	2.6499	1.6907	3.8532	4.4925	1.9941
Outlier	Yes(one)	No	Yes	Yes	No
Shapiro–Wilk Test p–value	0.1595	2.2e-16	1.655e-07	5.574e-15	2.33e-05
Method	WSmoonRF	SmoonRFM	WSmoonRFM	WSmoonRFM	WSmoonRFM

In none of the datasets is classical Random Forest the one with the smallest error. For Body Fat data that is normal with only one outlier, the proposed weighted random forest combined with sufficient bootstrap performs the best among the 8 methods. The distribution of response in Energy data has shorter tails with no outlier, and SmoonRFM is the one with smaller errors. For Forest Fire, there is a very small difference between the MPSE of SMoonRFM and WSmoonRFM, but WSmoonRFM has a smaller MAPE. WSmoonRFM is the one showing the best performance for Housing and Parkinson data.

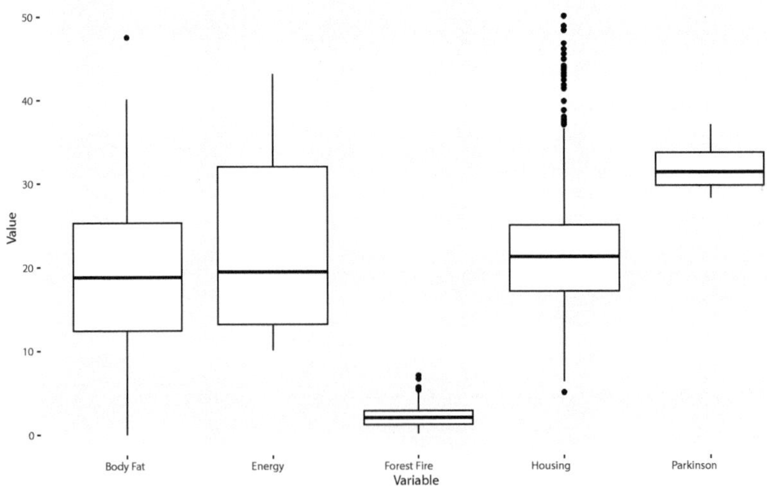

Figure 10.4 Box plots for real datasets.

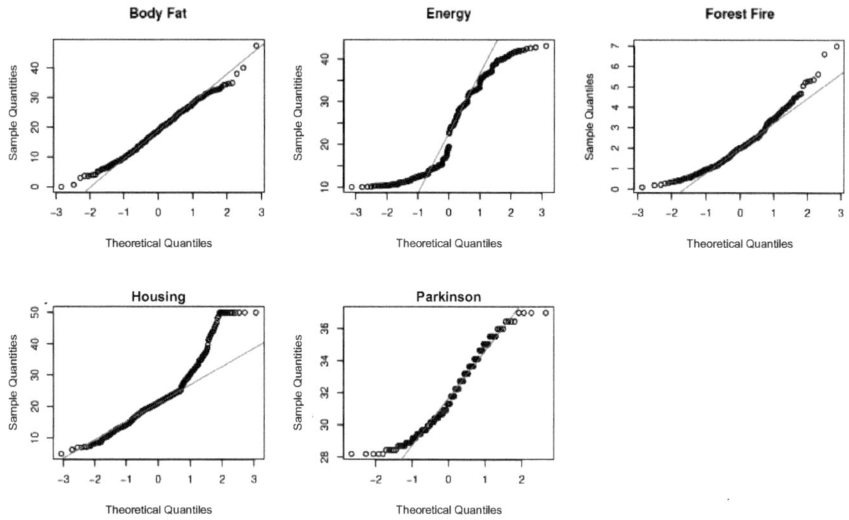

Figure 10.5 QQ norm plots for real datasets.

10.5 CONCLUSION

Outliers are a common problem in all fields. Robust methods are always needed to get more accurate analysis. Even though Random Forest Regression can tolerate the problems that original methods fail, it is not robust to outliers. Here, we mainly focus on amplifying the robustness by introducing weight function and alternative bootstrap method for bagging. We implement the sufficient moon bootstrap proposed by Alin et al. (2017) giving smaller and less complex trees. The other main contribution is to use the fair function to weight the response variable. All the numerical calculations reveal that the more the data heavily contaminated or depart from normality the more effective is the proposed weighted sufficient moon random forest regression. Even with RFM using sufficient moon bagging is a better choice compared to one with classical bagging.

REFERENCES

Ahmad, M.W., Mourshed, M. and Rezgui, Y. 2017. Trees vs neurons: Comparison between random forest and ANN for high-resolution prediction of building energy consumption. Energy and Buildings. 147: 77–89.

Alin, A., Martin, M., Beyaztas, U. and Pathak, K.P. 2017. Sufficient m-out-of-n (m/n) bootstrap. Journal of Statistical Computation and Simulation. 87: 1742–1753.

Breiman, L. 1996. Bagging predictors. Machine Learning. 24: 123–140.

Breiman, L. 2001. Random forests. Machine Learning. 45: 5–32.

Breiman, L. 2004. Consistency for a simple model of random forests. Technical Reports. Statistics Department University of California Berkeley.

Cano, G., Garcia-Rodriguez, J., Garcia-Garcia, A. et al. 2017. Automatic selection of molecular descriptors using random forest: Application to drug discovery. Expert Systems with Applications. 45: 151–159.

Chen, X. and Ishwaran, H. 2012. Random forests for genomic data analysis. Genomics. 99: 323–329.

Cortez, P. and Morais, A. 2007. A data mining approach to predict forest fires using meteorological data. URL: https://api.semanticscholar.org/CorpusID:36868619

Cummins, D. and Andrews, C. 1995. Iteratively re-weightted partial least squares: A performance analysis by Monte Carlo simulation. Journal of Chemometrics. 9: 489–507.

Cutler, A., Cutler, R. and Stevens, J.R. 2012. Ensemble Machine Learning. Chap. Random Forests. 157–175. Springer.

Denil, M., Matheson, D. and Freitas, N. (2013). Narrowing the gap: Random forests in theory and in practice. arXiv.

Dua, D. and Graff, C. 2017. UCI machine learning repository. URL http://archive.ics.uci.edu/ml

Díaz-Uriarte, R. and Alvarez de Andres, S. 2006. Gene selection and classification of´ microarray data using random forest. BMC Bioinformatics. 7.

Genuer, R., Poggi, J., Tuleau-Malot, C. et.al. 2017. Random forests for big data. Big Data Research. 9: 28–46.

Harrison, D. and Rubinfeld, D. 1978. Hedonic prices and the demand for clean air. Journal of Environmental Economics and Management. 5: 81–102.

Jones, P. 2019. A note on detecting statistical outliers in psychophysical data. Attention, Perception I & Psychophysics. 81: 1189–1196.

Krauss, C., Do, X. and Huck, N. 2017. Deep neural networks, gradient-boosted trees, random forests: Statistical arbitrage on the sp 500. European Journal of Operational Research. 259: 689–702.

Li, H. 2018. Robust Classification and Regression. PhD thesis, University of California. URL https://escholarship.org/uc/item/99b2s59x.

Little, M., McSharry, P., Roberts, S., Costello, D. and Moroz, M. 2007. Exploiting nonlinear recurrence and fractal scaling properties for voice disorder detection. BioMedical Engineering OnLine. 259: 6–23.

Mahapatra, D. 2014. Analyzing training information from random forests for improved image segmentation. IEEE Transactions on Image Processing. 23: 1504–1512.

Martin, M. and Roberts, S. 2014. Jackknife-after-bootstrap regression influence diagnostics. Journal of Nonparametric Statistics. 22: 257–269.

Meinshausen, M. 2006. Quantile regression forests. Journal of Machine Learning Research. 7: 983–999.

Meng, X. 2013. Scalable simple random sampling and stratified sampling. *In:* Proceedings of the 30th International Conference on Machine Learning. 28: 531–539.

Meng, X., Utkin, L., Kovalev, M. and Frank, C. 2019. Robust regression random forests by small and noisy training data. *In:* XXII International Conference on Soft Computing and Measurements. 134–137.

Pathak, P. 1962. On simple random sampling with replacement. Sankhya: The Indian Journal of Statistics, Series A. 24: 287–302.

Prasad, A., Iverson, L. and Liaw, A. 2006. Newer classification and regression tree techniques: Bagging and random forests for ecological prediction. Ecosystems. 9: 181–199.

Roger, J. 1995. Body fat data set. Journal of Statistics Education. 4(1). URL http://lib.stat.cmu.edu/datasets/bodyfat.

Roy, M. and Larocque, D. 2012. Robustness of random forests for regression. Journal of Nonparametric Statistics. 24: 993–1006.

Sage, A. 2018. Random Forest Robustness, Variable Importance and Tree Aggregation. PhD thesis, Iowa State University. URL https://dr.lib.iastate.edu/handle/20.500.12876/30636.

Sauer, S., Lemke, J., Zinn, W., Buettner, R. and Kohls, N. 2015. Mindful in a random forest: Assessing the validity of mindfulness items using random forests methods. Personality and Individual Differences. 81: 117–123.

Serneels, S., Croux, C., Filzmoser, P. and Van Espen, P.J. 2005. Partial robust M-regression. Chemometrics and Intelligent Laboratory Systems. 79: 55–64.

Shaikhina, T., Lowe, D., Dagad, S., Briggs, D., Higgins, R. and Khovanova, N. 2019. Decision tree and random forest models for outcome prediction in antibody incompatible kidney transplantation. Biomedical Signal Processing and Control. 52: 456–462.

Singh, S. and Sedory, S. 2011. Sufficient bootstrapping. Computational Statistics and Data Analysis. 55: 1629–1637.

Sinha, B. and Sen, P. 1989. On averaging over distinct units in sampling with replacement. Sankhya: The Indian Journal of Statistics, Series B. 51: 65–83.

Siroky, D. 2009. Navigating random forests and related advances in algorithmic modeling. Statistics Surveys. 3: 147–163.

Swanepoel, J. 1986. A note on proving that the (modified) bootstrap works. Communications in Statistics: Theory and Methods. 15: 3193–3203.

Tsanas, A. and Xifara, A. 2012. Accurate quantitative estimation of energy performance of residential buildings using statistical machine learning tools. Energy and Buildings. 49: 560–567.

Verikas, A., Gelzinis, A. and Bacauskiene, M. 2011. Mining data with random forests: A survey and results of new tests. Pattern Recognition. 44: 330–349.

Wei, B., Lee, S. and Wu, X. 2016. Stochastically optimal bootstrap sample size for shrinkage-type statistics. Statistics and Computing. 26: 249–262.

Williams, J. 2014. Using random forests to diagnose aviation turbulence. Machine Learning. 95: 51–70.

Xu, B., Ye, Y. and Nie, L. 2012. An improved random forest classifier for image classification. In: 2012 IEEE International Conference on Information and Automation. pp. 795–800.

Yu, P., Yang, T., Chen, S., et al. (2017. Comparison of random forests and support vector machine for real-time radar-derived rainfall forecasting. Journal of Hydrology. 552: 92–104.

Chapter **11**

Robust Remedies for Multiple Regression Model in the Presence of High-Leverage Collinearity-Influential Observations

Arezoo Bagheri[1*], Habshah Midi[2] and A.H.M. Rahmatullah Imon[3]

[1]Associate professor of applied statistics, Director of Statistical Methods and Population Modelling Department in National Population Studies and Comprehensive Management Institute, Tehran, Iran. Phone number: +1647-272-6714

[2]Laboratory of Applied and Computational Statistics, Institute for Mathematical Research University Putra Malaysia. Email: habshahmidi@gmail.com. Phone number: +60133925520

[3]Department of Mathematical Sciences, Ball State University, Muncie, IN 47306, USA. Email: rimon@bsu.edu. Phone number: +1765-285-8650

11.1 INTRODUCTION

The presence of outliers and multi-collinearity have an unduly effect on the parameter estimation of multiple linear regression models. It is

*For Correspondence: Arezoo Bagheri (abagheri_000@yahoo.com. arezoo.bagheri@nipr.ac.ir)

also evident that outliers in the x-direction or high-leverage points are another source of multi-collinearity (Bagheri et al. 2009, Bagheri 2015, Habshah et al. 2010). These leverage points may induce or hide near-linear dependency of explanatory variables in a dataset. Bagheri (2011) calls these leverages, high-leverage collinearity influential observations (HLCIO) either enhancing or reducing multi-collinearity. In this chapter, after introducing existing high-leverage diagnostic measures, first, a High-Leverage Collinearity Influential Measure (HLCIM) is proposed to study several criteria that cause these leverages to change the multi-collinearity pattern of collinear and non-collinear datasets (Habshah et al. 2011). Second, a new High-Leverage Collinearity-Influential Measure based on Diagnostic Robust Generalized Potentials (DRGP) with some non-parametric cutoff points is proposed (Bagheri et al. 2012). Third, three diagnostic plots which serve as powerful tools in separating outliers in the y-direction and the x-direction and are able to identify any HLCIO are recommended (Bagheri and Habshah 2015). Through applying some indicators on real and simulated datasets, the performance of several existing robust methods and the newly proposed methods are assessed and the results reveal that the proposed method outperforms the existing robust methods. This chapter is organized as follows.

In Section 11.2, existing high-leverage point diagnostic methods are reviewed. In Section 11.3, High-Leverage Collinearity-Influential Measure (HLCIM) and in Section 11.4, high-leverage collinearity-influential observation diagnostic measures based on a group deletion approach are introduced. In Section 11.4.1, new cutoff points for the existing collinearity influential measures are proposed. A new measure for the detection of multiple collinearity-influential observations is introduced in Section 11.4.2. The performance of the proposed measure is investigated in Section 11.4.2.1 through the use of some well-known datasets and Monte Carlo simulation studies. Section 11.5 focuses on new proposed high-leverage collinearity influential observation regression diagnostic plots. Some of the real datasets will be presented to evaluate the performance of the new proposed diagnostic plots in Section 11.5.1. Section 11.6 summarizes the contribution of the chapter and its significance results.

11.2 HIGH-LEVERAGE POINT DIAGNOSTIC METHODS

To explore high-leverage points in the dataset, some of the well-known detection methods are reviewed in the following sections.

11.2.1 Three-sigma Edit Rule

The three-sigma rule or empirical rule which is a traditional measure of the outlyingness of an observation x_i with respect to a sample, is the ratio between its distance to the sample mean, $\overline{x}$ and sample standard deviation s given as

$$T_i = \frac{x_i - \overline{x}}{s} \quad i = 1, 2, \ldots, n. \tag{1}$$

(Maronna et al. 2006). Any observations with $|T_i| > 3$ is suspected as an outlier. It is noticeable that a cutoff point of 3 has been chosen since $P(|x_i| \geq 3) = 0.003$ for a random variable $\mathbf{X}$ with a standard normal distribution. Maronna et al. (2006) noted the following drawbacks of applying the three-sigma rule diagnostic method:

1. In the large sample sizes with good observations, there are a large number of good observations which are swamped to detect as outliers.
2. In very small sample sizes, this rule is ineffective and always when $n \leq 10$ then $|T_i| \leq 3$.
3. In the presence of several outliers, some of them mask and remain unnoticed as a result of interacting their effects.

Detection procedures of outliers based on robust estimates are more reliable since outliers affect the classical estimators. Maronna et al. (2006) suggested to replace $\overline{x}$ and s in Eq. (1) by robust location and dispersion measures such as median and Normalized Median Absolute Deviation (*MADN*). Median can be defined as

$$t = median\ (\mathbf{x}),\ \text{if}\ \#\{x_i > t\} = \#\{x_i < t\},$$

where $\#\{\mathbf{A}\}$ denotes the number of elements of the set A. *MADN* which is a robust alternative to standard deviation s in Eq. (1) for $i = 1, 2, \ldots, n$ is defined as

$$MADN(\theta) = median\ |\theta_i - median\ (\theta_i)|/0.6745 = MAD(\theta)/0.6745. \tag{2}$$

The constant 0.6745 is used to make s an asymptotically unbiased estimator of s since 0.6745 is the *MAD* of a standard normal random variable. Hereafter, in this thesis, *MADN* will be used for the analysis and still refer to it as *MAD*. Hence, three-sigma edit rule may be presented as

$$T_i' = \frac{x_i - median(\mathbf{x}_i)}{MAD(\mathbf{x}_i)} \quad i = 1, 2, \ldots, n. \tag{3}$$

Any observations with $|T_i'| > 3$ is suspected as an outlier. This method can be used in univariate regression models as a diagnostic rule to detect high-leverage points. Since in most of the regression analysis, more than one explanatory variable exists in the regression model, investigating some useful methods in these cases seems necessary.

11.2.2 Hat Matrix

Hat matrix plays an important role in determining directly outlying **X** observations. This matrix is traditionally used as a measure of leverage points in regression analysis which is defined as

$$\mathbf{H} = \mathbf{X}(\mathbf{X}'\mathbf{X})^{-1}\mathbf{X}'. \tag{4}$$

The diagonal elements of the **H** matrix which are known as the leverage values are expressed by

$$h_{ii} = \mathbf{x}'_{\mathrm{i}}(\mathbf{X}'\mathbf{X})^{-1}\ \mathbf{x}_{\mathrm{i}} \quad i = 1, 2, \ldots, n. \tag{5}$$

The diagonal elements of the hat matrix, h_{ii} have some useful properties. In particular, their values are between 0 and 1 and their sum is equal to p. In addition, it can be shown that h_{ii} or leverage of the ith case is a measure of the distance between the **X** values for the ith case and the mean of **X** values for all n cases. Therefore, the large value of h_{ii} indicates that the ith case is distant from the center of all **X** observations (Cook and Weisberg 1999). The mean leverage value $\bar{h}$ is defined as

$$\bar{h} = \frac{\sum_{i=1}^{n} h_{ii}}{n} = \frac{p}{n}. \tag{6}$$

Therefore, leverage values greater than $2p/n$ are considered as outliers in **X**-direction. Hoaglin and Welsch (1978) called this rule as twice-the-mean-rule. Different cutoff points exist in the literature for the hat matrix to find high-leverage points such as thrice-the-mean-rule ($3p/n$) by Vellman and Welsch (1981). Huber (1981) introduced another cutoff point as a three interval range of Huber. He noted observations with $0.2 < h_{ii} < 0.5$ are risky to consider in analysis and those with $h_{ii} \geq 0.5 W_{ii} \leq 0.5$ should be avoided. Kutner et al. (2005) also noted that the existence of a gap between the leverage values for most of the cases and the unusually large leverage values is an additional evidence of an outlying case. It is noticeable that the entire above mentioned rules for identifying an outlying **X** value is intended for the datasets that are reasonably large, relative to the number of parameters in the regression model. However, Hadi (1992) pointed out that the hat matrix might fail to identify high-leverage points due to the effect of these points in the leverage structure.

11.2.3 Mahalanobis Distance

Mahalanobis Distance is a measure of leverage in the case of regression with a constant introduced by Rousseeuw and Leroy (1987). Its aim is to point to the observations with explanatory variables far from the bulk of the data. To introduce this measure, first split up the X_i is split up into the essential part t_i and the last coordinate 1, $\mathbf{x}_i = (x_{i1}\ x_{i1} \ldots x_{ik}\ 1) = (\mathbf{t}_i\ 1)$ where $\mathbf{t}_i = (x_{i1}\ x_{i2} \ldots x_{ik})$ is a k-dimensional row vector. Thus, the arithmetic mean and covariance matrix, $T(\mathbf{X})$ and $C(\mathbf{X})$, respectively, are defined as

$$T(\mathbf{X}) = \frac{1}{n}\sum_{i=1}^{n} \mathbf{t}_i \tag{7}$$

$$C(\mathbf{X}) = \frac{1}{n-1}\sum_{i=1}^{n} (\mathbf{t}_i - T(\mathbf{X}))'(\mathbf{t}_i - T(\mathbf{X})) \cdot \tag{8}$$

To find how far is $\mathbf{t}_i$ from $T(\mathbf{X})$ in the metric defined by $C(\mathbf{X})$, the Mahalanobis Distance for ith case is identified as

$$MD_i = \sqrt{(\mathbf{t}_i - T(\mathbf{X}))' C(\mathbf{X})^{-1} (\mathbf{t}_i - T(\mathbf{X}))} \quad i = 1, 2, \ldots, n. \tag{9}$$

Any observation with $MD_i > \sqrt{\chi^2_{(k,0.95)}}$ is considered a high-leverage point. Some authors for example, Habshah et al. (2009) prefer to use squared MD, MD^2 and compare MD^2 with $\chi^2_{(k,0.95)}$ to indicate high-leverage points which result in the same conclusion. Rousseeuw and Leroy (1987) proved that MD_i^2 can be written as a function of h_{ii} which present another clarification of the meaning of h_{ii} given as

$$MD_i^2 = (n-1)\left[h_{ii} - \frac{1}{n}\right] \quad i = 1, 2, \ldots, n. \tag{10}$$

Since MD_i^2 can be seen as a function of h_{ii}, MD_i^2 also will suffer from similar drawbacks as h_{ii} in identifying high-leverage points correctly.

11.2.4 Potential Measure

Hadi (1992) introduced a new type of high-leverage point diagnostic method, where the leverage of the ith case is computed based on a fit to the data after deletion of the ith case. Every possible subset of $(n - 1)$ observation is used to form the weight matrix, and weight of every deleted observation in turn is generated externally which is known as potentials. Although it seems that calculation of potentials will require construction of n weight matrices, it is possible to calculate them from

h_{ii} in a very simple way. Thus, a special type of single-case-deleted leverage measure, known as potential, is given by

$$p_i = \mathbf{x}_i'(\mathbf{X}_{(i)}'\mathbf{X}_{(i)})^{-1}\mathbf{x}_i \quad i = 1, 2, \ldots, n \tag{11}$$

where $\mathbf{X}_{(i)}$ is the data matrix $\mathbf{X}$ with the ith row deleted. It is straightforward to show that

$$p_i = \frac{h_{ii}}{1-h_{ii}} \quad i = 1, 2, \ldots, n. \tag{12}$$

Hadi (1992) proposed a cutoff point for p_i as

$$mean\ (p_i) + c\,\mathrm{SD}(p_i) \tag{13}$$

where SD stands for standard deviation and c is an appropriately chosen constant such as 2 or 3. The observations with p_i greater than this cutoff point value are said to be high-leverage points. This cutoff point is analogous to a confidence bound for a location parameter. But the weakness of this cutoff point is non-resistance of both *mean* and SD in the presence of a single extreme value. To avoid yielding a high cutoff point due to the extreme values, Hadi (1992) suggested a cutoff point for p_i as

$$median\ (p_i) + c\mathrm{MAD}(p_i) \tag{14}$$

where c can be taken as a constant value of 2 or 3. Any observation with p_i above the following non-parametric robust threshold is considered as a high-leverage point. Imon (2002) showed through simulated and real datasets that potentials are very successful in the identification of a single high-leverage point. However, they fail to identify multiple high-leverage points.

11.2.5 Generalized Potential Measure

To overcome the weakness of the potential measure, p_i, Imon (2002) proposed a group deletion version of potentials, known as Generalized Potentials (GP) which has been further studied by many scholars such as Imon (2005) and Habshah et al. (2009). First denote a set of good cases remaining in the analysis by R and a set of bad cases deleted by D. Also suppose that R contains $(n - d)$ cases after $d < (n - p)$ cases in D are deleted. Without loss of generality, assume that these observations are the last d rows of $\mathbf{X}$ and $\mathbf{y}$ so that the hat matrix defined in Eq. (4) can be partitioned as

$$\mathbf{H} = \begin{bmatrix} \mathbf{U_R} & \mathbf{V} \\ \mathbf{V}' & \mathbf{U_D} \end{bmatrix} \tag{15}$$

where $\mathbf{U}_R = \mathbf{X}_R(\mathbf{X}'\mathbf{X})^{-1}\mathbf{X}'_R$ and $\mathbf{U}_D = (\mathbf{X}'\mathbf{X})^{-1}\mathbf{X}'_D$ are symmetric matrices of $(n - d)$ and d, respectively, and $\mathbf{V} = \mathbf{X}_R(\mathbf{X}'\mathbf{X})^{-1}\mathbf{X}'_D$ is an $(n - d) \times d$ matrix. Based on a group of deleted cases indexed by D, can be defined as

$$h_{ii}^{(-D)} = \mathbf{x}'_i(\mathbf{X}'_R\mathbf{X}'_R)^{-1}\mathbf{x}_i \quad i = 1, 2, \ldots, n. \tag{16}$$

It should be noted that $h_{ii}^{(-D)}$ is the ith diagonal elements of $\mathbf{X}(\mathbf{X}'_R\mathbf{X}_R)^{-1}\mathbf{X}'$ matrix. When the size of R is $(n - 1)$ and $D = i$, it can be observed from Eqs. (11, 16) that

$$h_{ii}^{(-i)} = \mathbf{x}'_i(\mathbf{X}'_{(i)}\mathbf{X}_{(i)})^{-1}\mathbf{x}_i = p_i$$

which shows $h_{ii}^{(-D)}$ is a natural extension of p_i. Any ith case which is removed from the remaining subset R and joins the deletion subset D can be written as

$$h_{ii}^{-(D+i)} = \mathbf{x}'_i(\mathbf{X}'_R\mathbf{X}_R)^{-1}\mathbf{x}_i + \frac{(\mathbf{x}'_i(\mathbf{X}'_R\mathbf{X}_R)^{-1}\mathbf{x}_i)^2}{1-\mathbf{x}'_i(\mathbf{X}'_R\mathbf{X}_R)^{-1}\mathbf{x}_i} = \frac{h_{ii}^{-(D)}}{1-h_{ii}^{-(D)}}. \tag{17}$$

Imon (2002) introduced GP as

$$p_{ii} = \begin{cases} h_{ii}^{(-D)} & \text{for } i \in D \\ \dfrac{h_{ii}^{(-D)}}{1-h_{ii}^{(-D)}} & \text{for } i \in R \end{cases} \tag{18}$$

where D is any arbitrary deleted set of points. One could consider p_{ii} to be large if it is above the following threshold

$$median\,(p_{ii}) + cMAD(p_{ii}) \tag{19}$$

where c can be taken as a constant value of 2 or 3.

11.2.6 Diagnostic Robust Generalized Potential Based on Minimum Volume Ellipsoid (DRGP(MVE)) Measure

The commonly used diagnostic methods fail to identify high-leverage points correctly when a group of high-leverage points is present in a dataset due to the masking and/or swamping effects (Peña and Yohai 1995, Barnett and Lewis 1994, Rousseeuw and Leroy 1987). Rousseeuw (1985) introduced Robust Mahalanobis Distance (RMD), based on the Minimum Volume Ellipsoid (MVE), RMD(MVE), as

$$RMD_i(MVE) = \sqrt{(\mathbf{X}-T_R(\mathbf{X}))'C_R(\mathbf{X})^{-1}(\mathbf{X}-T_R(\mathbf{X}))} \quad i = 1, 2, \ldots, n \tag{20}$$

where $T_R(\mathbf{X})$ and $C_R(\mathbf{X})$ are robust locations and shape estimates of the MVE, respectively.

The robust alternative diagnostic methods such as RMD(MVE) can detect the high-leverage points correctly but they have a tendency to identify too many low leverage points as high-leverages which is not also desired. Habshah et al. (2009) attempt to make a compromise between these two approaches. They proposed an adaptive method where the suspected high-leverage points are identified by robust methods and then after diagnostic checking, the low leverage points (if any) are put back into the estimation dataset. Following the idea of Imon (2002) in developing GP, they proposed the diagnostic robust generalized potential (DRGP) based on MVE. DRGP(MVE) procedure can be summarized as follows:

Step 1: Calculate RMD(MVE) in Eq. (20) and consider a non-parametric cutoff point to determine the suspected high-leverage points (group D) as

$$median\ (RMD_i(MVE) + cMAD(RMD_i(MVE)) \quad i = 1, 2, \ldots, n \qquad (21)$$

where c is a constant value equals to 2 or 3. Hence, the members of group D are observations whose RMD(MVE) values exceeded the cutoff point in Eq. (21).

Step 2: Compute GP, p_{ii} in Eq. (18) to see whether all members of the deletion set have potentially high-leverages or not.

Step 3: If all members of the D set are greater than the rule in Eq. (16), they are declared as high-leverage points. Otherwise, those observations are put back into the estimation subset R.

The generalized potential values based on the final deletion set will be called DRGP(MVE) namely as p^*_{ii} and the points which are detected will be finally declared as high-leverage points. Leverage measures based on DRGP(MVE) have proven to be very effective in the identification of multiple high-leverage points (Habshah et al. 2009).

11.3 HIGH-LEVERAGE COLLINEARITY-INFLUENTIAL OBSERVATIONS

Outliers in **X**-direction or high-leverage points are the latest known source of multi-collinearity (Kamruzzaman and Imon 2002). Multi-collinearity is a non-orthogonality of two or more explanatory variables in multiple regression models which may have significant influential impacts on interpreting a fitted regression model (Gujarati 2002, Chatterjee and Hadi 2006). In this section, High-Leverage Collinearity Influential Measure (HLCIM) is proposed.

11.3.1 High-Leverage Collinearity-Influential Measure

The observations which create or hide the multi-collinearity pattern of the data are referred to as Collinearity-influential observations (Hadi 1988, Sengupta and Bhimasankaram 1997, Walker 1989, Gross 2003). Hadi (1988) introduced a collinearity-influential measure which is useful in detecting observations which cause a change in multi-collinearity pattern of the data. If Condition Number (CN) of **X** matrix, k defines as the largest value of k_j

$$k_j = \frac{\mu_{\max}}{\mu_j} \quad j = 1, 2, \ldots, k \tag{22}$$

and defines as the proportion of each singular value to the maximum singular value $\mu_{\max}$ of **X** matrix (Belsley et. al. 1980). Let consider Hadi's collinearity-influential measure as follows:

$$\delta_i = \frac{k_{(i)} - k}{k} \quad i = 1, 2, \ldots, n. \tag{23}$$

where k is the Condition Number of **X** matrix and $k_{(i)}$ is the Condition Number of **X** matrix when the ith row of **X** matrix has been deleted. Hadi's measure has the lack of symmetry which is due to the additive change in the Condition Number of **X**. To overcome the weakness of Hadi's measure, Sengupta and Behimasankaram (1997) introduced the following collinearity-influential measure as

$$l_i = \log\left(\frac{k_{(i)}}{k}\right) \quad i = 1, 2, \ldots, n. \tag{24}$$

It is important mentioning that a large negative or positive value of δ_i and l_i indicates that the ith observation is a collinearity-enhancing or collinearity-reducing observation. Bagheri (2011) proposed a new measure to study the influence of high-leverage points in the multi-collinearity pattern of a data based on the idea of Sengupta and Bhimasankaram in proposing l_i as collineaity-influential measure. If D is the group of high-leverage collinearity-influential observations, she defines High-leverage Collinearity-Influential Measure (HLCIM) as follows:

$$\text{HLCIM} = \log\left(\frac{k_{(D)}}{k}\right) \tag{25}$$

where $k_{(D)}$ is the Condition Number of **X** matrix when D rows of **X** matrix have been deleted. The HLCIM is used as an indicator to show whether these leverage points can cause multi-collinearity or not in the dataset. It is important to note that similar to Sengupta and Bhimasankaram's measure, high-leverage points can hide and induce a multi-collinearity pattern in two different situations. The first situation is when $\frac{k_{(D)}}{k} > 1$ and $k_{(D)} > k$ then, $\log\left(\frac{k_{(D)}}{k}\right) > 0$. Consequently, the degree of multi-collinearity increases due to the characteristics of high-leverages which hide the multi-collinearity pattern. In this situation, the D group of high-leverage points is referred to as high-leverage collinearity-reducing observations. Otherwise, the deletion of the D group high-leverage points may reduce the degree of multi-collinearity. Thus, in this situation these high-leverage points are referred as collinearity-enhancing observations and satisfy this inequality; $0 < \frac{k_{(D)}}{k} < 1$ and $k_{(D)} < k$ then $\log\left(\frac{k_{(D)}}{k}\right) < 0$. High-leverage points may cause a non-collinear data set to become multi-collinear. Habshah et al. (2011) call these high-leverage points as high-leverage collinearity-enhancing observations. They investigated the effect of the high-leverage collinearity-influential observations on the multi-collinearity pattern of non-collinear datasets through numerical and Monte Carlo simulation results.

Moreover, Bagheri et al. (2009) investigated the change in the degree of multi-collinearity caused by the high-leverage points in a collinear dataset. A collinear dataset may change to non-collinear dataset in the presence of high-leverage points. The high-leverage points which cause this change are called high-leverage collinearity-reducing observations by Bagheri et al. (2009). They studied the effect of high-leverage points in a multi-collinearity pattern of a real well-known collinear dataset and also the Monte Carlo simulation study.

11.4 HIGH-LEVERAGE COLLINEARITY-INFLUENTIAL OBSERVATION DIAGNOSTIC MEASURE BASED ON A GROUP DELETION APPROACH

Collinearity-influential observations are the points which might induce or disrupt the multi-collinearity pattern of a data (Hadi 1988, Sengupta and Behimasankaram 1997). Hadi (1988) noted that all high-leverage points are not collinearity-influential observations, whereas most collinearity-influential observations are points with high-leverages. In addition, he defined a measure for finding the influence of the ith row of $\mathbf{X}$ matrix on the condition index of $\mathbf{X}$ matrix (Hadi, 1988) as

$$\delta_i = \frac{k_{(i)} - k}{k} \quad i = 1, 2, \ldots, n. \tag{26}$$

where $k_{(i)}$ can be computed from the eigenvalues of $\mathbf{X}'_{(i)}\mathbf{X}_{(i)}$ or singular values of $\mathbf{X}_{(i)}$ when the ith row of $\mathbf{X}$ matrix has been deleted. A large negative value of δ_i indicates that the ith observation is a collinearity-enhancing observation, whereas a large positive δ_i value indicates a collinearity-reducing observation. In 1997, Sengupta and Behimasankaram identified that the weakness of Hadi's measure is in the lack of symmetry, which is due to the additive change in k. To overcome this problem, they suggested using this measure:

$$l_i = \log\left(\frac{k_{(i)}}{k}\right) \quad i = 1, 2, \ldots, n. \tag{27}$$

as a collinearity-influential measure for each row of the observations. Nonetheless, the use of δ_i or l_i possesses some drawbacks because no specific cutoff points were given to indicate which observations are collinearity-enhancing and which are collinearity-reducing.

11.4.1 New Cutoff Points for the Existing Collinearity-Influential Measures

Two of the most important collinearity-influential measures are Hadi's measure, δ_i in Eq. (26), and Sengupta and Bhimasankaram's measure, l_i in Eq. (27). According to Hadi (1988), a large negative value of δ_i indicates that the ith observation is a collinearity-enhancing observation, whereas a large positive δ_i value indicates a collinearity-reducing observation. However, he did not mention how large the δ_i should be. Sengupta and Bhimasankaram (1997) also did not introduce any specific cutoff points for l_i which correctly indicate collinearity-influential observations in the dataset. However, they suggested some easily computable upper bound and lower bound for l_i (Sengupta and Bhimasankaram 1997). Thus, the use of δ_i or l_i possesses some drawbacks because no specific cutoff points were given to indicate which observations are collinearity-enhancing and which are collinearity-reducing. In this respect, two non-parametric cutoff points are proposed for these two collinearity-influential measures following the proposed cutoff points of high-leverage diagnostic measures introduced by Hadi (1992), Imon (2002), and Habshah et al. (2009). These cutoff points may be defined as follows:

$$CEM - cut1(\theta) = median\ (\theta_i) - c\,MAD(\theta_i) \quad i = 1, 2 \ldots, n \tag{28}$$

$$CEM - cut2(\theta) = median\ (|\theta_i < 0|) - cMAD(\theta_i < 0|) \quad i = 1, 2 \ldots, n \tag{29}$$

where $CEM - cut$ is the Collinearity-Enhancing Measure cutoff point, θ_i may be δ_i or l_i and c is considered a constant value of 3. $|\theta_i| \geq |CEM - cut1(\theta)|$ and $|\theta_i| \geq |CEM - cut2(\theta)|$ for $\theta_i < 0$ is an indicator that the ith observation is a collinearity-enhancing observation.

11.4.2 High-leverage Collinearity-Influential Measure Based on DRGP(MVE)

Following the idea of Habshah et al. (2009) in proposing DRGP(MVE), p^*_{ii} defined in Section 11.2.6, Bagheri et al. (2012) utilized a similar concept in developing multiple high-leverage collinearity-influential measures. Hence, High-leverage Collinearity-Influential Measure based on DRGP(MVE), namely the HLCIM (DRGP) are proposed which is denoted as $\delta_i^{(D)}$. This new measure is defined as follows:

$$\delta_i^{(D)} = \begin{cases} \log\left(\dfrac{k_{(D)}}{k_{(D-i)}}\right) & \text{if } i \in D \text{ and } \#\{D\} \neq 1 \\ \log\left(\dfrac{k_{(i)}}{k}\right) & \text{if } \#\{D\} = 1 \text{ and } D = i, \quad i = 1, 2, \ldots, n \\ \log\left(\dfrac{k_{(D+i)}}{k_{(D)}}\right) & \text{if } i \in R \end{cases} \tag{30}$$

where D is the suspected group of multiple high-leverage collinearity-influential observations diagnosed by DRGP(MVE), p^*_{ii}, and $\#\{D\}$ is the number of elements in the D group. $k_{(D)}$ and $k_{(D-i)}$ indicate the condition number of the $\mathbf{X}$ matrix without the entire group of D high-leverage points and without the entire D group minus the ith high-leverage points where i belongs to the suspected D group, respectively. Furthermore, $k_{(D+i)}$ refers to the condition number of the $\mathbf{X}$ matrix without the entire group of D high-leverage points plus the ith additional observation of the remaining group. The $\delta_i^{(D)}$ value for the ith observation which belongs to the group D of high-leverage points, suspected as high-leverage collinearity-influential observations indicates the separate effect of each high-leverage point on the condition number under conditions where the influence of the other high-leverage points is ignored. On the other hand, the $\delta_i^{(D)}$ value for the observation that belongs to the remaining group describes the influence of this observation without taking into account the effect of the entire group of high-leverage points on the condition number. It is noticeable that when the suspected D group has just one member then the values of $\delta_i^{(D)}$ and l_i are equal. We propose the same cutoff points for $\delta_i^{(D)}$ as exemplified in Eqs. (28, 29).

Any observation that belongs to the suspected D group and which is less than the introduced cutoff points is put back to the R set and is not considered as high-leverage collinearity-influential observations.

The algorithm of HLCIM (DRGP) may be summarized as follows.

Step 1: Calculate DRGP(MVE), p^*_{ii}, for $i = 1, 2, \ldots, n$. Form D as high-leverage collinearity-influential suspected group for any observations which corresponding to p^*_{ii} which exceed the median $(p^*_{ii}) + 3$ MAD(p^*_{ii}). Obviously the rest of the observations belong to R, the remaining group.

Step 2: Compute high-leverage collinearity-influential values, $\delta_i^{(D)}$, as follows:

- If group D is single member, the size of R is $(n - 1)$, and $D = i$, calculate $\log\left(\frac{k_{(i)}}{k}\right)$ where $k_{(i)}$ indicates the condition number of the **X** matrix without the ith high-leverage points. In this way, $\delta_i^{(D)} = l_i$.
- If group D has more than one member, calculate $\log\left(\frac{k_{(D)}}{k_{(D-i)}}\right)$ where $k_{(D-i)}$ indicate the condition number of the **X** matrix without the entire D group minus the ith high-leverage points where i belongs to the suspected D group.
- For any observation in R group, compute $\log\left(\frac{k_{(D+i)}}{k_{(D)}}\right)$ where $k_{(D+i)}$ refers to the condition number of the **X** matrix without the entire group of D high-leverage points plus the ith additional observation of the remaining group.

Step 3: If any $\delta_i^{(D)}$ values for $i - 1, 2, \ldots, n$ does not exceed the cutoff points in Eqs. (28 or 29), put back the ith observation to R group. Otherwise, the D group is the high-leverage collinearity-enhancing observations.

11.4.2.1 *Numerical Results*

In this section, a few examples are considered that show how commonly used collinearity-influential measures fail in the presence of multiple high-leverage collinearity-enhancing observations and consider how successfully the proposed measure addresses this issue. Before proceeding to analysis of real datasets, it is important to mention that the explanatory variables are scaled before analysis. Because, doing so prevents the condition number from being dominated by large measurement units of some explanatory variables. To compute the condition number of the **X** matrix, the explanatory variables have been scaled by Stewart's Scaling Method. It is important to mention that in the leverage diagnostics tables for all of the real datasets in the following section, the bolded values in the brackets are the cutoff points of the leveraged diagnostic methods. Moreover, in collinearity-influential measures tables, the bolded and underlined cutoff points in the brackets are $CEM - cut1$ and $CEM - cut2$, respectively.

Real Data: The performance of the new proposed collinearity-influential measure, denoted as $\delta_i^{(D)}$ is studied in this section through two well-known datasets, namely, the Hawkins-Bradu-Kass data and the Commercial Properties data.

Hawkins-Bradu-Kass Dataset: The second example presented here is the well-known Hawkins et al. (1984) data. They generated an artificial three-predictor dataset containing 75 observations with 10 bad leverage points (cases 1–10) and 4 good leverage points (cases 11–14). Many authors (Hawkins et al. 1984, Rousseeuw and Leroy 1987, Imon 2005, Habshah et al. 2009) have reported that single-case deletion measures fail to identify high-leverage points correctly in this dataset. Figure 11.1 presents the scatter plots of Hawkins-Bradu-Kass Data with and without high-leverage points. Figure 11.1a illustrates the existence of 14 high-leverage points and Figure 11.1b shows the non-collinearity of the dataset without these high-leverage points. It seems that the first 13 high-leverage points are the cause of the multi-collinearity in the dataset, whereas the 14th observation changes the multi-collinearity pattern of the data (Figure 11.2a).

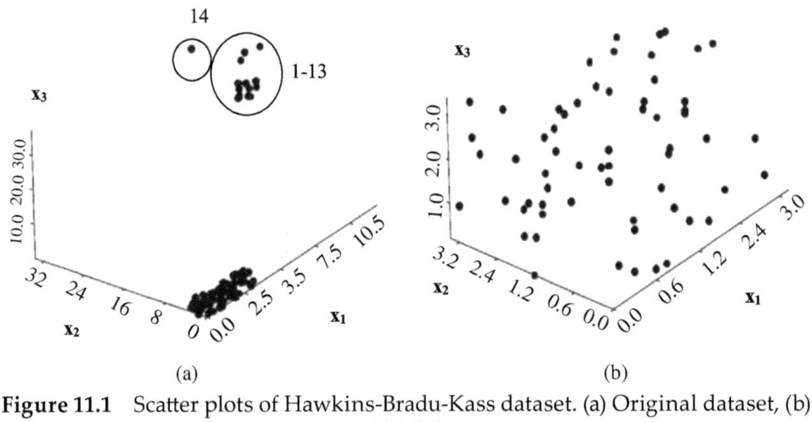

Figure 11.1 Scatter plots of Hawkins-Bradu-Kass dataset. (a) Original dataset, (b) Dataset without high-leverage points.

Table 11.1 presents leverages, potentials, and DRGP(MVE) values for this data. It is observed from the data in the table that leverages and potentials can only identify observation numbers 12–14 as high-leverage points, whereas the DRGP(MVE) can successfully identify all 14 high-leverage cases (to see the rest of the values for leverages, potentials, and DRGP(MVE), one can refer to Habshah et al., 2009). Next, collinearity-influential measures are computed for this data which are presented in Table 11.2.

Table 11.1 Leverage diagnostics for Hawkins-Bradu-Kass dataset

Case	h_{ii}(0.107)	p_i(0.109)	p^*_{ii}(0.552)
1	0.063	0.067	13.440
2	0.060	0.064	14.181
3	0.086	0.094	17.001
4	0.081	0.088	17.842
5	0.073	0.079	16.857
6	0.076	0.082	14.058
7	0.068	0.073	13.815
8	0.063	0.067	14.257
9	0.080	0.087	17.223
10	0.087	0.095	16.835
11	0.094	0.104	21.558
12	0.144	0.168	25.468
13	0.109	0.122	18.287
14	0.564	1.294	17.990

The results of Table 11.2 show that the existing measures δ_i and l_i can identify the first 13 high-leverage points as collinearity-enhancing observations based on $CEM - cut1$ (bolded values), while based on $CEM - cut2$ (underlined values), they failed to identify any collinearity-enhancing observation. However, the proposed $\delta_i^{(D)}$ measure can successfully identify the first 14 observations as high-leverage collinearity-enhancing observations using either cutoff point. Furthermore, the effect of the first 13 high-leverage points on the values of δ_i and l_i causes these two measures to wrongly indicate that observation 14 reduces multi-collinearity.

Figure 11.1a separately illustrates the effects of the 14th high-leverage point and the first 13 high-leverage points on the multi-collinearity pattern of the data. Figure 11.1a shows that, in the absence of the other high-leverage points, the 14th high-leverage point is a collinearity-enhancing observation. The plot shown in Figure 11.1b indicates that the first 13 high-leverage points bring multi-collinearity to the dataset in the absence of the 14th high-leverage point. In fact, the 14th observation disturbs the multi-collinearity pattern of the first 13 high-leverage points in this dataset (Figure 11.2a).

Table 11.2 Collinearity-influential measures for Hawkins-Bradu-Kass dataset.

Ind	$\delta_{i(-0.008)}$ (−0.048)	$l_{i(-0.004)}$ (−0.021)	$\delta^{(D)}_{i(-0.019)}$ (−0.022)	Ind	$\delta_{i(-0.008)}$ (−0.048)	$l_{i(-0.004)}$ (−0.021)	$\delta^{(D)}_{i(-0.019)}$ (−0.022)
1	**−0.027**	**−0.012**	**−0.228**	39	0.001	0.001	0.002
2	**−0.030**	**−0.013**	**−0.241**	40	0.003	0.001	−0.002
3	**−0.022**	**−0.010**	**−0.234**	41	0.000	0.000	−0.003
4	**−0.030**	**−0.013**	**−0.254**	42	0.002	0.001	0.005
5	**−0.029**	**−0.013**	**−0.248**	43	0.014	0.006	0.005
6	**−0.030**	**−0.013**	**−0.240**	44	0.006	0.003	0.010
7	**−0.031**	**−0.014**	**−0.248**	45	0.000	0.000	0.000
8	**−0.026**	**−0.011**	**−0.227**	46	0.005	0.002	0.008
9	**−0.026**	**−0.011**	**−0.242**	47	0.001	0.000	0.001
10	**−0.019**	**−0.008**	**−0.226**	48	0.000	0.000	−0.003
11	**−0.039**	**−0.017**	**−0.290**	49	0.004	0.002	−0.004
12	**−0.012**	**−0.005**	**−0.272**	50	0.001	0.000	−0.008
13	**−0.041**	**−0.018**	**−0.319**	51	0.005	0.002	0.003
14	0.260	0.101	**−0.391**	52	0.000	0.000	0.000
15	0.006	0.003	−0.005	53	0.009	0.004	0.017
16	0.012	0.005	0.010	54	0.017	0.007	0.010
17	0.004	0.002	0.002	55	0.006	0.003	0.001
18	0.000	0.000	−0.003	56	0.015	0.007	0.005
19	0.006	0.003	−0.002	57	0.000	0.000	0.000
20	0.009	0.004	−0.005	58	0.013	0.006	0.001
21	0.002	0.001	−0.004	59	0.000	0.000	−0.004
22	0.006	0.003	−0.004	60	0.002	0.001	0.005
23	0.008	0.003	0.002	61	0.002	0.001	−0.008
24	0.009	0.004	0.004	62	0.000	0.000	0.000
25	0.000	0.000	0.000	63	0.015	0.006	0.009
26	0.004	0.002	−0.007	64	0.002	0.001	−0.009
27	0.001	0.000	−0.008	65	0.003	0.001	0.004
28	0.001	0.000	0.000	66	0.003	0.001	0.000
29	0.000	0.000	−0.001	67	0.003	0.001	−0.001
30	0.006	0.003	0.011	68	0.001	0.001	0.002
31	0.002	0.001	0.000	69	0.001	0.000	0.001
32	0.010	0.004	0.005	70	0.000	0.000	−0.004
33	0.009	0.004	−0.001	71	0.002	0.001	−0.006
34	0.000	0.000	0.001	72	0.002	0.001	−0.003
35	0.000	0.000	−0.008	73	0.000	0.000	−0.002
36	0.000	0.000	−0.005	74	0.002	0.001	−0.002
37	0.002	0.001	−0.006	75	0.005	0.002	0.009
38	0.003	0.001	0.002				

To investigate the effect of collinearity-influential observations on the multi-collinearity structure of the data, multi-collinearity diagnostics are computed including pair-wise correlation coefficients, variance inflation factors, and condition indices. These results are presented in Table 11.3.

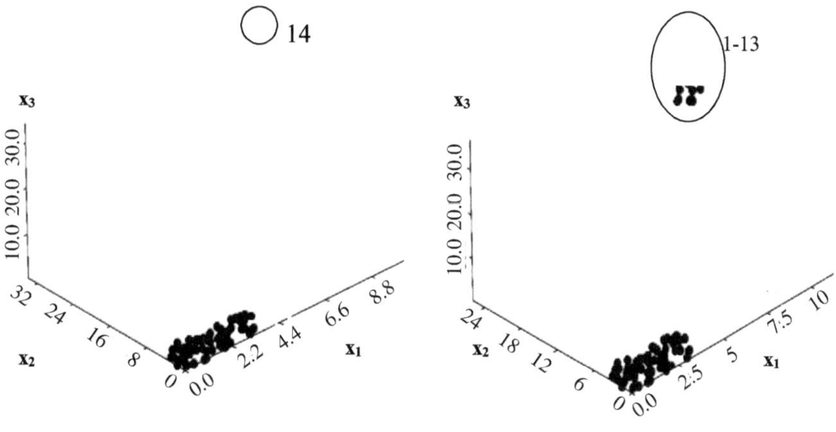

(a) (b)

Figure 11.2 Scatter plots of Hawkins-Bradu-Kass data. (a) Dataset without the first 13 high-leverage points, (b) Data set without the 14th high-leverage point.

The data in the table show that the multi-collinearity problem of this data is reflected in the pair-wise Pearson correlation coefficients, Variance Inflation Factors (VIF) and CN values. Since scaling reduces some of the multi-collinearity between explanatory variables, the CN value (13.586) may be affected. Due to this, the CN fails to detect the correct degree of multi-collinearity in this dataset. It can be seen from the result of Table 11.3 and Figure 11.2b that after the deletion of 14 high-leverage points, this dataset has no multi-collinearity problem, confirming that these leverage points are also collinearity-influential observations. Moreover, after deletion of the first 13 high-leverage points, the 14th observation still induces multi-collinearity for the dataset.

Table 11.3 Multi-collinearity diagnostics for Hawkins-Bradu-Kass dataset.

Diagnostics	Status	1	2	3
Pearson correlation coefficient	Original data	$r_{12}=0.946$	$r_{13}=0.962$	$r_{23}=0.979$
	Without observations 1–14	$r_{12}=0.044$	$r_{13}=0.107$	$r_{23}=0.127$
	Without observations 1–13	$r_{12}=0.734$	$r_{13}=0.746$	$r_{23}=0.947$
	Without observation no. 14	$r_{12}=0.954$	$r_{13}=0.959$	$r_{23}=0.988$
VIF > 5	Original data	13.432	23.853	33.432
	Without observations 1–14	1.012	1.017	1.027
	Without observations 1–13	2.295	9.795	10.170
	Without observation no. 14	12.795	41.916	47.155
Condition index of **X** matrix > 10	Original data	13.586	7.839	1.000
	Without observations 1–14	3.275	2.946	1.000
	Without observations 1–13	8.055	3.210	1.000
	Without observation no. 14	17.125	7.848	1.000

This multi-collinearity is especially evident from the large values of VIF. Because scaling reduces some of the multi-collinearity between explanatory variables, it may cause the CN value (8.055) to fail in detecting the presence of multi-collinearity in this dataset. Furthermore, in the absence of the 14th observation, the dataset still has multi-collinearity problem, indicated by the fact that the values of CN and VIF are larger than their cutoff points. It can be concluded that the effects of the first 13 high-leverage points on δ_i and l_i cause these two measures to wrongly indicate that the 14th observation reduces collinearity. In addition, the dataset suffers from multi-collinearity problems without the 14th high-leverage point. The results also indicate that the degree of the multi-collinearity problem has been reduced in the presence of just the 14th high-leverage point.

Commercial Properties Data-set: The next example is the commercial properties data introduced in Section 11.3.3. Nineteen high-leverage points (observations 1, 2, 3, 6, 7, 8, 17, 21, 26, 29, 37, 45, 53, 54, 58, 61, 62, 72, and 79) are identified in this dataset when the DRGP(MVE) is applied which is also denoted as p^*_{ii}. As mentioned earlier, Hadi (1988) pointed out that collinearity-influential observations are usually points with high-leverages. He also noted that not all high-leverage points are collinear-influential. It will be presented in the following results (Figure 11.3 and Table 11.4) that these high-leverage points are not collinearity-influential. To make the dataset collinear in each of three explanatory variables, the values of high-leverage points at observations 1, 2, 3, 6, 7, and 8 were modified to 300, 200, 100, 300, 200, and 100, respectively. Figure 11.3 presents the original and modified commercial

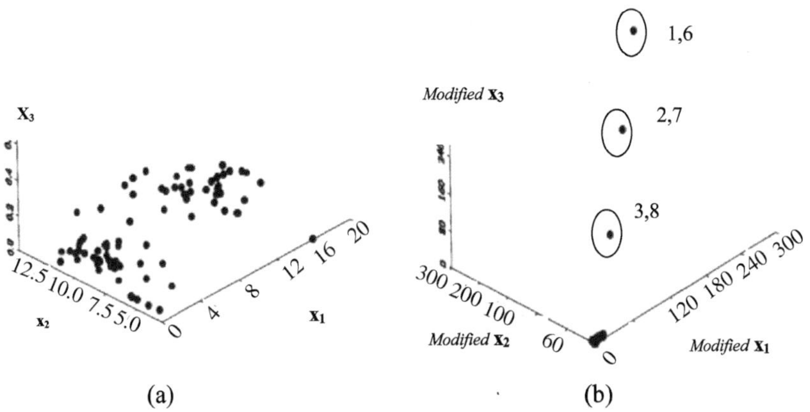

Figure 11.3 Scatter plots of the original and the modified commercial properties data. (a) Original data, (b) Modified data.

Table 11.4 Collinearity-influential measures for the original commercial properties dataset.

Ind	$\delta_{i(-0.030)}$ (−0.022)	$l_{i(-0.013)}$ (−0.010)	$\delta^{(D)}_{i(-0.018)}$ (−0.010)	Ind	$\delta_{i(-0.013)}$ (−0.022)	$l_{i(-0.013)}$ (−0.010)	$\delta^{(D)}_{i(-0.018)}$ (−0.010)
1	0.000	0.000	0.005	42	−0.003	−0.001	0.000
2	0.001	0.000	0.007	43	0.001	0.001	0.001
3	0.019	0.008	0.010	44	−0.007	−0.003	−0.003
4	0.004	0.002	0.002	45	−0.002	−0.001	0.005
5	−0.006	−0.003	−0.003	46	0.012	0.005	0.006
6	−0.003	−0.001	0.004	47	−0.001	0.000	0.000
7	−0.001	0.000	0.007	48	0.004	0.002	0.002
8	0.002	0.001	0.037	49	0.009	0.004	0.004
9	0.004	0.002	0.002	50	0.012	0.005	0.006
10	−0.003	−0.001	−0.002	51	0.019	0.008	0.010
11	−0.011	−0.005	−0.006	52	0.015	0.006	0.007
12	0.013	0.006	0.007	53	0.003	0.001	0.036
13	0.009	0.004	0.005	54	0.001	0.000	0.012
14	−0.011	−0.005	−0.006	55	0.008	0.004	0.004
15	0.015	0.006	0.007	56	−0.009	−0.004	−0.005
16	0.021	0.009	0.011	57	−0.009	−0.004	−0.003
17	0.002	0.001	0.005	58	0.001	0.001	0.005
18	−0.009	−0.004	−0.003	59	−0.011	−0.005	−0.006
19	−0.008	−0.003	−0.003	60	0.007	0.003	0.005
20	0.005	0.002	0.002	61	0.015	0.007	0.043
21	−0.002	−0.001	0.001	62	0.001	0.001	0.011
22	0.013	0.006	0.007	63	−0.010	−0.004	−0.004
23	0.006	0.003	0.003	64	−0.012	−0.005	−0.005
24	0.006	0.003	0.003	65	−0.011	−0.005	−0.006
25	−0.009	−0.004	−0.004	66	−0.007	−0.003	−0.001
26	−0.001	0.000	0.021	67	−0.009	−0.004	−0.004
27	−0.007	−0.003	−0.004	68	−0.013	−0.006	−0.007
28	−0.009	−0.004	−0.005	69	0.002	0.001	0.001
29	−0.001	−0.001	0.010	70	0.008	0.003	0.004
30	−0.008	−0.004	−0.004	71	−0.007	−0.003	−0.003
31	0.006	0.002	0.003	72	0.000	0.000	0.011
32	0.005	0.002	0.003	73	−0.013	−0.006	−0.007
33	0.011	0.005	0.005	74	0.004	0.002	0.002
34	−0.002	−0.001	−0.001	75	−0.006	−0.002	−0.001
35	−0.008	−0.003	−0.004	76	0.008	0.004	0.004
36	−0.009	−0.004	−0.004	77	0.009	0.004	0.005
37	−0.001	0.000	0.010	78	0.002	0.001	0.002
38	−0.004	−0.002	−0.002	79	−0.012	−0.005	−0.005
39	0.003	0.002	0.002	80	−0.008	−0.003	−0.004
40	−0.003	−0.001	0.000	81	−0.011	−0.005	−0.006
41	0.004	0.002	0.002				

properties dataset. Figure 11.3a shows that the original dataset follows no special pattern, while the multi-collinearity pattern of the dataset depicted in Figure 11.3b is obvious. Table 11.4 demonstrates the collinearity-influential measures for the original Commercial Properties dataset. The results of this table show that none of the high-leverage points cause multi-collinearity for this dataset, evidenced by the small values of δ_i, l_i, and $\delta_i^{(D)}$. However, modification of the dataset changes the results dramatically, as shown in Figure 11.3b. This change in the dataset results in it becoming collinear. Table 11.4 presents the leverage diagnostics for the modified Commercial Properties dataset.

As in the results shown in Table 11.5, 19 high-leverage points (observations 1, 2, 3, 6, 7, 8, 17, 21, 26, 29, 37, 45, 53, 54, 58, 61, 62, 72, and 79) are detected based on p^*_{ii}, while h_{ii} and p_i detect fewer high-leverage points for modified dataset (Table 11.5). Table 11.6 presents the collinearity-influential measures for modified Commercial Properties dataset. It can be seen from Table 11.6 that the values of δ_i and l_i exceed the $CEM - cut1$ value for all the 6 added high-leverage collinearity-influential observations. Hence, δ_i and l_i detect observations 1, 2, 3, 6, 7, and 8 as high-leverage collinearity-enhancing observations with the proposed cutoff point, $CEM - cut1$. While, δ_i and l_i fail to detect all modified observations as high-leverage collinearity-enhancing observations when $CEM - cut2$ is applied. The new proposed collinearity-influential measure $\delta_i^{(D)}$ correctly identifies all the added high-leverage points as collinearity-enhancing observations, regardless of which cutoff point is used. Moreover, the values of $\delta_i^{(D)}$ are larger in magnitude compared to δ_i and l_i. Hence, the new proposed measure strongly proves that these points are high-leverage collinearity-enhancing observations.

Multi-collinearity diagnostic methods, including correlation matrix, variance inflation factors, and condition indices of the scaled explanatory variables for the original and the modified Commercial Properties dataset are presented in Table 11.7. These measures show that although the original dataset contains 19 high-leverage points, these leverage points may not cause a multi-collinearity problem. These results agree reasonably well with the results shown in Figure 11.3a and Table 11.6. This confirms that these 19 high-leverage points are not collinearity-enhancing observations. From Table 11.6, it can be seen that the values of VIF and CN (24.152) for the modified data indicate a multi-collinearity problem. Thus, these new high-leverage points are collinearity-enhancing observations.

Table 11.5 Leverage diagnostics for the modified commercial properties dataset.

Ind	h_{ii} (0.099)	p_i (0.093)	p^*_{ii} (0.197)	Ind	h_{ii}(0.099)	p_i (0.093)	p^*_{ii} (0.197)
1	0.337	0.509	1293746.496	42	0.048	0.050	0.075
2	0.147	0.172	575027.609	43	0.074	0.079	0.185
3	0.098	0.109	143778.906	44	0.032	0.033	0.054
4	0.023	0.024	0.038	45	0.053	0.056	0.457
5	0.021	0.022	0.058	46	0.037	0.039	0.054
6	0.337	0.509	1293746.496	47	0.050	0.052	0.116
7	0.147	0.172	575027.609	48	0.044	0.046	0.104
8	0.098	0.109	143778.906	49	0.044	0.046	0.052
9	0.039	0.040	0.090	50	0.049	0.051	0.075
10	0.022	0.023	0.025	51	0.061	0.065	0.088
11	0.056	0.059	0.098	52	0.040	0.042	0.056
12	0.042	0.044	0.049	53	0.049	0.052	4.725
13	0.025	0.026	0.040	54	0.038	0.040	0.918
14	0.033	0.034	0.040	55	0.025	0.026	0.039
15	0.048	0.051	0.057	56	0.029	0.030	0.033
16	0.055	0.059	0.076	57	0.039	0.041	0.070
17	0.028	0.029	0.288	58	0.033	0.034	0.315
18	0.019	0.019	0.149	59	0.028	0.029	0.032
19	0.035	0.036	0.061	60	0.027	0.028	0.061
20	0.021	0.021	0.032	61	0.068	0.073	7.824
21	0.037	0.038	0.215	62	0.035	0.036	0.714
22	0.044	0.046	0.064	63	0.046	0.048	0.086
23	0.026	0.027	0.043	64	0.040	0.041	0.082
24	0.021	0.021	0.036	65	0.030	0.031	0.036
25	0.026	0.027	0.052	66	0.059	0.063	0.103
26	0.061	0.064	1.686	67	0.045	0.047	0.052
27	0.033	0.034	0.039	68	0.040	0.041	0.066
28	0.034	0.035	0.041	69	0.058	0.062	0.142
29	0.025	0.026	0.668	70	0.026	0.026	0.037
30	0.035	0.036	0.078	71	0.036	0.037	0.043
31	0.024	0.025	0.050	72	0.028	0.029	0.736
32	0.020	0.020	0.037	73	0.039	0.041	0.049
33	0.043	0.045	0.050	74	0.022	0.023	0.031
34	0.054	0.057	0.068	75	0.061	0.064	0.101
35	0.035	0.036	0.037	76	0.025	0.025	0.039
36	0.022	0.022	0.091	77	0.027	0.028	0.041
37	0.051	0.053	0.724	78	0.068	0.073	0.118
38	0.039	0.040	0.050	79	0.030	0.031	0.206
39	0.020	0.020	0.040	80	0.018	0.019	0.019
40	0.049	0.052	0.078	81	0.032	0.033	0.040
41	0.040	0.041	0.092				

Table 11.6 Collinearity-influential measures for the modified commercial properties data.

Ind	δ_i(−0.013) (−0.095)	l_i(−0.005) (−0.042)	$\delta_i^{(D)}$(−0.019) (−0.012)	Ind	δ_i(−0.013) (−0.095)	l_i(−0.005) (−0.042)	$\delta_i^{(D)}$(−0.019) (−0.012)
1	**−0.172**	**−0.082**	**−0.603**	42	0.008	0.003	0.000
2	**−0.073**	**−0.033**	**−0.441**	43	0.002	0.001	0.001
3	**−0.018**	**−0.008**	**−0.203**	44	0.004	0.002	−0.003
4	0.008	0.003	0.002	45	0.001	0.000	0.005
5	0.001	0.000	−0.003	46	0.013	0.006	0.006
6	**−0.172**	**−0.082**	**−0.603**	47	0.000	0.000	0.000
7	**−0.073**	**−0.033**	**−0.441**	48	0.004	0.002	0.002
8	**−0.018**	**−0.008**	**−0.203**	49	0.013	0.005	0.004
9	0.005	0.002	0.002	50	0.013	0.006	0.006
10	0.002	0.001	−0.002	51	0.021	0.009	0.010
11	0.001	0.000	−0.006	52	0.015	0.007	0.007
12	0.016	0.007	0.007	53	0.004	0.002	0.036
13	0.010	0.004	0.005	54	0.015	0.007	0.012
14	0.000	0.000	−0.006	55	0.009	0.004	0.004
15	0.018	0.008	0.007	56	0.003	0.001	−0.005
16	0.022	0.009	0.011	57	0.005	0.002	−0.003
17	0.008	0.004	0.005	58	0.006	0.003	0.005
18	0.000	0.000	−0.003	59	0.000	0.000	−0.006
19	0.005	0.002	−0.003	60	0.011	0.005	0.005
20	0.007	0.003	0.002	61	0.003	0.001	0.043
21	0.008	0.003	0.001	62	0.009	0.004	0.011
22	0.015	0.006	0.007	63	0.006	0.003	−0.004
23	0.010	0.004	0.003	64	0.001	0.001	−0.005
24	0.007	0.003	0.003	65	0.000	0.000	−0.006
25	0.001	0.000	−0.004	66	0.012	0.005	−0.001
26	0.003	0.001	0.021	67	0.007	0.003	−0.004
27	0.004	0.002	−0.004	68	0.001	0.001	−0.007
28	0.004	0.002	−0.005	69	0.002	0.001	0.001
29	0.006	0.003	0.010	70	0.010	0.005	0.004
30	0.005	0.002	−0.004	71	0.005	0.002	−0.003
31	0.006	0.003	0.003	72	0.009	0.004	0.011
32	0.006	0.003	0.003	73	0.000	0.000	−0.007
33	0.014	0.006	0.005	74	0.007	0.003	0.002
34	0.010	0.004	−0.001	75	0.014	0.006	−0.001
35	0.005	0.002	−0.004	76	0.009	0.004	0.004
36	0.001	0.000	−0.004	77	0.010	0.004	0.005
37	0.003	0.002	0.010	78	0.015	0.007	0.002
38	0.005	0.002	−0.002	79	0.001	0.001	−0.005
39	0.006	0.003	0.002	80	0.000	0.000	−0.004
40	0.008	0.003	0.000	81	0.000	0.000	−0.006
41	0.005	0.002	0.002				

Table 11.7 Multi-collinearity diagnostics for commercial properties data.

Diagnostics	Status	1	2	3
Pearson correlation coefficient	original data	$r_{12}=0.389$	$r_{13}=-0.253$	$r_{23}=-0.380$
	modified data	$r_{12}=0.995$	$r_{13}=0.993$	$r_{23}=0.999$
VIF > 5	original data	1.196	1.309	1.186
	modified data	95.295	559.959	441.488
Condition index of **X** matrix > 10	original data	3.389	1.597	1.000
	modified data	24.152	13.466	1.000

11.4.2.2 *Monte Carlo Simulation Study*

In this section, a Monte Carlo simulation study that is designed to assess the performance of the new proposed measure, $\delta_i^{(D)}$, in detecting multiple high-leverage collinearity-influential observations are reported. Moreover, its performance with that of other commonly used measures is compared. Simulated datasets with 3 and 5 independent regressors were generated. The first 100% $(1 - \alpha)$ observations of the independent regressor variables for sample sizes $n = 20, 40, 60, 100$, and 300 were generated as Uniform (0, 1), and the first $100\alpha\%$ observations were constructed as high-leverage points. Several cases with different percentages of high-leverages (5, 10, 15, 20, and 25) with unequal weights are considered. Following the idea of Habshah et al. (2009) to generate the high-leverage points with unequal weights, the values corresponding to the first high-leverage point are kept fixed at 10 and those of the successive values are created by multiplying the observations index, i by 10. The three diagnostic measures namely, δ_i, l_i, and $\delta_i^{(D)}$, with two different cutoff points in each case, are then applied to the simulated datasets. 10,000 simulations are run for each of the five sample sizes and the results based on their averages are presented in Tables 11.8 to 11.10. In these tables, α and HLCIO indicates the percentages and the numbers of added high-leverage collinearity-enhancing observations to the simulated datasets. Furthermore, the number of high-leverage points which are identified by DRGP(MVE) is denoted as HL. The CN^* and also CN^{**} indicates the condition number of **X** matrix with and without high-leverage collinearity-enhancing observations, respectively. Moreover, $\text{Cut}(\theta_i)^1$ and $\text{Cut}(\theta_i)^2$ signifies $CEM - cut1$ and $CEM - cut2$ for θ_i.

Tables 11.8 to 11.10 clearly show the merit of the new proposed multiple high-leverage collinearity-influential measure. Furthermore, no other measures that were considered in this experiment performed satisfactorily. It is interesting to point out that when $\alpha\%$ high-leverage colinearity-enhancing observations are added to the simulated dataset, the percentage of the points indicated by p_{ii}^* as high-leverages is more than $\alpha\%$. This confirms that there are some more high-leverage points in the simulated datasets than the added high-leverage points. The added multiple high-leverage collinearity-enhancing observations may cause multi-collinearity for all of the simulated datasets; this multi-collinearity may result from the large values of condition number, CN*. Obviously, these observations were the cause of multi-collinearity, which confirmed by the small values of the condition number of the **X** matrix without these points, CN** for the simulated datasets.

Two different cutoff points which were introduced in Eqs. (28, 29) were considered for the proposed and the existing collinearity-influential measures. The results in Tables 11.8 to 11.10 indicate that when there is only a single high-leverage collinearity-influential observation, the existing δ_i and l_i measures seem to be more accurate than the new proposed measure $\delta_i^{(D)}$ with $CEM - cut1$. However, by increasing the number of explanatory variables in the simulated datasets, the accuracy of the new proposed measure $\delta_i^{(D)}$ with $CEM - cut1$ in detecting a single

Table 11.8 Collinearity-influential measures for the simulated datasets with p = 3 and 5, n = 20 and 40.

$p = 3$	n									
Measures	**20**					**40**				
α	5	10	15	20	25	5	10	15	20	25
HLCIO	1	2	3	4	5	2	4	6	8	10
HL	1.821	2.580	3.394	4.280	5.185	2.633	4.393	6.222	8.12	10.065
CN^*	16.845	38.109	65.870	100.338	141.610	24.613	61.678	110.681	171.52	244.745
CN^{**}	4.254	4.207	4.197	4.193	4.212	3.707	3.690	3.675	3.68	3.691
$Cut(\delta_i^{(D)})^1$	1.042	2.021	3.005	4.002	5.001	2.040	4.019	6.006	8.00	10.000
$Cut(\delta_i^{(D)})^2$	1.191	2.079	3.025	3.957	4.604	2.210	4.063	6.016	8.00	9.911
$Cut\ (\delta_i)^1$	1.000	1.872	2.065	2.261	2.208	2.050	3.160	4.208	4.66	4.584
$Cut\ (\delta_i)^2$	0.813	0.353	0.122	0.055	0.023	1.533	0.547	0.174	0.07	0.002
$Cut\ (l_i)^1$	1.000	1.870	2.083	2.319	2.318	2.053	3.171	4.233	4.72	4.676
$Cut\ (l_i)^2$	0.813	0.747	0.468	0.055	0.023	1.646	0.547	0.174	0.07	0.031
$p = 5$	n									
Measures	**20**					**40**				
α	5	10	15	20	25	5	10	15	20	25
HLCIO	1	2	3	4	5	2	4	6	8	10
HL	1.958	2.707	3.486	4.322	5.205	2.759	4.508	6.295	8.197	10.088
CN^*	27.142	62.349	109.026	166.930	237.470	36.485	92.138	165.893	260.264	369.972
CN^{**}	7.279	7.188	7.209	7.302	7.253	5.511	5.509	5.494	5.581	5.536
$Cut\ (\delta_i^{(D)})^1$	1.009	2.007	3.002	4.000	5.000	2.011	4.003	6.001	8.000	10.000
$Cut\ (\delta_i^{(D)})^2$	1.147	2.060	2.997	3.859	4.078	2.164	4.051	6.011	7.973	9.766
$Cut\ (\delta_i)^1$	1.000	1.907	1.981	2.050	1.794	2.020	3.102	4.098	4.497	4.281
$Cut\ (\delta_i)^2$	0.766	0.322	0.100	0.031	0.014	1.465	0.488	0.152	0.054	0.002
$Cut\ (l_i)^1$	1.000	1.945	2.016	2.134	1.966	2.023	3.111	4.123	4.555	4.400
$Cut\ (l_i)^2$	0.766	0.714	0.412	0.031	0.014	1.591	0.488	0.152	0.054	0.022

Table 11.9 Collinearity-influential measures for simulated datasets with p = 3 and 5, n = 60 and 100.

$p = 3$	n									
Measures	**60**					**100**				
α	5	10	15	20	25	5	10	15	20	25
HLCIO	3	6	9	12	15	5	10	15	20	25
HL	3.530	6.316	9.162	12.082	15.033	5.476	10.240	15.103	20.030	25.012
CN*	32.660	85.356	155.886	243.257	348.059	48.870	132.916	245.849	386.111	554.660
CN**	3.546	3.539	3.537	3.542	3.545	3.427	3.427	3.424	3.419	3.431
Cut $(\delta_i^{(D)})^1$	3.039	6.012	9.003	12.001	15.000	5.030	10.007	15.001	20.000	25.000
Cut $(\delta_i^{(D)})^2$	3.214	6.053	9.010	11.999	14.975	5.215	10.035	15.003	20.000	24.991
Cut $(\delta_i)^1$	3.045	4.963	6.350	7.048	6.962	4.689	7.979	10.592	11.795	11.725
Cut $(\delta_i)^2$	2.176	0.618	0.050	0.011	0.003	2.971	0.264	0.045	0.000	0.000
Cut $(l_i)^1$	3.050	4.967	6.369	7.092	7.049	4.702	7.981	10.614	11.826	11.804
Cut $(l_i)^2$	2.255	0.618	0.206	0.011	0.003	2.971	0.678	0.045	0.000	0.000
$p = 5$	n									
Measures	**60**					**100**				
α	5	10	15	20	25	5	10	15	20	25
HLCIO	3	6	9	12	15	5	10	15	20	25
HL	3.581	6.427	9.226	12.106	15.039	5.549	10.302	15.124	20.054	25.013
CN*	46.832	123.344	225.394	354.007	508.453	68.391	186.617	346.681	544.213	785.278
CN**	5.020	5.077	5.071	5.096	5.096	4.721	4.731	4.756	4.746	4.780
Cut$(\delta_i^{(D)})^1$	3.004	6.003	9.000	12.000	15.000	5.011	10.002	15.000	20.000	25.000
Cut$(\delta_i^{(D)})^2$	3.159	6.060	9.001	12.000	14.947	5.194	10.040	15.004	20.000	24.945
Cut$(\delta_i)^1$	2.899	4.921	6.228	6.818	6.654	4.619	7.948	10.507	11.564	11.387
Cut$(\delta_i)^2$	2.086	0.554	0.035	0.006	0.003	2.818	0.228	0.043	0.001	0.000
Cut$(l_i)^1$	2.890	4.932	6.247	6.874	6.755	4.635	7.955	10.529	11.616	11.484
Cut$(l_i)^2$	2.181	0.554	0.169	0.006	0.003	2.818	0.624	0.043	0.001	0.000

high-leverage collinearity-enhancing observation increase. It is important to note that by increasing the number of added high-leverage collinearity-influential observations, the new proposed measure $\delta_i^{(D)}$ becomes more precise in detecting the exact number of multiple high-leverage collinearity-influential observations in the datasets. The result of Tables 11.8 to 11.10 reveal that the use of $CEM - cut1$ for $\delta_i^{(D)}$ by increasing the sample size and also the number of the regressor variables is more practical because it can detect exactly the number of added multiple high-leverage collinearity-influential observations. Moreover $CEM - cut2$, is not as efficient as $CEM - cut1$ to be applied as a cutoff point for high leverage collinearity-influential measure. Hence, the results of Tables 11.8 to 11.10 clearly reveal the merit of the new proposed measure $\delta_i^{(D)}$ with $CEM - cut1$ in detecting multiple high-leverage collinearity-influential observations.

Table 11.10 Collinearity-influential measures for simulated datasets with $p = 3$ and 5, $n = 300$.

Measures	$p = 3$					$p = 5$				
α	5	10	15	20	25	5	10	15	20	25
HLCIO	15	30	45	60	75	15	30	45	60	75
HL	15.466	30.137	45.036	60.007	75.000	15.598	30.258	45.09	60.018	75.001
CN^*	129.659	368.829	691.435	1093.824	1578.910	175.014	497.506	934.98	1481.497	2136.034
CN^{**}	3.290	3.295	3.294	3.300	3.306	4.367	4.365	4.37	4.380	4.387
$Cut(\delta_i^{(D)})^1$	15.017	30.000	45.000	60.000	75.000	15.003	30.000	45.00	60.000	75.000
$Cut(\delta_i^{(D)})^2$	15.287	30.011	45.000	60.000	75.000	15.328	30.042	45.00	60.000	74.925
$Cut(\delta_i)^1$	12.984	23.457	31.734	35.487	35.602	12.974	23.416	31.79	35.470	35.499
$Cut(\delta_i)^2$	2.837	0.023	0.000	0.000	0.000	2.642	0.032	0.000	0.000	0.000
$Cut(l_i)^1$	12.984	23.467	31.748	35.526	35.674	12.974	23.425	31.81	35.507	35.581
$Cut(l_i)^2$	2.837	0.023	0.000	0.000	0.000	2.642	0.032	0.000	0.000	0.000

11.5 REGRESSION DIAGNOSTIC PLOTS FOR HIGH-LEVERAGE COLLINEARITY-INFLUENTIAL OBSERVATIONS

Identifying outliers and high-leverage points is a fundamental step in the least squares regression model-building process. Graphical tools are one of the easiest ways to detect the abnormal points in a dataset. For reviewing the existing influential observations diagnostic plots, one can refer to Andersen (2008).

There are two most practical plots in this area which were proposed by Rousseeuw and van Zomeren (1990) and Rousseeuw and van Driessen (1999). An influence plot of the robust residuals against robust distances or Residual-Distance (RD) plot (also denoted as Residual-Leverage plot by Habshah et al. 2009) which detects multiple outliers more accurately than the traditional methods is proposed by Rousseeuw and van Zomeren (1990) (see also Cook and Hawkins 1990, Ruppert and Simpson 1990, Kempthorne and Mendel 1990, Olive 2008, Andersen 2008 for a debate about this topic). This plot classifies the dataset to clean data, bad and good leverage points, and vertical outliers. Since efficiency is not a concern for the regression diagnostic plots such as Rousseeuw and van Zomeren's RD plot, robust residuals which come from the highly resistant Least Median of Squares (LMS) or Least Trimmed Squares (LTS) regressions are often employed (Andersen 2008). The absolute robust residuals which are more than 2.5 will be considered as vertical outliers. The Robust Mahalanobis Distance based on MVE, that is RMD(MVE), is defined by Rousseeuw (1985) as

$$RMD_i(MVE) = \sqrt{(\mathbf{X} - T_R(\mathbf{X}))' C_R(\mathbf{X})^{-1} (X - T_R(\mathbf{X}))} \quad i = 1, 2, ..., n \quad (31)$$

where $T_R(\mathbf{X})$ and $C_R(\mathbf{X})$ are robust locations and shape estimates of the MVE is introduced by Rousseeuw (1983, 1984), respectively. The squared value of Robust Mahalanobis Distance based on MVE in Equation (31), that is $\text{RMD}^2(\text{MVE})$, is used in the RD plot. The $\text{RMD}^2(\text{MVE})$ which exceeds 0.95% point of the chi-square distribution with degrees of freedom equal to the number of explanatory variables in the model are considered as high-leverage points. The DD plot which is introduced by Rousseeuw and Van Driessen (1999), plots the squared values of the classical Mahalanobis distance, introduced by Rousseeuw and Leroy (1987), against the squared values of Robust Mahalanobis distance based on Minimum Covariance Determinant (MCD), $\text{RMD}^2(\text{MCD})$. The $T_R(\mathbf{X})$ and $C_R(\mathbf{X})$ are robust locations and shape estimates of the MCD in the $\text{RMD}^2(\text{MCD})$ which is introduced in Equation (31). The points which values for both classical and robust Mahalanobis distance exceed 0.95%

point of the chi-squared distribution with degrees of freedom equal to the number of explanatory variables in the model are considered as high-leverage points.

In this section, three different plots are proposed following the idea of Rousseeuw and van Zomeren (1990) and Rousseeuw and van Driessen (1999). Habshah et al. (2009) proposed the latest high-leverage diagnostic method, namely the DRGP(MVE). Since it is more accurate than the RMD^2(MVE) in diagnosing multiple high-leverage points (Habshah et al. 2009, Norazan 2008) the first and the second proposed plots would be based on DRGP(MVE). Following the idea of Rousseeuw and Leroy (1987), the LMS residuals are employed in developing the new proposed diagnostic plots. We call the first proposed plot as SR(LMS)-DRGP plot. Firstly, the LMS residuals, r_i for $i = 1, 2, \ldots, n$, are standardized by $\hat{\tau} = 1.4826 \times \left(\left(1 + \frac{5}{n-p-1} \right) median\,(\,|r_i|\,) \right)$, following the idea of Wilcox (2005) in developing the GM6 algorithm. The SR(LMS)-DRGP plots the standardized LMS residuals against the DRGP(MVE). In SR(LMS)-DRGP plot, any observation which exceed the **y**-axis boundaries (±2.5), **x**-axis boundaries (median (p^*_{ii}) + cMAD$(p^*_{ii})$where p^*_{ii} is the value of DRGP(MVE)), or both **y**-axis and **x**-axis boundaries are called vertical outliers, good leverage points, and bad leverage points, respectively.

The second proposed plot is based on the new developed diagnostic measure for identification of multiple high-leverage collinearity-influential observations, HLCIM(DRGP), denoted as $\delta_i^{(D)}$ introduced in Eq. (30). This plot is called the DRGP-HLCIM plot. It plots the DRGP(MVE) against the High-Leverage Collinearity-Influential Measure.

The third proposed plot is also based on HLCIM(DRGP). This plot is called the SR(LMS)-HLCIM plot. In this plot, the Standardized LMS Residuals are plotted against the High-Leverage Collinearity-Influential Measure.

Figures 11.4, 11.5, and 11.6 show the Venn diagram or Ballentine view of the SR(LMS)-DRGP, the DRGP-HLCIM, and the SR(LMS)-HLCIM plots, respectively. It is important to note that the proposed cutoff points are as follows:

$$cut_{DRGP(MVE)} = median(p^*_{ii}) + c\,MAD(p^*_{ii})$$

where p^*_{ii} is DRGP(MVE). If the proposed $\delta_i^{(D)}$ in Equation (30) is employed, then

$$cut^1_{HLCIM(DRGP)} = median(\delta_i^{(D)}) - c\,MAD(\delta_i^{(D)})$$

where $\delta_i^{(D)} < 0$ and

$$cut^2_{HLCIM(DRGP)} = median(\delta_i^{(D)}) - c\,MAD(\delta_i^{(D)})$$

where $\delta_i^{(D)} > 0$. The $\text{cut}^1{}_{\text{HLCIM(DRGP)}}$ is the $CEM - cut1$ for $\delta_i^{(D)}$ while $\text{cut}^2{}_{\text{HLCIM(DRGP)}}$ is the cutoff point which can be obtained from the results of Section 11.4. It is obvious that any observations with absolute $\delta_i^{(D)}$ values exceeding the absolute value of $\text{cut}^1{}_{\text{HLCIM(DRGP)}}$ or absolute value of $\text{cut}^2{}_{\text{HLCIM(DRGP)}}$ will be considered as high-leverage collinearity-enhancing or collinearity-reducing observations.

Figure 11.4 separates the dataset to different groups, specifically clean data, vertical (or regression) outliers, and good or bad leverage points. This figure clusters the dataset according to whether the observation is the high-leverage point and/or vertical outliers. Nevertheless, it does not take care of the multi-collinearity pattern of a dataset.

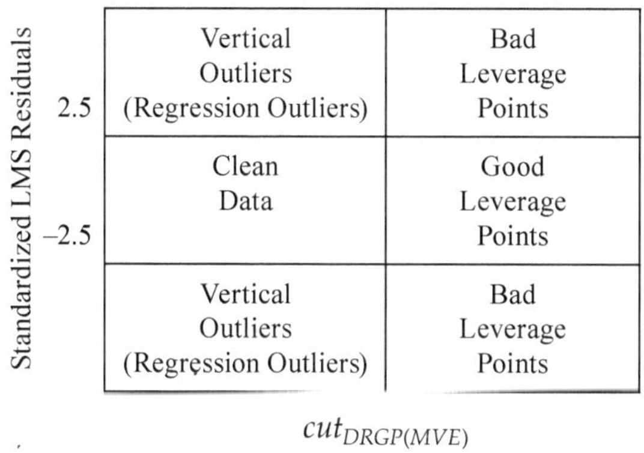

Figure 11.4 The Venn Diagram or ballentine view of SR(LMS)-DRGP plot.

Figure 11.5 separates the dataset according to whether the observation is high-leverage point or collinearity-influential observation. Hence, it classifies the dataset to clean data, high-leverage points, high-leverage collinearity-enhancing or collinearity-reducing observations, and collinearity-enhancing or collinearity-reducing observations.

This figure does not take into consideration whether the observation is abnormal in the y-direction. Finally, Figure 11.5 classifies the data into clean data, vertical outliers, good leverage collinearity-enhancing or collinearity-reducing observations/or collinearity-enhancing or collinearity-reducing observations and bad leverage collinearity-enhancing or collinearity-reducing observations/or collinearity-enhancing or collinearity-reducing observations with large residuals. One of the attractive features of this figure is that it takes into account the good leverage points which are also collinearity-influential observations. Most statisticians believe that good leverage points are not problematic since they are in the same fitted regression line as the other dataset and they decrease the standard error of the parameter estimations due to make widen the spread of **X** (see for instant Moller et al. 2005, Andersen 2008). However, these points may be collinearity-influential observations and may be destructive as much as bad leverage points for the regression analysis. The DRGP-HLCIM and SR(LMS)-HLCIM plots jointly give the most important view about the outlyingness of any points in **X**-direction or **y**-direction and also the multi-collinearity pattern of a dataset. The performance of the new proposed diagnostic plots is investigated by applying these plots to diagnose influential cases in the real well-known datasets in the following section.

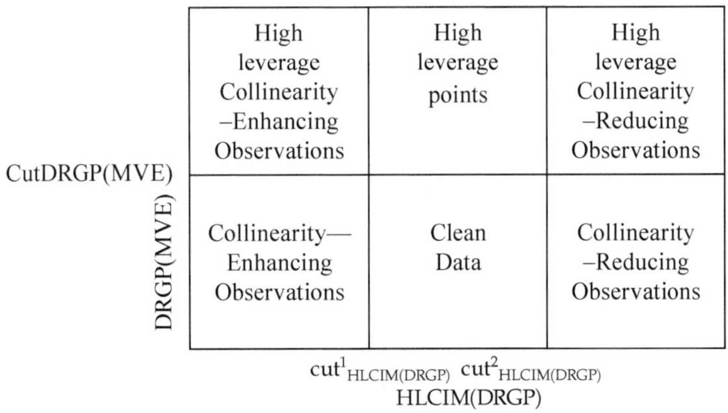

Figure 11.5 The Venn diagram or Ballentine view of DRGP-HLCIM plot.

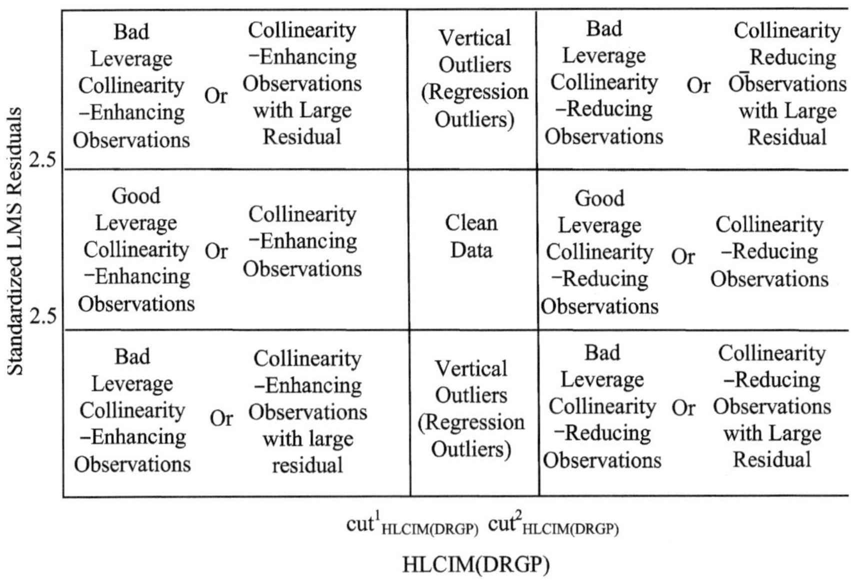

Figure 11.6 The Venn diagram or Ballentine view of SR(LMS)-HLCIM plot.

11.5.1 Numerical Results

In this section, the performance of the new proposed diagnostic plots, namely the SR(LMS)-DRGP, the DRGP-HLCIM, and the SR(LMS)-HLCIM are investigated through some well-referred datasets such as Hawkins-Bradu-Kass data, Commercial Properties data, and Body Fat dataset.

Real Data: Hawkins-Bradu-Kass data is an artificial dataset with 10 bad leverage points (cases 1–10) and 4 good leverage points (cases 11–14) (Rousseeuw and Leroy 1987, Habshah et al. 2009). Figure 11.7 presents new proposed diagnostic plots for Hawkins-Bradu-Kass dataset. According to Figure 11.7a and 11.7c, cases 11–14 not only are good leverage points but also are good leverage collinearity-enhancing observations. Moreover, cases 1–10 are bad leverage points and bad leverage collinearity-enhancing observations. It is important to note that cases 1–14 are all high-leverage collinearity-enhancing observations (Figure 11.7(b). It is important to mention that however cases 11–14 are good leverage points, they are collinearity-enhancing observations. Hence, they need more attention in the estimation of the parameters.

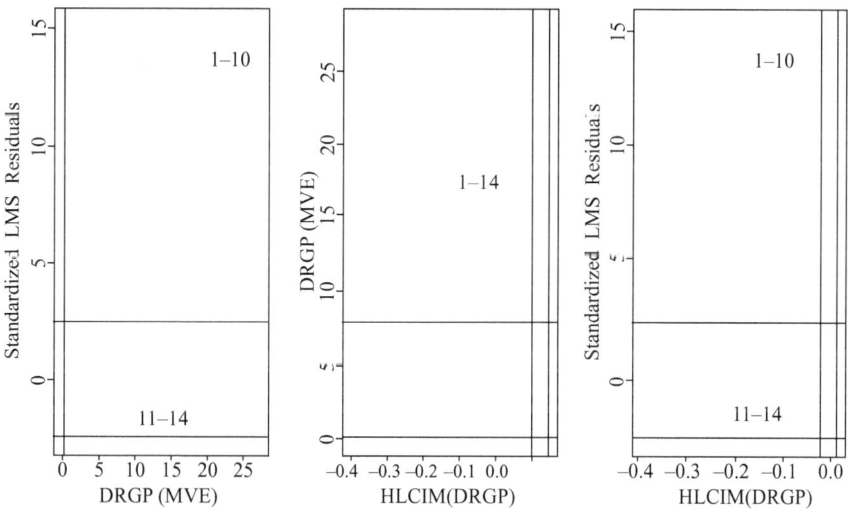

Figure 11.7 Diagnostic Q of Hawkins-Bradu-Kass dataset.

Figure 11.8 presents diagnostic plots for Hawkins-Bradu-Kass dataset without the first 14 observations. It can be observed from Figure 11.8a,b that this dataset does not have vertical outlier and high-leverage points. Nonetheless, it has one collinearity-reducing observation (case 53) which was masked in the presence of the first 14 observations.

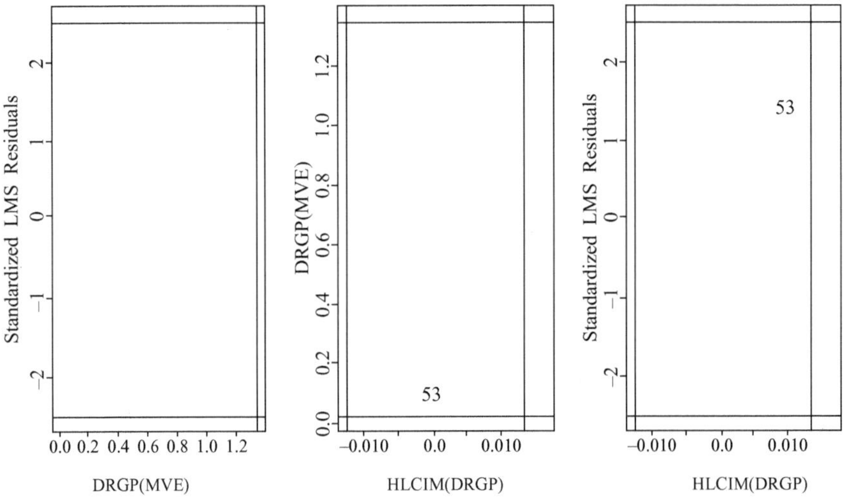

(a) SR(LMS)-DRGP (b) DRGP-HLCIM (c) SR(LMS)-HLCIM

Figure 11.8 Diagnostic plots of Hawkins-Bradu-Kass dataset without the first 14 obse.rvations.

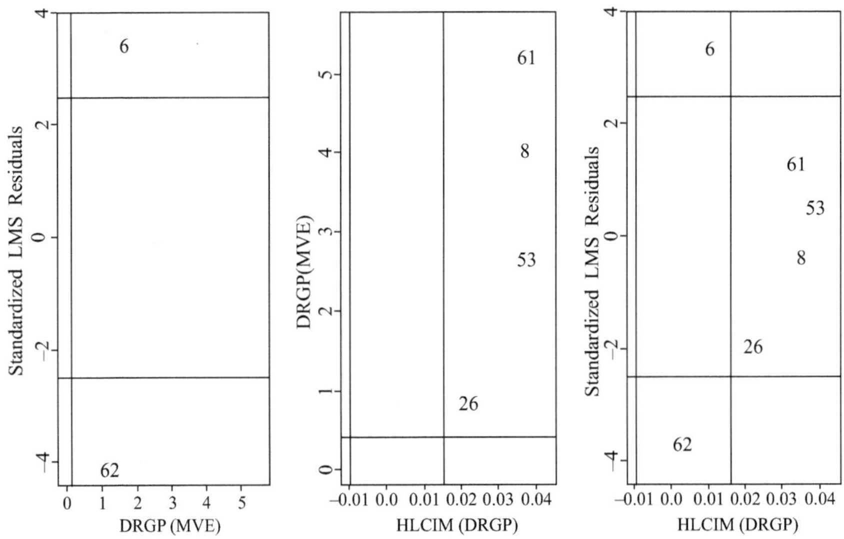

Figure 11.9 Diagnostic plots of commercial properties dataset.

Diagnostic plots for the original and modified Commercial Properties dataset are presented in Figures 11.9 and 11.10, respectively. The original dataset has 19 leverage points (observations 1, 2, 3, 6, 7, 8, 17, 21, 26, 29, 37, 45, 53, 54, 58, 61, 62, 72, and 79) while just cases 6 and 62 are bad leverage points (Figure 11.9a). Moreover, cases 9, 63, 64, 65, and

68 are vertical outliers. This dataset does not have any high-leverage collinearity-enhancing observations (Figure 11.9b). Figure 11.9b,c reveal that cases 8, 26, 53, and 61 are high-leverage collinearity-reducing observations and good leverage collinearity-reducing observations, respectively. After modifying the Commercial Properties dataset by fixing the values of high-leverage points at observations 1, 2, 3, 6, 7, and 8 to be 300, 200, 100, 300, 200, and 100 cases 1, 2, 3, 6, and 7 become good leverage points (Figure 11.10a).

Figure 11.10a also indicates case 8 as a bad leverage point. All the modified cases 1, 2, 3, 6, 7, and 8 are high-leverage collinearity-enhancing observations (Figure 11.10b). According to Figure 11.10c, case 8 is a bad leverage collinearity-enhancing observation and cases 1, 2, 3, 6, and 7 are good leverage collinearity-enhancing observations. Hence, cases 1, 2, 3, 6, and 7 deserve more attention to prevent misleading conclusions, although they are good leverage points.

Figures 11.11 to 11.12 are diagnostic plots for the original and the modified Body Fat dataset. Figure 11.11a shows that the original Body fat dataset has two good leverage points (cases 5 and 15), and two bad leverage points (cases 1 and 3). Moreover, it has three vertical outliers (cases 1, 3, and 19). However, only case 15 is high-leverage collinearity-reducing observation. Furthermore, case 13 which is not high-leverage is also collinearity-reducing observation (Figure 11.7.8b). Additionally, cases 15 and 13 are good leverage collinearity-reducing and collinearity-reducing observations, respectively (Figure 11.11c). Figures 11.12 and 11.13

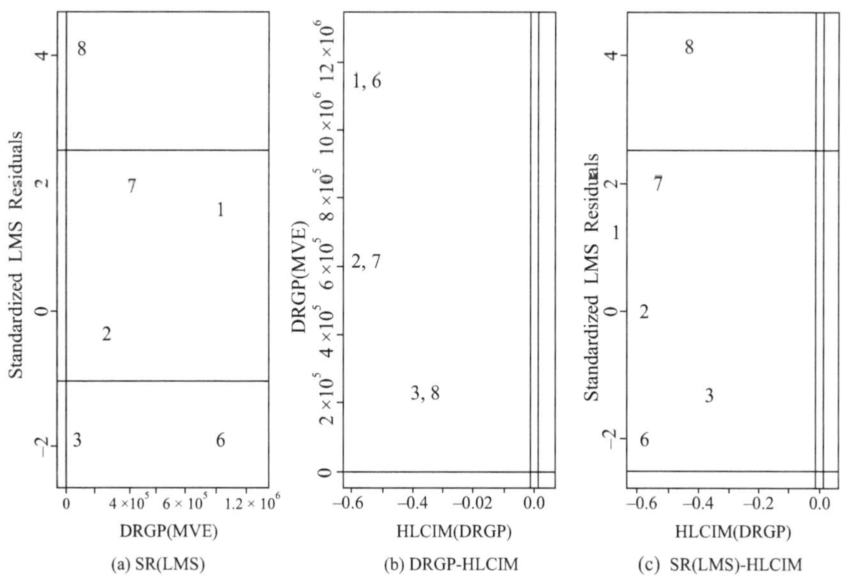

Figure 11.10 Diagnostic plots of modified commercial properties dataset.

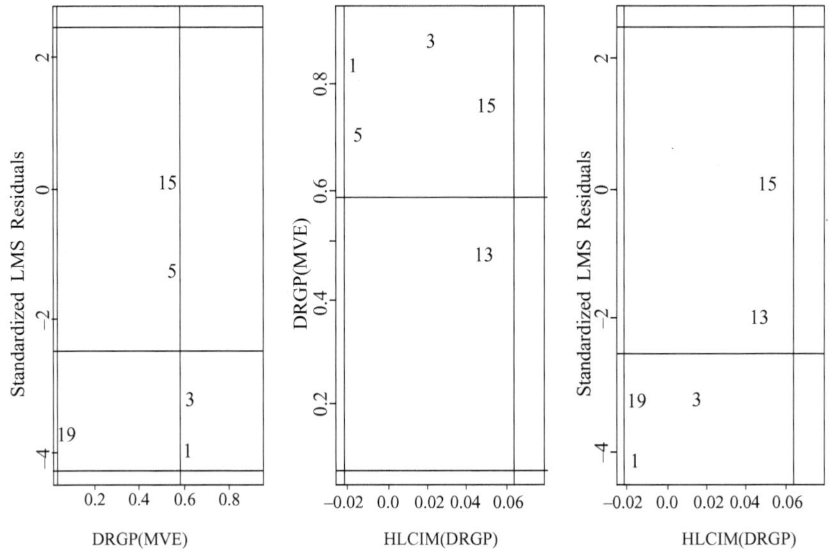

(a) SR(LMS) (b) DRGP-HLCIM (c) SR(LMS)-HLCIM

Figure 11.11 Diagnostic plots of original body fat dataset.

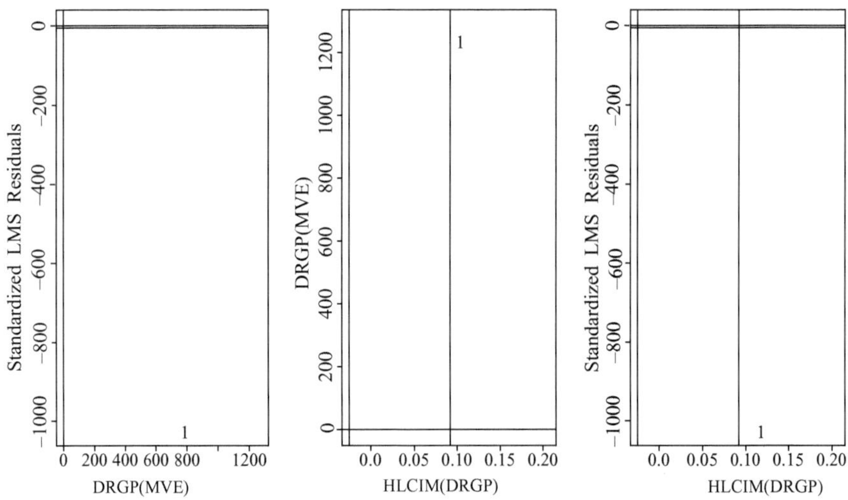

(a) SR(LMS) (b) DRGP-HLCIM (c) SR(LMS)-HLCIM

Figure 11.12 Diagnostic plots of modified x_1 body fat dataset.

illustrate the modified Body Fat dataset when the first observation of $\mathbf{x}_1$ is fixed to be 300 and when the first observation of $\mathbf{x}_1$ and $\mathbf{x}_2$ is fixed to be 300, respectively. The results of Figures 11.12 and 11.13a reveal that the added contaminated point is bad leverage point.

Moreover, when the high-leverage point only exists in $\mathbf{x}_1$, case 1 becomes high-leverage collinearity-reducing observation (Figure 11.12b). But, according to Figure 11.13b, when modification is in case 1 for $\mathbf{x}_1$ and $\mathbf{x}_2$ in the same position, it becomes high-leverage collinearity-enhancing observation. Furthermore, case 1 in Figure 11.12c is bad leverage collinearity-reducing and in Figure 11.13c is bad leverage collinearity-enhancing observation.

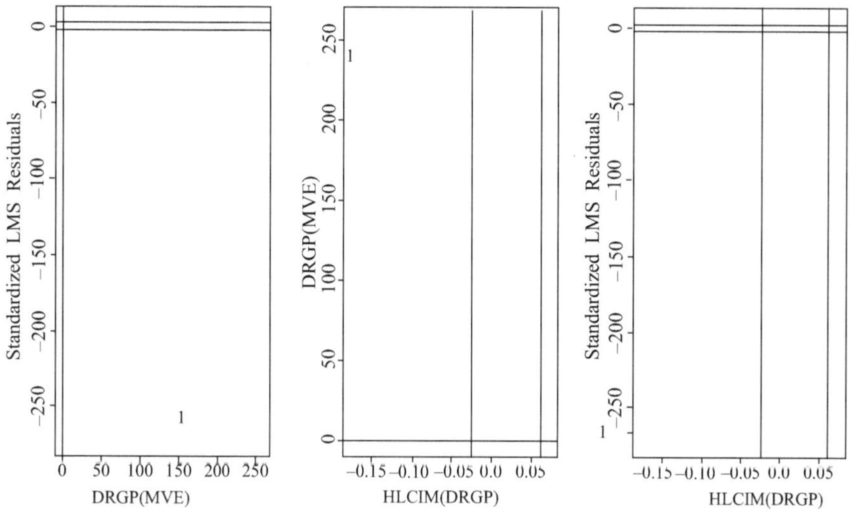

(a) SR(LMS) (b) DRGP-HLCIM (c) SR(LMS)-HLCIM

Figure 11.13 Diagnostic plots of modified x_1 and x_2 in the same positions body fat data set.

11.6 CONCLUSION

High-leverage points are abnormal points in the explanatory variables which may cause multi-collinearity in the dataset that result in interpretive problems such as wrong sign problems, producing unstable and inconsistent estimates of parameters, and insignificant regression coefficients. Thus, it is very essential to investigate and detect the presence of multi-collinearity caused by high-leverage collinearity-influential observations to reduce the destructive effects of these observations on the regression estimates. Little work has been explored in the conditions which cause any leverage points change to be collinearity-influential observations. High-leverage collinearity-influential observations which induce or hide multi-collinearity are called High-leverage collinearity-enhancing or collinearity-reducing observations in this study. One of the objectives of this chapter was to introduce High-leverage Collinearity-

Influential measures. Since, all the high-leverage points are not collinearity-influential observations, the diagnostics measure, namely the High-leverage Collinearity-Influential Measure based on DRGP(MVE), is proposed. It is denoted as HLCIM (DRGP) and finds the exact amount of multiple high-leverage collinearity-influential observations in a dataset. In addition, some practical cutoff points for the new proposed and also the existing collinearity-influential measures, were formulated. Well-referred real datasets and Monte Carlo simulation study were employed to verify the merit of the new proposed diagnostic measure. The results indicate that the new proposed measure can successfully detect the exact multiple high-leverage collinearity-enhancing observations in the datasets. Another objective of this chapter was defining three different kinds of regression diagnostic plots namely the SR(LMS)-DRGP, the DRGP-HLCIM and the SR(LMS)-HLCIM were proposed. The merit of the proposed diagnostic plots is confirmed through applying them to different real datasets. The results of the proposed regression diagnostic plots agree reasonably well with the result of the proposed High-leverage Collinearity-influential Measure, denoted as, ($\delta_i^{(D)}$) for diagnosing high-leverage collinearity-enhancing observations.

REFERENCES

Andersen, R. 2008. Modern Methods for Robust Regression. The United States of America: Sara Miller McCune. Sage Publications.

Bagheri, A. 2011. Robust Estimation Methods and Robust Multicollinearity Diagnostics for Multiple Regression Model in the Presence of High Leverage Collinearity-Influential Observations. Ph.D. thesis. Institute for Mathematical Research, Universiti Putra Malaysia (UPM), Kuala Lumpur, Malaysia.

Bagheri, A. 2015. Sample size impacts on high leverage collinearity-enhancing observations. Economics Computation and Economic Cybernetics Studies and Research, 3.

Bagheri, A. and Habshah, M. 2015. Diagnostic plot for the identification of high leverage collinearity-influential observations. Statistics and Operations Research Transactions, SORT. 39(1): January–June 2015, 51–70.

Bagheri, A., Habshah, M. and Imon, A.H.M.R. 2009. Two-Step robust diagnostic method for identification of multiple high leverage points. Journal of Mathematics and Statistics. 5(2): 97–106.

Bagheri, A., Habshah, M. and Imon, A.H.M.R. 2012. A novel collinearity-influential observation diagnostic measure based on a group deletion approach. Journal of Communication and Statistics, Simulation and Computation. 41(8): 1379–1396.

Barnett, V. and Lewis, T. 1994. Outliers in Statistical Data. 3rd Edn. New York: Wiley.

Belsley, D.A., Kuh, E. and Welsch, R.E. 1980. Regression Diagnostics: Identifying Influential Data and Sources of Collinearity. New York: Wiley.

Chatterjee, S. and Hadi, A.S. 2006. Regression Analysis by Example. 4th Edn. New York: Wiley.

Cook, R.D. and Hawkins, D.M. 1990. Unmasking multivariate outliers and leverage points: Comment. Journal of the American Statistical Association. 85: 640–644.

Cook, R.D. and Weisberg, S. 1999. Applied Regression Including Computing and Graphics. John Wiley and Sons.

Gross, J. 2003. Linear Regression. 1st Edn. New York: Springer-Verlag.

Gujarati, D.N. 2002. Basic Econometrics. 4th Edn. New York: Macgraw-Hill.

Habshah, M., Norazan, M.R. and Imon, A.H.M.R. 2009. The performance of Diagnostic-Robust Generalized Potentials for the identification of multiple high leverage points in linear regression. Journal of Applied Statistics. 36(5): 507–520.

Habshah, M., Bagheri, A. and Imon, A.H.M.R. 2010. The effect of collinearity—influential observations on collinear dataset: a monte carlo simulation study. Journal of Applied Sciences. 10(18): 2086–2093.

Habshah, M., Bagheri, A. and Imon, A.H.M.R. 2011. High leverage Collinearity—Enhancing observation and its effect on multicollinearity pattern: monte carlo simulation study. Sains Malaysiana. 40(12): 1437–1447.

Hadi, A.S. 1988. Diagnosing collinearity-influential observations. Computational Statistics and Data Analysis. 7: 143–159.

Hadi, A.S. 1992. A new measure of overall potential influence in linear regression. Computational and Statistical Data Analysis. 14: 1–27.

Hawkins, D.M., Bradu, D. and Kass, G.V. 1984. Location of several outliers in multiple regression data using elemental sets. Technometrics. 26: 197–208.

Hoaglin, D.C. and Welsch, R.E. 1978. The hat matrix in regression and ANOVA. American Statistician. 32: 17–22.

Huber, P.J. 1981. Robust Statistics. New York: Wiley.

Imon, A.H.M.R. 2002. Identifying multiple high leverage points in linear regression. Journal of Statistical Studies. Special Volume in Honour of Professor Mir Masoom Ali. 3: 207–218.

Imon, A.H.M.R. 2005. Identifying multiple influential observations in linear regression. Journal of Applied Statistics. 32: 929–946.

Kamruzzaman, M.D. and Imon, A.H.M.R. 2002. High leverage point: Another source of multicollinearity. Pakistani Journal of Statistics. 18: 435–448.

Kempthorne, P.J. and Mendel, M.B. 1990. Unmasking multivariate outliers and leverage points: Comment. Journal of the American Statistical Association. 85: 647–648.

Kutner, M.H., Nachtsheim, C.J., Neter, J. and Li, W. 2005. Applied linear regression models. 5th Edn. New York: Mcgraw-Hill.

Maronna, R.A., Martin, R.D. and Yohai, V.J. 2006. Robust Statistics Theory and Methods. New York: Wiley and Sons.

Norazan, M.R. 2008. Weighted Maximum Median Likelihood Estimation for Parameters in Multiple Linear Regression Model, Unpublished Ph.D. Thesis, Universiti Putra Malaysia, Malaysia.

Olive, D.J. 2008. Applied Robust Statistics. Southern Illinois University. http://www.math.siu.edu/olive/run.pdf.

Peña, D. and Yohai, V.J. 1995. The detection of influential subsets in linear regression by using an influence matrix. Journal of Royal Statistical Society, B. 57: 18–44.

Rousseeuw, P.J. 1983. Multivariate estimation with high breakdown point. Proceedings of the 4th Pannonian Symposium on Mathematical Statistics, Bad Tatzmannsdorf, Austria, September 4–10.

Rousseeuw, P.J. 1984. Least median of squares regression. Journal of the American Statistical Association. 79: 871–880.

Rousseeuw, P.J. 1985. Multivariate estimation with high breakdown point. Mathematical Statistics and Applications, B: 283–297.

Rousseeuw, P.J. and Leroy, A.M. 1987. Robust Regression and Outlier Detection. New York: Wiley.

Rousseeuw, P. and Van Zomeren, B. 1990. Unmasking multivariate outliers and leverage points. Journal of American Statistical Associations. 85: 633–639.

Rousseeuw P.J. and Van Driessen, K. 1999. A fast algorithm for the minimum covariance determinant estimator. Technometrics. 41: 212–223.

Ruppert, D. and Simpson, D.G. 1990. Unmasking multivariate outliers and leverage points: Comment. Journal of the American Statistical Association, 85: 644–646.

Sengupta, D. and Bhimasankaram, P. 1997. On the roles of observations in collinearity in the linear model. Journal of American Statistical Association, 92: 1024–1032.

Vellman, P.F. and Welsch, R.E. 1981. Efficient computing of regression diagnostics. American Statistician. 27: 234–242.

Walker, E. 1989. Detection of collinearity-influential observations. Communications in Statistics: Theory and Methodology. 18: 1675–1690.

Wilcox, R.R. 2005. Introduction to Robust Estimation and Hypothesis Testing. 2nd Edn. The United States of America: Elsevier Academic Press.

Chapter 12

Mixture Modeling of Car Insurance Claim in Zimbabwe

Nyasha Mangava[1], Saralees Nadarajah[2*] and M. Masoom Ali[3]

[1]Department of Statistics, University of Zimbabwe, Harare, Zimbabwe. Email: nyashamangava8@gmail.com

[2]Department of Mathematics, University of Manchester, Manchester M13 9PL, UK. Email: mbbsssn2@manchester.ac.uk

[3]Department of Mathematical Sciences, Ball State University, Muncie, IN, USA. Email: mali@bsu.edu

12.1 INTRODUCTION

Many papers on modeling of insurance data have been published mainly by Europeans and North Americans using data from Europe and North America (for example, McNeil 1997, Resnick 1997, Geyer and Peter 1999, Richaudeau 1999, Scollnik 2007, Nasvadi and Wister 2009, Aizcorbe et al. 2012, Rieck et al. 2014, Brazauskas and Kleefeld 2016 and Donadio et al. 2021). Good quality insurance data are available in many African countries, including some of the poorest countries. Africa also has many talented scholars, but lacking guidance and resources. No attention has been paid to the data or the scholars. Europeans and North Americans appear to have looked to Africa only for slaves, minerals to steal, and

*For Correspondence: Saralees Nadarajah (mbbsssn2@manchester.ac.uk)

countries to colonize among others. Nadarajah and Kwofie (2022) were the first to provide modeling of automobile insurance claim data from Ghana. Except for Nadarajah and Kwofie (2022), there are no papers on insurance modeling of data from Africa.

The aim of this paper is to provide statistical modeling of insurance claims by car users in Zimbabwe. There have been no such papers published for Zimbabwe. We determine how the claim amount depends on various factors and covariates. The data on claim amounts (in Zimbabwe dollars) were obtained from a leading insurance company in Harare. The data also contained the following:

- age of driver,
- region (highly rural/rural versus highly urban/urban),
- motor vehicle record score (higher the score the poorer the driver),
- car license revoked previously or not,
- red claim flag or not (false claims about damage to car),
- car type (minivan, panel truck, pickup, sports car, SUV, or van),
- number of traffic inspection failures,
- blue book value (value of the car according to guidelines in the Kelly Blue Book),
- car use (commercial or private),
- travel time,
- occupation (blue collar, clerical, doctor, homemaker, lawyer, manager, professional, or student),
- education (bachelors, high school, masters, or PhD),
- gender (male or female),
- marital status (married or not),
- home value,
- single parent or not,
- income,
- years since joining insurance,
- number of kids at home,
- age of car.

Region, car license revoked previously or not, red claim flag or not, car type, car use, occupation, education, gender, marital status, single parent or not and the number of kids at home are factors. Age of driver, motor vehicle record score, number of traffic inspection failures, blue book value, travel time, home value, income, years since joining insurance and age of car are covariates. The results of this paper could be useful to car users as well as insurance companies in Zimbabwe and surrounding countries.

A histogram of the claim amount is shown in Figure 12.1. Some summary statistics are minimum = 30, first quartile = 2588, median = 4127, mean = 5505, third quartile = 5686, maximum = 85524, variance = 50875651, skewness = 5.369857, and kurtosis = 39.70605.

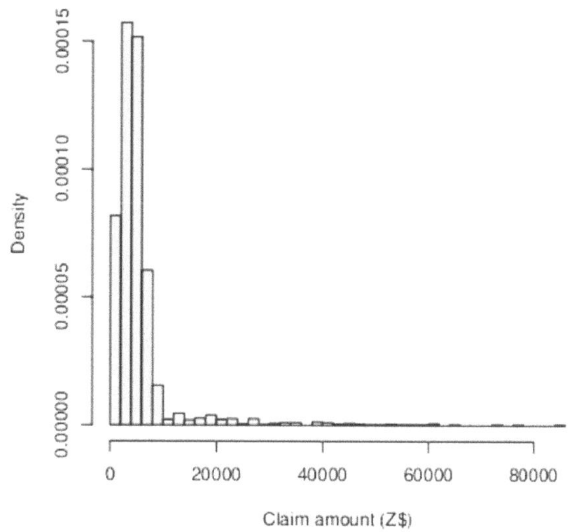

Figure 12.1 Histogram of the claim amount.

We see that the histogram is multimodal. In particular, the tail appears bimodal. One way to model the data is to adopt the approach in Nadarajah and Kwofie (2022) and fit a composite lognormal distribution. That is, fit a lognormal distribution to the body and a mixture distribution to the tail. The fitting of such a composite distribution comes at a cost, the cost of two parameters for the lognormal distribution plus additional parameters for the mixture distribution. In this paper, we show that data in Figure 12.1 can be modeled by a simpler class of mixture distributions, see Section 12.3.

Another aspect of modeling is regression to account for the effects of the various factors and covariates. Nadarajah and Kwofie (2022) built regression models based on the "cutoff point" of composite lognormal distributions. Regression models for the data in Figure 12.1 can be built more easily using the scale parameters of mixture distributions, see Section 12.4.

The contents of the remainder of this paper are organized as follows. Section 12.2 performs some exploratory analysis to identify features in the data. Section 12.3 states models used for the claim amount. Section 12.4 describes fitting of the models as well as results and discussion. Finally, some conclusions are reported in Section 12.5.

12.2 EXPLORATORY DATA ANALYSIS

The relationship of claim amount versus the factors and covariates is shown in Figures 12.2, 12.3 and 12.4.

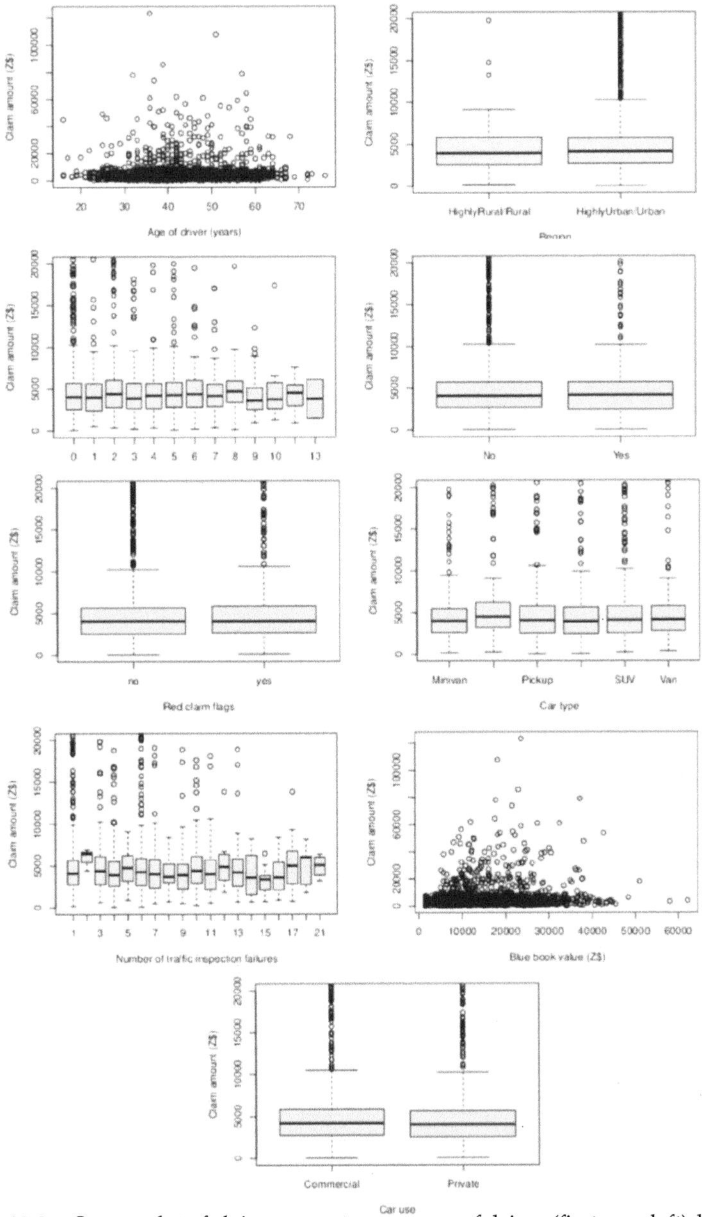

Figure 12.2 Scatter plot of claim amount versus age of driver (first row, left); box plot of claim amount versus region (first row, middle); box plot of claim amount versus motor vehicle record score (first row, right); box plot of claim amount versus license revoked previously or not (second row, left); box plot of claim amount versus red claim flag or not (second row, middle); box plot of claim amount versus car type (first row, right); box plot of claim amount versus number of traffic inspection failures (third row, left); scatter plot of claim amount versus blue book value (third row, middle); box plot of claim amount versus car use (third row, right).

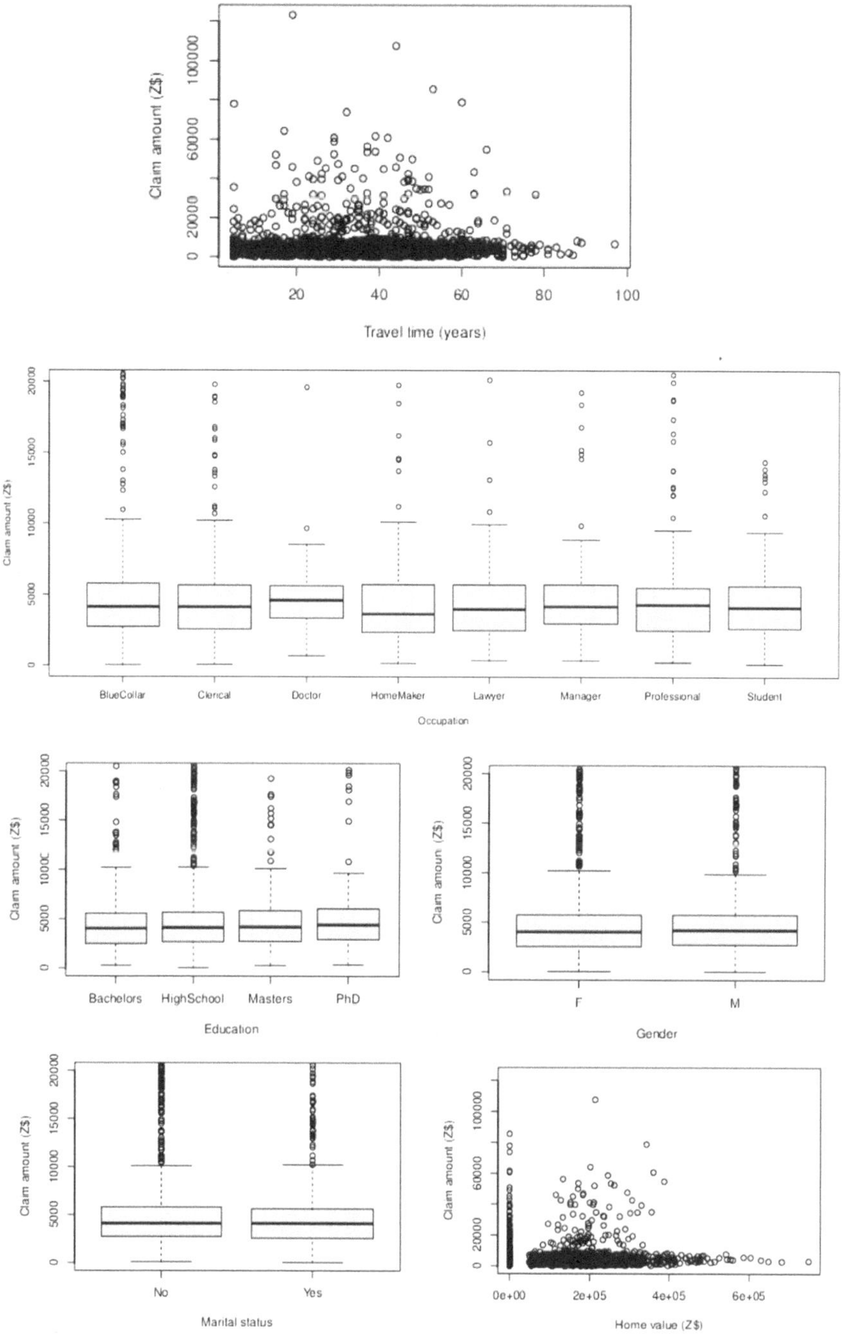

Figure 12.3 (*Contd.*)

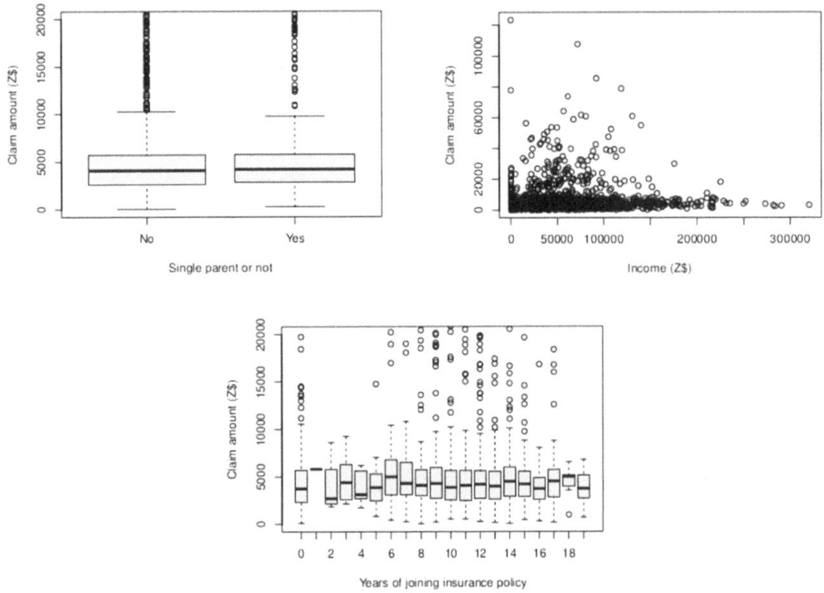

Figure 12.3 Scatter plot of claim amount versus travel time (first row, left); box plot of claim amount versus occupation (first row, middle); box plot of claim amount versus education (first row, right); box plot of claim amount versus gender (second row, left); box plot of claim amount versus marital status (second row, middle); scatter plot of claim amount versus home value (second row, right); box plot of claim amount versus single parent or not (third row, left); scatter plot of claim amount versus income (third row, middle); box plot of claim amount versus years of joining insurance policy (third row, right).

The relationship of claim amount versus most of the factors or covariates is not clear. However, the claim amount appears to decrease with respect to the blue book value, travel time, income, and age of car.

The pairwise correlations among the covariates are reported in Table 12.1. None of the correlations are strong. The motor vehicle record score is negatively correlated with all of the covariates. The number of traffic inspection failures is negatively correlated with all of the covariates but the blue book value. The remaining pairwise correlations are positive.

The correlations between the claim amount and the covariates (in the order listed in Table 12.1) are 0.010, 0.031, 0.059, 0.009, 0.011, 0.135, 0.013, −0.013, and −0.001. These correlations are once again not strong.

12.3 MODELS

One way to model the data in Figure 12.1 is to use composite lognormal distributions as in Nadarajah and Kwofie (2022). But the fitting of

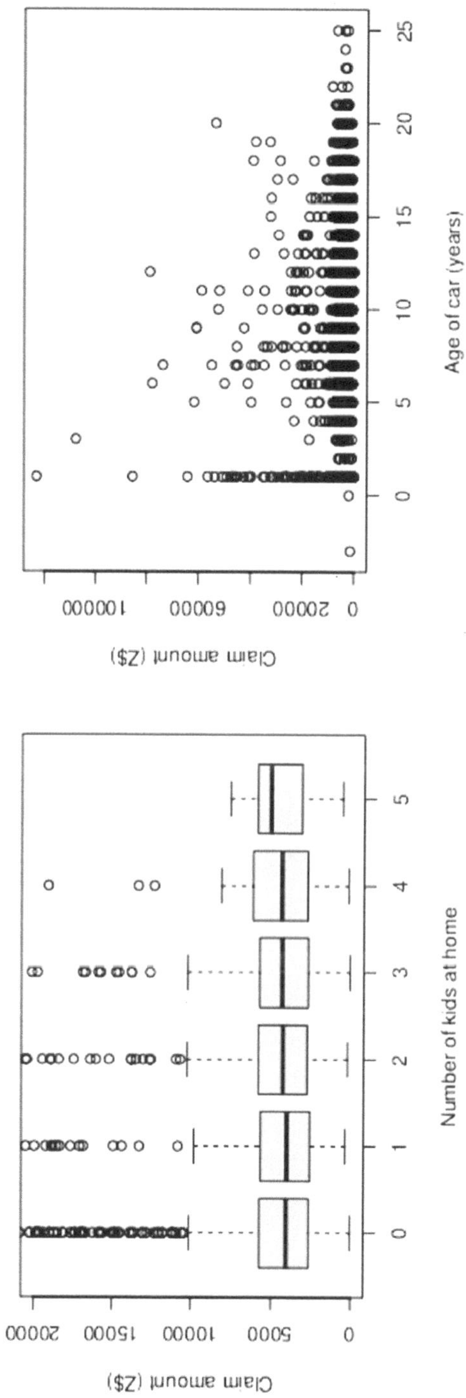

Figure 12.4 Box plot of claim amount versus number of kids at home (left); box plot of claim amount versus age of car (middle).

Table 12.1 Pairwise correlation among the covariates.

	Age of driver	Years since insurance	Income	Home value	Travel time	Blue book value	Motor vehicle record score	Car age	No. of traffic inspection failures
Age of driver	1	0.156	0.190	0.212	0.048	0.154	−0.054	0.156	−0.009
Years since insurance		1	0.366	0.322	0.030	0.172	−0.030	0.058	−0.004
Income			1	0.475	0.028	0.432	−0.074	0.422	−0.016
Home value				1	0.039	0.207	−0.093	0.186	−0.031
Travel time					1	0.005	−0.005	0.049	−0.028
Blue book value						1	−0.054	0.182	0.009
Motor vehicle record score							1	−0.069	−0.025
Car age								1	−0.009
No. of traffic inspection failures									1

composite lognormal distributions comes at the cost of a large number of parameters. We shall see later in Section 12.4 that a mixture of two heavy- tailed distributions can provide adequate fits to the claim amount. So, we take the probability density function of the claim amount as

$$f(x;\Theta) = \alpha f_1(x;\Theta_1) + (1-\alpha) f_2(x;\Theta_2) \quad (1)$$

where $0 < \alpha < 1$, $f_i(x;\Theta_i)$ are valid probability density functions and $\Theta = (\Theta_1, \Theta_2, \alpha)$.

We chose $f_i(x;\Theta_i)$, $i = 1, 2$ to belong to the same parametric family. It was taken to correspond to the most popular heavy tailed distributions: Fréchet distribution, inverse gamma distribution, inverse paralogistic distribution, log-logistic distribution, lognormal distribution, paralogistic distribution, Pareto distribution, and the Weibull distribution. We have considered only two parameters of heavy-tailed distributions for the mixture components as distributions. Having a higher number of parameters did not provide significantly better fits.

We fitted Eq. (1) for each of the 8 choices for $f_i(x;\Theta_i)$, $i = 1, 2$. The method of maximum likelihood and the EM algorithm were used for the fitting. All computations were performed using the R software (R Core Team 2022). Discrimination among the fitted models was performed using the two criteria:

- the Akaike information criterion due to Akaike (1974) defined by

  $$AIC = 2(2m+1) - 2\ln L(\widehat{\Theta});$$

  where m denotes the length of each Θ_i, $\widehat{\Theta}$ denotes the maximum likelihood estimate of Θ, and $\ln L(\widehat{\Theta})$ denotes the log-likelihood value;
- the Bayesian information criterion due to Schwarz (1978) defined by

  $$BIC = (2m+1)\ln n - 2\ln L(\widehat{\Theta}),$$

  where n denotes the sample size.

The smaller the values of these criteria the better the fit.

12.4 RESULTS AND DISCUSSION

12.4.1 Mixture Modeling

The model given by (1) was fitted for the eight choices for $f_i(x;\Theta_i)$, $i = 1, 2$. The values of AIC and BIC are given in Table 12.2. We also fitted the composite lognormal distribution in Nadarajah and Kwofie (2022) with the tail distribution given by (1). The corresponding values of AIC and BIC are given in Table 12.3.

Table 12.2 Values of AIC and BIC when $f_i(x; \Theta_i)$ are the two-parameter densities

Component Distribution	AIC	BIC
Weibull	−8944.186	−8916.036
Fréchet	−8680.244	−8652.094
Log-logistic	−8879.014	−8850.864
Inverse gamma	−8749.262	−8721.112
Pareto	−7965.393	−7937.243
Paralogistic	−8883.32	−8855.17
Inverse paralogistic	−8868.008	−8839.858
Lognormal	−2560.814	−2532.664

Table 12.3 Values of AIC and BIC when the composite lognormal distribution was fitted with the given tail distribution

Tail Distribution	AIC	BIC
Two-component Weibull	−8277.418	−8238.009
Two-component Fréchet	−8087.199	−8047.79
Two-component log-logistic	−8185.151	−8145.741
Two-component inverse gamma	−8112.036	−8072.626
Two-component Pareto	−7494.948	−7455.538
Two-component paralogistic	−8125.127	−8085.717
Two-component inverse paralogistic	−8448.456	−8409.046
Two-component lognormal	−2120.207	−2080.797

We see that Eq. (1) fits better than the composite lognormal distribution for all the choices. The smallest AIC and BIC values in Table 12.2 are when $f_i\ (x; \Theta_i)$ are Weibull densities. That is

$$f_i\left(x;\Theta_i\right)=\frac{a_i}{b_i}\left(\frac{x}{b_i}\right)^{a_i-1}\exp\left[-\left(\frac{x}{b_i}\right)^{a_i}\right], \tag{2}$$

where $x > 0$, $a_i > 0$ are the shape parameters, $b_i > 0$ are the scale parameters and $\Theta_i = (a_i, b_i)$. The parameter estimates and standard errors are $\widehat{a_1} = 0.979(0.112)$, $\widehat{b_1} = 0.169(0.013)$, $\widehat{a_2} = 2.181(1.712)$, $\widehat{b_2} = 0.045(0.011)$ and $\widehat{\alpha} = 0.127\left(0.043\right)$. The fitted probability density function of the claim amount given by

$$f\left(x\right)=\frac{\widehat{\alpha}\widehat{a_1}}{\widehat{b_1}}\left(\frac{x}{\widehat{b_1}}\right)^{\widehat{a_1}-1}\exp\left[-\left(\frac{x}{\widehat{b_1}}\right)^{\widehat{a_1}}\right]+\frac{\left(1-\widehat{\alpha}\right)\widehat{a_2}}{\widehat{b_2}}\left(\frac{x}{\widehat{b_2}}\right)^{\widehat{a_2}-1}\exp\left[-\left(\frac{x}{\widehat{b_2}}\right)^{\widehat{a_2}}\right] \tag{3}$$

is shown in Figure 12.5. The corresponding probability and quantile plots are shown in Figure 12.6. These figures show that the fitted distribution adequately describes the data.

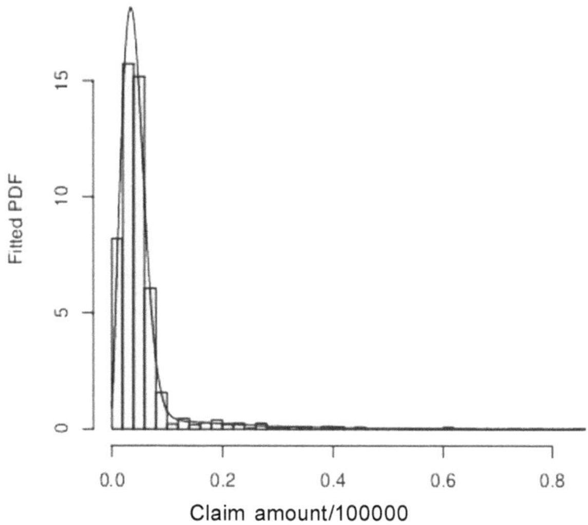

Figure 12.5 Histogram of the claim amount and the fitted probability density function Eq. (3).

12.4.2 Regression Modeling

We now incorporate the factors (region, car license revoked previously or not, red claim flag or not, car type, car use, occupation, education, gender, marital status, single parent or not, and the number of kids at home), and covariates (age of driver, motor vehicle record score, number of traffic inspection failures, blue book value, travel time, home value, income, years since joining insurance, and age of car) into modeling to see how they affect the claim amount. We let the scale parameters in Eq. (2) depend on the factors and covariates as follows:

$$
\begin{aligned}
b_i = \exp(&b_{intercept,i} + b_{region,i} \cdot I\{region = 1\} \\
&+ b_{car\ license\ revoked,i} \cdot I\{car\ license\ revoked = 1\} \\
&+ b_{red\ claim\ flag,i} \cdot I\{red\ claim\ flag = 1\} + \sum_{j=1}^{6} b_{car\ type,i,j} \cdot I\{\{car\ type = j\} \\
&+ b_{car\ use,i} \cdot I\{car\ use = 1\} + b_{region,i} \cdot I\{region = 1\} \\
&+ \sum_{j=1}^{6} b_{occupation,i,j} \cdot I\{occupation = j\} \\
&+ \sum_{j=1}^{3} b_{education,i,j} \cdot I\{education = 1\} + b_{gender,i} \cdot I\{gender = 1\} \\
&+ b_{marital\ status,i} \cdot I\{marital\ status = 1\} \\
&+ b_{single\ parent\ or\ not,i} \cdot I\{single\ parent\ or\ not = 1\}
\end{aligned}
$$

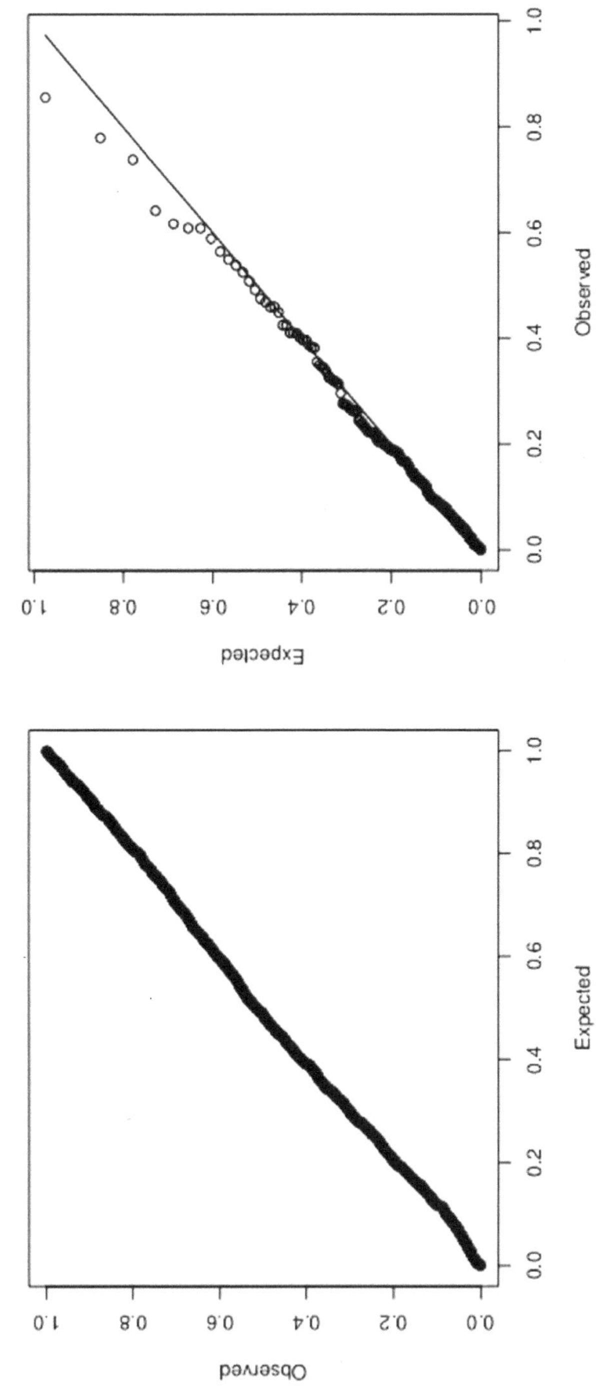

Figure 12.6 Probability plot (left) and quantile plot (right) corresponding to Eq. (3).

$$+\sum_{j=1}^{5} b_{no\ of\ kids,i,j} \cdot I\{no\ of\ kids = j\} + b_{age\ of\ driver,i} \cdot age\ of\ driver$$
$$+ b_{motor\ vehicle\ record\ score,i} \cdot motor\ vehicle\ record\ score$$
$$+ b_{number\ of\ traffic\ inspection\ failures,i} \cdot number\ of\ traffic\ inspection\ failures$$
$$+ b_{blue\ book\ value,i} \cdot blue\ book\ value + b_{travel\ time,i} \cdot travel\ time$$
$$+ b_{home\ value,i} \cdot home\ value$$
$$+ b_{income,i} \cdot income + b_{years\ since\ joining\ insurance,i} \cdot years\ since\ joining\ insurance$$
$$+ b_{age\ of\ car,i} \cdot age\ of\ car)$$

for i = 1, 2, where $I\{\cdot\}$ denotes the indicator function. We begin by fitting the constant model $b_i = \exp(b_{(intercept,i)})$ for i = 1, 2. Then, we fit the term that is most significant among the factors and covariates. Thereafter, we fit the term that is most significant among the remaining factors and covariates. We continue this process until the model cannot be improved further. In the final model, none of the factors and none of the covariates turned out to be significant except for the blue book value. That is

$$\widehat{b_i} = \exp\left(\widehat{b_{intercept,i}} + \widehat{b_{blue\ book\ value,i}} \cdot blue\ book\ value\right) \tag{4}$$

for i = 1, 2. The parameter estimates were

$$\widehat{b_{intercept,1}} = 2.494, \quad \widehat{b_{blue\ book\ value,1}} = 4.769 \cdot 10^{-5}, \quad \widehat{b_{intercept,2}} = -3.128$$

and

$$\widehat{b_{blue\ book\ value,2}} = 2.014 \cdot 10^{-6}.$$

So, the blue book value has a positive effect on the scale parameters.

According to the fitted probability density function in Eq. (3), the first four moments of the claim amount are

$$E(claim\ amount) = \widehat{\alpha}\widehat{b_1}\Gamma\left(1+\frac{1}{\widehat{a_1}}\right) + \left(1-\widehat{\alpha}\right)\widehat{b_2}\Gamma\left(1+\frac{1}{\widehat{a_2}}\right),$$

$$E(claim\ amount^2) = \widehat{\alpha}\widehat{b_1}^2\Gamma\left(1+\frac{2}{\widehat{a_1}}\right) + (1-\widehat{\alpha})\,\widehat{b_2}^2\Gamma\left(1+\frac{2}{\widehat{a_2}}\right),$$

$$E(claim\ amount^3) = \widehat{\alpha}\widehat{b_1}^3\Gamma\left(1+\frac{3}{\widehat{a_1}}\right) + \left(1-\widehat{\alpha}\right)\widehat{b_2}^3\Gamma\left(1+\frac{3}{\widehat{a_2}}\right)$$

and

$$E(claim\ amount^4) = \widehat{\alpha}\widehat{b_1}^4\Gamma\left(1+\frac{4}{\widehat{a_1}}\right) + \left(1-\widehat{\alpha}\right)\widehat{b_2}^4\Gamma\left(1+\frac{4}{\widehat{a_2}}\right),$$

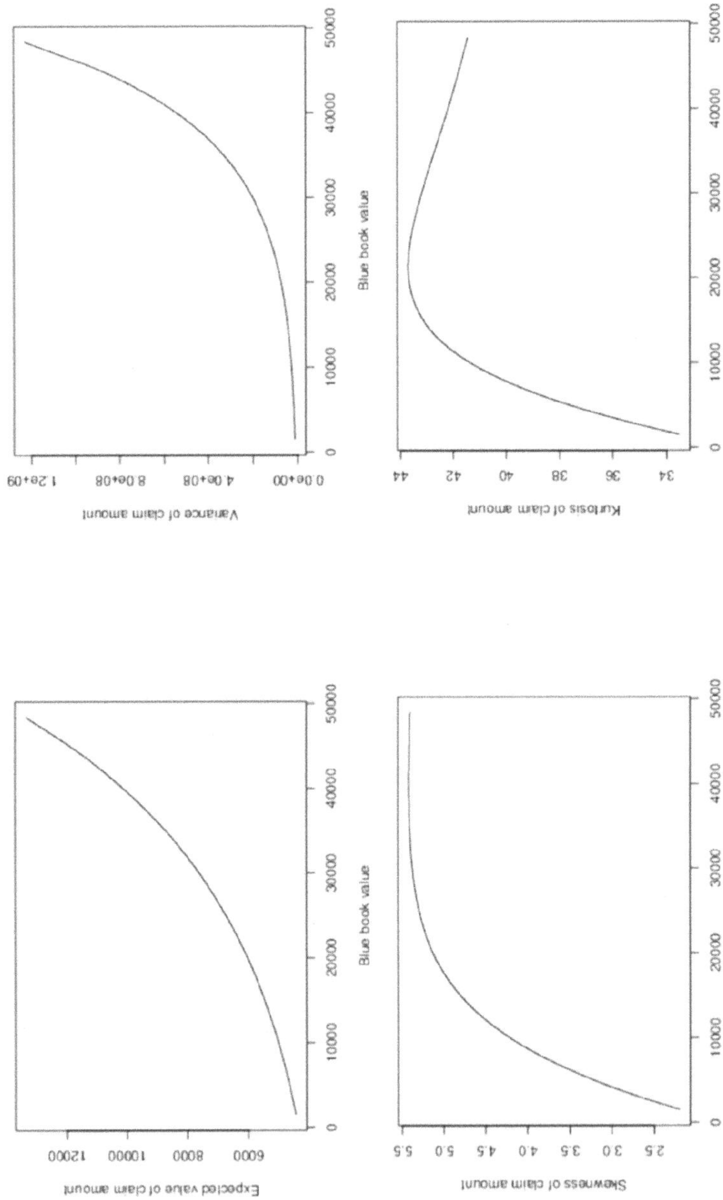

Figure 12.7 Expectation of the claim amount versus the blue book value (top left); variance of the claim amount versus the blue book value (top right); skewness of the claim amount versus the blue book value (bottom left); kurtosis of the claim amount versus the blue book value (bottom right).

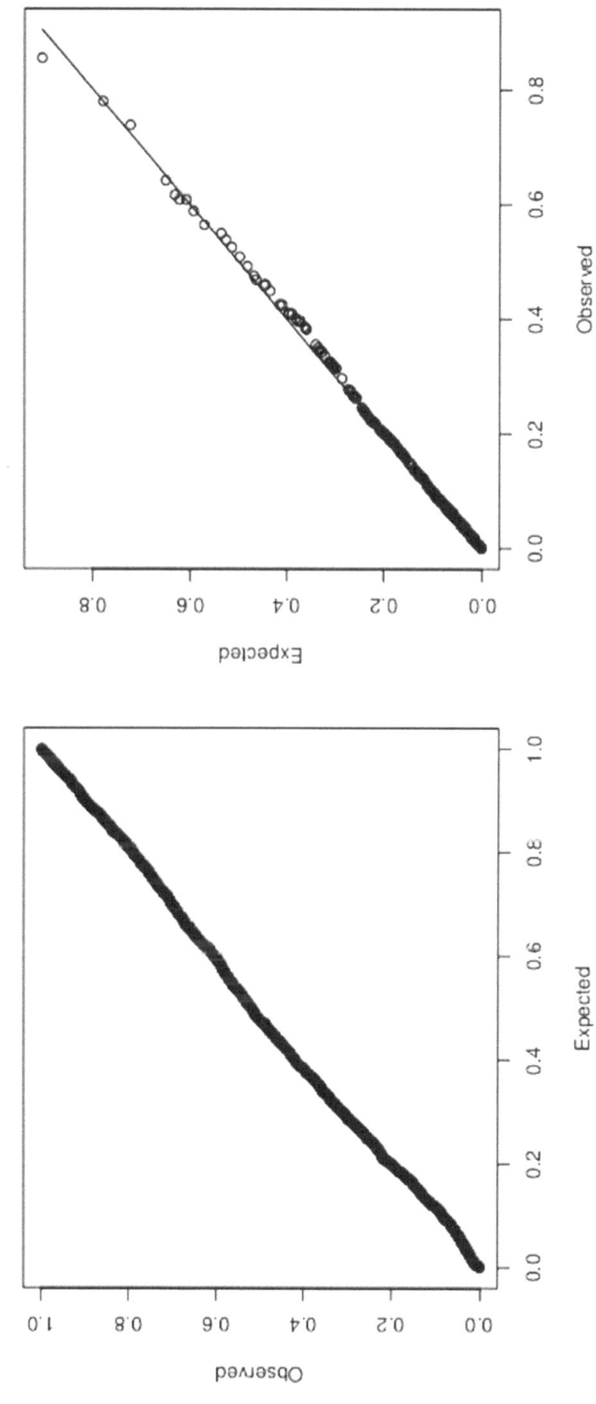

Figure 12.8 Probability plot (left) and quantile plot (right) corresponding to Eq. (4).

where $\Gamma(\cdot)$ denotes the gamma function. Figure 12.7 plots the expectation, variance, skewness, and kurtosis of the claim amount versus the blue book value.

We see that the expectation, variance, and skewness are increasing with increasing blue book value. Kurtosis initially increases before decreasing. The maximum kurtosis occurs when the blue book value is around 20000.

The probability and quantile plots corresponding to Eq. (4) are shown in Figure 12.8. These plots show that the fitted regression adequately describes the data.

12.4.3 Risk Estimation

We now give risk estimates associated with car insurance claim. The most popular risk measure is the value at risk (Jorion 2006). The value at risk with probability p corresponding to Eq. (3) is the root of

$$1-\widehat{\alpha}\exp\left[-\left(\frac{\widehat{VaR_p}}{\widehat{b_1}}\right)^{\widehat{a_1}}\right]-\left(1-\widehat{\alpha}\right)\exp\left[-\left(\frac{\widehat{VaR_p}}{\widehat{b_2}}\right)^{\widehat{a_2}}\right]=p. \quad (5)$$

Equation (5) was solved using the uniroot function in R. The estimates of value at risk versus blue book value are shown in Tables 12.4 and 12.5.

Table 12.4 Estimates of Eq. (5) versus p = 0.9, 0.95, 0.99, 0.995, 0.999, 0.9995, 0.9999, and blue book value

Blue Book Value	$\widehat{VaR}_{0.9}$	$\widehat{VaR}_{0.95}$	$\widehat{VaR}_{0.99}$	$\widehat{VaR}_{0.995}$	$\widehat{VaR}_{0.999}$	$\widehat{VaR}_{0.9995}$	$\widehat{VaR}_{0.9999}$
1000	7268	9072	19670	24551	35510	40112	50580
2000	7331	9254	20629	25747	37245	42070	53052
3000	7394	9456	21636	27005	39062	44123	55645
4000	7459	9684	22693	28325	40970	46279	58364
5000	7524	9939	23802	29709	42974	48542	61215
6000	7591	10231	24965	31160	45073	50913	64205
7000	7659	10567	26184	32685	47275	53402	67340
8000	7729	10949	27464	34279	49585	56010	70634
9000	7799	11384	28805	35951	52009	58744	74081
10000	7876	11874	30212	37711	54547	61614	77697
11000	7949	12414	31687	39550	57212	64625	81495
12000	8026	12999	33233	41484	60007	67781	85479
13000	8100	13622	34859	43510	62938	71090	89650
14000	8180	14283	36562	45633	66012	74565	94031
15000	8262	14979	38348	47865	69237	78209	98625
16000	8345	15711	40222	50203	72620	82025	103444
17000	8433	16478	42186	52656	76167	86035	108497

(Contd...)

(Contd...)

Blue Book Value	$\widehat{VaR}_{0.9}$	$\widehat{VaR}_{0.95}$	$\widehat{VaR}_{0.99}$	$\widehat{VaR}_{0.995}$	$\widehat{VaR}_{0.999}$	$\widehat{VaR}_{0.9995}$	$\widehat{VaR}_{0.9999}$
18000	8518	17283	44244	55228	79888	90236	113797
19000	8609	18127	46407	57925	83790	94646	119358
20000	8700	19013	48673	60753	87881	99270	125184
21000	8796	19942	51053	63721	92177	104118	131301
22000	8894	20916	53548	66836	96679	109204	137714
23000	8998	21937	56163	70102	101402	114540	144441
24000	9101	23009	58907	73525	106356	120134	151499
25000	9209	24133	61787	77117	111552	126003	158898
26000	9328	25312	64803	80885	117000	132159	166663
27000	9447	26548	67969	84836	122716	138615	174805
28000	9571	27845	71285	88980	128711	145386	183342
29000	9703	29205	74769	93327	134999	152487	192299
30000	9843	30632	78423	97886	141591	159937	201693
31000	9993	32128	82254	102665	148510	167750	211545
32000	10153	33698	86271	107682	155765	175944	221879
33000	10329	35344	90485	112945	163374	184537	232717
34000	10514	37070	94908	118461	171354	193553	244086
35000	10725	38880	99543	124247	179723	203010	256009
36000	10953	40778	104406	130316	188506	212925	268513
37000	11218	42775	109507	136682	197713	223325	281632
38000	11516	44864	114856	143359	207371	234238	295389
39000	11863	47054	120465	150362	217499	245679	309819
40000	12268	49355	126351	157707	228127	257680	324956

Table 12.5 Estimates of Eq. (5) versus p = 0.9, 0.95, 0.99, 0.995, 0.999, 0.9995, 0.9999 and blue book value

Blue Book Value	$\widehat{VaR}_{0.9}$	$\widehat{VaR}_{0.95}$	$\widehat{VaR}_{0.99}$	$\widehat{VaR}_{0.995}$	$\widehat{VaR}_{0.999}$	$\widehat{VaR}_{0.9995}$	$\widehat{VaR}_{0.9999}$
41000	12738	51764	132524	165411	239270	270270	340828
42000	13273	54293	138998	173489	250958	283472	357478
43000	13872	56944	145788	181967	263218	297318	374940
44000	14525	59726	152910	190856	276077	311842	393257
45000	15224	62643	160379	200180	289562	327078	412467
46000	15964	65705	168213	209958	303709	343055	432616
47000	16743	68913	176430	220214	318543	359811	453749
48000	17560	72279	185051	230972	334104	377389	475913
49000	18418	75809	194091	242255	350425	395823	499162
50000	19318	79516	203569	254091	367541	415160	523546
51000	20261	83398	213514	266501	385499	435440	549123
52000	21251	87472	223942	279521	404329	456711	575947
53000	22289	91745	234884	293175	424081	479020	604082
54000	23378	96227	246359	307496	444798	502421	633591
55000	24520	100929	258392	322517	466527	526963	664543
56000	25717	105860	271015	338274	489315	552705	697005

(Contd...)

(Contd...)

Blue Book Value	$\widehat{VaR}_{0.9}$	$\widehat{VaR}_{0.95}$	$\widehat{VaR}_{0.99}$	$\widehat{VaR}_{0.995}$	$\widehat{VaR}_{0.999}$	$\widehat{VaR}_{0.9995}$	$\widehat{VaR}_{0.9999}$
57000	26974	111032	284254	354796	513219	579709	731055
58000	28291	116450	298140	372128	538291	608026	766766
59000	29673	122140	312704	390304	564586	637724	804220
60000	31122	128108	327978	409373	592165	668880	843509
61000	32643	134366	344003	429371	621092	701555	884715
62000	34237	140930	360806	450346	651433	735824	927932
63000	35909	147814	378431	472345	683255	771771	973262
64000	37663	155035	396918	495419	716631	809472	1020806
65000	39503	162608	416307	519619	751640	849015	1070672
66000	41433	170552	436643	545006	788355	890491	1122974
67000	43457	178883	457975	571628	826867	933990	1177832
68000	45579	187621	480344	599550	867260	979615	1235369
69000	47806	196787	503809	628839	909626	1027468	1295719
70000	50141	206400	528421	659558	954059	1077660	1359012
71000	52590	216482	554233	691777	1000667	1130304	1425399
72000	55159	227058	581309	725571	1049551	1185520	1495029
73000	57853	238149	609707	761016	1100822	1243433	1568064
74000	60679	249783	639487	798190	1154596	1304175	1644662
75000	63643	261984	670730	837182	1210998	1367884	1725005
76000	66752	274782	703494	878078	1270153	1434703	1809270
77000	70013	288205	737859	920973	1332201	1504789	1897653
78000	73433	302283	773906	965962	1397279	1578298	1990356
79000	77020	317049	811710	1013149	1465533	1655398	2087580
80000	80783	332536	851362	1062641	1537127	1736262	2189561

The numbers in Tables 12.4 and 12.5 can be useful for car users in Zimbabwe. For example, a vehicle with a blue book value of 80000 will have a claim amount less than 2189561 with probability 0.9999. A vehicle with a blue book value of 1000 will have a claim amount less than 7268 with probability 0.99.

12.5 CONCLUSIONS

We have modeled car insurance claim data from Zimbabwe. A mixture of two Weibull distributions was shown to provide the best fit. The best fit was assessed in terms of the Akaike Information Criterion, the Bayesian Information Criterion, probability plots, and quantile plots.

We have also fitted a regression model based on the best fitting distribution to see if the factors and covariates have any effect on the claim amount. The regression model showed that only the blue book value was significant. It has a positive effect on the expected value, variance, and skewness of the claim amount.

REFERENCES

Aizcorbe, A., Liebman, E., Pack, S., Cutler, D.M., Chernew, M.E. and Rosen, A.B. 2012. Measuring health care costs of individuals with employer-sponsored health insurance in the U.S.: A comparison of survey and claims data. Statistical Journal of IAOS. 28: 43–51.

Akaike, H. 1974. A new look at the statistical model identification. IEEE Transactions on Automatic Control. 19: 716–723.

Brazauskas, V. and Kleefeld, A. 2016. Modeling severity and measuring tail risk of Norwegian fire claims. North American Actuarial Journal. 20: 1–16.

Donadio, L., Schifanella, R., Binder, C.R. and Massaro, E. 2021. Leveraging insurance customer data to characterize socioeconomic indicators of Swiss municipalities. PLoS ONE. 16: Article number e0246785.

Geyer, S. and Peter, R. 1999. Occupational status and all-cause mortality: A study with health insurance data from Nordrhein-Westfalen, Germany. European Journal of Public Health. 9: 114–118.

Jorion, P. 2006. Value at Risk: The New Benchmark for Managing Financial Risk. McGraw-Hill, New York.

McNeil, A.J. 1997. Estimating the tails of loss severity distributions using extreme value theory. ASTIN Bulletin, 27: 117–137.

Nadarajah, S. and Kwofie, C. 2022. Heavy tailed modeling of automobile claim data from Ghana. Journal of Computational and Applied Mathematics, 405: Article number 113947.

Nasvadi, G.C. and Wister, A. 2009. Do restricted driver's licenses lower crash risk among older drivers? A survival analysis of insurance data from British Columbia. The Gerontologist. 49: 474–484.

R Core Team. 2022. R: A language and environment for statistical computing. R Foundation for Statistical Computing. Vienna, Austria.

Resnick, S.I. 1997. Discussion of the Danish data on large fire insurance losses. ASTIN Bulletin. 27: 139–151.

Richaudeau, D. 1999. Automobile insurance contracts and risk of accident: An empirical test using French individual data. The Geneva Papers on Risk and Insurance Theory. 24: 97–114.

Rieck, T., Feig, M., Eckmanns, T., Benzler, J., Siedler, A. and Wichmann, O. 2014. Vaccination coverage among children in Germany estimated by analysis of health insurance claims data. Human Vaccines and Immunotherapeutics. 10: 476–484.

Schwarz, G.E. 1978. Estimating the dimension of a model. Annals of Statistics, 6: 461–464.

Scollnik, D.P.M. 2007. On composite lognormal-Pareto models. Scandinavian Actuarial Journal. 20–33.

Chapter **13**

Examining Research in Statistics and Computer Science Literatures: The Need to Cross the Borders and Adapt and Modernize the Statistics Discipline

Ali S. Hadi*[1] and A.H.M. Rahmatullah Imon[2]

[1]Department of Mathematics and Actuarial Science,
The American University in Cairo,
Egypt and Department of Statistics and Data Science, Cornell University, USA.
Emails: ahadi@aucegypt.edu and ali-hadi@cornell.edu

[2]Department of Mathematical Sciences, Ball State University,
Muncie, Indiana, USA.
E-mail: rimon@bsu.edu

13.1 INTRODUCTION

The focus of this book is on *Statistical Outliers and Related Topics*. This chapter goes beyond the statistical outliers and into the related topics particularly the statistics and computer science literatures. As can be seen from the list of references, an abundant number of papers

*For Correspondence: Ali S. Hadi (ahadi@aucegypt.edu)

about outliers (also known as anomaly or novelty detection in the computer science literature) have been published in each of these two literatures. Unfortunately, the two areas seem to be disconnected. That is, researchers in one area seem to be unaware of the work done by researchers in the other area. This is evident by the few incidences of cross citations (papers in one area cite papers in the other area). Another evidence of this disconnect is the use of diverse terminologies for the same concepts. Table 13.1 provides some examples of the terminologies commonly found in the statistics literature and the corresponding terminologies in the computer science literature. Table 13.1 also shows that some terminologies in the computer science literature have no corresponding terminologies in the statistics literature. Consequently, the first point of this chapter is that there is a great need for researchers in the statistics and computer science areas to cross borders and to unify terminologies. Although the list of references in this chapter is by no means exhaustive, the list contains many references from both the statistics and computer science literatures. It is hoped that this will facilitate more collaboration among researchers in both areas.

Table 13.1 Examples of terminologies in statistics literature and the corresponding terminologies in computer science literature

No.	Statistics	Computer Science
1	Outlier or Aberrant	Anomaly or Novelty
2	Variables	Attributes or Features
3	Observations	Objects or Cases
4	Modeling	Structural Learning
5	Parameter Estimation	Parametric Learning
6	Classification	Supervised Learning
7	Clustering	Unsupervised Learning
8		Semi-supervised Learning
9		Neural Networks
10		Data Mining
11		Text Mining
12		Deep Learning
13		Visualization
14		Machine Learning
15		Reinforcement Learning
16		Image Processing
17		Bagging
18		Boosting, etc.

13.2 A MOTIVATING SIMPLE EXAMPLE

Another goal of this chapter is to point out that if we, statisticians, do not adapt ourselves to the needs of the world in which we live, we will go into extinction! Whether we like it or not, Statistics or Data Analysis are now being replaced by Data Science! To demonstrate this point, we use the following very simple example. Suppose we show our readers examples of four tools that have been used by farmers in agriculture for centuries. The agricultural tools given in Figure 13.1 were very useful tools for farmers until very recently. At the time they were invented, they were called technology, just like today's tools that we call technology. On the other hand, Figure 13.2 shows an example of a nowadays wheat harvesting, threshing, and winnowing machine, all in one. The tools that were once considered as the technology of their time, now many, if not all of them, have gone into extinction!

(a) Water lifting, a tool used for irrigation

(b) A tool for wheat threshing

(c) Two winnowing tools known as Pitchforks.

Figure 13.1 Examples of four tools that have been used by farmers in agriculture for centuries until only recently. *Source*: Obtained from the Internet by searching, for example, for "Winnowing Tools".

Suppose further that we measured the following variables from each of our readers of this chapter:

X1: Have you lived in a developing or underdeveloped country?

X2: Have you lived in a rural area?

X3: Are you older than, say, 50 years?

Y: Have you seen the tools in Figure 13.1 used in practise?

Clearly, all the variables are binary taking either Yes or No values. We wish to predict (or classify) the response variable Y based on some or all of the predictor or factor variables X1, X2, and X3. This is a classification (also known as supervised learning) problem because the response variable Y is known (given in our data).

Figure 13.2 An example of a nowadays wheat harvesting, threshing, and winnowing machine all in one.

This is a simple classification problem. Which statistical method should we use to solve this classification problem? Obviously, we cannot use Fisher (1936) Linear Discriminant Method because:

1. The data are not multivariate normal; not even quantitative.
2. Even if the predictor variables X1, X2, and X3 are numerical, the categories of the response Y may or may not have equal covariance matrices. The means, variances, covariances, and correlations are not even defined for categorical variables!

The answer to this classification (prediction) question, however, can be provided by a regression method. Some of these methods are given in Table 13.2. This raises two questions:

1. Which of these methods is suitable here? and
2. How many of them do we teach in our regression course?

Of course, the answer to these questions will vary from one academic statistician to another. The current authors, for example, would use the Classification and Regression Trees (CART) or Neural Networks, to choose only two. Two examples of CART models for this dataset

Table 13.2 Examples of regression methods

No.	Method
1	Linear Regression
2	General Linear Models
3	Generalized Linear Models
4	Polynomial Regression
5	Logistic Regression
6	Quantile Regression
7	Ridge Regression
8	Lasso Regression
9	Ordinal Regression
10	Poisson Regression
11	Elastic Net Regression
12	Cox Regression
13	Nonlinear Regression
14	Neural Networks
15	Support Vector Regression
16	Principal Component Regression
17	Partial Least Square Regression
18	Negative Binomial Regression
19	Classification and Regression Trees (CART)

are given in Figure 13.3. Other examples could be obtained using different sequences of the variables. Do we usually teach this method in my regression course? Our honest answer is no. This is because one regression course is not enough to teach even the majority, not to mention all, of these methods.

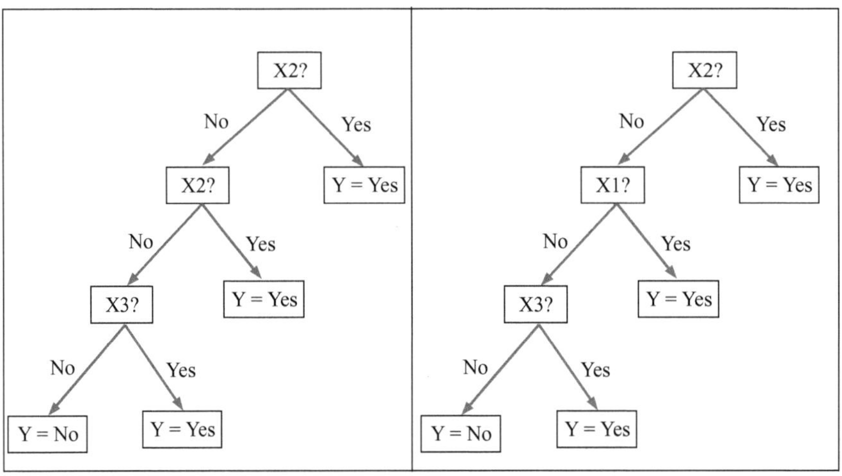

Figure 13.3 Two examples of CART models.

Another reason is that statistics teachers tend to emphasize theory more than application, which is, of course, time consuming. We tend to focus heavily on theoretical (mathematical) aspects and continue to discuss things that are no longer necessary. For example:

- Different formulas to compute the variance of just one variable.
- What is the optimal number of bars in a histogram?
- How to compute the angles when constructing a Pie Chart?

Our readings of both the statistics and computer science literatures indicate a very unbalanced interest in and, hence, coverage of, methods for dealing with numerical versus categorical variables. Although the most commonly observed real-world datasets are of the type mixed datasets (some variables are numerical, while others are categorical in the same dataset), the research on numerical only datasets far exceeds that on the categorical datasets. We explain here 4 of the topics demonstrating this imbalance of treatment: (a) statistical inference, (b) identification of outliers, (c) measures of association between variables, and (d) problems associated with Big data.

13.3 ISSUES RELATED TO STATISTICAL INFERENCE

Most of the statistical inference procedures assume that the data are not only numerical but also normally distributed. These lead to the most commonly used test statistics such as the normal, the Student T, and the F distributions. By comparison, tests for categorical data, such as the Chi-square tests of independence, receive much less coverage. Look also at the interest in investigating the statistical properties of the arithmetic mean as opposed to other summary measures such as the median and mode.

When we explain statistical inference, we tend to assume difficulties away by using methods that depend on ridiculous assumptions such as the predictor variables in a regression equation are nonrandom and also measured without error! We also tend to use these methods:

- Without explicitly mentioning many of these assumptions,
- Without discussing sensitivity analysis (the impact of violations), and
- Without discussing tools for the validation of these assumptions (e.g., graphs, tests, etc.)

We often concentrate very heavily on continuous data, normal distribution, equal variances, etc. How about methods for the analysis of other types of data such as categorical data, nominal data, text data, images (photos, MRI), video, voice data, or even unstructured mixed data, and data streams?

We also tend not to teach our students how to "think critically". For example:

- Why do we choose H_0 as the no-effect hypothesis?
- What would happen if we switched H_0 and H_1?
- Why do we control Type I error probability (α) and only hope that Type II error probability (β) will be small?
- Would the least squares' estimators continue to be BLUE (Best Linear Unbiased Estimators) when some assumptions are violated?

13.4 IDENTIFICATION OF OUTLIERS

Outlier identification, in the statistics literature (also known as anomaly or novelty detection in the computer science literature) has a venerable history. Outliers are a minority of observations that are inconsistent with the pattern suggested by the majority of observations in the same dataset (Taha and Hadi 2019). Most of the attention, however, was paid to the identification of outliers in numerical data. See, for example, Chatterjee and Hadi (1986), Rousseeuw and Leroy (1987), Hadi (1992a, 1992b, 1994), Hadi and Simonoff (1993), Barnett and Lewis (1994), Rousseeuw and van Driessen (1999), Angiulli, Basta and Pizzuti (2006), Jiang et al. (2006), Cheng et al. (2009), Kondylis, Hadi and Werner (2012), Bouguessa (2014), Filzmoser, Ruiz-Gazen and Thomas-Agnan (2014), Hadi and Imon (2019), and Touny, Moussa and Hadi (2020).

In addition, the following authors, for example, provide extensive surveys for the identification of outliers in quantitative data: Hodge and Austin (2004), Agyemang, Barker and Alhajj (2006), Lin and Brown (2006), Chandola et al. (2009), Hadi et al. (2009), Aggarwal (2013), Gupta et al. (2014), Kalinichenko, Shanin and Taraban (2014).

Consequently, numerous methods have been suggested in the domain of outlier detection in numerical, but three methods seemed to have emerged from this research. The first and by far the most common is the Mahalanobis (1936) distance,

$$d_i(\underline{x}, S) = \sqrt{(x_i - \underline{x})^T S^{-1} (x_i - \underline{x})}, \quad i = 1, \ldots, n,$$

where x_i is the i-th observation in the numeric $n \times p$ data matrix $\mathbf{X}$, $\underline{x}$ are the sample mean, and S is the sample covariance matrix. Values of $d_i^2 > \chi^2_{(p,\alpha)}$ are considered outliers. Mahalanobis distance is fast to compute (efficiently), even for large datasets. The problem with the Mahalanobis distance, however, is that it is not robust because the sample mean and covariance matrix are not robust. Robust versions of the Mahalanobis distance can be obtained by substituting more robust estimates of the mean and covariance matrix. Two such estimators are the microcolony detection method (MCD) suggested by Rousseeuw and van Driessen (1999) and the BACON method suggested by Billor, Hadi and Velleman (2000). BACON has been implemented in some software packages such as Stata (Weber 2010) and R (the functions BACON and mvBACON in the package robustX). Versions of BACON that can handle missing values and different observation weights are provided by Béguin and Hulliger (2008): the function BEM (block, element, modifier) in the package MODI (multivariate outlier detection and imputation) and Schoch (2021) in the package wBACON, respectively.

Much less attention has been paid to the problem of the identification of outliers in categorical data. Taha and Hadi (2019), however, provide an extensive survey on the identification of outliers in categorical data. They categorized existing algorithms into different approaches based on the conceptual definition of outliers they use. For each approach, they survey outlier identification algorithms, and then show the similarities and differences among them.

It is interesting to note here that most published work on the topic outlier/anomaly detection comes from the computer science/machine learning area. We conducted an Internet search for the terms "anomaly detection" and "outlier detection", restricted to articles published during the first three months of 2023 and we found 55 articles all of which are published in computer science/machine learning journals/conferences.

13.5 MEASURES OF ASSOCIATION

Another topic which receives an disproportionate coverage for numeric and categorical data is: How to measure relationship between variables? Most of the time is spent on the case where the two variables are quantitative. In this case we teach our students to calculate the correlation coefficient between the two variables and, possibly, but not always augment this numeric summary by the corresponding scatter plot to validate the two assumptions (the linearity and no outliers) behind a correct interpretation of the correlation coefficient. By contrast, we spend relatively less time, if any, about measuring association between two categorical variables. This is despite the fact that such measures are abundant in the statistical literature. Examples of such measures are:

1. The Uncertainty coefficient (Theil 1970),
2. The Cramer's (1999) V^2,
3. The Concentration association coefficient, τ (Goodman and Kruskal 1954),

4. The simple correspondence analysis association coefficient (Benzecri 1973, and Blasius and M. Greenacre 1994),
5. The maximal information coefficient (Reshef et al. 2011), and
6. The two association measures η_1 and η_2 proposed by Taha and Hadi (2016), who also provide a review and comparisons among the above association measures.
7. The multiple association measure that can be used with categorical, numerical, and/or mixed datasets proposed by Taha et al. (2023).

13.6 THE PROBLEMS OF BIG DATA

"Statisticians have missed several opportunities in the past to meet the needs of analyzing big and complex datasets." (Hadi 2017.) Even when the two variables are quantitative, how many of us discuss the problems associated with large datasets not to mention big data. In our teaching of statistics, we continue to insist on using very small datasets, while big data represent many challenges for both statisticians and computer programmers:

- Computing Speed (Complexity, waiting time): Cloud computing and Parallel programming,
- Statistical Inference (when n is large, everything is significant) and, hence the null hypothesis will very likely be rejected (recall, for example, that the standard error of the sample mean tends to zero as the sample size tends to infinity), and
- The Cluttering problem: Cluttering occurs when the sample size is large causing points to be over-plotted. For example, as can be seen in Figure 13.4, the relationship between the variables Y and X is clearly positive when the sample size is n = 200. This relationship cannot be seen because of cluttering when n = 1,000,000. This size does not even classify the dataset to be called big data. Big data comes from different sources and are characterized by the so-called 5 Vs: volume, value, variety, velocity, and veracity; see for example, Minelli, Chambers and Dhiraj (2013) and Acharya and Chellappan (2015). Some possible solutions to the cluttering problem have been proposed in the literature. See, for example, the research on the Parallel Coordinate Plots (e.g., Inselberg, 1985, and Wegman 1990) and the associated enveloping methods proposed by Inselberg, Chomut and Reif (1987), and Moustafa, Hadi and Symanzik (2011) and the references therein.

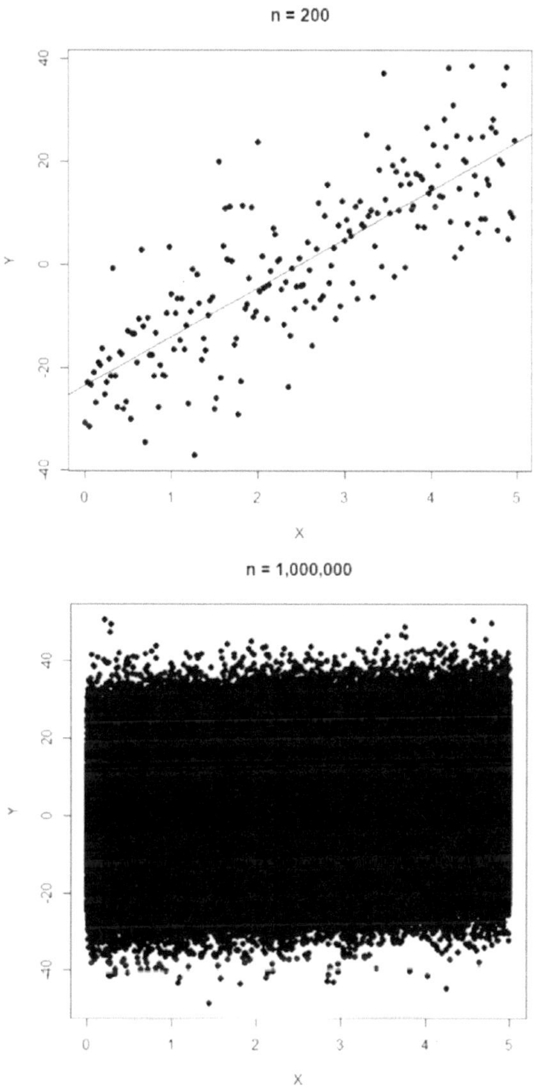

Figure 13.4 An example of cluttering.

13.7 SUMMARY AND CONCLUSIONS

We start this concluding section with the following quotation: "The demand for data science education is surging and traditional courses offered by statistics departments are not meeting the needs of those seeking training." (Hicks and Irizarry 2017.) Consequently, we need to adapt and modernize our discipline (and ourselves) to meet the continually changing needs of the world in which we live, see e.g., Cleveland (2001) and Hardin et al. (2015). If we, statisticians, do not adapt ourselves to the needs of the world in which we live, we will go into extinction! We have no other viable option but to adopt, embrace, and contribute to the newly developing interdisciplinary/multidisciplinary field of Data Science, which is defined as the study of generalizable extractions of knowledge from data. What do we need to do to adapt?

- We need to change our thinking about our profession (in other words, we need a cultural change), that is, we need to adapt and modernize our discipline (and ourselves) to meet the continually changing needs of the world in which we live.
- We need to revolutionize our statistics curricula and the way we teach them.
- When we teach statistics, we should distinguish between the different types of our audience (students), statistics majors and non-majors. The first audience is more interested in theory but the second is most interested in application.
- We should move away from using obsolete and very expensive software packages and instead teach our students (including majors) the more modern, free, and multiplatform languages/packages such as R and Python.

REFERENCES

Acharya, S. and Chellappan, S. 2015. Big Data Analytics., John Wiley & Sons, New York.

Aggarwal, C.C. 2013. Outlier Analysis. Springer-Verlag, New York, NY.

Agyemang, M., Barker, K. and Alhajj, R. 2006. A comprehensive survey of numeric and symbolic outlier mining techniques. Intelligent Data Analysis. 10(6): 521–538.

Angiulli, F., Basta, S. and Pizzuti, C. 2006. Distance-based detection and prediction of outliers. IEEE Transaction on Knowledge and Data Engineering. 18(2): 145–160.

Barnett, V. and Lewis, T.B. 1994. Outliers in Statistical Data. Wiley, New York.

Béguin, C. and Hulliger, B. 2008. The BACON-EEM algorithm for multivariate outlier detection in incomplete survey data. Survey Methodology. 34(1): 91–103.

Billor, N., Hadi, A.S. and Velleman, P.F. 2000. BACON: Blocked adaptive computationally efficient outlier nominators. Computational Statistics and Data Analysis. 34: 279–298.

Benzecri, J. 1973. "Lánalyse des donn´ees: Lánalyse des correspondances. Lánalyse des donnees: Lecons sur lánalyse factorielle et la reconnaissance des formes et travaux. Dunod, Paris.

Blasius, J. and Greenacre, M. 1994. Computation of correspondence analysis. Correspondence Analysis in the Social Sciences, 53–75.

Bouguessa, M. 2014. A mixture model-based combination approach for outlier detection. International Journal on Artificial Intelligence Tools. 23(4): 1–21.

Chandola, V., Banerjee, A. and Kumar, V. 2009. Anomaly detection: A survey. ACM Computing Surveys (CSUR). 41(3): 1–58.

Chatterjee, S. and Hadi, A.S. 1986. Influential observations, high leverage points, and outliers in linear regression (with discussions). Statistical Science. 1: 379–416.

Cheng, H., Tan, P.-N., Potter, C. and Klooster, S.A. 2009. Detection and characterization of anomalies in multivariate time series. In: Proceedings of the SIAM International Conference on Data Mining, SDM. 413–424. Lisbon, Portugal.

Cleveland, W. 2001. Data science: An action plan for expanding the technical areas of the field of statistics. International Statistics Review. 69: 21–26.

Cramer, H. 1999. Mathematical Methods of Statistics. Princeton University Press, Princeton, N.J.

Filzmoser, P., Ruiz-Gazen, A. and Thomas-Agnan, C. 2014. Identification of local multivariate outliers. Statistical Papers. 55: 29–47.

Fisher, R.A. 1936. The use of multiple measurements in taxonomic problems. Annals of Eugenics. Part II. 7: 179–188.

Goodman, L. and Kruskal, W. 1954. Measures of associations for cross classifications. Journal of the American Statistical Association. 49: 732–764.

Gupta, M., Gao, J., Aggarwal, C.C., and Han, J. 2014. Outlier detection for temporal data: A survey. IEEE Transactions on Knowledge and Data Engineering. 26(9): 2250–2267.

Hadi, A.S. 1992a. A new measure of overall potential influence in linear regression. Computational Statistics and Data Analysis. 14: 1–27.

Hadi, A.S. 1992b. Identifying multiple outliers in multivariate data. Journal of the Royal Statistical Society, Series (B). 54: 761–771.

Hadi, A.S. 1994. A modification of a method for the detection of outliers in multivariate samples. Journal of the Royal Statistical Society, Series (B). 56: 393–396.

Hadi. A.S. 2017. Big data: Many challenges and some solutions. Presented at the 61st ISI World Statistics Congress, Marrakech, Morocco.

Hadi, A.S. and Imon, A.H.M.R. 2019. Identification of multiple outliers in spatial data. International Journal of Statistical Sciences. 16: 87–96.

Hadi, A.S. and Simonoff, J.S. 1993. Procedures for the identification of multiple outliers in linear models. Journal of the American Statistical Association. 88: 1264–1272.

Hadi, A.S., Imon, A.H.M.R. and Werner, M. 2009. Detection of outliers. Wiley Interdisciplinary Reviews: Computational Statistics. 1: 57–70.

Hardin, J., Hoerl, R., Horton, N.J., Nolan, D., Baumer, B., Hall-Holt, O., Murrell, P. et al. 2015. Data science in statistics curricula: Preparing students to think with data. American Statistician. doi: 10.1080/00031305.2015.1077729.

Hicks, S.C. and Irizarry, R.A. 2017. A Guide to teaching data science. American Statistician, Accepted manuscript. Online Journal homepage: http://www.tandfonline.com/loi/utas20.

Hodge, V. and Austin, J. 2004. A survey of outlier detection methodologies. Artificial Intelligence Review. 22: 85–126.

Inselberg, A. 1985. The plane with parallel coordinates. The Visual Computer. 1: 69–91.

Inselberg, A., Chomut, T. and Reif, M. 1987. Convexity algorithms in parallel coordinates. Journal of the ACM. 34: 765–801.

Jiang, S., Song, X., Wang, H., Han, J.-J. and Li, Q.-H. 2006. A clustering-based method for unsupervised intrusion detections. Pattern Recognition Letters. 27: 802–810.

Kalinichenko, L., Shanin, I. and Taraban, I. 2014. Methods for anomaly detection: A survey. In: All-Russian Conference Digital Libraries: Advanced Methods and Technologies, Digital Collections, RCDL. 20–25. Dubna, Russia.

Knorr, E. and Ng, R. 1997. A unified notion of outliers: properties and computation. In: Proceedings of the International Conference on Knowledge Discovery and Data Mining. 219–222.

Kondylis, A., Hadi, A.S. and Werner, M. 2012. The Bacon approach for rank-deficient data. Pakistan Journal of Statistics and Operation Research, Statistics in the Twenty-First Century: Special Volume in honor of Dr. Mir Masoom Ali on the Occasion of his Seventy-Fifth Birthday. 8: 359–379.

Lin, S. and Brown, D.E. 2006. An outlier-based data association method for linking criminal incidents. Decision Support Systems. 41(3): 604–615.

Mahalanobis, P.C. 1936. On the generalised distance in statistics. In: Proceedings of the National Institute of Statistics of India. 2(1): 49–55.

Minelli, M., Chambers, M. and Dhiraj, A. 2013. Big Data, Big Analytics: Emerging Business Intelligence and Analytic Trends for Today's Businesses. John Wiley & Sons, New York.

Moustafa, R.E., Hadi, A.S. and Symanzik, J. 2011. Multi-class data exploration using space transformed visualization plots. Journal of Computational and Graphical Statistics. 20: 298–315.

Reshef, D.N., Reshef, Y.A., Finucane, H.K., Grossman, S.R., McVean, G., Turnbaugh, P.J., Lander, E.S. et al. 2011. Detecting novel associations in large data sets. Science. 334: 1518–1524.

Rousseeuw, P.J. and Leroy, A. 1987. Robust Regression and Outlier Detection. Wiley, New York.

Rousseeuw, P.J. and van Driessen, K.V. 1999. A fast algorithm for the minimum covariance determinant estimator. Technometrics. 41: 212–223.

Schoch, T. 2021. WBACON: Weighted BACON algorithms for multivariate outlier nomination (detection) and robust linear regression. The Journal of Open Source Software. doi: 10.21105/joss.03238.

Taha, A. and Hadi, A.S. 2016. Pair-wise association for categorical and mixed attributes. Information Sciences. 346: 73–89. doi: http://dx.doi.org/10.1016/j.ins.2016.01.022.

Taha, A. and Hadi, A.S. 2019. Anomaly detection for categorical data: A review. ACM Computing Surveys. Article 38, 52(2): 1–35. https://doi.org/10.1145/3312739.

Taha, A., Hadi, A.S., Cosgrave, B. and McKeever, S. 2023. A multiple association-based unsupervised feature selection algorithm for mixed data sets. Expert Systems with Applications, 212, 118718. https://doi.org/10.1016/j.eswa.2022.118718.

Theil, H. 1970. On the estimation of relationships involving qualitative variables. American Journal of Sociology. 76: 103–154.

Touny, H.M., Moussa, A.S. and Hadi, A.S. 2020. Fuzzy multivariate outliers with application on the BACON algorithm. In: 2020 IEEE International Conference on Fuzzy Systems (FUZZ-IEEE). Glasgow, United Kingdom. 1–7. doi: 10.1109/FUZZ48607.2020.9177652.

Weber, S. 2010. BACON: An effective way to detect outliers in multivariate data using Stata (and Mata). The Stata Journal. 10(3): 331–338.

Wegman, E.J. 1990. Hyperdimensional data analysis using parallel coordinates. Journal of the American Statistical Association. 85: 664–675.

Chapter **14**

On a Unit-EEG Quantile Regression Analysis

Mohamed El Moustapha Sidi[1], Coşkun Kuş[2], Mustafa Çağatay Kokmaz[3], Ahmet Pekgör[4] and Kadir Karakaya*[5]

[1,2,5] Department of Statistics, Selçuk University, Konya, Turkey. Emails: [1]moustafabb@gmail.com; [2]coskun@selcuk.edu.tr; [5]kkarakaya@selcuk.edu.tr

[3]Department of Measurement and Evaluation, Artvin Çoruh University, Artvin, Turkey. Email: mustafacagataykorkmaz@gmail.com,

[4]Department of Statistics, Necmettin Erbakan University, Konya, Turkey. Email: istatistik@hotmail.com,

14.1 INTRODUCTION

Many statistical distributions have been introduced in the past two decades. However, offering distribution with unit domain is becoming popular and quantile regression analysis is provided to demonstrate the applicability of introduced unit distributions. Unit distributions are used for modeling the recovery rate, mortality rate, proportion of the educational measurements, etc. The distribution with bounded support can be used for modeling the lifetime data. A failure time distribution with upper limit does not mean that all failures in a life test can be observed. From an engineering point of view, the bounded lifetime

*For Correspondence: Kadir Karakaya (kkarakaya@selcuk.edu.tr)

distribution is more appropriate than the lifetime distribution with infinite support because life is always finite. There are many practical situations that lead to bounded support in the areas of lifetime study and reliability engineering. For example, when failure or reliability life test data is collected by an automatic recorder, the lifetimes below L and above T ($L < T$) may not be measured at all due to the device's resolving power or other environmental influences (Zhang and Xie 2011). From this perspective, new distributions with bounded support such as the Beta or Kumaraswamy distribution are required for modeling such lifetime data.

Some recent studies in this context may be listed: Mazucheli et al. (2018), Ghitany et al. (2019), Mazucheli et al. (2019a), Mazucheli et al. (2019b), Mazucheli et al. (2020a), Mazucheli et al. (2020b), Mazucheli et al. (2021), Korkmaz and Korkmaz (2021), Korkmaz and Chesneau (2021), Korkmaz et al. (2021a), Korkmaz et al. (2021b), and Korkmaz et al. (2021c).

On the other hand, there is a problem with quantile regression analysis in the papers listed above in obtaining maximum likelihood estimates (MLEs). The R functions such as maxLik, optim, or nlm are used to optimize the likelihood functions. These functions use the Nelder-Mead, Newton-Raphson, BFGS, CG, and other numerical methods. However, these methods are sensitive to the initial values of parameters. In the example section, as an alternative to the existing methods, we propose an algorithm that is not influenced by the initial values to maximize the likelihood function.

In this study, we introduce a new distribution with the unit domain and construct the new quantile regression analysis. In addition, a new algorithm is proposed to get the maximum likelihood estimates which work well with arbitrary initial values. In Section 14.2, a new distribution is proposed with support (0, 1). Some estimation issues are discussed for the distribution parameters. In Section 14.3, quantile regression analysis is discussed based on introduced distribution and maximum likelihood estimation is studied for the regression model parameters and an optimization algorithm and some numerical examples are provided to close the paper.

14.2 UNIT EXTENDED EXPONENTIAL-GEOMETRIC DISTRIBUTION

Adamidis et al. (2005) has introduced extended exponential-geometric (EEG) distribution via a transformation of the exponential-geometric distribution introduced by Adamidis and Loukas (1998). Let the random variable Z follow an EEG distribution with probability density function (pdf) and cumulative distribution function (cdf)

$$f_{EEG}(z,\alpha,\beta)=\frac{\alpha(\beta+1)\exp(-\alpha z)}{(1+\beta\exp(-\alpha z))^2} \tag{1}$$

and

$$F_{EEG}(z,\alpha,\beta)=\frac{1-\exp(-\alpha z)}{1+\beta\exp(-\alpha z)}, \tag{2}$$

respectively, where $z > 0$, $\alpha, \beta > 0$. Let us consider transformation $X = Z/(1 + Z)$, where Z is a random variable with pdf (Eq. 1). The distribution of X is called unit extended exponential-geometric (UEEG) distribution. The pdf and cdf of the UEEG distribution are given, respectively, by

$$f(x,\alpha,\beta)=\frac{\alpha(\beta+1)\exp\left(-\frac{\alpha x}{1-x}\right)}{(1-x)^2\left(1+\beta\exp\left(-\frac{\alpha x}{1-x}\right)\right)^2} \tag{3}$$

and

$$F(x,\alpha,\beta)=\frac{1-\exp\left(-\frac{\alpha x}{1-x}\right)}{1+\beta\exp\left(-\frac{\alpha x}{1-x}\right)}, \tag{4}$$

where $0 < x < 1$, $\alpha > 0$, $\beta > 0$ are the parameters. Hereafter, it is denoted by $UEEG(\alpha, \beta)$.

The approximate rth moment of the $UEEG(\alpha, \beta)$ distribution is obtained as follows:

$$E(X^r)=\int_0^1 x^r\frac{\alpha(\beta+1)\exp\left(-\frac{\alpha x}{1-x}\right)}{(1-x)^2\left(1+\beta\exp\left(-\frac{\alpha x}{1-x}\right)\right)^2}dx$$

$$=\int_{-1}^{1}\left(\frac{y+1}{2}\right)^r\frac{\alpha(\beta+1)\exp\left(-\frac{\alpha(y+1)}{1-y}\right)}{2\left(\frac{1-y}{2}\right)^2\left(1+\beta\exp\left(-\frac{\alpha(y+1)}{1-y}\right)\right)^2}dy$$

$$\approx\sum_{\ell=1}^{N}\varpi_\ell\left(\frac{y_\ell+1}{2}\right)^r\frac{\alpha(\beta+1)\exp\left(-\frac{\alpha(y_\ell+1)}{1-y_\ell}\right)}{2\left(\frac{1-y_\ell}{2}\right)^2\left(1+\beta\exp\left(-\frac{\alpha(y_\ell+1)}{1-y_\ell}\right)\right)^2},$$

where y_ℓ and ϖ_ℓ are the zeros and the corresponding Christoffel numbers of the Legendre-Gauss quadrature formula on the interval (–1, 1) respectively (Canuto et al. 2007) and N is the degree of the Legendre polynomial. It should be noted that y_ℓ and ϖ_ℓ can be easily calculated with an R function **legendre.quadrature** from the **gaussquad** library (Novomestky and Novomestky 2013). Note that a large N value is not necessary to approximate the true value.

The quantile function of the UEEG distribution can be written by

$$x_u(\alpha,\beta)=\frac{\log\left(\dfrac{1-u}{1+u\beta}\right)}{\log\left(\dfrac{1-u}{1+u\beta}\right)-\alpha},$$

where $0 < u < 1$. This function is used to construct the quantile regression in Section 14.3.

Figure 14.1 shows the head map of skewness and kurtosis. It can be seen from Figure 14.1 that as α increases, the skewness coefficient increases, and the kurtosis coefficient decreases. Furthermore, as β increases, the skewness coefficient decreases, and the kurtosis coefficient increases.

Stochastic ordering is a fundamental measure for evaluating the comparative behavior of random variables in distribution theory. The following theorem indicates that the UEEG random variables can be ordered in terms of likelihood ratio.

Theorem 1. Let $X \sim UEEG(\alpha, \beta_1)$ and $Y \sim UEEG(\alpha, \beta_2)$. If $\beta_1 < \beta_2$ then the random variable X is smaller than the random variable Y in the likelihood ratio order.

Proof. For any $x \in (0, 1)$, the ratio of the two densities is given by

$$g(x)=\frac{(\beta_1+1)\left(1+\beta_2\exp\left(-\dfrac{\alpha x}{1-x}\right)\right)^2}{(\beta_2+1)\left(1+\beta_1\exp\left(-\dfrac{\alpha x}{1-x}\right)\right)^2}.$$

Consider the derivative of logarithm of the $g(x)$ in x

$$\frac{\partial\log(g(x))}{\partial x}=\frac{2\alpha(\beta_1-\beta_2)\exp\left(\dfrac{-\alpha x}{1-x}\right)}{(1-x)^2\left(1+\beta_2\exp\left(-\dfrac{\alpha x}{1-x}\right)\right)\left(1+\beta_1\exp\left(-\dfrac{\alpha x}{1-x}\right)\right)}<0$$

for $\beta_1 < \beta_2$ and this completes the proof.

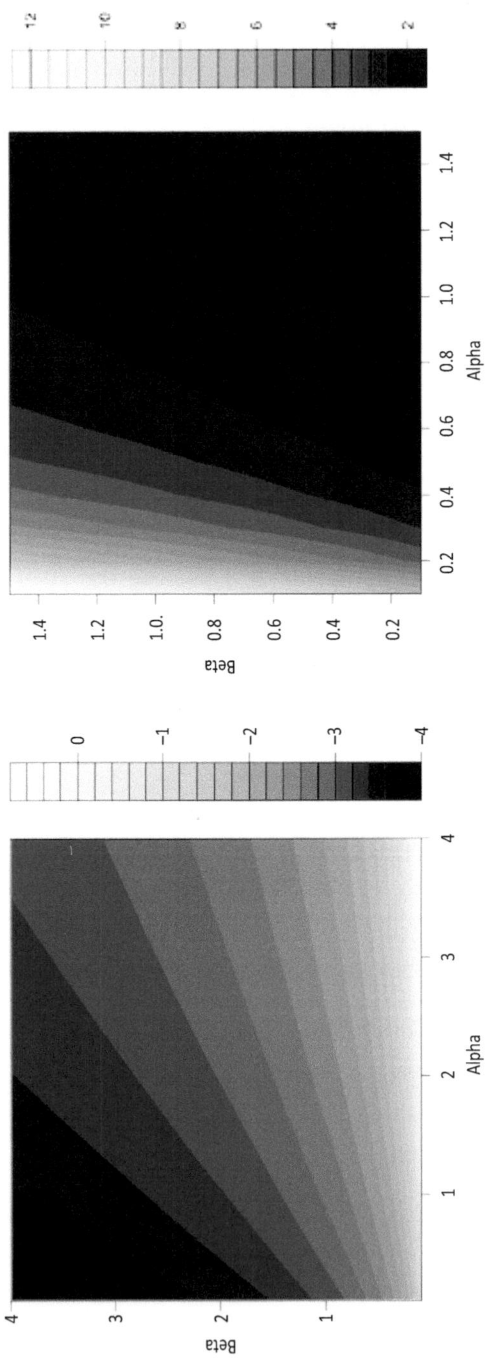

Figure 14.1 Skewness (on the left) and kurtosis (on the right) of the UEEG distribution.

14.2.1 Point Estimation of UEEG Distribution Parameters

In this section, different estimators of distribution parameters are discussed for the UEEG distribution. Let $\boldsymbol{X} = (X_1, X_2, \ldots, X_n)$ be a random sample from the $UEEG(\alpha, \beta)$ and $\boldsymbol{x} = (x_1, x_2, \ldots, x_n)$ the observed values of the sample. Then, the log likelihood function can be written by

$$\ell(\Xi) = n\log(\alpha) + n\log(\beta+1) - \alpha\sum_{i=1}^{n}\frac{x_i}{1-x_i}$$
$$-2\sum_{i=1}^{n}\log(1-x_i) - 2\sum_{i=1}^{n}\log\left(1+\beta\exp\left(-\frac{\alpha x_i}{1-x_i}\right)\right),$$

where $\Xi = (\alpha, \beta)$. Hence, the MLE of Ξ is given by

$$\hat{\Xi} = \arg\max_{\Xi} \; \ell(\Xi), \tag{5}$$

where $\hat{\Xi} = (\hat{\alpha}, \hat{\beta})$.

Let us define $X_{(1)}, X_{(2)}, \ldots, X_{(n)}$ be the ordered statistics based on sample $\boldsymbol{X}$ with the realization $x_{(1)}, x_{(2)}, \ldots, x_{(n)}$. Let us consider the following functions:

$$LS(\Xi) = \sum_{i=1}^{n}\left(F(x_{(i)};\Xi) - \frac{i}{n+1}\right)^2, \tag{6}$$

$$WLS(\Xi) = \sum_{i=1}^{n}\frac{(n+2)(n+1)^2}{i(n-i+1)}\left(F(x_{(i)};\Xi) - \frac{i}{n+1}\right)^2, \tag{7}$$

$$ADD(\Xi) = -n - \sum_{i=1}^{n}\frac{2i-1}{n}\left[\log F(x_{(i)};\Xi) + \log\{1 - F(x_{(n+1-i)};\Xi)\}\right] \tag{8}$$

and

$$CVM(\Xi) = \frac{1}{12n} + \sum_{i=1}^{n}\left[F(x_{(i)};\Xi) - \frac{2i-1}{2n}\right]^2. \tag{9}$$

The least square estimate (LSE), the weighted least square estimate (WLSE), the Anderson-Darling type estimate (ADE), and the Cramer-von Mises type estimate (CvME) are obtained by minimizing Eqs. (6)–(9), respectively.

It is noted that the LSE, WLSE, ADE, and CvME are denoted by $(\hat{\alpha}_2, \hat{\beta}_2)$, $(\hat{\alpha}_3, \hat{\beta}_3)$, $(\hat{\alpha}_4, \hat{\beta}_4)$ and $(\hat{\alpha}_5, \hat{\beta}_5)$, respectively in the Tables 14.1 and 14.2.

In the simulation study, the bias and MSEs of MLE, LSE, WLE, ADE, and CvME of distribution parameters are simulated with 5000 trials. The simulation results are presented in Tables 14.1 and 14.2 for the true parameter vectors Ξ = (0.5, 0.7), (0.9, 2), (1.5, 0.3) and (2, 2), respectively. These tables show that the bias and MSEs of MLEs, LSEs, WLEs, ADEs, and CvMEs decrease to zero when the sample size increases. MLE of α and ADE of β are better than the others in terms of MSEs.

Table 14.1 Average bias for the estimators.

			MLE		LSE		WLSE		ADE		CvME	
n			$\hat{\alpha}$	$\hat{\beta}$	$\hat{\alpha}_2$	$\hat{\beta}_2$	$\hat{\alpha}_3$	$\hat{\beta}_3$	$\hat{\alpha}_4$	$\hat{\beta}_4$	$\hat{\alpha}_5$	$\hat{\beta}_5$
10			0.2599	4.1784	0.0463	1.2479	0.0881	2.6080	0.1497	2.5138	0.2161	2.8507
20			0.1326	1.5914	0.0397	0.8777	0.0582	1.0852	0.0831	1.1697	0.1301	1.6805
30			0.0880	0.9683	0.0272	0.5791	0.0415	0.6689	0.0552	0.7305	0.0874	1.0546
40			0.0640	0.6695	0.0193	0.4170	0.0319	0.4749	0.0398	0.5099	0.0640	0.7455
50			0.0513	0.5079	0.0154	0.3139	0.0265	0.3666	0.0328	0.3930	0.0510	0.5599
60			0.0427	0.4156	0.0128	0.2627	0.0226	0.3053	0.0267	0.3207	0.0424	0.4608
70			0.0352	0.3435	0.0093	0.2127	0.0182	0.2511	0.0208	0.2584	0.0344	0.3765
80			0.0283	0.2768	0.0077	0.1785	0.0153	0.2111	0.0172	0.2154	0.0298	0.3188
90			0.0255	0.2456	0.0064	0.1574	0.0140	0.1901	0.0153	0.1911	0.0259	0.2799
100	0.5	0.7	0.0248	0.2370	0.0073	0.1524	0.0141	0.1817	0.0154	0.1850	0.0248	0.2617
150			0.0131	0.1255	0.0010	0.0677	0.0065	0.0934	0.0067	0.0915	0.0126	0.1367
200			0.0096	0.0912	0.0008	0.0507	0.0050	0.0699	0.0049	0.0671	0.0095	0.1016
250			0.0058	0.0663	−0.0012	0.0338	0.0023	0.0494	0.0021	0.0471	0.0057	0.0740
500			0.0031	0.0329	−0.0003	0.0181	0.0014	0.0250	0.0012	0.0231	0.0032	0.0379
1000			0.0011	0.0128	−0.0007	0.0044	0.0003	0.0091	0.0002	0.0079	0.0010	0.0141
2500			0.0005	0.0074	−0.0003	0.0038	0.0002	0.0061	0.0001	0.0054	0.0004	0.0077
5000			0.0006	0.0053	0.0001	0.0028	0.0003	0.0042	0.0003	0.0038	0.0004	0.0048
10			0.2976	5.7948	−0.0357	0.9429	0.0431	3.4015	0.1385	3.2621	0.1885	2.7753
20			0.1395	2.2990	−0.0100	0.8209	0.0272	1.3913	0.0672	1.5313	0.1129	1.8511
30			0.0787	1.2495	−0.0116	0.5996	0.0113	0.7925	0.0325	0.8938	0.0734	1.2958
40			0.0612	0.8739	−0.0054	0.4512	0.0132	0.5531	0.0259	0.6172	0.0594	0.9660
50			0.0445	0.6564	−0.0083	0.3398	0.0082	0.4245	0.0174	0.4710	0.0432	0.7325
60			0.0365	0.5103	−0.0075	0.2416	0.0072	0.3234	0.0137	0.3531	0.0355	0.5551
70			0.0305	0.4419	−0.0059	0.2325	0.0072	0.3004	0.0115	0.3188	0.0308	0.4981
80			0.0289	0.3777	−0.0033	0.2049	0.0087	0.2617	0.0121	0.2737	0.0288	0.4341
90			0.0236	0.3162	−0.0043	0.1640	0.0062	0.2136	0.0088	0.2230	0.0242	0.3644
100	0.9	2	0.0246	0.3075	−0.0010	0.1704	0.0091	0.2207	0.0113	0.2287	0.0246	0.3500
150			0.0139	0.1903	−0.0031	0.0991	0.0045	0.1359	0.0050	0.1352	0.0139	0.2145
200			0.0120	0.1440	−0.0010	0.0743	0.0053	0.1076	0.0055	0.1057	0.0117	0.1594
250			0.0090	0.1227	−0.0006	0.0719	0.0039	0.0945	0.0039	0.0919	0.0096	0.1397
500			0.0028	0.0395	−0.0013	0.0192	0.0010	0.0308	0.0008	0.0281	0.0038	0.0522
1000			0.0030	0.0305	0.0009	0.0200	0.0021	0.0259	0.0018	0.0240	0.0034	0.0364
2500			0.0009	0.0125	−0.0003	0.0061	0.0004	0.0093	0.0003	0.0084	0.0007	0.0126
5000			0.0005	0.0066	−0.0002	0.003	0.0002	0.0053	0.0002	0.0049	0.0003	0.0064
10			0.9951	3.7955	0.2796	1.4300	0.3692	2.2423	0.6159	2.3592	0.9142	3.2250
20			0.5363	1.4670	0.2238	0.8704	0.2842	1.0365	0.3644	1.0994	0.5331	1.5657
30			0.3672	0.9098	0.1769	0.6270	0.2246	0.6908	0.2601	0.7130	0.3806	1.0412
40			0.2972	0.6796	0.1509	0.4730	0.1893	0.5168	0.2166	0.5434	0.3025	0.7544
50			0.2462	0.5500	0.1326	0.4007	0.1668	0.4393	0.1848	0.4560	0.2533	0.6147
60			0.2076	0.4537	0.1127	0.3314	0.1439	0.3665	0.1547	0.3739	0.2125	0.5003

(Contd.)

(Contd...)

n			MLE $\hat{\alpha}$	$\hat{\beta}$	LSE $\hat{\alpha}_2$	$\hat{\beta}_2$	WLSE $\hat{\alpha}_3$	$\hat{\beta}_3$	ADE $\hat{\alpha}_4$	$\hat{\beta}_4$	CvME $\hat{\alpha}_5$	$\hat{\beta}_5$
70			0.1775	0.3779	0.0974	0.2783	0.1264	0.3111	0.1337	0.3141	0.1826	0.4173
80			0.1603	0.3374	0.0838	0.2446	0.1117	0.2753	0.1177	0.2775	0.1579	0.3632
90			0.1458	0.3057	0.0813	0.2267	0.1055	0.2529	0.1109	0.2555	0.1470	0.3303
100	1.5	0.3	0.1350	0.2836	0.0766	0.2155	0.0991	0.2398	0.1031	0.2409	0.1356	0.3078
150			0.0857	0.1720	0.0449	0.1262	0.0626	0.1447	0.0624	0.1416	0.0840	0.1838
200			0.0638	0.1319	0.0304	0.0930	0.0454	0.1096	0.0456	0.1082	0.0596	0.1351
250			0.0496	0.1025	0.0269	0.0774	0.0374	0.0885	0.0368	0.0864	0.0503	0.1106
500			0.0208	0.0447	0.0103	0.0337	0.0161	0.0395	0.0150	0.0375	0.0220	0.0497
1000			0.0089	0.0182	0.0017	0.0100	0.0057	0.0145	0.0049	0.0133	0.0075	0.0179
2500			0.0037	0.0067	0.0024	0.0056	0.0033	0.0066	0.0030	0.0060	0.0048	0.0088
5000			0.0013	0.0028	0.0007	0.0022	0.0011	0.0027	0.0009	0.0024	0.0018	0.0038
10			0.6541	5.7149	−0.0684	1.0691	0.0929	3.3262	0.2946	3.1529	0.4579	3.1514
20			0.3092	2.2574	−0.0225	0.8023	0.0667	1.4006	0.1441	1.5017	0.2505	1.8160
30			0.1949	1.3164	−0.0059	0.6589	0.0454	0.8499	0.0914	0.9497	0.1844	1.3754
40			0.1261	0.8649	−0.0222	0.4325	0.0242	0.5761	0.0498	0.6140	0.1214	0.9434
50			0.1049	0.6396	−0.0092	0.3372	0.0256	0.4241	0.0457	0.4621	0.1062	0.7308
60			0.0844	0.5584	−0.0165	0.2810	0.0177	0.3655	0.0316	0.3951	0.0787	0.5986
70			0.0707	0.4336	−0.0132	0.2235	0.0173	0.2902	0.0270	0.3060	0.0685	0.4891
80			0.0716	0.3995	−0.0010	0.2195	0.0280	0.2851	0.0349	0.2950	0.0704	0.4497
90			0.0529	0.3258	−0.0123	0.1632	0.0122	0.2166	0.0185	0.2284	0.0509	0.3632
100	2	2	0.0490	0.3024	−0.0079	0.1658	0.0140	0.2128	0.0186	0.2193	0.0491	0.3452
150			0.0379	0.2214	−0.0008	0.1269	0.0156	0.1636	0.0181	0.1670	0.0370	0.2431
200			0.0221	0.1337	−0.0039	0.0736	0.0081	0.0993	0.0087	0.0983	0.0244	0.1587
250			0.0182	0.1187	−0.0033	0.0670	0.0073	0.0917	0.0070	0.0888	0.0193	0.1347
500			0.0128	0.0574	0.0013	0.0312	0.0077	0.0456	0.0069	0.0423	0.0126	0.0643
1000			0.0041	0.0272	−0.0042	0.0067	−0.0001	0.0164	−0.0005	0.0148	0.0014	0.0230
2500			0.0028	0.0164	0.0009	0.0118	0.0023	0.0155	0.0021	0.0147	0.0032	0.0183
5000			0.0001	0.0030	−0.0011	−0.0001	−0.0003	0.0018	−0.0004	0.0014	0.0001	0.0031

Table 14.2 Average MSEs for the estimators.

n			MLE $\hat{\alpha}$	$\hat{\beta}$	LSE $\hat{\alpha}_2$	$\hat{\beta}_2$	WLSE $\hat{\alpha}_3$	$\hat{\beta}_3$	ADE $\hat{\alpha}_4$	$\hat{\beta}_4$	CvME $\hat{\alpha}_5$	$\hat{\beta}_5$
10			0.1937	80.2638	0.1146	11.4082	0.1641	82.5802	0.1398	34.7024	0.1828	25.0557
20			0.0609	10.8483	0.0598	6.6126	0.0589	16.0371	0.0541	8.4875	0.0836	11.6651
30			0.0336	4.4866	0.0393	3.5862	0.0355	4.2536	0.0326	3.8628	0.0510	5.5757
40			0.0219	2.0812	0.0272	2.0866	0.0234	1.9617	0.0218	1.8651	0.0337	3.0905
50			0.0167	1.3753	0.0220	1.4597	0.0187	1.3837	0.0175	1.3016	0.0260	2.0143
60			0.0132	1.0116	0.0179	1.1069	0.0149	1.0158	0.0141	0.9783	0.0207	1.4662
70			0.0110	0.7755	0.0156	0.9097	0.0128	0.8079	0.0120	0.7580	0.0176	1.1578
80			0.0091	0.5818	0.0136	0.7280	0.0109	0.6492	0.0103	0.6087	0.0150	0.9022
90			0.0079	0.4964	0.0122	0.6364	0.0097	0.5629	0.0091	0.5254	0.0133	0.7711
100	0.5	0.7	0.0076	0.4817	0.0110	0.5907	0.0089	0.5218	0.0086	0.4999	0.0120	0.7025
150			0.0048	0.2576	0.0072	0.3213	0.0057	0.2848	0.0055	0.2738	0.0075	0.3592
200			0.0037	0.1908	0.0057	0.2492	0.0045	0.2162	0.0044	0.2091	0.0059	0.2706
250			0.0028	0.1464	0.0043	0.1924	0.0034	0.1655	0.0033	0.1616	0.0044	0.2051

			MLE		LSE		WLSE		ADE		CvME	
n			$\hat{\alpha}$	$\hat{\beta}$	$\hat{\alpha}_2$	$\hat{\beta}_2$	$\hat{\alpha}_3$	$\hat{\beta}_3$	$\hat{\alpha}_4$	$\hat{\beta}_4$	$\hat{\alpha}_5$	$\hat{\beta}_5$
500			0.0015	0.0726	0.0023	0.0971	0.0018	0.0822	0.0018	0.0815	0.0023	0.1003
1000			0.0007	0.0339	0.0011	0.0456	0.0009	0.0384	0.0009	0.0381	0.0011	0.0463
2500			0.0003	0.0131	0.0005	0.0182	0.0004	0.0151	0.0003	0.0151	0.0005	0.0183
5000			0.0001	0.0065	0.0002	0.0087	0.0002	0.0073	0.0002	0.0073	0.0002	0.0088
10			0.3313	159.7071	0.2017	16.2960	0.2819	173.6028	0.2421	63.6888	0.2609	28.7059
20			0.1225	33.5640	0.1125	11.5762	0.1211	28.6643	0.1100	20.4638	0.1351	18.3961
30			0.0699	12.5533	0.0836	8.6619	0.0784	11.9525	0.0707	10.8282	0.0949	12.7121
40			0.0496	6.2622	0.0629	5.4795	0.0547	5.5641	0.0510	5.4004	0.0710	7.8385
50			0.0385	3.9104	0.0513	3.8833	0.0438	3.7556	0.0414	3.6675	0.0563	5.2493
60			0.0313	2.8663	0.0401	2.7746	0.0349	2.7386	0.0332	2.6212	0.0434	3.5770
70			0.0266	2.4336	0.0365	2.6048	0.0309	2.4727	0.0294	2.3857	0.0390	3.2398
80			0.0236	1.9541	0.0316	2.2393	0.0269	2.0631	0.0256	1.9691	0.0336	2.7184
90			0.0206	1.6346	0.0282	1.8465	0.0236	1.6878	0.0227	1.6405	0.0297	2.1992
100	0.9	2	0.0187	1.4771	0.0254	1.6979	0.0213	1.5645	0.0208	1.5310	0.0267	1.9916
150			0.0123	0.8500	0.0169	1.0040	0.0139	0.9001	0.0136	0.8783	0.0174	1.1183
200			0.0088	0.5891	0.0122	0.7214	0.0100	0.6427	0.0098	0.6272	0.0125	0.7825
250			0.0069	0.4708	0.0096	0.5737	0.0079	0.5096	0.0078	0.5002	0.0098	0.6141
500			0.0034	0.2089	0.0047	0.2621	0.0038	0.2277	0.0038	0.2251	0.0048	0.2703
1000			0.0017	0.1037	0.0024	0.1324	0.0019	0.1149	0.0019	0.1142	0.0024	0.1348
2500			0.0007	0.0412	0.0010	0.0518	0.0008	0.0449	0.0008	0.0448	0.0010	0.0522
5000			0.0003	0.0202	0.0005	0.0259	0.0004	0.0223	0.0004	0.0223	0.0005	0.0260
10			2.3507	65.7180	1.4901	14.0267	1.6057	60.2570	1.6107	28.5274	2.8332	37.7410
20			0.7578	7.4073	0.7110	4.6823	0.7090	8.3813	0.6420	5.6091	1.0751	8.4653
30			0.3921	2.5852	0.4630	2.6357	0.4246	2.7822	0.3690	2.1930	0.6384	4.2657
40			0.2752	1.4208	0.3311	1.4761	0.2885	1.4133	0.2663	1.2568	0.4347	2.2315
50			0.2028	0.9693	0.2582	1.1086	0.2237	1.0067	0.2071	0.9356	0.3267	1.5789
60			0.1562	0.6687	0.2072	0.7677	0.1735	0.6950	0.1612	0.6452	0.2544	1.0485
70			0.1260	0.5030	0.1712	0.5959	0.1418	0.5372	0.1325	0.4981	0.2054	0.7844
80			0.1050	0.4038	0.1487	0.5063	0.1203	0.4414	0.1119	0.4064	0.1745	0.6471
90			0.0929	0.3478	0.1271	0.4164	0.1044	0.3673	0.0996	0.3475	0.1479	0.5225
100	1.5	0.3	0.0838	0.3078	0.1216	0.3901	0.0986	0.3446	0.0940	0.3270	0.1391	0.4788
150			0.0480	0.1524	0.0757	0.2085	0.0585	0.1761	0.0558	0.1661	0.0828	0.2410
200			0.0359	0.1094	0.0554	0.1455	0.0425	0.1235	0.0414	0.1195	0.0592	0.1628
250			0.0289	0.0862	0.0455	0.1162	0.0348	0.0980	0.0340	0.0951	0.0480	0.1273
500			0.0142	0.0395	0.0236	0.0556	0.0175	0.0457	0.0173	0.0449	0.0242	0.0581
1000			0.0076	0.0199	0.0121	0.0275	0.0092	0.0229	0.0092	0.0228	0.0122	0.0280
2500			0.0032	0.0082	0.0052	0.0116	0.0039	0.0095	0.0039	0.0094	0.0052	0.0117
5000			0.0015	0.0040	0.0025	0.0056	0.0019	0.0046	0.0019	0.0046	0.0025	0.0056
10			1.6179	159.0407	1.1594	21.6142	1.4625	153.6532	1.1815	61.6586	1.6337	46.2437
20			0.6027	31.5853	0.5768	11.2716	0.6291	31.3008	0.5455	20.5203	0.6881	17.5644
30			0.3688	12.3300	0.4288	8.8606	0.4046	12.2830	0.3690	10.7049	0.4958	13.3317
40			0.2460	6.6307	0.3096	5.7295	0.2765	7.8026	0.2537	5.8801	0.3466	8.1538
50			0.1906	3.8175	0.2499	3.9967	0.2184	3.9813	0.2038	3.6744	0.2769	5.4033
60			0.1567	3.2274	0.2036	3.1862	0.1753	3.1612	0.1666	3.0538	0.2199	4.1148
70			0.1314	2.3596	0.1802	2.6029	0.1515	2.4115	0.1433	2.2952	0.1928	3.2453
80			0.1159	2.0328	0.1607	2.3350	0.1351	2.1621	0.1285	2.0450	0.1720	2.8546
90			0.0998	1.5667	0.1337	1.7429	0.1130	1.6032	0.1086	1.5648	0.1407	2.0807

(Contd.)

(*Contd.*)

n			MLE $\hat{\alpha}$	$\hat{\beta}$	LSE $\hat{\alpha}_2$	$\hat{\beta}_2$	WLSE $\hat{\alpha}_3$	$\hat{\beta}_3$	ADE $\hat{\alpha}_4$	$\hat{\beta}_4$	CvME $\hat{\alpha}_5$	$\hat{\beta}_5$
100	2	2	0.0912	1.4900	0.1266	1.7333	0.1057	1.5736	0.1022	1.5270	0.1328	2.0305
150			0.0612	0.8869	0.0824	1.0215	0.0693	0.9342	0.0677	0.9148	0.0854	1.1427
200			0.0439	0.5956	0.0614	0.7246	0.0505	0.6454	0.0496	0.6337	0.0629	0.7857
250			0.0344	0.4646	0.0482	0.5639	0.0394	0.5020	0.0387	0.4941	0.0491	0.6031
500			0.0173	0.2112	0.0244	0.2691	0.0198	0.2348	0.0196	0.2318	0.0247	0.2782
1000			0.0084	0.1025	0.0117	0.1297	0.0095	0.1132	0.0095	0.1128	0.0118	0.1316
2500			0.0034	0.0415	0.0048	0.0525	0.0038	0.0456	0.0037	0.0455	0.0048	0.0530
5000			0.0017	0.0210	0.0024	0.0264	0.0019	0.0228	0.0019	0.0228	0.0024	0.0264

14.3 AN ALTERNATIVE QUANTILE REGRESSION MODEL

In this section, we introduce an alternative quantile regression based on the UEEG distribution. Firstly, the pdf of the UEEG distribution can be given with a re-parameterization based on its quantile function. Let $\mu = x_u(\alpha, \beta)$, then we have following relation:

$$\alpha = \log\left(\frac{1-u}{1+u\beta}\right)\frac{(\mu-1)}{\mu}.$$

Then, the cdf and pdf given in Eqs. (4)–(5) are re-parameterized by

$$G(y,\beta,\mu) = \frac{1-\exp\left(-\dfrac{\log\left(\dfrac{1-u}{1+u\beta}\right)\dfrac{y(\mu-1)}{\mu}}{1-y}\right)}{1+\beta\exp\left(-\dfrac{\log\left(\dfrac{1-u}{1+u\beta}\right)\dfrac{y(\mu-1)}{\mu}}{1-y}\right)}$$

and

$$g(y,\beta,\mu) = \frac{\log\left(\dfrac{1-u}{1+u\beta}\right)\dfrac{(\mu-1)}{\mu}(\beta+1)\exp\left(-\dfrac{\log\left(\dfrac{1-u}{1+u\beta}\right)\dfrac{y(\mu-1)}{\mu}}{1-y}\right)}{(1-y)^2\left(1+\beta\exp\left(-\dfrac{\log\left(\dfrac{1-u}{1+u\beta}\right)\dfrac{y(\mu-1)}{\mu}}{1-y}\right)\right)^2}, \tag{10}$$

respectively, where $\beta > 0$ is the parameter, the parameter $\mu \in (0, 1)$ represents the quantile parameter, and $u \in (0, 1)$ is given or known. It is noted that the random variable Y is denoted by $Y \sim QUEEG(\beta, \mu, u)$.

After defining the QUEEG distribution, it can be interested in the quantile regression model based on the QUEEG distribution with pdf in Eq. (10). Let $y_1, y_2, \ldots, y_n$ such that y_i is a realization of $Y_i \sim QUEEG(\beta, \mu_i, u)$ for $i = 1, \ldots, n$, where μ_i and β are the unknown parameters, and the u is given or known. Hence, the proposed quantile regression model is defined as

$$g(\mu_i) = \mathbf{x}_i\boldsymbol{\gamma}^T,$$

where $\boldsymbol{\gamma} = (\gamma_0, \gamma_1, \gamma_2, \ldots, \gamma_p)$, $\mathbf{x}_i = (1, x_{i1}, x_{i2}, x_{i3}, \ldots, x_{ip})$ are the unknown regression model parameter vector and the function $g(\cdot)$ is a link function. In the simulation study and real data analysis $u = 0.5$ is used in which the covariates are linked to the conditional median of the response variable. Since the QUEEG distribution has domain (0, 1) interval, we use the following logit-link function

$$g(\mu_i) = \log\left(\frac{\mu_i}{1-\mu_i}\right) = \mathbf{x}_i\boldsymbol{\gamma}^T, \quad i = 1, 2, \ldots, n. \tag{11}$$

It is noticed that the probit and log-log link functions can be used for linking to conditional quantiles of the response variable.

14.3.1 Estimation for the Quantile Regression Model Parameters

From Eq. (11), one can write

$$\mu_i = \frac{\exp(\mathbf{x}_i\boldsymbol{\gamma}^T)}{1+\exp(\mathbf{x}_i\boldsymbol{\gamma}^T)}. \tag{12}$$

Let $Y_1, Y_2, \ldots, Y_n$ be a random sample of size n from the $QUEGG(\beta, \mu_i, u)$ distribution with realizations $y_1, y_2, \ldots, y_n$, where the μ_i is given by Eq. (12) for $i = 1, \ldots, n$. Then the associated log-likelihood function is given by

$$\ell(\Theta) = n\log(\beta+1) + n\log\left(\log\left(\frac{1-u}{1+u\beta}\right)\right) + \sum_{i=1}^{n}\log\left(\frac{\mu_i-1}{\mu_i}\right) - 2\sum_{i=1}^{n}\log(1-y_i)$$

$$-\sum_{i=1}^{n}\frac{\log\left(\frac{1-u}{1+u\beta}\right)\frac{y_i(\mu_i-1)}{\mu_i}}{1-y_i} - 2\sum_{i=1}^{n}\log\left(1+\beta\exp\left(-\frac{\log\left(\frac{1-u}{1+u\beta}\right)\frac{y_i(\mu_i-1)}{\mu_i}}{1-y_i}\right)\right), \tag{13}$$

where $\Theta = (\beta, \gamma)$ is the unknown parameter vector. The MLE of Θ, say $\widehat{\Theta} = (\widehat{\beta}, \widehat{\boldsymbol{\gamma}})$, is obtained by maximizing $\ell(\Theta)$ with respect to Θ. The **maxLik** function in R can be used to maximize Eq. (13) directly, where $\widehat{\boldsymbol{\gamma}} = (\widehat{\gamma}_0, \widehat{\gamma}_1, \ldots, \widehat{\gamma}_p)$.

Based on some regularity conditions, the asymptotic distribution of $(\widehat{\Theta} - \Theta)$ is multivariate normal $N_{p+1}(0, J^{-1})$, where J is the expected information matrix. In practise, it is generally used the observed information matrix instead of J. The observed information matrix can be calculated by any software. In R, the **maxLik** function gives the asymptotic standard errors of the estimates based on observed Fisher information matrix.

The likelihood equations corresponding to the likelihood function in Eq. (13) are given below:

$$\frac{\partial \ell(\Theta)}{\partial \beta} = \sum_{i=1}^{n} \frac{\xi_1}{\xi_3} = 0, \tag{14}$$

$$\frac{\partial \ell(\Theta)}{\partial \gamma_0} = \sum_{i=1}^{n} \frac{\xi_4}{\xi_5} = 0, \tag{15}$$

and

$$\frac{\partial \ell(\Theta)}{\partial \gamma_j} = \sum_{i=1}^{n} \frac{\xi_4}{\xi_5} xij = 0, \quad j = 1, 2, \ldots, p, \tag{16}$$

where

$$\xi_1 = \xi_2 \times \exp\left(- \frac{\log\left(\frac{1-u}{1+u\beta}\right) y_i}{(\exp(\mathbf{x}_i \boldsymbol{\gamma}^T))(-1+y_i)} \right)$$

$$+ \left((-1+y_i)(1+u\beta)\exp(\mathbf{x}_i \boldsymbol{\gamma}^T) + u y_i (\beta+1) \right) \log\left(\frac{1-u}{1+u\beta}\right)$$

$$- u \exp(\mathbf{x}_i \boldsymbol{\gamma}^T)(\beta+1)(-1+y_i),$$

$$\xi_2 = \left(-(\beta+2)(-1+y_i)(1+u\beta)\exp(\mathbf{x}_i \boldsymbol{\gamma}^T) - y_i u \beta(\beta+1) \right) \log\left(\frac{1-u}{1+u\beta}\right)$$

$$- u\beta \exp(\mathbf{x}_i \boldsymbol{\gamma}^T)(\beta+1)(-1+y_i),$$

$$\xi_3 = \left(\exp(\mathbf{x}_i \boldsymbol{\gamma}^T)\right) \left(1 + \beta \exp\left(- \frac{\log\left(\frac{1-u}{1+u\beta}\right) y_i}{(\exp(\mathbf{x}_i \boldsymbol{\gamma}^T))(-1+y_i)} \right) \right)$$

$$\times (-1+y_i)(1+u\beta)\left(\log\left(\frac{1-u}{1+u\beta}\right) \right)(\beta+1),$$

$$\xi_4 = -\beta\left(\log\left(\frac{1-u}{1+u\beta}\right)y_i + \exp(\mathbf{x}_i\boldsymbol{\gamma}^T)(-1+y_i)\right)\exp\left(-\frac{\log\left(\frac{1-u}{1+u\beta}\right)y_i}{(\exp(\mathbf{x}_i\boldsymbol{\gamma}^T))(-1+y_i)}\right)$$

$$+\log\left(\frac{1-u}{1+u\beta}\right)y_i - \exp(\mathbf{x}_i\boldsymbol{\gamma}^T)(-1+y_i),$$

and

$$\xi_5 = (\exp(\mathbf{x}_i\boldsymbol{\gamma}^T))\left(1+\beta\exp\left(-\frac{\log\left(\frac{1-u}{1+u\beta}\right)y_i}{(\exp(\mathbf{x}_i\boldsymbol{\gamma}^T))(-1+y_i)}\right)\right)(-1+y_i).$$

These gradients can be used to achieve the MLEs. See Algorithm 1 in the next section.

14.3.2 Simulation Study for Quantile Regression Model

In this subsection, a simulation study is conducted to observe the MLE behaviors of the parameters of the QUEEG regression model. The simulation is done with β = 0.5, 1, 2, γ_0 = 1, γ_1 = 1, γ_2 = 1, γ_3 = 1, n = 100, 200, ..., 1000 and correlation matrix

$$\boldsymbol{\rho}_1 = \begin{pmatrix} 1 & 0 & 0 \\ & 1 & 0 \\ & & 1 \end{pmatrix}, \boldsymbol{\rho}_2 = \begin{pmatrix} 1 & 0.25 & 0.25 \\ & 1 & 0.25 \\ & & 1 \end{pmatrix} \text{ and } \boldsymbol{\rho}_3 = \begin{pmatrix} 1 & 0.75 & 0.75 \\ & 1 & 0.75 \\ & & 1 \end{pmatrix}$$

The covariates are generated from multivariate normal distribution. The simulated bias and MSEs of the MLEs of the QUEEG regression model and the coverage probabilities and mean lengths of the MLEs based confidence intervals are plotted Figures 14.2 to 14.8.

From Figures 14.2 to 14.8 it is observed that the $\hat{\gamma}_0, \hat{\gamma}_1, \hat{\gamma}_2$ and $\hat{\gamma}_3$ are unbiased and MSEs of these estimates decrease to 0 for a large sample of size. This indicates that the $\hat{\gamma}_0, \hat{\gamma}_1, \hat{\gamma}_2$ and $\hat{\gamma}_3$ seem to be consistent. The $\hat{\beta}$ is biased but it is asymptotically unbiased. In addition, the MSE of $\hat{\beta}$ decreases to zero when the sample of size increases. The approximate confidence intervals for all model parameters work quite well. The coverage probabilities for all parameters are almost equal to the nominal level (0.95) and the mean length are decreases to zero when the sample of size is increased.

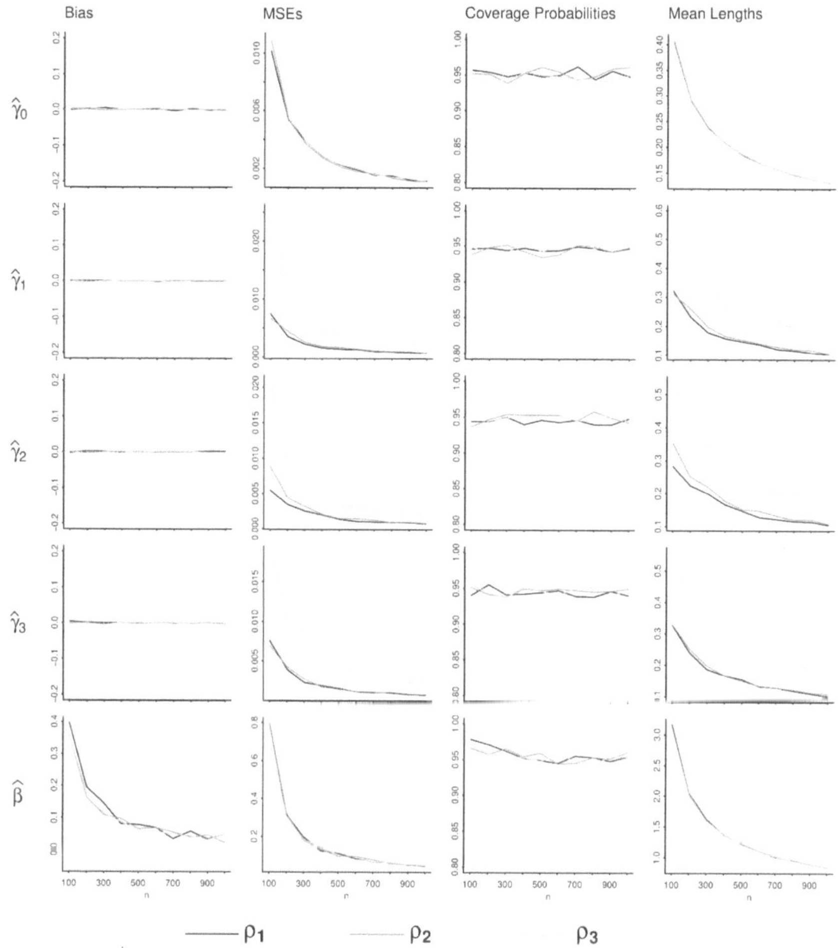

Figure 14.2 Bias and MSEs of the MLEs of the QUEEG regression model and the coverage probabilities and mean lengths of the MLE based approximately confidence intervals for $\beta = 0.5$.

14.3.3 Illustrative Examples with a New Searching Algorithm

In this subsection, two illustrative examples are presented. An algorithm is also provided to achieve the MLEs of the model parameters.

Let us discuss the first example with one dependent and one covariate. The data is generated by settings $n = 30$, $\gamma_0 = 1$, $\gamma_1 = 0.5$ and $\beta = 3$. The following R code is used to get the simulated data:

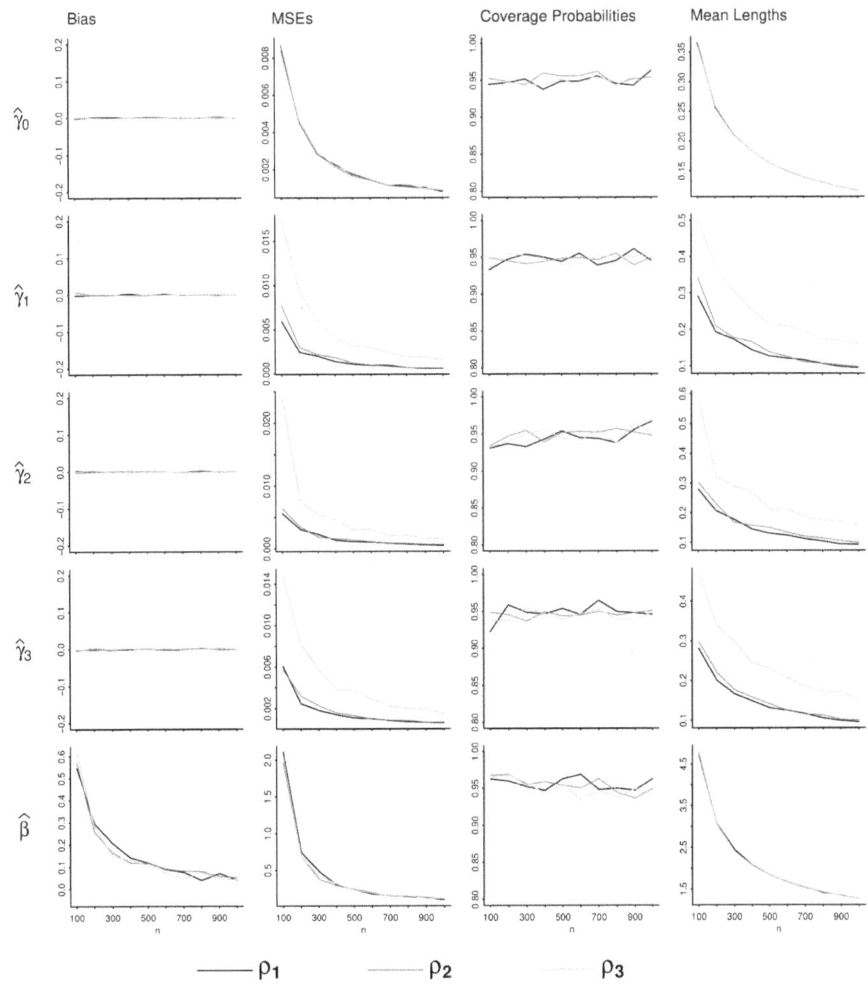

Figure 14.3 Bias and MSEs of the MLEs of the QUEEG regression model and the coverage probabilities and mean lengths of the MLE based approximately confidence intervals for $\beta = 1$.

```
set.seed(3)
n=30
x1=rnorm(n)
gam0=1;gam1=0.5
xxgamma=gam0+gam1*x1
mu=exp(xxgamma)/(1+exp(xxgamma))
beta=3
uu=runif(n)
u=0.5
alpha=log((1-u)/(1+u*beta))*(mu-1)/mu
y=log(-(-1+uu)/(1+uu*beta))/(-alpha+log(-(-1+uu)/(1+uu*beta)))
```

The generated data are given below:

x_1: –0.96193342, –0.29252572, 0.25878822, –1.15213189, 0.19578283, 0.03012394, 0.08541773, 1.11661021, –1.21885742, 1.26736872, –0.74478160, –1.13121857, –0.71635849, 0.25265237, 0.15204571, –0.30765643, –0.95301733, –0.64824281, 1.22431362, 0.19981161, –0.57848372, –0.94230073, –0.20372818, –1.66647484, –0.48445511, –0.74107266, 1.16061578, 1.01206712, –0.07207847, –1.13678230.

y: 0.75372679, 0.24197206, 0.84564075, 0.26649304, 0.83147945, 0.63461920, 0.82436912, 0.83717883, 0.52125874, 0.52131999, 0.75259441, 0.71531681, 0.81149615, 0.90102591, 0.75139999, 0.71950652, 0.37754608, 0.41514432, 0.89563900, 0.82751229, 0.77626051, 0.69376186, 0.65289861, 0.02433737, 0.85534884, 0.78226489, 0.68921333, 0.81690074, 0.77205617, 0.78711495.

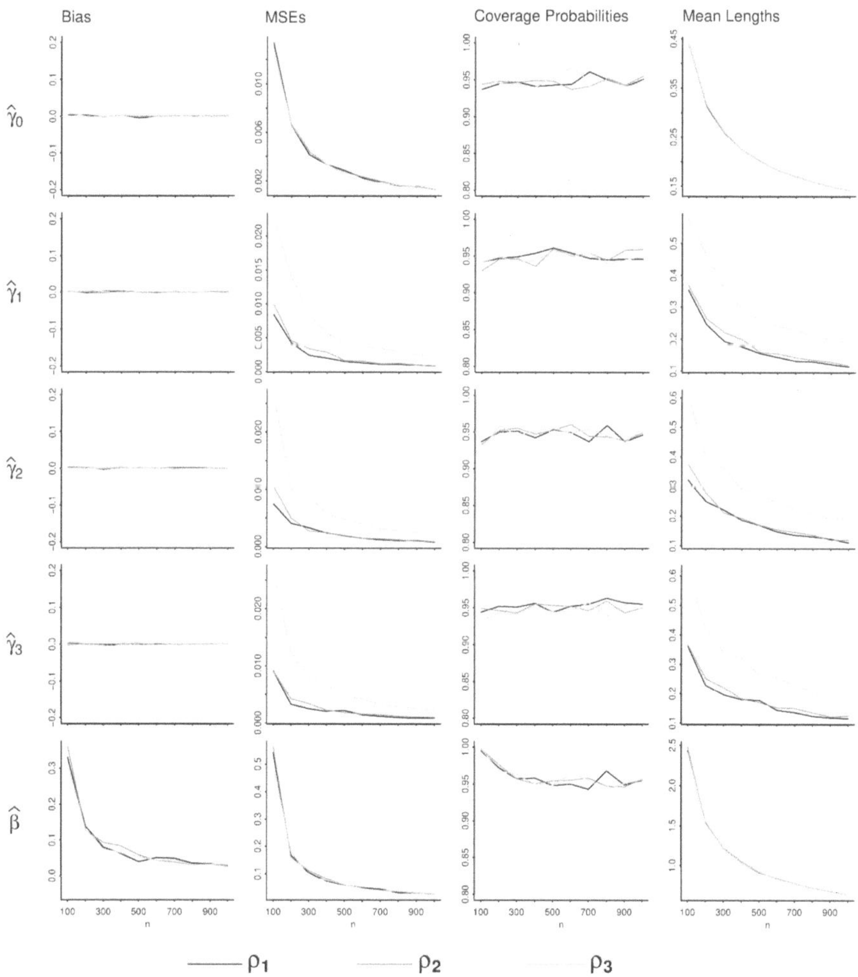

Figure 14.4 Bias and MSEs of the MLEs of the QUEEG regression model and the coverage probabilities and mean lengths of the MLE based approximately confidence intervals for $\beta = 2$.

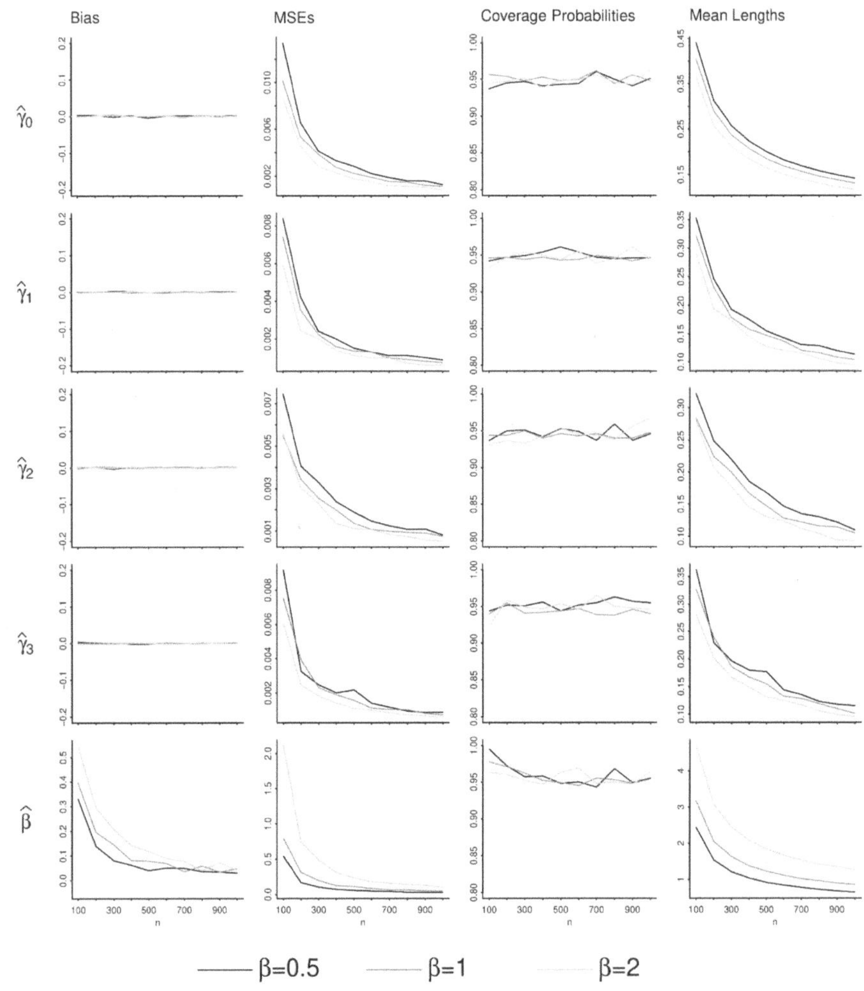

Figure 14.5 Bias and MSEs of the MLEs of the QUEEG regression model and the coverage probabilities and mean lengths of the MLE based approximately confidence intervals for ρ_1.

The MLEs with standard errors (SE) and p-values associated with the significance tests are given in Table 14.3 for the Beta, Kw, and QUEEG regression analyses, respectively. The p-values corresponding to Anderson Darling normality tests are 0.2835, 0.1101, and 0.7623 for the residuals of the Beta, Kw, and QUEEG regression analyses.

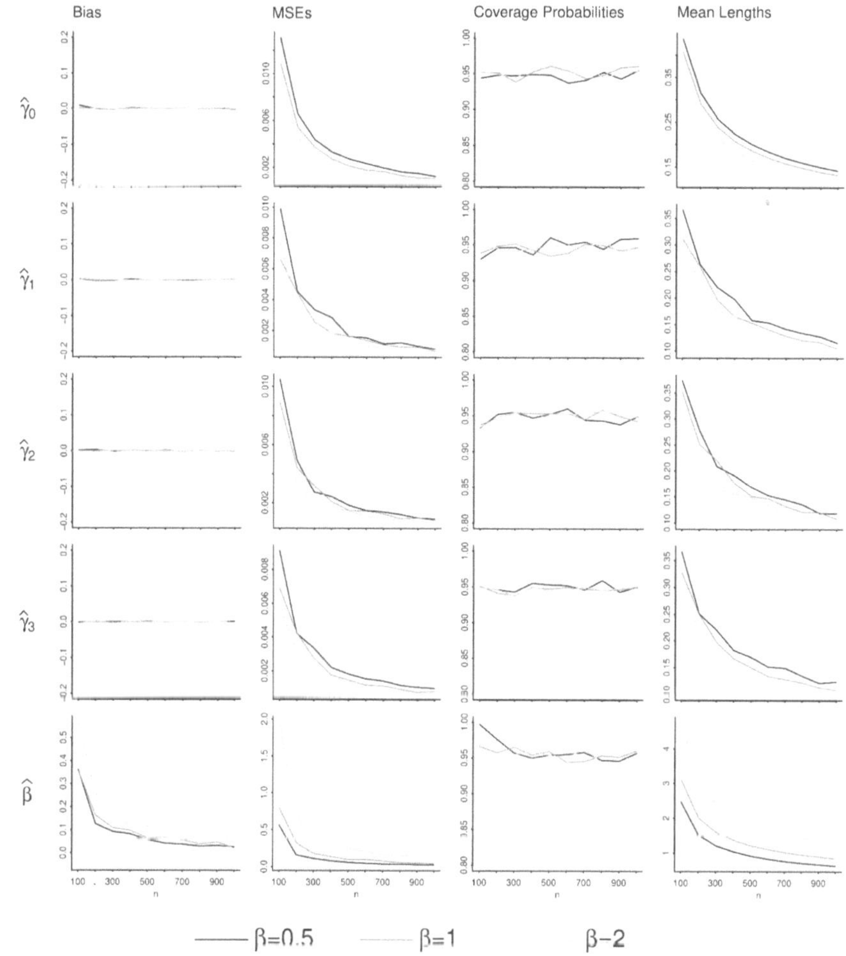

Figure 14.6 Bias and MSEs of the MLEs of the QUEEG regression model and the coverage probabilities and mean lengths of the MLE based approximately confidence intervals for ρ_2.

Table 14.3 Beta, Kw, and QUEEG regression outputs for simulated dataset 1.

	Beta			Kw			QUEEG		
Parameters	**Estimate**	**SE**	***p*-value**	**Estimate**	**SE**	***p*-value**	**Estimate**	**SE**	***p*-value**
Intercept	0.7936	0.1664	<0.0001	0.8274	0.2098	<0.0001	1.1698	0.1330	<0.0001
x_1	0.6142	0.1975	0.0018	0.4536	0.2688	0.0915	0.3998	0.1527	0.0088
ϕ	5.262	1.275	<0.0001	–	–	–	–	–	–
β	–	–	–	2.8118	0.6410	<0.0001	6.9181	5.0731	0.1726
Log-likelihood		10.28			8.31			18.92	
AIC (BIC)	−14.56(−10.35)				−10.62(−6.41)			−31.84(−27.63)	

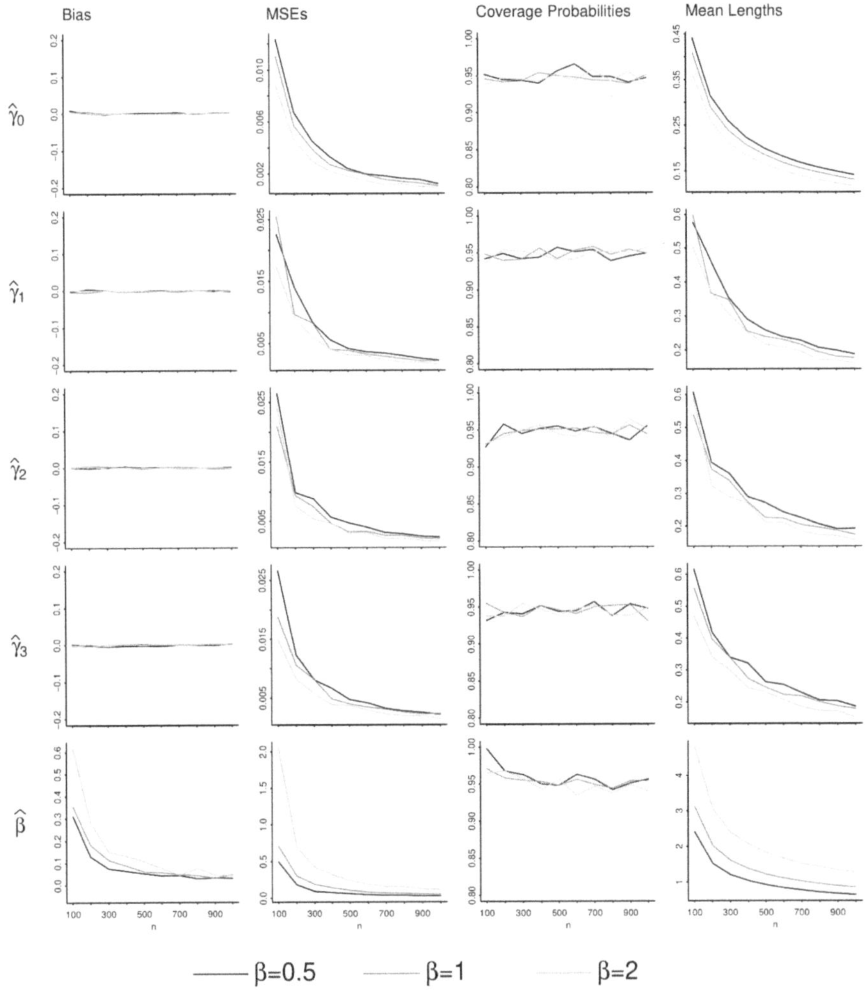

Figure 14.7 Bias and MSEs of the MLEs of the QUEEG regression model and the coverage probabilities and mean lengths of the MLE based approximately confidence intervals for ρ_3.

QUEEG regression results are obtained by the maxLik function in R with the initials values of parameters $\beta^{(0)} = 1$, $\gamma_0^{(0)} = 1$ and $\gamma_1^{(0)} = 1$. If the initial values are selected as $\beta^{(0)} = 5$, $\gamma_0^{(0)} = -5$ and $\gamma_1^{(0)} = -5$, there is no solution to the likelihood equations. We give a new algorithm to achieve the MLE of the parameters with the following algorithm.

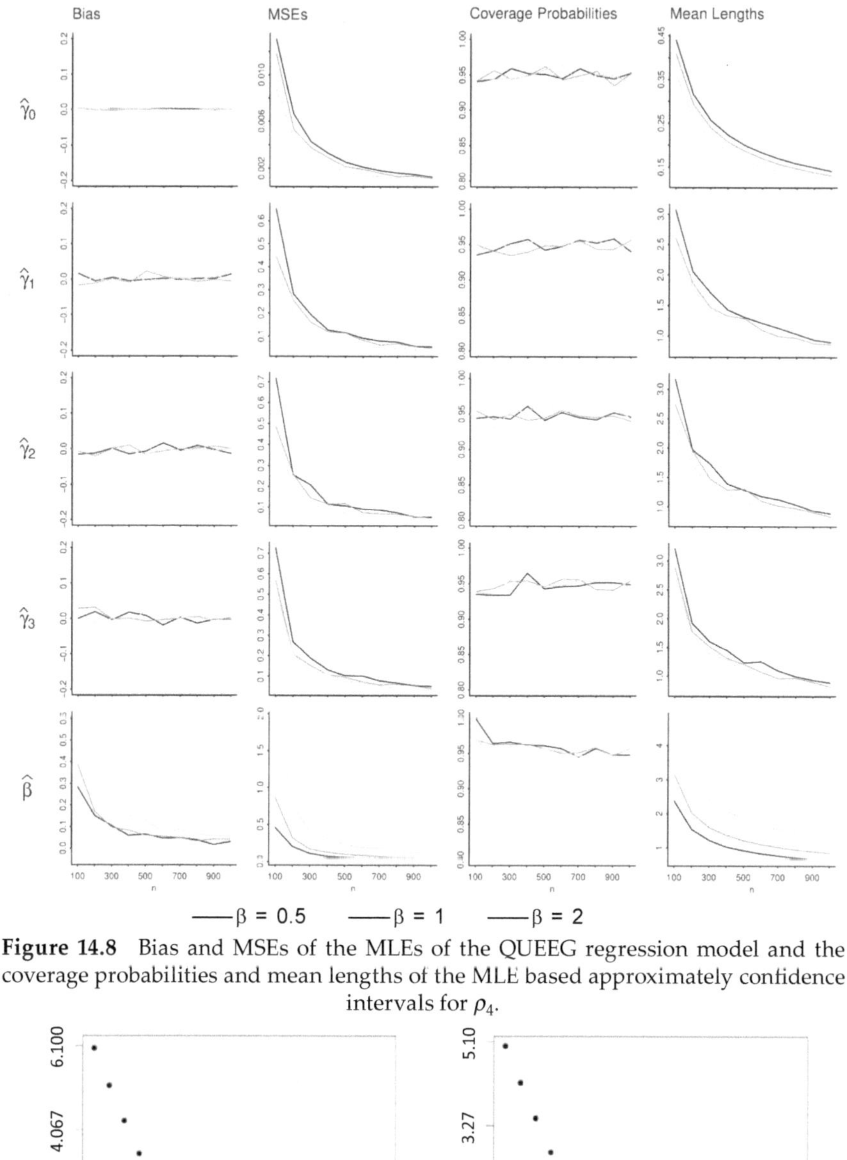

Figure 14.8 Bias and MSEs of the MLEs of the QUEEG regression model and the coverage probabilities and mean lengths of the MLE based approximately confidence intervals for ρ_4.

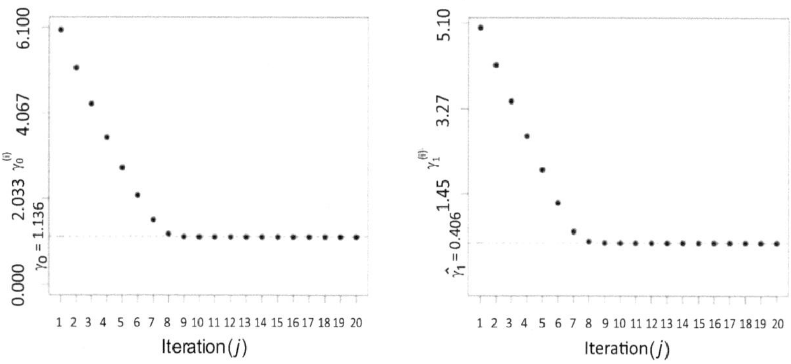

Figure 14.9 Trace plot for (j) under Algorithm 1 (Step A2) (simulated data).

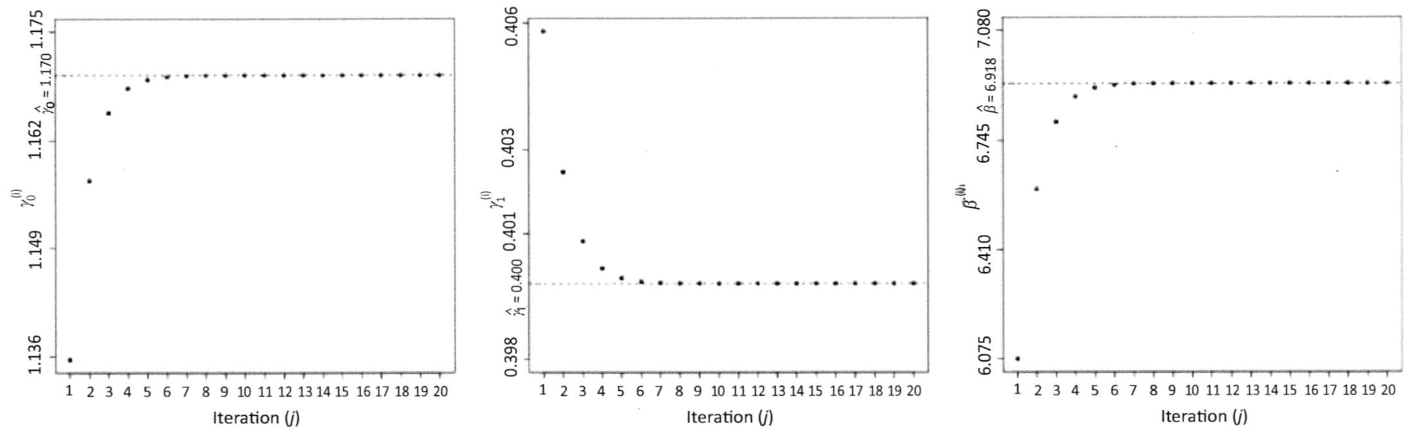

Figure 14.10 Trace plot for (j) under Algorithm 1 (Step A4) (simulated data).

Algorithm 1:

A1. Give initial values of parameters $\beta^{(0)}$, $\gamma_0^{(0)}$ and $\gamma_1^{(0)}$ and define *n_iter*1 and *n_iter*2.

A2. Calculate the roots of Eq. (14) at $\beta^{(0)}$ and $\gamma_1^{(j-1)}$ and it is denoted by $\gamma_0^{(j)}$; Calculate the roots of Eq. (15) at $\beta^{(0)}$ and $\gamma_0^{(j)}$ and it is denoted by $\gamma_1^{(j)}$, $j = 1, \ldots, n\_iter1$.

A3. Update the initial values as $\beta^{(0)} = \beta^{(0)}$, $\gamma_0^{(0)} = \gamma_0^{(n\_iter1)}$ and $\gamma_1^{(0)} = \gamma_1^{(n\_iter1)}$.

A4. Calculate the roots of Eq. (14) at $\beta^{(j-1)}$ and $\gamma_1^{(j-1)}$ and it is denoted by $\gamma_0^{(j)}$; Calculate the roots of Eq. (15) at $\beta^{(j-1)}$ and $\gamma_0^{(j)}$ and it is denoted by $\gamma_1^{(j)}$; Calculate the roots of Eq. (16) at $\gamma_0^{(j)}$ and $\gamma_1^{(j)}$ and it is denoted by $\beta^{(j)}$, $j = 1, \ldots, n\_iter2$.

Using Algorithm 1, we achieve the MLEs of the parameters for almost all parameter settings including the choice $\beta^{(0)} = 5$, $\gamma_0^{(0)} = -5$ and $\gamma_1^{(0)} = -5$. The first and second iterations are plotted in Figures 14.9–14.10. From these figures, it is observed that a few iterations (about 20 times) are enough to achieve the MLEs of the parameters. In addition, it should be pointed out that Algorithm 1 is not affected by the initial values to give true MLEs.

A second data set is provided to demonstrate the Algorithm 1 works quite well for two-covariate cases. One dependent and two covariates are generated for the numerical example. The data generated by settings $n = 30$, $\gamma_0 = 1$, $\gamma_1 = 0.5$, $\gamma_2 = 2$ and $\beta = 3$. The simulated data is given below:

x_1: −0.84085548, 1.38435934, −1.25549186, 0.07014277, 1.71144087, −0.60290798, −0.47216639, −0.63537131, −0.28577363, 0.13810822, 1.22763034, −0.80177945, −1.08039260, −0.15753436, −1.07176004, −0.13898614, −0.59731309, −2.18396676, 0.24081726, −0.25935541, 0.90051195, 0.94186939, 1.46796190, 0.70676109, 0.81900893, −0.29348185, 1.41858907, 1.49877383, −0.65708209, −0.85279544.

x_2: 0.31591504, 1.10969417, 2.21546057, 1.21710364, 1.47922179, 0.95157383, −1.00953265, −2.00047274, −1.76218587, −0.14260813, 1.55006037, −0.80242318, −0.07457892, 1.89566795, −0.45656894, 0.56222336, −0.88700851, −0.46024458, −0.72432849, −0.06921116, 1.46324856, 0.18772610, 1.02202286, −0.59183483, −0.11220066, −0.92495309, 0.75330480, −0.11260907, −0.06409093, 0.23327529.

y: 0.49054461, 0.99405410, 0.99539669, 0.98512928, 0.98633823, 0.57632866, 0.28886502, 0.05585406, 0.03201018, 0.59947030, 0.98809278, 0.12383778, 0.05589164, 0.99538960, 0.33021151, 0.09633235, 0.07350479, 0.24088027, 0.36211348, 0.71378101, 0.98532009, 0.76852738, 0.97311048, 0.72629838, 0.80847603, 0.35890237, 0.96236226, 0.85760601, 0.69913215, 0.80349312.

QUEEG regression results are obtained by the maxLik function in R with the initials values of parameters $\beta^{(0)} = 1$, $\gamma_0^{(0)} = 1$, $\gamma_1^{(0)} = 1$ and $\gamma_2^{(0)} = 2$ (Table 14.4).

Table 14.4 QUEEG regression output for simulated dataset 2

Parameters	Estimate	SE	p-value
Intercept	0.9277	0.1512	< 0.0001
x_1	0.5853	0.1311	< 0.0001
x_2	2.0804	0.1047	< 0.0001
β	3.8153	2.9179	0.1910
Log-likelihood		42.21	

Table 14.5 Iteration for the first stage of the Algorithm 1

Iteration	$\gamma_0^{(j)}$	$\gamma_1^{(j)}$	$\gamma_2^{(j)}$
1	9.4770	−4.7851	1.6474
2	8.2775	−3.7731	1.3428
3	7.0493	−2.7198	1.0126
4	5.8276	−1.6223	0.6200
5	4.6798	−0.4545	0.2447
6	3.7649	0.0434	0.6860
7	2.7412	0.2704	1.2964
8	1.6797	0.5196	1.9469
9	0.9911	0.6012	2.0763
10	0.9611	0.5818	2.0811
11	0.9592	0.5810	2.0813
12	0.9591	0.5809	2.0813
13	0.9591	0.5809	2.0813
14	0.9591	0.5809	2.0813
15	0.9591	0.5809	2.0813
16	0.9591	0.5809	2.0813
17	0.9591	0.5809	2.0813
18	0.9591	0.5809	2.0813
19	0.9591	0.5809	2.0813
20	0.9591	0.5809	2.0813

Using Algorithm 1 with 20 iterations for the two stages and initials values $\beta^{(0)} = 5$, $\gamma_0^{(0)} = 5$, $\gamma_1^{(0)} = 5$ and $\gamma_2^{(0)} = 5$, the MLEs of parameters are obtained as $\gamma_0^{(20)} = 0.9277$, $\gamma_1^{(20)} = 0.5853$, $\gamma_2^{(20)} = 2.0804$, $\beta^{(20)} = 3.8139$.

The iterations are given in Tables 14.5–14.6. From Tables 14.5–14.6, it is observed that the iteration is converged in a few steps. The Algorithm 1 is not affected by these initial values. However, if we use these initial values in the maxLik function, we have the following error message in R:

"Error in optim(par = c(gama0 = 5, gama1 = 5, gama2 = 5, beta = 5), fn = function (theta, : function cannot be evaluated at initial parameters))".

We can say that the maxLik is dramatically affected by initial values. Of course, it is only revealed by a numerical study but we observed that

Algorithm 1 works almost all initial values but maxLik fail to achieve the MLE for some initial values. The second example also shows that Algorithm 1 can be used for more than 3 parameters.

Table 14.6 Iteration for the second stage of the Algorithm 1

Iteration	$\gamma_0^{(j)}$	$\gamma_1^{(j)}$	$\gamma_2^{(j)}$	$\beta^{(j)}$
1	0.9591	0.5809	2.0813	4.2113
2	0.9391	0.5836	2.0808	3.9564
3	0.9319	0.5847	2.0805	3.8664
4	0.9293	0.5851	2.0804	3.8334
5	0.9283	0.5852	2.0804	3.8212
6	0.9279	0.5853	2.0804	3.8167
7	0.9278	0.5853	2.0804	3.8150
8	0.9277	0.5853	2.0804	3.8143
9	0.9277	0.5853	2.0804	3.8141
10	0.9277	0.5853	2.0804	3.8140
11	0.9277	0.5853	2.0804	3.8140
12	0.9277	0.5853	2.0804	3.8139
13	0.9277	0.5853	2.0804	3.8139
14	0.9277	0.5853	2.0804	3.8139
15	0.9277	0.5853	2.0804	3.8139
16	0.9277	0.5853	2.0804	3.8139
17	0.9277	0.5853	2.0804	3.8139
18	0.9277	0.5853	2.0804	3.8139
19	0.9277	0.5853	2.0804	3.8139
20	**0.9277**	**0.5853**	**2.0804**	**3.8139**

Furthermore, a real data application as a thirds example is conducted to observe the applicability of the QUEGG regression model. The considered data set is obtained from OECD.Stat with link https://stats.oecd.org/. Detailed information about the data is given and also analyzed by Korkmaz et al. (2021c). This data is used in some papers with different covariates Korkmaz et al. (2021d), Korkmaz et al. (2021e), and Korkmaz and Korkmaz (2021b). The data can be briefly described as follows: The dependent variable y is the percentage of the educational attainment values of the OECD countries and the covariates are x_1 is the percentage of the voter turnout, x_2 is homicide rate, x_3 is life satisfaction. The results for Beta, Kw, and QUEGG regression analyses are given in Table 14.7. From Table 14.7, it is observed that the QUEGG regression model is better than all models studied by Korkmaz et al. (2021c). The half normal plots for the randomized quantile residuals is given in Figure 14.11. It shows that the QUEEG regression has good performance.

Table 14.7 Beta, Kw, and QUEEG regression outputs for OECD data

	Beta			Kw			QUEEG		
Parameters	**Estimate**	**SE**	***p*-value**	**Estimate**	**SE**	***p*-value**	**Estimate**	**SE**	***p*-value**
Intercept	0.9615	0.9685	0.3208	1.6247	1.1740	0.1664	2.9536	1.0409	0.0045
x_1	−2.9211	1.0176	0.0041	−4.1197	1.3892	0.0030	−3.6845	1.0538	0.0004
x_2	−0.0470	0.0178	0.0084	−0.0404	0.0168	0.0159	0.0097	0.0412	0.8137
x_3	0.3794	0.1492	0.0110	0.4237	0.2546	0.0960	0.1857	0.1951	0.3412
ϕ	11.5900	2.6100	<0.0001	–	–	–	–	–	–
β	–	–	–	6.2167	1.0787	<0.0001	7.7410	6.1438	0.2076
Log-likelihood		30.90			29.43			33.44	
AIC (BIC)		−55.80(−50.88)			−52.86(−47.94)			−60.88(−55.96)	

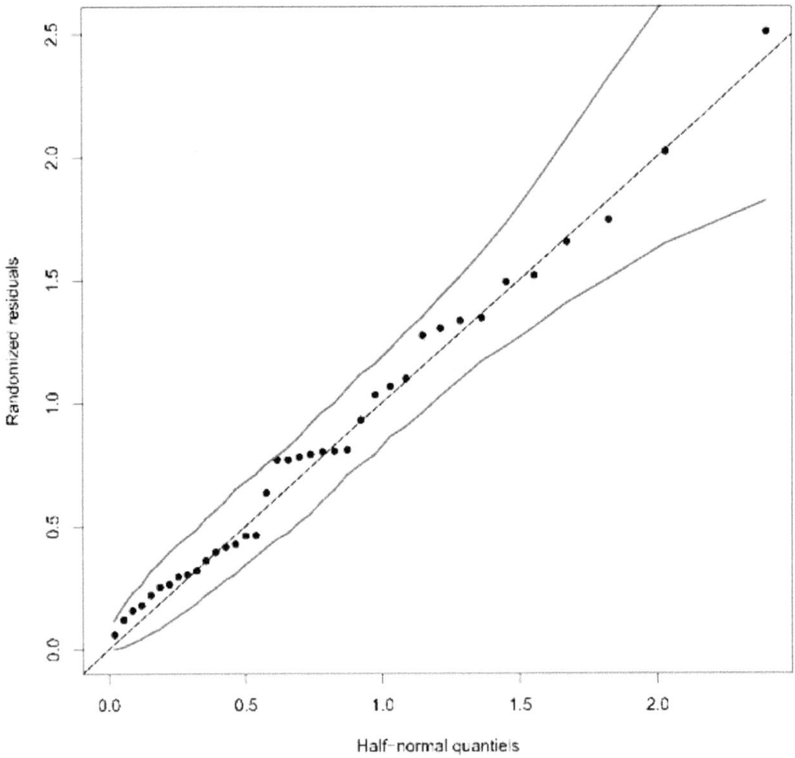

Figure 14.11 The half-normal plots for the randomized quantile residuals for OECD data.

14.4 CONCLUSION

In this paper, a new unit distribution is introduced and it is also used to construct the quantile regression analysis. The simulation study shows that the MLEs of the distribution and regression model parameters have desired performance. In the regression model, a new algorithm is also presented to achieve the MLEs, numerically. It is observed that our algorithm works for arbitrary initial values. Our algorithm can easily be extended for all lifetime regression and quantile regression analysis.

REFERENCES

Adamidis, K. and Loukas, S. 1998. A lifetime distribution with decreasing failure rate. *Statistics & Probability Letters*. 39(1): 35–42.

Adamidis, K., Dimitrakopoulou, T. and Loukas, S. 2005. On an extension of the exponential-geometric distribution. Statistics & Probability Letters. 73(3): 259–269.

Canuto, C., Hussaini, M.Y., Quarteroni, A. and Zang, T.A. 2007. Spectral Methods: Fundamentals in Single Domains. Springer-Verlag, New York.

Ghitany, M.E., Mazucheli, J., Menezes, A.F.B. and Alqallaf, F. 2019. The unit-inverse Gaussian distribution: A new alternative to two-parameter distributions on the unit interval. Communication Statisics Theory Methods. 48: 3423–3438.

Korkmaz, M.Ç. and Chesneau, C. 2021. On the unit Burr-XII distribution with the quantile regression modeling and applications. Computational and Applied Mathematics. 40(1): 1–26.

Korkmaz, M.Ç. and Korkmaz, Z.S. 2021. The unit log–log distribution: A new unit distribution with alternative quantile regression modeling and educational measurements applications. Journal of Applied Statistics. 1–20.

Korkmaz, M.Ç., Chesneau, C. and Korkmaz, Z.S. 2021a. On the arcsecant hyperbolic normal distribution: Properties, quantile regression modeling and applications. Symmetry. 13: 117.

Korkmaz, M.Ç. and Korkmaz, Z.S. 2021b. The unit log–log distribution: A new unit distribution with alternative quantile regression modeling and educational measurements applications. Journal of Applied Statistics. 2021. DOI: https://doi.org/10.1080/02664763.2021.2001442.

Korkmaz, M.Ç., Chesneau, C. and Korkmaz, Z.S. 2021c. Transmuted unit Rayleigh quantile regression model: Alternative to beta and Kumaraswamy quantile regression models. Univ. Politeh. Buchar. *Sci. Bull. Ser. Appl. Math. Phys.* 83(3): 149–158.

Korkmaz, M.Ç., Chesneau, C. and Korkmaz, Z.S. 2021d. A new alternative quantile regression model for the bounded response with educational measurements applications of OECD countries. Journal of Applied Statistics. 1–24.

Korkmaz, M.Ç., Altun, E., Alizadeh, M. and El-Morshedy, M. 2021e. The log exponential-power distribution: properties, estimations and quantile regression model. Mathematics. 9(21): 2634.

Mazucheli, J., Menezes, A.F.B. and Ghitany, M.E. 2018. The unit-Weibull distribution and associated inference. Journal of Applied Probability Statistics. 13: 1–22.

Mazucheli, J., Menezes, A.F.B. and Chakraborty, S. 2019a. On the one parameter unit-Lindley distribution and its associated regression model for proportion data. Journal of Applied Statistics. 46: 700–714.

Mazucheli, J., Menezes, A.F.B. and Dey, S. 2019b. Unit-Gompertz distribution with applications. Statistica. 79(1): 25–43.

Mazucheli, J., Bapat, S.R. and Menezes A.F.B. 2020a. A new one-parameter unit-Lindley distribution. Chilean Journal of Statistics. 11(1).

Mazucheli, J., Menezes, A.F.B., Fernandes, L., de Oliveira, R. and Ghitany, M. 2020b. The unit-Weibull distribution as an alternative to the Kumaraswamy distribution for the modeling of quantiles conditional on covariates. Journal of Applied Statistics. 47: 954–974.

Mazucheli, J., Leiva, V., Alves, B. and Menezes A.F.B. 2021. A new quantile regression for modeling bounded data under a unit Birnbaum–Saunders distribution with applications in medicine and politics. Symmetry. 13: 682.

Novomestky, F. and Novomestky, M.F. 2013. R Package 'gaussquad'.

Zhang, T. and Xie, M. 2011. On the upper truncated Weibull distribution and its reliability implications. Reliability Engineering & System Safety. 96: 194–200.

Chapter 15

The Cos-Modified Lindley Distribution: Properties, Estimation, and Applications

Lishamol Tomy[1], Veena G.[2] and Christophe Chesneau[3*]

[1]Department of Statistics, Deva Matha College, Kuravilangad, Kerala-686633, India. Email: lishatomy@gmail.com

[2]Department of Statistics, St. Thomas College, Palai, Kerala-686574, India. Email: veenagpillai@hotmail.com

[3]Department of Mathematics, Université de Caen Basse-Normandie, UFR de Sciences, F-14032, Caen, France. Email: christophe.chesneau@unicaen.fr

15.1 INTRODUCTION

The study of trigonometric distributions and their applications in statistical literature has gained prominence in recent years. In this chapter, we contribute to the subject by using the concept of weighted trigonometric distributions. The basic concept of weighted distributions was developed by Fisher (1934) to model the ascertainment bias. Later, Rao (1965) formalized it in a unified manner while modeling statistical data. Based on this key work, weighted distributions were applied quite successfully in various research areas related to biomedicine, reliability, ecology, and branching processes, where standard distributions showed

*For Correspondence: Christophe Chesneau (christophe.chesneau@unicaen.fr)

some functional limitations. Polynomial weight functions are the most commonly used. Recently, a few research articles have investigated trigonometric weight functions. We may cite Bakouch (2018) who developed a three-parameter distribution, named the new exponential trigonometric (NET) distribution. The NET distribution is considered an oscillatory generalization of the exponential distribution. One of its many interesting features is the presence of periodicity in its hazard rate function (hrf). The supremacy of the NET distribution over the exponential, Weibull, and gamma distributions is proved by using the lifetime data of fatigue fractures of Kevlar 373/epoxy that are subject to constant pressure at the 90% stress level until all have failed. On the other hand, the cos-Lindley (CL) distribution, proposed by Chesneau (2023), is also another remarkable weighted trigonometric distribution. On the practical aspect, it is able to win against comparable competitors, such as the Lindley, modified Lindley, and NET distributions. This was proved with two real-world datasets, which represent the service and failure times of the aircraft windshield, to study the applicability of the distribution. Like the modified Lindley (ML) distribution developed by Chesneau (2021), it is found to be an alternative to the Lindley and exponential distributions. With this thought in mind, we introduce a new lifetime distribution defined with a weighted trigonometric function and the functionalities of the ML distribution.

In this article, we thus emphasize the weighted cos-modified Lindley (Wc-ML) distribution. It is generated using the ML distribution and the trigonometric two-parameter weight function: $w_t(x;\alpha,\beta) = 1 - \alpha\cos(\beta x)$. We recall that the ML distribution, as introduced in Chesneau (2021), is defined with the probability density function (pdf) indicated as

$$h(y;\theta) = \frac{\theta}{1+\theta} e^{-2\theta y}[(1+\theta)e^{\theta y} + 2\theta y - 1],\ y > 0, \tag{1}$$

with $h(y;\theta) = 0$ for $y \leq 0$. Based on a random variable Y having this pdf and the concept of weighted distributions, we define Wc-ML distribution by the following pdf:

$$f_{Wc\text{-}ML}(y;\alpha,\beta,\theta) = \frac{w_t(y;\alpha,\beta)h(y;\theta)}{E[w_t(Y;\alpha,\beta)]},\ y > 0,$$

where $\alpha \in [-1, 1]$ and $\beta, \theta \geq 0$, and $f_{Wc\text{-}ML}(y;\alpha,\beta,\theta) = 0$ for $y \leq 0$. With this trigonometrically weighted operation, the Wc-ML distribution logically exhibits oscillation and periodicity in the shapes of its pdf and hrf. Such features are rare, and in view of the behavior of the ML distribution, we hope for a significant gain in terms of fit for some data. Also, in view of its construction, the Wc-ML distribution can be considered an alternative to the NET and CL distributions in modeling periodic data.

The structure of this chapter is as follows: Section 15.2 gives a more detailed expression of the pdf as well as the cumulative density function (cdf) and hrf related to the Wc-ML distribution, along with their shape properties. Section 15.3 investigates the moments of the Wc-ML distribution and their analysis. The estimation of the parameters using the maximum likelihood estimation (MLE) method is carried out in Section 15.4. The Wc-ML distribution and other competitive distributions are tested on two real-world datasets in Section 15.5. Section 15.6 highlights the results and concludes the chapter.

15.2 THE WC-ML DISTRIBUTION

15.2.1 Pdf

Using Eq. (1) and the weight function $w_t(y;\ \alpha,\ \beta) = 1 - \alpha\cos(\beta y)$, we obtain

$$E[w_t(Y;\alpha,\beta)] = \frac{\theta}{1+\theta}\left[\frac{\theta+1}{\theta} - \frac{\alpha\theta(1+\theta)}{\theta^2+\beta^2} + \frac{4\theta\alpha\beta^2}{(4\theta^2+\beta^2)^2}\right].$$

Hereafter, to lighten the notations, $E[w_t(Y;\ \alpha,\ \beta)]$ is denoted as W_c. Then the pdf of the Wc-ML distribution with parameters α, β and θ is developed as follows:

$$f_{Wc\text{-}ML}(y;\alpha,\beta,\theta) = W_c^{-1}\frac{\theta}{1+\theta}[1-\alpha\cos(\beta y)]e^{-2\theta y} \times [(1+\theta)e^{\theta y}+2\theta y-1],\ \ y>0, \tag{2}$$

with $f_{Wc\text{-}ML}(y;\ \alpha,\ \beta,\ \theta) = 0$ for $y \leq 0$.

By using the series expansion of the cosine function, for $y > 0$, it can be expanded as follows:

$$f_{Wc\text{-}ML}(y;\alpha,\beta,\theta) = W_c^{-1}\frac{\theta}{1+\theta}\sum_{k=0}^{+\infty} a_k(\alpha,\beta)y^{2k}e^{-2\theta y}[(1+\theta)e^{\theta y}+2\theta y-1], \tag{3}$$

where the sequence $(a_m(\alpha,\beta))_{m\in\mathbb{N}}$ is defined as

$$a_m(\alpha,\beta) = \begin{cases} 1-\alpha, & \text{if } m=0 \\ \alpha(-1)^{m+1}\dfrac{\beta^{2m}}{(2m)!}, & \text{if } m\geq 1. \end{cases} \tag{4}$$

15.2.2 Properties

Some fundamental properties are now investigated, beginning with a confirmation that $f_{Wc\text{-}ML}(y;\ \alpha,\ \beta,\ \theta)$ is indeed a pdf.

Proposition 1: *The properties of a pdf are satisfied by* $f_{Wc\text{-}ML}(y;\ \alpha,\ \beta,\ \theta)$.

Proof. We know that $\cos(\beta y) \in [-1, 1]$. Since $\alpha \in [-1, 1]$, it is clear that $f_{Wc\text{-}ML}(y;\ \alpha,\ \beta,\ \theta) \geq 0$.

Now, we need to prove that $\int_0^{+\infty} f_{Wc\text{-}ML}(y; \alpha, \beta, \theta)dy = 1$.

We have

$$\int_0^{+\infty} f_{Wc\text{-}ML}(y; \alpha, \beta, \theta)dy = \frac{\theta}{1+\theta}\int_0^{+\infty} [1-\alpha\cos(\beta y)]e^{-2\theta y} \times [(1+\theta)e^{\theta y} + 2\theta y - 1]dy.$$

After some integrations, this integral is equal to

$$\frac{\theta}{1+\theta}\left[\left(1+\frac{1}{\theta}\right) - \frac{\theta\alpha(1+\theta)}{\theta^2+\beta^2} + \frac{1}{2\theta} - \frac{2\theta\alpha(4\theta^2-\beta^2)}{(4\theta^2+\beta^2)^2} - \frac{1}{2\theta} + \frac{2\theta\alpha}{4\theta^2+\beta^2}\right]$$

$$= \frac{\theta}{1+\theta}\left[\left(1+\frac{1}{\theta}\right) - \frac{\alpha\theta(1+\theta)}{\theta^2+\beta^2} + \frac{4\theta\alpha\beta^2}{(4\theta^2+\beta^2)^2}\right] = W_c.$$

Especially, we used the following integral formulas for the development of the proof: we have

$$\int_0^{+\infty} 2\theta y e^{-2\theta y}\alpha\cos(\beta y)dy = \frac{2\theta\alpha(4\theta^2-\beta^2)}{(4\theta^2+\beta^2)^2}$$

and

$$\int_0^{+\infty} e^{-2\theta y}\alpha\cos(\beta y)dy = \frac{2\theta\alpha}{4\theta^2+\beta^2}.$$

□

This ends the proof.

An illustration of the shapes of the pdf of the Wc-ML distribution for various parameter values is seen in Figure 15.1.

The following are some observations about the shapes of $f_{Wc\text{-}ML}(y;\ \alpha,\ \beta,\ \theta)$:

1. The overall tendency of the curves seems to increase or decrease as α varies.
2. As β decreases, the overall tendency of the pdf becomes more spread out, with less visible signs of periodicity.
3. As θ increases, the shape of the pdf is visible with a peak, single modes, and no periodicity.

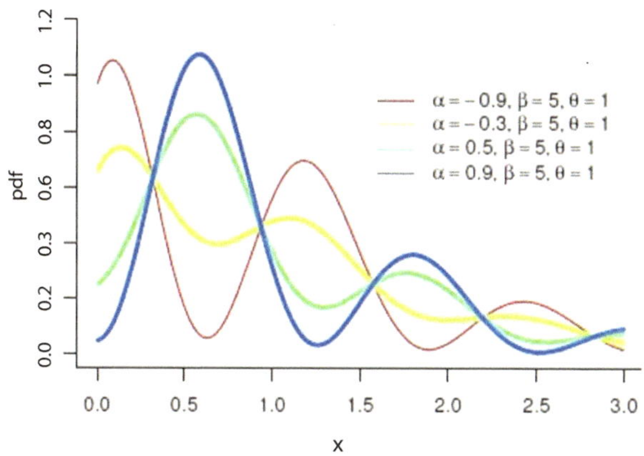

(a) Oscillatory shapes with a decreasing tendency

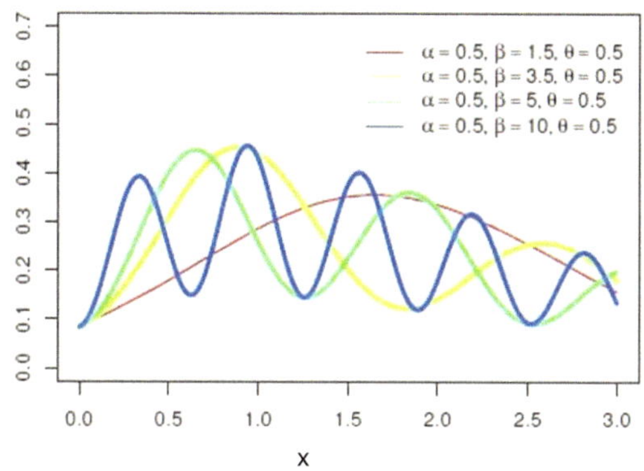

(b) Oscillatory shapes with a non-monotonic tendency

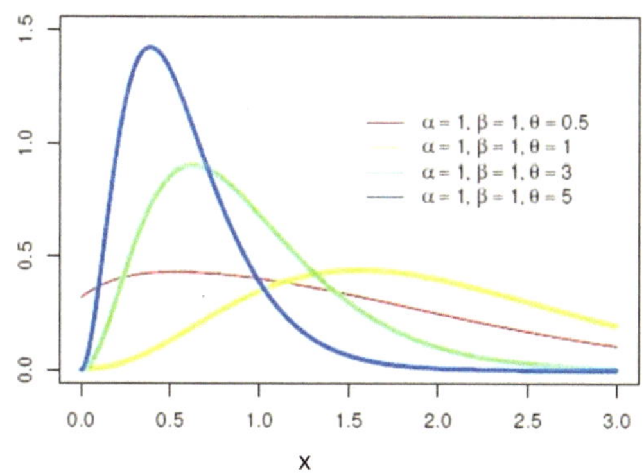

(c) Unimodal shape

Figure 15.1 Shapes of the pdf of the Wc-ML distribution with varying parametric values: (a) oscillatory shapes with a decreasing tendency, (b) oscillatory shapes with a non-monotonic tendency, and (c) Unimodal shape.

Next, we discuss the cdf of the Wc-ML distribution.

Proposition 2: *The cdf of the Wc-ML distribution has a closed form expression, and it is given by*

$$F_{Wc\text{-}ML}(t;\alpha,\beta,\theta)=$$

$$W_c^{-1}\frac{\theta}{1+\theta}\left[\frac{1+\theta}{\theta}\left(1-e^{-\theta t}\right)-te^{-2\theta t}-\frac{\alpha(1+\theta)e^{-\theta t}}{\theta^2+\beta^2}\left\{\beta\sin(\beta t)-\theta\cos(\beta t)+\theta e^{\theta t}\right\}\right.$$

$$-\frac{2\theta\alpha e^{-2\theta t}}{\left(4\theta^2+\beta^2\right)^2}\{(4\theta^2-\beta^2)e^{2\theta t}+\beta(\beta^2 t+4\theta^2 t+4\theta)\sin(\beta t)$$

$$+\left[\beta^2(1-2\theta t)-4\theta^2(2\theta t+1)\right]\cos(\beta t)\}$$

$$\left.+\frac{\alpha e^{-2\theta t}}{4\theta^2+\beta^2}\{\beta\sin(\beta t)-2\theta\cos(\beta t)+2\theta e^{2\theta t}\}\right],\ t>0$$

and $F_{Wc\text{-}ML}(t;\ \alpha,\ \beta,\ \theta)=0$*, for* $t\le 0$.

Proof. The proof is mainly based on several integrations. We have

$$F_{Wc\text{-}ML}(t;\alpha,\beta,\theta)=\int_0^t f_{Wc\text{-}ML}(y;\alpha,\beta,\theta)dy$$

$$=W_c^{-1}\frac{\theta}{1+\theta}\int_0^t\left[1-\alpha\cos(\beta y)\right]e^{-2\theta y}\left[(1+\theta)e^{\theta y}+2\theta y-1\right]dy$$

$$=W_c^{-1}\frac{\theta}{1+\theta}\left[\frac{1+\theta}{\theta}\left(1-e^{-\theta t}\right)+te^{-2\theta t}-\left\{\int_0^{t\theta}\frac{\alpha}{\theta}(1+\theta)\cos\left(\frac{\beta y}{\theta}\right)e^{-y}dy\right.\right.$$

$$\left.\left.+\frac{\alpha}{2\theta^2}\int_0^{2t\theta}\cos\left(\frac{\beta y}{2\theta}\right)e^{-y}ydy-\int_0^{2t\theta}\frac{\alpha}{2\theta}\cos\left(\frac{\beta y}{2\theta}\right)e^{-y}dy\right\}\right]$$

$$=W_c^{-1}\frac{\theta}{1+\theta}\left[\frac{1+\theta}{\theta}\left(1-e^{-\theta t}\right)-te^{-2\theta t}\right.$$

$$-\frac{\alpha(1+\theta)e^{-\theta t}}{\theta^2+\beta^2}\left\{\beta\sin(\beta t)-\theta\cos(\beta t)+\theta e^{\theta t}\right\}$$

$$-\frac{2\theta\alpha e^{-2\theta t}}{(4\theta^2+\beta^2)^2}\left\{(4\theta^2-\beta^2)e^{2\theta t}+\beta(\beta^2 t+4\theta^2 t+4\theta)\sin(\beta t)\right.$$

$$\left.+(\beta^2(1-2\theta)-4\theta^2(2\theta t+1))\cos(\beta t)\right\}$$

$$\left.+\frac{\alpha e^{-2\theta t}}{4\theta^2+\beta^2}\left\{\beta\sin(\beta t)-2\theta\cos(\beta t)+2\theta e^{2\theta t}\right\}\right].$$

We used the following formulas: we have

$$\int_0^t \cos(ax)e^{-x}dx = e^{-t}\,\frac{a\sin(at)-\cos(at)+e^t}{1+a^2}$$

and

$$\int_0^t x\cos(ax)e^{-x}dx = e^{-t}\,\frac{(1-a^2)e^t + a(a^2t+t+2)\sin(at)-(a^2(t-1)+t+1)\cos(at)}{(1+a^2)^2}.$$

This ends the proof. □

Figure 15.2 illustrates the shapes of the cdf and survival function (sf) of the Wc-ML distribution.

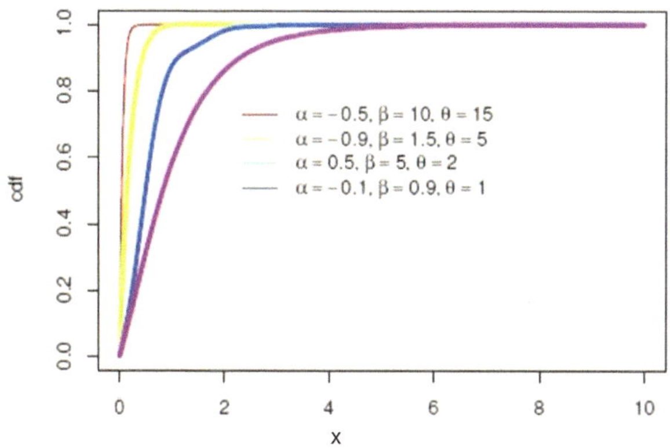

(a) The cdf display an increasing function

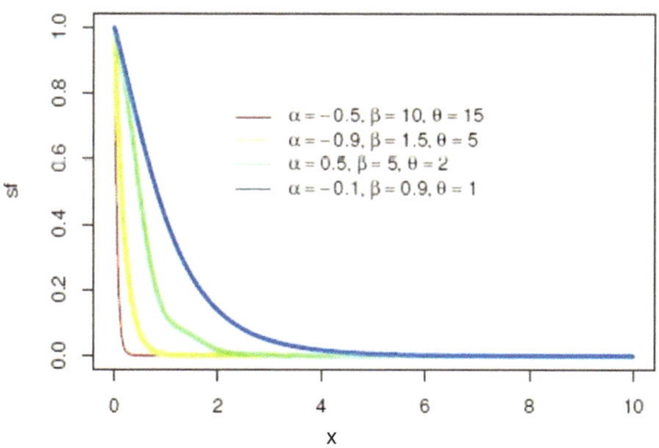

(b) The sf displays a decreasing function

Figure 15.2 Shapes of the (a) cdf and (b) sf of the Wc-ML distribution.

It is clear from Figure 15.2 that the cdf is valid in the mathematical sense. The hrf of the Wc-ML distribution is developed by taking the ratio of $f_{Wc\text{-}ML}(y;\,\alpha,\,\beta,\,\theta)$ to $1 - f_{Wc\text{-}ML}(y;\,\alpha,\,\beta,\,\theta)$.

Figure 15.3 illustrates the shapes of the hrf for different parametric values.

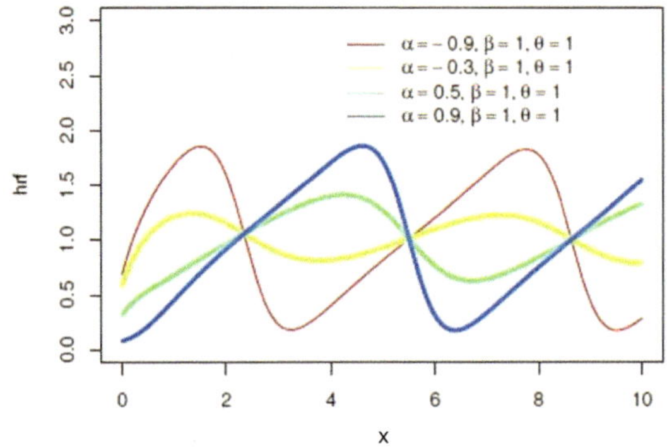

(a) Shapes of the hrf with various values of α

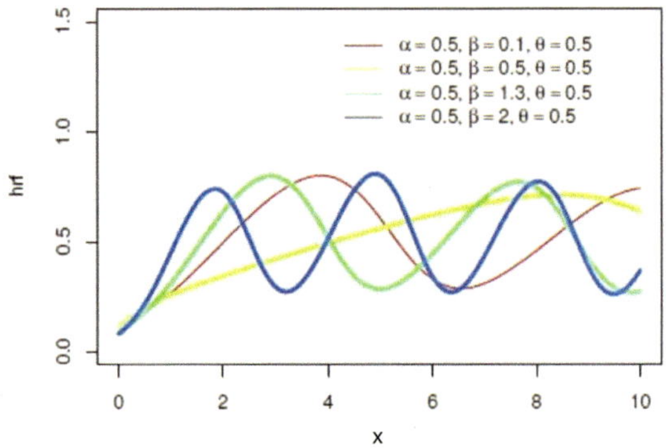

(b) Shapes of the hrf with various values of β

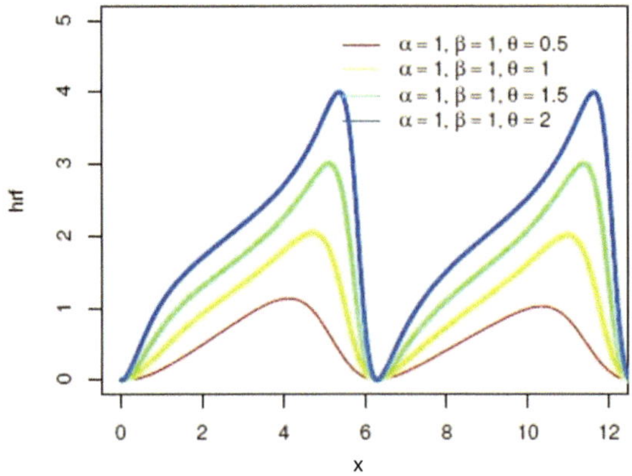

(c) Shapes of the hrf with various values of θ

Figure 15.3 Shapes of the hrf of the Wc-ML distribution for different parameter values with (a) varying values of α, (b) varying values of β and (c) varying values of θ.

From Figure 15.3, the following observations hold:

1. With constant values for β and θ, and varying α, we can see periodicity in the shapes of the hrf. As α increases, it is more evident that the overall tendency of the hrf begins with lower probability values.
2. With different values for β, the overall tendency of the hrf begins with lower probability, and shows periodic nature.
3. With same values for α and β, and different values of θ, the overall tendency of the hrf displays prominent peaks, indicating that it can be bimodal or multimodal.

15.2.3 Shape Analysis

The tail behavior of the pdf of the Wc-ML distribution can be studied as follows:

$$\lim_{y\to 0} f_{Wc\text{-}ML}(y;\alpha,\beta,\theta) = \frac{\theta^2}{1+\theta}(1-\alpha)$$

and $\lim_{y\to +\infty} f_{Wc\text{-}ML}(y;\alpha,\beta,\theta) = 0$. Thus, the parameters θ and α have a significant effect on the initial value but no effect on the infinite limit.

To obtain the modes of the Wc-ML distribution, we take the first derivative of $f_{Wc\text{-}ML}(y;\ \alpha,\ \beta,\ \theta)$ with respect to y, and equate it to zero. This gives the following sophisticated non-linear equation

$$\begin{aligned} &e^{-2\theta y}[2\theta+\theta(1+\theta)e^{\theta y}][1-\alpha\cos(\beta y)] - 2\theta e^{-2\theta y}[(1+\theta)e^{\theta y}+2\theta y-1] \\ &\quad \times[1-\alpha\cos(\beta y)] + \alpha\beta e^{-2\theta y}[-1+(1+\theta)e^{\theta y}+2\theta y]\sin(\beta y) = 0. \end{aligned} \tag{5}$$

Due to the trigonometric functions, this equation can be very complicated to solve, because we can have a large number of modes, as observed graphically. Our distribution can be "exaggeratedly multimodal".

15.3 MATHEMATICAL RESULTS

In this section, we investigate diverse moment properties of the Wc-ML distribution. Thanks to them, one can easily analyze the measures of central tendency, skewness, and kurtosis.

15.3.1 Moment Analysis

The moment properties of the Wc-ML distribution are examined in this subsection. To this aim, we consider a random variable Y following the

Wc-ML distribution with parameters α, β and θ. The raw moments of Y are provided in the result below.

Proposition 1: *For a positive integer r, the rth raw moment of Y is given by*

$$\mu_r' = E(Y^r) = W_c^{-1}\frac{\theta}{1+\theta}$$

$$\times\left[\theta^{-(r+1)}r!(\theta+1+2^{-(r+1)}r) - \frac{\alpha r!(1+\theta)}{(\theta^2+\beta^2)^{\frac{r+1}{2}}}\left\{\cos\left((r+1)\arctan\left(\frac{\beta}{\theta}\right)\right)\right\}\right.$$

$$-\frac{\alpha r!}{(4\theta^2+\beta^2)^{\frac{r+1}{2}}}\left\{\frac{2\theta(r+1)}{(4\theta^2+\beta^2)^{\frac{1}{2}}}\cos\left((r+2)\arctan\left(\frac{\beta}{2\theta}\right)\right)\right.$$

$$\left.\left.-\cos\left((r+1)\arctan\left(\frac{\beta}{2\theta}\right)\right)\right\}\right].$$

Proof. We can express the rth raw moment of Y as

$$\mu_r' = E(Y^r) = \int_0^{+\infty} y^r f_{Wc\text{-}ML}(y;\alpha,\beta,\theta)dy$$

$$= W_c^{-1}\frac{\theta}{1+\theta}\int_0^{+\infty} y^r\,(1-\alpha\cos(\beta y))e^{-2\theta y}\left[(1+\theta)e^{\theta y}+2\theta y-1\right]dy$$

$$= W_c^{-1}\frac{\theta}{1+\theta}\left[\int_0^{+\infty} y^r e^{-\theta y}dy + \int_0^{+\infty}\theta y^r e^{-\theta y}dy + \int_0^{+\infty}2\theta y^{r+1}e^{-2\theta y}dy\right.$$

$$-\int_0^{+\infty} y^r e^{-2\theta y}dy - \int_0^{+\infty} y^r\alpha\cos(\beta y)e^{-\theta y}dy - \int_0^{\infty}\theta y^r\alpha\cos(\beta y)e^{-\theta y}dy$$

$$\left.-\int_0^{+\infty}2\theta y^{r+1}\alpha\cos(\beta y)e^{-2\theta y}dy + \int_0^{\infty} y^r\alpha\cos(\beta y)e^{-2\theta y}dy\right].$$

After several integral developments, we establish that

$$\mu_r' = W_c^{-1}\frac{\theta}{1+\theta}$$

$$\times\left[\theta^{-(r+1)}r!(\theta+1+2^{-(r+1)}r) - \frac{\alpha r!(1+\theta)}{(\theta^2+\beta^2)^{\frac{r+1}{2}}}\left\{\cos\left((r+1)\arctan\left(\frac{\beta}{\theta}\right)\right)\right\}\right.$$

$$-\frac{\alpha r!}{(4\theta^2+\beta^2)^{\frac{r+1}{2}}}\left\{\frac{2\theta(r+1)}{(4\theta^2+\beta^2)^{\frac{1}{2}}}\cos\left((r+2)\arctan\left(\frac{\beta}{2\theta}\right)\right)\right.$$

$$\left.\left.-\cos\left((r+1)\arctan\left(\frac{\beta}{2\theta}\right)\right)\right\}\right].$$

The following integral formula was used:

$$\int_0^{+\infty} y^r \cos(\beta y)e^{-\theta y}dy = \frac{r!}{(\theta^2+\beta^2)^{\frac{r+1}{2}}}\cos\left((r+1)\arctan\left(\frac{\beta}{\theta}\right)\right).$$

Hence, we conclude the proof. □

From Proposition 1, we can obtain the expression of the mean of Y as follows:

$$\mu_1' = E(Y) = W_c^{-1}\frac{\theta}{1+\theta}\left[\theta^{-2}(\theta+1+2^{-2)})-\frac{\alpha(1+\theta)}{\theta^2+\beta^2}\cos\left(2\arctan\left(\frac{\beta}{\theta}\right)\right)\right.$$

$$\left.-\frac{\alpha}{(4\theta^2+\beta^2)}\left\{\frac{4\theta}{(4\theta^2+\beta^2)^{1/2}}\cos\left(3\arctan\left(\frac{\beta}{2\theta}\right)\right)-\cos\left(2\arctan\left(\frac{\beta}{2\theta}\right)\right)\right\}\right]$$

$$= W_c^{-1}\frac{\theta}{1+\theta}\left[\frac{4\theta+5}{4\theta^2}-\frac{\alpha(1+\theta)}{(\theta^2+\beta^2)}\left(\frac{\theta^2-\beta^2}{\theta^2+\beta^2}\right)-\frac{\alpha}{(4\theta^2+\beta^2)}\right.$$

$$\left.\times\left\{\frac{16\theta^2(4\theta^2-3\beta^2)}{(4\theta^2+\beta^2)^2}-\frac{4\theta^2-\beta^2}{4\theta^2+\beta^2}\right\}\right].$$

Also, the second raw moment of Y is given by

$$\mu_2' = W_c^{-1}\frac{\theta}{1+\theta}\left[\frac{5}{\theta^5}-\frac{2\alpha(1+0)}{(\theta^2+\beta^2)^{3/2}}\cos\left(3\arctan\left(\frac{\beta}{\theta}\right)\right)\right.$$

$$\left.-\frac{2\alpha}{(4\theta^2+\beta^2)^{\frac{3}{2}}}\left[\frac{6\theta}{(4\theta^2+\beta^2)^{1/2}}\cos\left(4\arctan\left(\frac{\beta}{2\theta}\right)\right)-\cos\left[3\arctan\left(\frac{\beta}{2\theta}\right)\right]\right]\right]$$

$$= W_c^{-1}\frac{\theta}{1+\theta}\left[\frac{5}{\theta^5}-\frac{2\alpha\theta(1+\theta)(\theta^2-3\beta^2)}{(\theta^2+\beta^2)^3}-\frac{2\alpha}{(4\theta^2+\beta^2)^{\frac{3}{2}}}\right.$$

$$\left.\times\left\{\frac{6\theta}{(4\theta^2+\beta^2)^{1/2}}\left(\frac{\beta^4-24\theta^2\beta^2+16\theta^4}{(4\theta^2+\beta^2)^2}\right)-\frac{4\theta(4\theta^2-3\beta^2)}{(4\theta^2+\beta^2)^{3/2}}\right\}\right].$$

The following formulas were used:

$$\cos\{2\arctan(x)\}=\frac{1-x^2}{1+x^2},\ \cos\{3\arctan(x)\}=\frac{1-3x^2}{(1+x^2)^{3/2}},\cos\{4\arctan(x)\}$$
$$=\frac{x^4-6x^2+1}{(1+x^2)^2}.$$

From the first and second raw moments, we can define the variance of Y by

$$\mu_2 = \mu'_2 - \mu_1'^2 .$$

With the development of the raw moments, the central moments of the Wc-ML distribution can also be developed. One can further develop the general formula for skewness and kurtosis of the Wc-ML distribution. Their equations are not derived in context since they are quite complex.

15.3.2 Incomplete Moments

In this subsection, an analysis on the incomplete moments of the Wc-ML distribution is carried out. A manageable expansion is given in the proposition below.

Proposition 2: *For a positive integer r, the rth incomplete moment of a random variable Y following the Wc-ML distribution is expressed as*

$$m_r(x)=E(Y^r 1_{\{Y\le x\}})=W_c^{-1}\frac{\theta}{1+\theta}\sum_{k=0}^{+\infty}a_k(\alpha,\beta)\Big[(1+\theta)\frac{1}{\theta^{2k+r+1}}\gamma(2k+r+1,\theta x)$$
$$+\frac{1}{(2\theta)^{2k+r+1}}\gamma(2k+r+2,2\theta x)-\frac{1}{(2\theta)^{2k+r+1}}\gamma(2k+r+1,2\theta x)\Big],$$

where $\gamma(w,z)=\int_0^z t^{w-1}e^{-t}dt, w,z>0$, *is the incomplete gamma function and the sequence* $(a_m(\alpha,\beta))_{m\in\mathbb{N}}$ *is defined in Eq. (4).*

Proof. The incomplete moment of Y is defined by

$$m_r(x)=E(Y^r 1_{\{Y\le x\}})=\int_0^x y^r f_{Wc\text{-}ML}(y;\alpha,\beta,\theta)dy.$$

It follows from Eq. (4) and several integrations that

$$m_r(x)=W_c^{-1}\frac{\theta}{1+\theta}\sum_{k=0}^{+\infty}a_k(\alpha,\beta)\int_0^x y^{2k+r}e^{-2\theta y}\left[(1+\theta)e^{\theta y}+2\theta y-1\right]dy$$

$$=W_c^{-1}\frac{\theta}{1+\theta}\sum_{k=0}^{+\infty}a_k(\alpha,\beta)\Bigg[(1+\theta)\frac{1}{\theta^{2k+r+1}}\gamma(2k+r+1,\theta x)+\frac{1}{(2\theta)^{2k+r+1}}$$

$$\times\gamma(2k+r+2,2\theta x)-\frac{1}{(2\theta)^{2k+r+1}}\gamma(2k+r+1,2\theta x)\Bigg].$$

Hence, the proof ends here. □

When only a portion of the data is available or it is challenging to compute complete moments, incomplete moments can be used as approximations of population moments. Also, data can be represented concisely and with fewer dimensions using incomplete moments, which is helpful for data display and further analysis.

15.4 INFERENTIAL ANALYSIS

In this section, we develop the estimates of the parameters α, β and θ based on the MLE method. Suppose that the vector $y = (y_1, y_2, \ldots, y_n)$ contained observations from a random sample $(Y_1, \ldots, Y_n)$ drawn from a random variable Y following the Wc-ML distribution with parameters α, β and θ. The likelihood function is then obtained as

$$L(y;\alpha,\beta,\theta)=\prod_{i=1}^{n}f_{Wc\text{-}ML}(y_i;\alpha,\beta,\theta)$$

$$=\prod_{i=1}^{n}\left[\frac{1}{W_c}(1-\alpha\cos(\beta y_i))\frac{\theta}{1+\theta}e^{-2\theta y_i}\times\left\{(1+\theta)e^{\theta y_i}+2\theta y_i-1\right\}\right]$$

$$=\frac{1}{(W_c)^n}\left(\frac{\theta}{1+\theta}\right)^n\left[\prod_{i=1}^{n}(1-\alpha\cos(\beta y_i))\right]e^{-2\theta\sum_{i=1}^{n}y_i}$$

$$\prod_{i=1}^{n}\left\{(1+\theta)e^{\theta y_i}+2\theta y_i-1\right\}.$$

As a result, the log likelihood (LL) function is given by

$$LL=-n\log\left[\frac{\theta+1}{\theta}-\frac{\alpha\theta(1+\theta)}{\theta^2+\beta^2}+\frac{4\theta\alpha\beta^2}{(4\theta^2+\beta^2)^2}\right]+\sum_{i=1}^{n}\log(1-\alpha\cos(\beta y_i))$$

$$-2\theta\sum_{i=1}^{n}y_i+\sum_{i=1}^{n}\log\left[(1+\theta)e^{\theta y_i}+2\theta y_i-1\right].$$

We aim to maximize this function with respect to the parameters α, β, and θ, respectively. To this aim, one can investigate the derivatives of the LL function with respect to the parameters α, β, θ. These derivatives are obtained as follows:

$$\frac{dLL}{d\alpha}=-\frac{n\left(\frac{4\beta^2\theta}{(\beta^2+4\theta^2)^2}-\frac{\theta(\theta+1)}{\beta^2+\theta^2}\right)}{\frac{4\beta^2\alpha\theta}{(4\theta^2+\beta^2)^2}-\frac{\alpha\theta(\theta+1)}{\beta^2+\theta^2}+\frac{1}{\theta}+1}-\sum_{i=1}^{n}\frac{\cos(\beta y_i)}{1-\alpha\cos(\beta y_i)},$$

$$\frac{dLL}{d\beta}=-\frac{n\left(\frac{2\beta\alpha\theta(\theta+1)}{(\beta^2+\theta^2)^2}+\frac{8\beta\alpha\theta}{(\beta^2+4\theta^2)^2}-\frac{16\beta^3\alpha\theta}{(\beta^2+4\theta^2)^3}\right)}{\frac{4\beta^2\alpha\theta}{(4\theta^2+\beta^2)^2}-\frac{\alpha\theta(\theta+1)}{\beta^2+\theta^2}+\frac{1}{\theta}+1}+\sum_{i=1}^{n}\frac{\alpha y_i\sin(\beta y_i)}{1-\alpha\cos(\beta y_i)},$$

$$\frac{dLL}{d\theta}=$$

$$-\frac{n\left(\frac{2(\theta+1)\theta^2\alpha)}{(\theta^2+\beta^2)^2}-\frac{64\theta^2\beta^2\alpha}{(4\theta^2+\beta^2)^3}-\frac{\theta\alpha}{\theta^2+\beta^2}-\frac{(\theta+1)\alpha}{\theta^2+\beta^2}+\frac{4\beta^2\alpha}{(4\theta^2+\beta^2)^2}-\frac{\theta+1}{\theta^2}+\frac{1}{\theta}\right)}{\frac{4\beta^2\alpha\theta}{(4\theta^2+\beta^2)^2}-\frac{\alpha\theta(\theta+1)}{\beta^2+\theta^2}+\frac{1}{\theta}+1}$$

$$-2\sum_{i=1}^{n}y_i+\sum_{i=1}^{n}\frac{2y_i+e^{\theta y_i}(1+y_i+\theta y_i)}{(1+\theta)e^{\theta y_i}+2\theta y_i-1}.$$

These derivatives are then equated to zero to obtain the estimates, $\hat{\alpha}$, $\hat{\beta}$ and $\hat{\theta}$ of α, β and θ, respectively. These estimates cannot be obtained by solving the equations manually. Here, we use the R software to compute their values (Team 2005).

15.5 DATA ANALYSIS

This section is relevant to prove the applicability of the newly proposed distribution over other comparable distributions, which are the NET, CL, and exponential distributions.

Table 15.1 displays the definitions of the pdfs of the considered competitive distributions.

Table 15.1 Presentation of the pdfs of the competitive distributions

Model	pdf
NET	$\frac{\theta(\beta^2+\theta^2)}{\beta^2+(1-\alpha)\theta^2}[1-\alpha\cos(\beta y)]e^{-\theta y}$
CL	$\left\{\frac{1+\theta}{\theta^2}-\frac{\alpha\theta}{\theta^2+\beta^2}-\frac{\alpha(\theta^2-\beta^2)}{(\theta^2+\beta^2)^2}\right\}^{-1}[1-\alpha\cos(\beta y)](1+y)e^{-\theta y}$
Expo	$\theta e^{-\theta y}$

15.5.1 Methodology

To begin, we study the characteristics of the data sets in terms of basic statistical descriptive, which include mean, standard deviation (SD), skewness, and kurtosis.

A box plot and a total time on test (TTT) plot are illustrated to help understand the datasets better. We redirect the readers to Aarset (1987) for a more in-depth interpretation of the TTT plot.

The estimates $\widehat{\alpha}$, $\widehat{\beta}$ and $\widehat{\theta}$ along with their standard errors (SEs) of the proposed and competitive distributions are obtained.

The goodness-of-fit (gof) criteria measures, such as the Kolmogorov-Smirnov (KS) statistic and its p-value, the Anderson-Darling (AD) statistic, and the Cramér von-Mises (CVM) statistic, are obtained. For more information on these gof criteria, one can refer to Anderson (1954) and Darling (1957).

The adequacy measures, which include the Akaike information criterion (AIC) and the Bayesian information criterion (BIC), are used.

The distribution having the highest p-value and the lowest values for the gof criteria measures and the adequacy measures is considered the best distribution when modeling the datasets.

For a visual depiction of the best distribution and to support the numerical results obtained, the estimated pdfs (epdfs) and estimated cdfs (ecdfs) are plotted.

15.5.2 Datasets

In this section, we provide the datasets considered for the study.

15.5.2.1 Failure Dataset

The first dataset refers to the time between failures for repairable items, it has been received from Murthy (2004). The data are: 1.43, 0.11, 0.71, 0.77, 2.63, 1.49, 3.46, 2.46, 0.59, 0.74, 1.23, 0.94, 4.36, 0.40, 1.74, 4.73, 2.23, 0.45, 0.70, 1.06, 1.46, 0.30, 1.82, 2.37, 0.63, 1.23, 1.24, 1.97, 1.86, 1.17.

15.5.2.2 *Animal Migration Dataset*

Observed target distances from transect lines are frequently used in line-transect distance sampling to estimate population densities and simulate detectability. Significant wild animal populations in a particular environment are a contributing factor to the current situation. This approach's fundamental premise is that all organisms can be found where they first appear. The natural food chain of the community could be considerably disrupted by animal migration that is not under the observer's and transects' control. This dataset, which was collected from Patil (1987), gives the positions of the 68 snakes that were discovered while paddling $L = 1000m$ and scanning $w = 20m$ on either side of the transect line.

The data are: 2.0, 0.5, 10.4, 3.6, 0.9, 1.0, 3.4, 2.9, 8.2, 6.5, 5.7, 3.0, 4.0, 0.1, 11.8, 14.2, 2.4, 1.6, 13.3, 6.5, 8.3, 4.9, 1.5, 18.6, 0.4, 0.4, 0.2, 11.6, 3.2, 7.1, 10.7, 3.9, 6.1, 6.4, 3.8, 15.2, 3.5, 3.1, 7.9, 18.2, 10.1, 4.4, 1.3, 13.7, 6.3, 3.6, 9.0, 7.7, 4.9, 9.1, 3.3, 8.5, 6.1, 0.4, 9.3, 0.5, 1.2, 1.7, 4.5, 3.1, 3.1, 6.6, 4.4, 5.0, 3.2, 7.7, 18.2, 4.1.

15.5.3 Statistical Analysis

Failure dataset: The mean of the failure data of components is 1.5427, with an SD of 1.127686. The corresponding skewness and kurtosis are obtained as 1.295462 and 4.31917, respectively, depicting that the shape of the data is positively skewed and leptokurtic in nature.

Figure 15.4 illustrates the box plot and the TTT plot for the failure dataset.

From Figure 15.4, the box plot supports the positive skewness of the data. The colored TTT line in the TTT plot is concave, meaning that the empirical hrf of the data is increasing, which is also a property satisfied by the proposed model.

Tables 15.2 and 15.3 provide the estimates, SEs, and gof measures for the considered dataset.

Table 15.2 Estimates and SEs of the models for the failure dataset

Model	$\hat{\alpha}$	$\hat{\beta}$	$\hat{\theta}$
NET	0.7578	2.0027	0.7034
	(0.1551)	(0.1757)	(0.2142)
CL	0.5850	2.0156	1.0554
	(0.1639)	(0.2006)	(0.2484)
Wc-ML	0.4852	2.0049	0.7827
	(0.1338)	(0.2336)	(0.2702)
Exp	–	–	0.6482
			(0.1183)

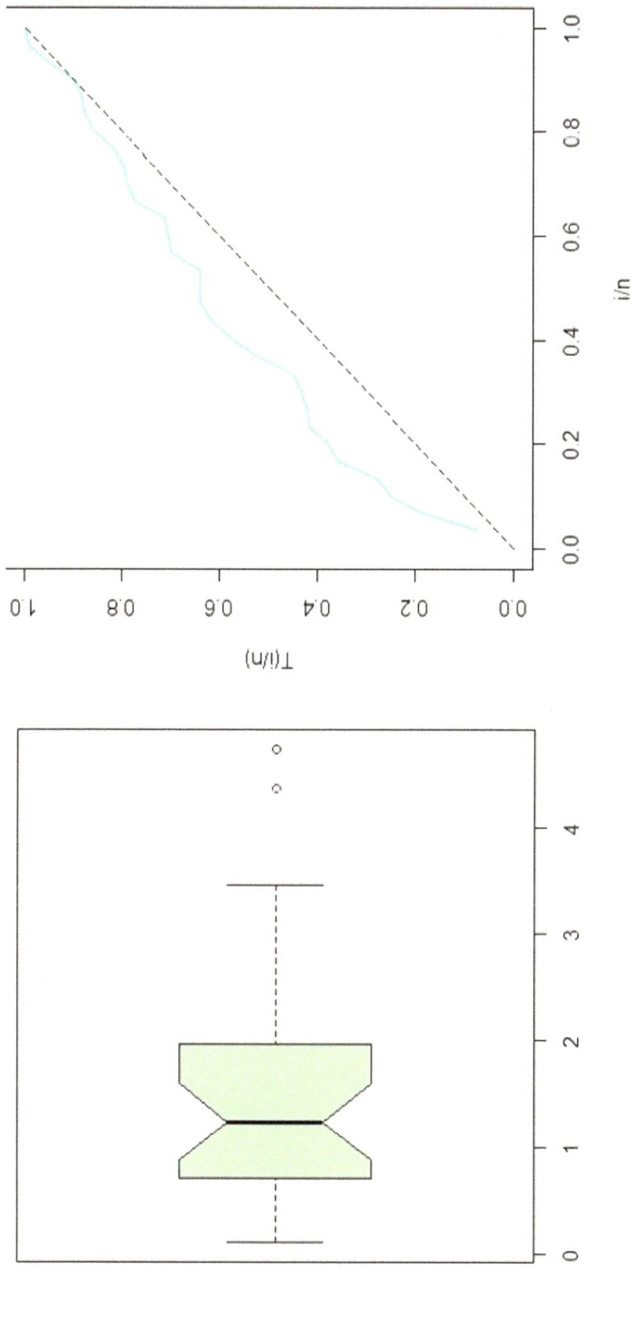

(a) Positively skewed data

(b) Increasing hrf

Figure 15.4 (a) Box plot and (b) TTT plot of the failure dataset.

Table 15.3 Adequacy measures and the gof metrics for the failure dataset

Model	AIC	BIC	AD	CVM	KS	p-value
NET	84.9629	89.1665	0.2006	0.0298	0.0860	0.9794
CL	84.6197	88.8233	0.1784	0.02754	0.0936	0.9551
Wc-ML	84.6075	88.8111	0.16302	0.02545	0.0881	0.9738
Exp	88.0108	89.411	1.329	0.232	0.184	0.258

From Table 15.3, it can be observed that the NET distribution has the highest p-value of 0.9794, but the Wc-ML distribution has a competitive p-value of 0.9738 and the lowest AIC and BIC values compared to the other distributions. Figure 15.5 then provides a visual illustration of the epdfs and ecdfs of the considered models to support Table 15.3.

In terms of fit efficiency, Figure 15.5 supports the findings presented in Table 15.3.

Animal migration dataset: The mean of the distance traveled by 68 snakes from the transect line is 5.853, with a SD of 4.61278. The corresponding skewness and kurtosis are obtained as 1.0436 and 3.575, respectively, stating that the shape of the data is positively skewed and is mesokurtic in nature.

Figure 15.6 displays the box plot and TTT plot for the animal migration dataset.

The box plot in Figure 15.6 supports the positive skewness of the data, as indicated by the skewness value. Additionally, the portion to the left of the median is short compared to the other portion, indicating that smaller values are more closely spaced out than the greater ones. The typical normal distribution is insufficient for modelling these data, as a quick observation shows. The colored TTT line in the TTT plot is concave, which means that the corresponding empirical hrf is increasing. Our proposed model has this characteristic.

Tables 15.4 and 15.5 provide the obtained estimates and SEs along with the gof metrics for the animal migration dataset.

Table 15.4 Estimates and SEs of the models for the animal migration dataset

Model	$\hat{\alpha}$	$\hat{\beta}$	$\hat{\theta}$
NET	0.1715	2.5643	0.4282
	(0.02084)	(0.03897)	(0.16830)
CL	0.4058	2.56460	0.3015
	(0.02623)	(0.04027)	(0.17102)
Wc-ML	−0.502	1.725	0.208
	(0.152)	(0.038)	(0.019)
Exp	–	–	0.17085
			(0.02072)

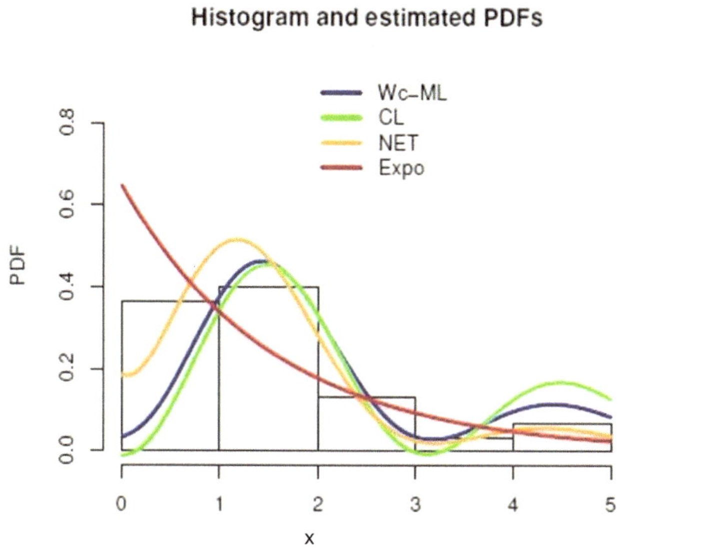

(a) The epdfs of the considered models over the corresponding the histogram of the data

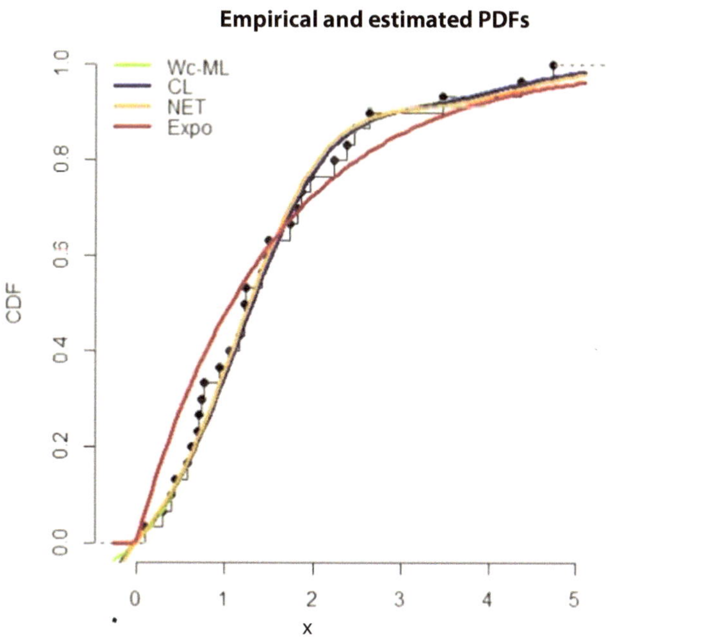

(b) The ecdfs of the considered models along with the corresponding empirical cdf of the data

Figure 15.5 Illustration of the (a) epdfs and (b) ecdfs for the failure dataset.

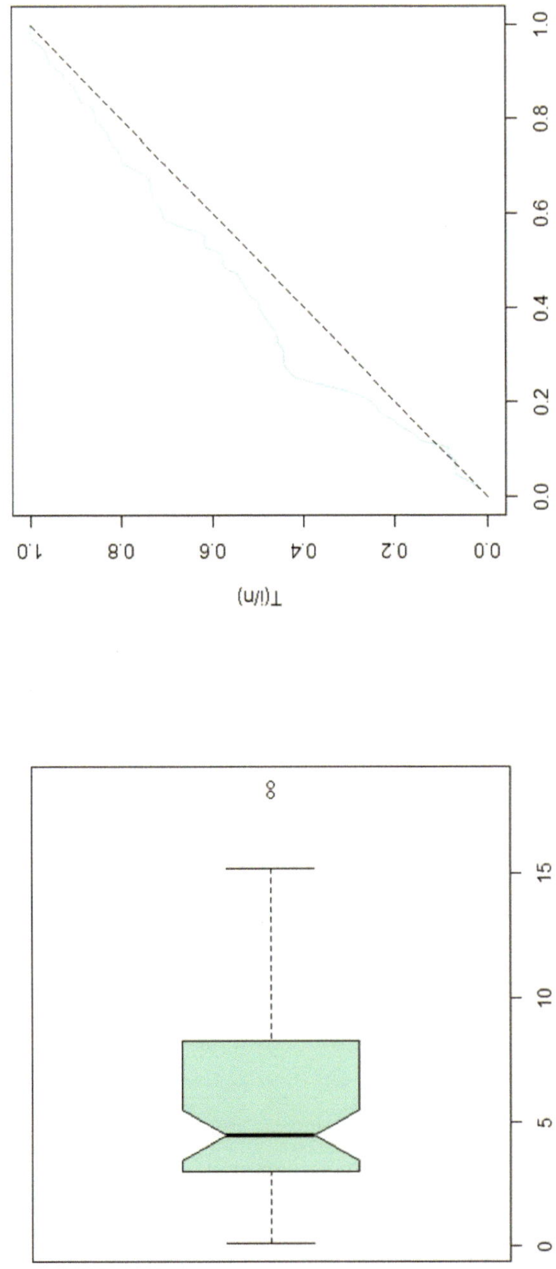

(a) Positively skewed data

(b) Increasing hrf

Figure 15.6 (a) Box plot and (b) TTT plot of the animal migration dataset.

Table 15.5 Adequacy measures and the gof metrics for the animal migration dataset

Model	AIC	BIC	AD	CVM	KS	p-value
NET	376.5597	383.2182	1.01278	0.197	0.1400	0.139
CL	373.5631	380.2216	0.466	0.03317	0.0746	0.842
Wc-ML	372.6238	379.2823	0.3270	0.03621	0.0605	0.9642
Exp	378.3044	380.5239	1.0652	0.2123	0.15546	0.0748

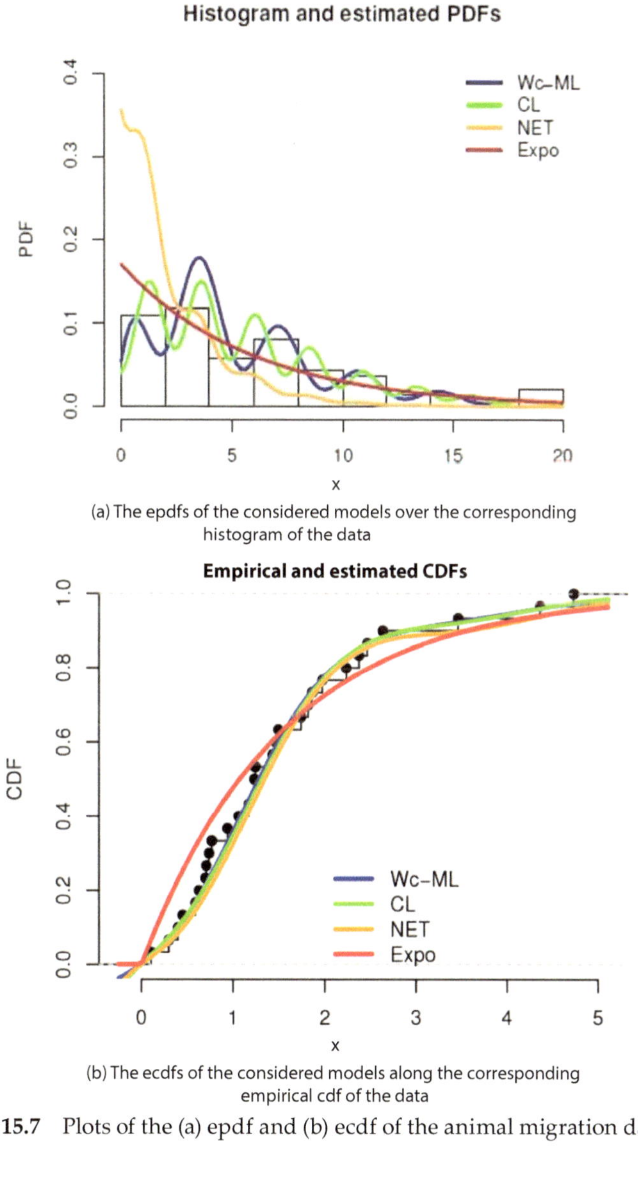

(a) The epdfs of the considered models over the corresponding histogram of the data

(b) The ecdfs of the considered models along the corresponding empirical cdf of the data

Figure 15.7 Plots of the (a) epdf and (b) ecdf of the animal migration dataset.

From Table 15.5, it can be seen that the Wc-ML distribution provides the highest KS p-value of 0.9642, with the least values for the AIC, BIC, AD statistic, and CVM statistics. Thus, we can conclude that the Wc-ML distribution provides the best fit to the animal migration dataset. We support the numerical findings by the illustration of the epdfs and ecdfs in Figure 15.7.

From Figure 15.7, we can observe that the Wc-ML distribution gives a better fit to the dataset compared to the other distributions.

15.6 CONCLUSION

In this article, we developed a trigonometric weighted distribution by merging a trigonometric weight function with the functionalities of the modified Lindley distribution. The corresponding probability density function and the hazard rate function exhibit periodic shapes, which is a rare and highly demanded feature of distribution theory. The basic functions associated with the distribution, shape analysis of the probability density function, and moments were also studied. In comparison to the new exponential trigonometric function and cos-Lindley models, the applicability of the proposed model to two real-world data sets is examined. The results of the comparison demonstrate that, for some criteria, the suggested model is superior to the competing distributions.

REFERENCES

Aarset, M.V. 1987. How to identify a bathtub hazard rate. IEEE Transactions on Reliability. 106–108.

Anderson, T.W. 1954. A test of goodness of fit. Journal of the American Statistical Association. 49(268): 765–769.

Bakouch, H.S. 2018. A new lifetime model with a periodic hazard rate and an application. Journal of Statistical Computation and Simulation. 88(11): 2048–2065.

Chesneau, C. 2021. A new modified Lindley distribution with properties and applications. Journal of Statistics and Management Systems. 24(7): 1383–1403.

Chesneau, C. 2023. The Cos-Lindley model with modeling to service times and failure times data. International Journal of Modelling and Simulation.

Darling, D.A. 1957. The Kolmogorov-Smirnov, Cramér-von Mises tests. The Annals of Mathematical Statistics. 28(4): 823–838.

Fisher, R.A. 1934. The effect of methods of ascertainment upon the estimation of frequencies. Annals of eugenics. 6(1): 13–25. doi:10.1111/j.1469-1809.1934.tb02105.x.

Murthy, D. 2004. Weibull Models. John Wiley & Sons, Inc. New Jersey, NJ, USA.

Patil, G.P. 1987. Weighted Distributions. Center for Statistical Ecology and Environmental Statistics, Pennsylvania State University.

Rao, C.R. 1965. On discrete distributions arising out of methods of ascertainment. Sankhyā: The Indian Journal of Statistics, Series A. doi:10.1111/j.1469-1809.1934.tb02105.x.

Team, R.D. 2005. R Foundation for Statistical Computing. R: A Language and Environment for Statistical Computing.

Chapter 16

The Two-Sided Generalized Lomax Distribution: Estimation and Application

Kadir Karakaya[1*], Yunus Akdoğan[2], Mustafa Çağatay Korkmaz[3] and Fatih Şahin[4]

[1,2,4]Department of Statistics, Selçuk University, Konya, Turkey. Emails: [1]kkarakaya@selcuk.edu.tr, [2]yakdogan@selcuk.edu.tr, [4]fatihsahin0644@gmail.com

[3]Department of Measurement and Evaluation, Artvin Çoruh University, Artvin, Turkey. Email: mustafacagataykorkmaz@gmail.com

16.1 INTRODUCTION

Lomax (Lo) distribution was studied by Lomax (1954) for modeling business failure data. The *Lo* distribution is known as Pareto II or the second type of beta distributions. *Lo* distribution has a wide range of applications, including income and wealth inequality, engineering, lifetime modeling, biological and medical sciences, actuarial science, and reliability. The probability density function (pdf) and cumulative distribution function (cdf) of the *Lo* distribution are presented, respectively, by

$$f_{Lo}(x,\lambda,\theta)=\frac{\theta}{\lambda}(1+x/\lambda)^{(-\theta-1)}, \tag{1}$$

*For Correspondence: Kadir Karakaya (kkarakaya@selcuk.edu.tr)

and

$$F_{Lo}(x, \lambda, \theta) = 1 - (1 + x/\lambda)^{-\theta}, x > 0 \tag{2}$$

where $\lambda > 0$ is the scale parameter and $\theta > 0$ is shape parameter. The pdf of *Lo* distribution is reverse-J shaped. The hazard rate function (hrf) of the *Lo* model is equal to $h(x, \theta, \lambda) = \theta/(\lambda[1 + x/\lambda])$ which is a decreasing function of x. These properties of the *Lo* distribution cannot be adequate for data modeling. On the other hand, several family generators have been presented in the literature to increase flexibility by adding one or more parameters to a baseline model. As an example the two-sided generalized family ($TSG - G$) of the model is introduced by Korkmaz and Genç (2017) following cdf:

$$F_{TSG-G}(x, \alpha, \beta, \xi) = \{\beta^{(1-\alpha)}[G(x; \xi)]^{\alpha} ,$$
$$-\infty < x < G^{-1}_{(x,\xi)}(\beta) 1 - (1-\beta)^{(1-\alpha)}[1 - G(x; \xi)]^{\alpha} , [G^{-1}_{(x,\xi)}(\beta)] \leq x < \infty \tag{3}$$

where $\alpha > 0$ is shape and $\beta \in (0, 1)$ is the location parameters, which plays a role of the reflection of the distribution, $G(x, \xi)$ which is a baseline cdf and is affected by a parameter vector ξ, and $G^{(-1)}_{(x, \xi)}(\cdot)$ is inverse of baseline cdf. The authors have shown that this family brings flexible pdf and hrf shapes to the baseline model to extend its data modeling area.

In this study, a new generalization of the *Lo* model with a two-sided idea defined by Eq. (3) is investigated. The cdfs and pdfs of the new generalized *Lo* model are presented, respectively, by

$$F_{TSG-Lo}(x; \alpha, \beta, \theta, \lambda) = \{\beta^{1-\alpha}[1 - (1 + x/\lambda)^{-\theta}]^{\alpha},$$
$$0 < x < \tau 1 - (1-\beta)^{(1-\alpha)}(1 + x/\lambda)^{-\theta\alpha}, \tau \leq x < \infty \tag{4}$$

and

$$f_{TSG-Lo}(x; \alpha, \beta, \theta, \lambda) = \{\alpha\beta^{1-\alpha}\theta\lambda^{-1}[1 - (1 + x/\lambda)^{-\theta}]^{\alpha-1}(1 + x/\lambda)^{-\theta-1} ,$$
$$0 < x < \tau \alpha(1-\beta)^{1-\alpha}\theta\lambda^{-1}(1 + x/\lambda)^{-\theta(\alpha-1)}(1 + x/\lambda)^{-\theta-1} , \tau \leq x < \infty \tag{5}$$

where $\tau = \lambda[(1-\beta)^{(-1/\theta)} - 1]$. We call it two-sided generalized Lomax ($TSG - Lo$) distribution. For $\alpha = 1$, the ordinary *Lo* distribution is obtained. When $\beta \to 0$, the distribution is reduced to exponentiated *Lo* (Abdul-Moniem and Abdel-Hameed 2012). A $TSG - Lo$ model will be a more flexible model than the ordinary *Lo* model in relation to pdf, hrf, kurtosis, and skewness.

In this study, some basic properties of $TSG - Lo$ such as moments, skewness, and kurtosis are examined. The maximum likelihood (ML) estimation method is evaluated to estimate four model parameters. A simulation experiment is studied to observe behavior of ML estimators. Finally, we fit the new model to real data to evaluate it with current models.

16.1.1 The Density Shapes

Figure 16.1 illustrates some possible plots of the *TSG – Lo* density for various parameters. From Figure 16.1, one can see that pdf of the proposed model is extremely flexible in shape. It has unimodal or decreasing pdfs.

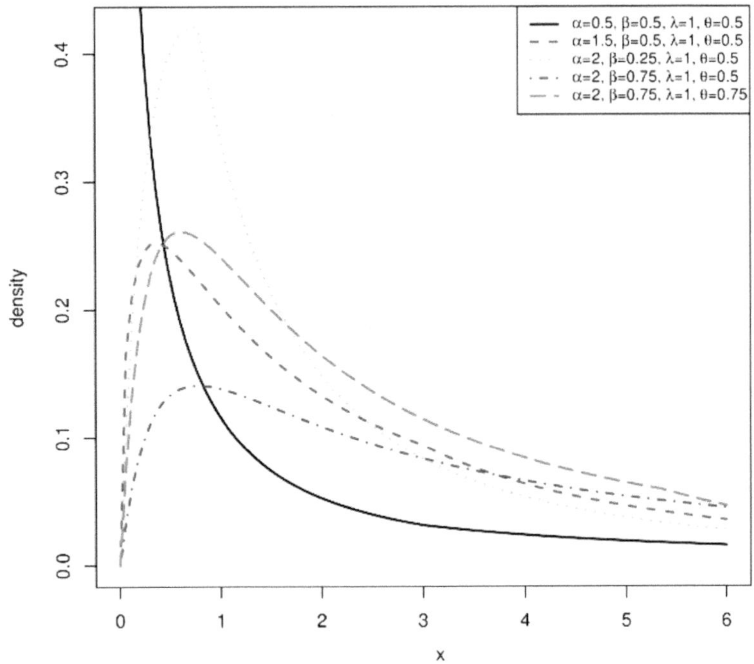

Figure 16.1 The plots of pdf of *TSG – Lo* distributions.

16.1.2 Moments

Let the random variable X have *TSG – Lo* with pdf in Eq. (5). The rth moment of X is given as follows:

$$\mu_r = \sum_{l=1}^{N} \varpi_l y^r f_1(y) + \sum_{l=1}^{N} \varpi_l \frac{1}{y^2}\left(\frac{1}{y}\right)^r f_2\left(\frac{1}{y}\right), \tag{6}$$

where $f_1(\cdot)$ and $f_2(\cdot)$ are pdfs given in Eq. (5), y_l is zero and ϖ_l denotes the Legendre-Gauss quadrature formula's associated Christoffel numbers on the interval (–1, 1), as defined by Canuto et al. (2006). It is also noted that ϖ_l is calculated by

$$\varpi_l = \frac{2}{(1-y_l)^2 [L'_{N+1}(y_l)]^2}, \tag{7}$$

where

$$L'_{N+1}(y_l) = \frac{dL_{N+1}(y)}{dy} \quad (8)$$

at $y = y_l$ and $L_{(N+1)}$ (·) is the Legendre polynomial of degree N. The expected value (μ_1), variance (V), skewness (S), kurtosis (K), and first four moments of a *TSG – Lo* model are reported in Tables 16.1–16.3. N = 50 is used in numerical computations. We observe the following:

Table 16.1 The μ_1 and V of *TSG – Lo* distribution for different values of λ, θ, β and α

		(λ, β)							
		(1,2)		**(2,1)**		**(3,5)**		**(4,6)**	
β	α	μ_1	V	μ_1	V	μ_1	V	μ_1	V
0.3	2	0.4514	0.2701	2.7538	84.1184	0.4472	0.1540	0.4874	0.1742
0.5	3.5	0.4670	0.0764	2.4589	4.3913	0.4840	0.0584	0.5295	0.0675
0.9	5	1.3048	0.4064	9.4595	42.4234	1.1498	0.2107	1.2370	0.2334

Table 16.2 The S and K of *TSG – Lo* distribution for different values of λ, θ, β and α

		(λ, β)							
		(1,2)		**(2,1)**		**(3,5)**		**(4,6)**	
β	α	S	K	S	K	S	K	S	K
0.3	2	6.9130	492.3262	254.795	150,854.428	2.6898	16.9605	2.5154	14.7073
0.5	3.5	2.6238	20.7217	10.2680	4,941.4938	1.6295	9.0131	1.5478	8.4164
0.9	5	0.6379	3.7119	1.7886	15.7442	0.1902	2.7275	0.1433	2.6857

Table 16.3 The first four moments of *TSG – Lo* distribution for different values of λ, θ, β and α

β	α	λ	θ	μ_1	μ_2	μ_3	μ_4
0.3	2	1	2	0.4514	0.4739	1.4283	38.0447
		2	1	2.7538	91.7019	197,291.1585	1,069,601,232.4765
		3	5	0.4472	0.35407	0.4587	0.9183
		4	6	0.4874	0.4118	0.5535	1.1081
0.5	3.5	1	2	0.4670	0.2945	0.2644	0.3723
		2	1	2.4589	10.4376	141.7525	96,418.5663
		3	5	0.4840	0.2927	0.2212	0.2124
		4	6	0.5295	0.3479	0.2829	0.2881
0.9	5	1	2	1.3048	2.1092	3.9785	8.5287
		2	1	9.4595	131.9071	2,544.6486	77,821.5812
		3	5	1.1498	1.5329	2.2657	3.6258
		4	6	1.2370	1.7638	2.7757	4.7118

16.2 ESTIMATION

Let $X_1, X_2, \ldots, X_n$ be a random sample of size n from the $TSG - Lo$ model with parameter vector $(\beta, \alpha, \lambda, \theta)$ and let $X_{1:n} \leq X_{2:n} \leq \ldots \leq X_{n:n}$ represent related-order statistics. The log-likelihood function is obtained as

$$\ell(\beta,\alpha,\lambda,\theta) = n\log\alpha + \sum_{i=1}^{r} \log\left(\alpha\beta^{1-\alpha}\theta\lambda^{-1}[1-(1+x_i/\lambda)^{-\theta}]^{\alpha-1}(1+x_i/\lambda)^{-\theta-1}\right)$$

$$+\sum_{i=r+1}^{n} \log\left(\alpha(1-\beta)^{1-\alpha}\,\theta\lambda^{-1}(1+x_i/\lambda)^{-\theta(\alpha-1)}(1+x_i/\lambda)^{-\theta-1}\right)$$

$$+(\alpha-1)\log\left(\frac{\prod_{i=1}^{r}\beta^{1-\alpha}\left[1-(1+x_i/\lambda)^{-\theta}\right]^{\alpha}\prod_{i=r+1}^{n}(1-\beta)^{(1-\alpha)}(1+x_{ip}/\lambda)^{-\theta\alpha}}{\beta^{r}(1-\beta)^{n-r}}\right) \tag{9}$$

where $x(r) \leq \lambda[(1-\beta)^{(-1/\theta)} - 1] \leq x(r+1)$ for $r = 1, 2, \ldots, n$ and $x_{(0)} = 0$, $x_{(n+1)} = \infty$. The ML estimates of α and β, based on van Dorp and Kotz (2002) are provided by

$$\widehat{\alpha} = \frac{-n}{\log\log M(\hat{r},\theta,\lambda)} \tag{10}$$

and

$$\widehat{\beta} = G(x_{\hat{r}};\theta,\lambda) \tag{11}$$

respectively, where $\hat{r} = M(r, \beta, \alpha, \lambda, \theta)$, $r \in (1, 2, \ldots, n)$ with

$$M(r,\lambda,\theta) = \prod_{i=1}^{r-1}\frac{G(x_i;\theta,\lambda)}{G(x_r;\theta,\lambda)}\prod_{i=r+1}^{n}\frac{1-G(x_i;\theta,\lambda)}{1-G(x_r;\theta,\lambda)}. \tag{12}$$

For parameter vector $\xi = (\lambda, \theta)$ of $TSG - Lo$ distribution, the associated gradients are obtained to be

$$\frac{\partial l}{\partial \xi_k} = \sum_{i=1}^{n}\frac{\partial g(x_i;\xi)/\partial\xi_k}{g(x_i;\xi)} + (\alpha-1)\left(\sum_{i=1}^{\hat{r}}\frac{\partial G(x_i;\xi)/\partial\xi_k}{G(x_i;\xi)} - \sum_{i=\hat{r}+1}^{n}\frac{\partial G(x_i;\xi)/\partial\xi_k}{1-G(x_i;\xi)}\right). \tag{13}$$

The above equations cannot be achieved analytically, hence statistical software with an optimization procedure can be utilized to solve them numerically using an iterative approach such as Newton-Raphson. For instance, the optim function in R can be used.

16.3 SIMULATION EXPERIMENTS

In this section, two simulation experiments to observe the behavior of ML estimates of the $TSG - Lo$ distribution are conducted. All results are obtained via CG method in function in the R program.

16.3.1 Simulation Experiment I

The graphical findings are presented in the first simulation study. $N = 1000$ samples of size $n = 20, 25, \ldots, 1000$ from $TSG - Lo$ model with a parameter setting $\alpha = 5$, $\beta = 0.75$, $\lambda = 1$, and $\theta = 0.5$ are generated. In the simulation experiment, the mean, standard deviation (SD), bias, and mean square error (MSE) of the ML estimates are computed as empirically. The bias and MSE are computed below, respectively,

$$\widehat{Bias}_h = \frac{1}{1000} \sum_{i=1}^{1000} (\hat{h}_i - h), \tag{14}$$

and

$$\widehat{MSE}_h = \frac{1}{1000} \sum_{i=1}^{1000} (\hat{h}_i - h)^2 \tag{15}$$

where $h = \alpha, \beta, \lambda, \theta$. The simulation Experiment-I illustrated in Figure 16.2 demonstrated that when the sample size increases, the empirical means approach the true parameter value while all biases, SD, and MSEs go to zero.

16.3.2 Simulation Experiment II

In this sub-section, we generate $N = 1000$ samples of sizes 20, 60, and 100 from $TSG - Lo$ distribution. We achieve the empirical means and SD of the ML estimates in the simulation study. The results are reported in Table 16.4 for five parameter settings. Table 16.4 indicates that as the sample of size increases, the SD decreases whereas the empirical means approach true parameter values.

16.4 REAL DATA ANALYSIS

In this section, an application to the real dataset to prove empirically the potentiality of $TSG - Lo$ distribution is provided. Dataset shows remission rates (in months) of randomly selected 128 bladder cancer patients (Lee and Wang 2003). This data is also investigated by Lemonte and Cordeiro (2013). On this data set, we compare it with McDonald

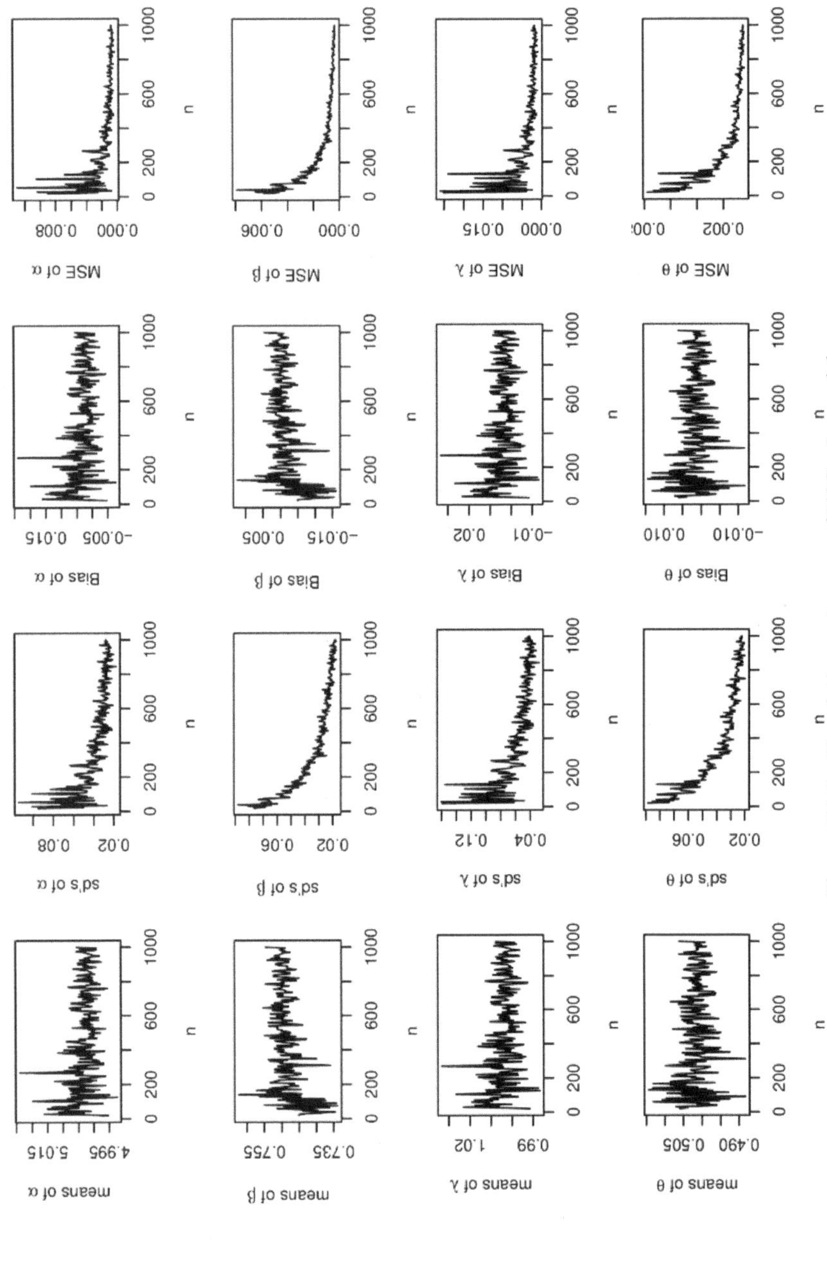

Figure 16.2 Simulation outcomes of $TSG - Lo$ model.

Table 16.4 Empirical means (SD) for the *TSG – Lo* distribution

Parameters					
$\alpha, \beta, \lambda, \theta$	n	$\hat{\alpha}$	$\hat{\beta}$	$\hat{\lambda}$	$\hat{\theta}$
0.5,0.5,0.5,0.5	20	0.6680 (0.3287)	0.4765(0.4537)	0.7073(0.8293)	0.5640(0.4217)
	60	0.5533(0.1932)	0.4740(0.3908)	0.5605(0.4204)	0.5198(0.2286)
	100	0.5415(0.1568)	0.4506(0.4699)	0.4709(0.3043)	0.5100(0.2385)
2.5,0.5,0.5,0.5	20	2.6045(0.2563)	0.5366(0.1590)	0.6185(0.2357)	0.5866(0.1714)
	60	2.5698(0.1627)	0.5184(0.1228)	0.5595(0.1647)	0.5515(0.1047)
	100	2.5609(0.1607)	0.5008(0.1060)	0.5505(0.1490)	0.5239(0.0806)
5,0.25,5,5	20	5.0175(0.2593)	0.2574(0.0313)	4.9907(0.1814)	5.0128(0.1837)
	60	5.0071(0.1989)	0.2471(0.0185)	5.0116(0.1019)	4.9867(0.1129)
	100	4.9910(0.1531)	0.2519(0.0161)	4.9941(0.0891)	5.0065(0.0988)
10,0.9,10,10	20	10.0763(1.3400)	0.8982(0.0296)	10.0264(0.6134)	9.9789(0.5958)
	60	10.0184(0.4465)	0.8975(0.0270)	10.0308(0.5651)	9.9674(0.5409)
	100	10.0154(0.3757)	0.8993(0.0098)	10.0026(0.1595)	9.9940(0.2306)
0.5,0.7,10,0.2	20	0.5144(0.1473)	0.6170(0.6727)	9.9041(0.8852)	0.2507(0.2465)
	60	0.5018(0.0968)	0.6869(0.6403)	9.8430(0.8627)	0.2318(0.1463)
	100	0.4984(0.0574)	0.6965(0.1099)	10.0068(0.1405)	0.2038(0.0323)

Lomax (Mc-Lx), Beta Lomax (B-Lx), Kumaraswamy Lomax (Kw-Lx), Exponentiated Lomax (E-Lx), and *Lo* distributions. The Mc-Lx, B-Lx, and Kw Lx distributions have been introduced by Lemonte and Cordeiro (2013) and E-Lx distribution has been introduced by Abdul-Moniem and Abdel-Hameed (2012). The cdfs of the above distribution where $x > 0$; $\alpha, \beta, c, \lambda, \theta > 0$ are presented as

$$F_{Mc\text{-}Lx}\left(x;\alpha,\beta,c,\lambda,\theta\right)=\frac{c}{B(a,b)}\int_{0}^{\left[1-(1+x/\lambda)^{-\theta}\right]^{c}} w^{\alpha-1}\left(1-w\right)^{\beta-1}dw, \tag{16}$$

$$F_{B\text{-}Lx}\left(x;\alpha,\beta,\lambda,\theta\right)=F_{Mc\text{-}Lx}\left(x;\alpha,\beta,1,\lambda,\theta\right), \tag{17}$$

$$F_{Kw\text{-}Lx}\left(x;\beta,c,\lambda,\theta\right)=F_{Mc\text{-}Lx}\left(x;1,\beta,c,\lambda,\theta\right) \tag{18}$$

and

$$F_{E\text{-}Lx}\left(x;\beta,c,\lambda,\theta\right)=F_{Mc\text{-}Lx}\left(x;1,1,c,\lambda,\theta\right). \tag{19}$$

We calculate the log-L values $(\hat{l})$, Akaike Information Criteria (AIC), Kolmogorov-Smirnov (KS), Cramer von Mises (W^*), and Anderson-Darling (A^*) goodness-of-fit statistics for all models to determine the best model. In Chen and Balakrishnan (1995), the statistics W^* and A^* are discussed in detail. The model with the smaller values of the AIC,

KS, W^* and A^* statistics and the larger values of $\hat{\ell}$ may generally be considered to be the best one. All computations of the ML estimates are achieved by the maxLik function and all goodness-of-fits statistics are achieved by the gofTest function in R.

On the other hand, choosing a certain model may be aided by the shape of hrf. The total time on test (TTT) plot in Aarset (1987) can be utilized to achieve this aim. The TTT plot is achieved by plotting $T\left(\frac{i}{n}\right)$ against i/n where $T\left(\frac{i}{n}\right)=\left[\sum_{r=1}^{n} y_{(r)}+(n-i)y_{(i)}\right]/\sum_{r=1}^{n} y_{(i)}$, $i = 1,\ldots, n$ and y_i are the order statistics of the sample. For decreasing hrf, it has a convex form, and for increasing hrf, a concave shape. The TTT plot for the dataset is given in Figure 16.3. The dataset refers to concave-convex shapes, as seen in Figure 16.3. That is, it has a unimodal-shape.

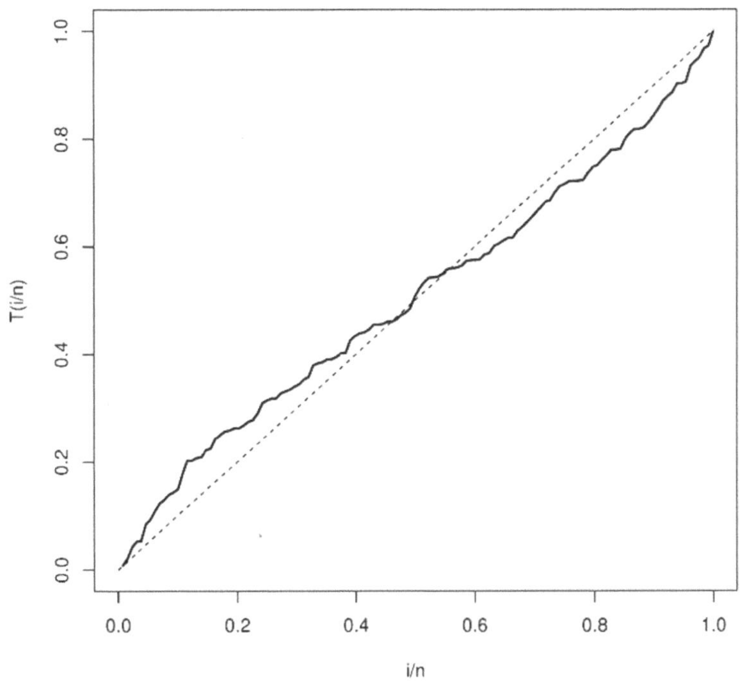

Figure 16.3 The TTT plot for the bladder cancer data.

Table 16.5 lists the ML estimates for above distributions, standard errors of their parameters, $\hat{\ell}$ and goodness-of-fits statistics from the fitted models. Table 16.5 indicates that the $TSG-Lo$ distribution can be selected as the best since these models have the lowest values of the KS, W^*, and A^* statistics, and have the biggest $\hat{\ell}$ values. The E-Lx model

Table 16.5 ML estimates, standard errors of the estimates (in parentheses), $\hat{\ell}$ and goodness-of-fits statistics for the bladder cancer datasets

Model	$\hat{\alpha}$	$\hat{\beta}$	$\hat{\lambda}$	$\hat{\theta}$	$\hat{c}$	$-\hat{\ell}$	AIC	KS	A^*	W^*
TSG-Lx	1.4767	0.5600	19.6072	2.6002		409.6502	827.3005	0.0363	0.1229	0.0180
	(0.1422)	(0.0010)	(3.4771)	(0.4317)						
Lo			120.1652	13.8392		413.8329	831.6658	0.0966	1.3776	0.2127
			(4.3990)	(1.3263)						
E-Lx			24.7638	4.5890	1.5861	410.0718	826.1436	0.0405	0.1797	0.0262
			(3.3346)	(0.6024)	(0.2058)					
Kw-Lx		17.0779	11.5996	0.2923	1.5154	409.9412	827.8825	0.0390	0.1616	0.0235
		(0.7381)	(3.9363)	(0.0578)	(0.1647)					
B-Lx	1.6601	0.2786	38.1245	20.6599		410.3330	828.6660	0.0443	0.2321	0.0345
	(0.2625)	0.0277	(0.1491)	(0.1459)						
Mc-Lx	0.7385	6.1245	10.9292	0.7021	2.0382	409.9130	829.8259	0.0391	0.1585	0.0231
	(0.6042)	(1.3330)	(5.4210)	(0.2055)	(1.5034)					

has the lowest AIC value whereas the *TSG* – *Lo* model comes forward according to others goodness-of-fits statistics.

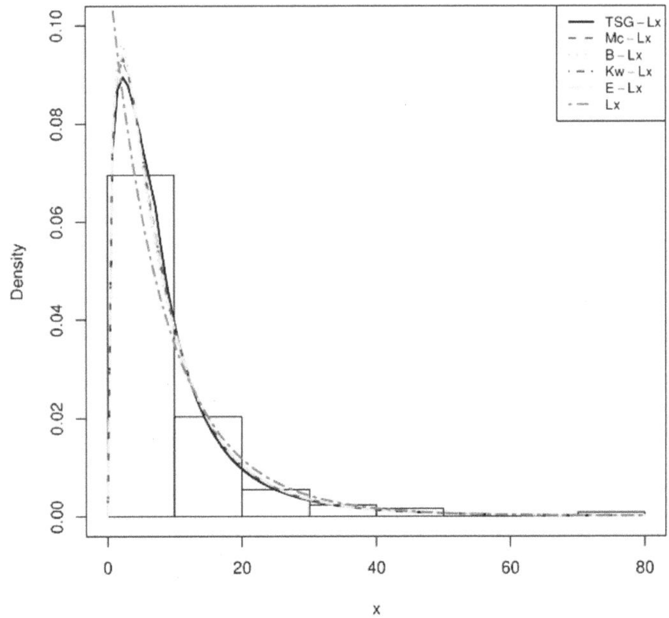

Figure 16.4 The fitted pdfs for all models.

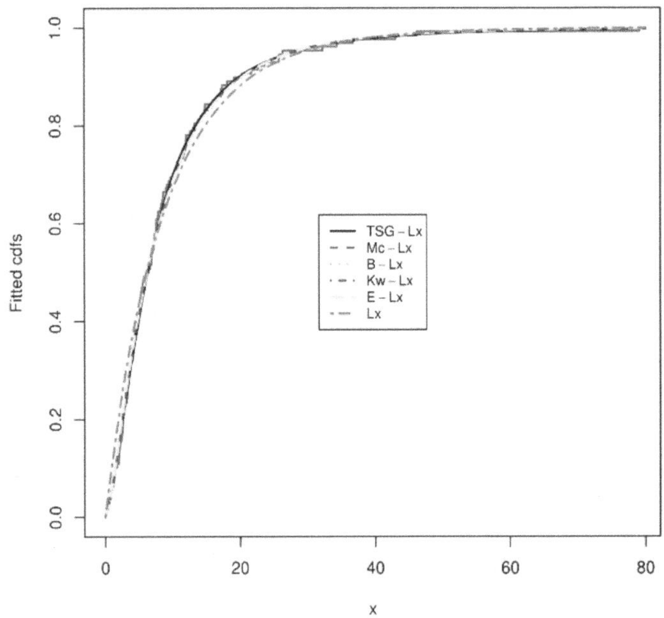

Figure 16.5 The fitted cdfs for all models.

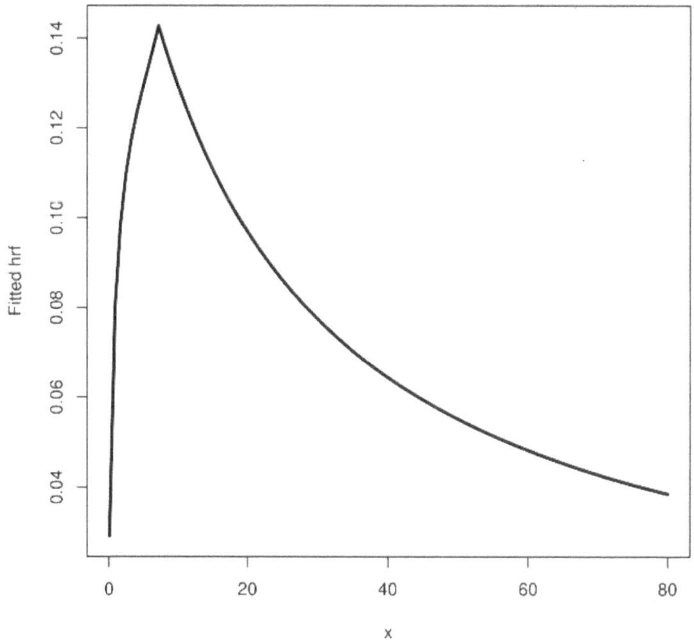

Figure 16.6 The fitted hrf of $TSG - Lo$ distribution.

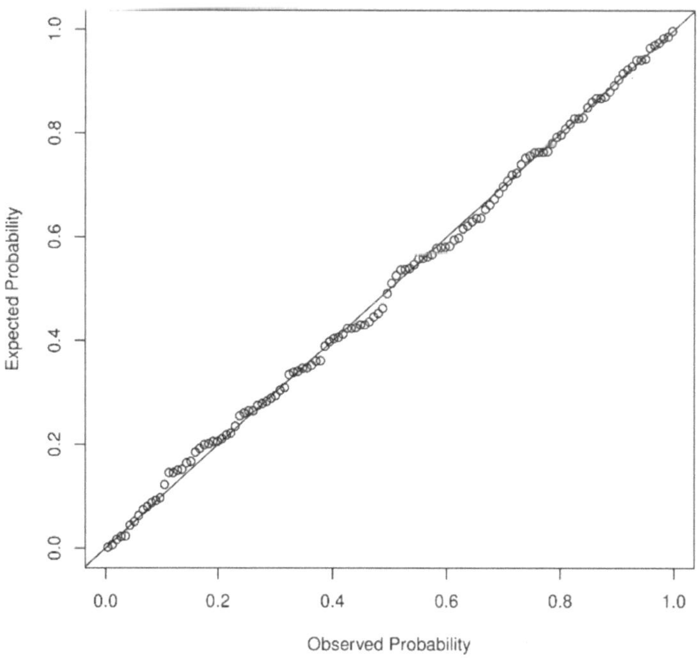

Figure 16.7 The P-P plot of $TSG - Lo$ distribution.

Figures 16.4 and 16.5 show fitted densities and cdfs for all models. We also give a fitted hrf and probability-probability (P-P) plot of $TSG - Lo$ model in Figures 16.6 and 16.7. These plots show that the $TSG - Lo$ model provides a good fit to these data compared to the other models. The fitted hrf shape of $TSG - Lo$ model has firstly an increasing and then a decreasing shape, similar to Figure 16.3 depicting TTT.

REFERENCES

Abdul-Moniem, B.I. and Abdel-Hameed, H.F. 2012. On exponentiated lomax distribution. International Journal of Mathematical Archive. 3(5): 2144–2150.

Aarset, M.V. 1987. How to identify a bathtub hazard rate. IEEE Transactions on Reliability. 36(1): 106–108.

Chen, G. and Balakrishnan, N. 1995. A general-purpose approximate goodness-of-fit test. Journal of Quality Technology. 27(2): 154–161.

Canuto, C., Hussaini, M.Y., Quarteroni, A. and Zang, T.A. 2006. Spectral Methods: Fundamentals in Single Domains. Springer-Verlag, New York.

Korkmaz, M.Ç. and Genç, A.İ. 2017. A new generalized two-sided class of distributions with an emphasis on two-sided generalized normal distribution. Communication Statistical Simulation and Computation. 46(2): 1441–1460.

Lee, E.T. and Wang, J.W. 2003. Statistical Methods for Survival Data Analysis, 3rd Edn. Wiley, New York.

Lemonte, A.J. and Cordeiro, G.M. 2013. An extended Lomax distribution. Statistics. 47(4): 800–816.

Lomax, K.S. 1954. Business failures: Another example of the analysis of failure data. Journal of the American Statistical Association. 49: 847–852.

van Dorp, J.R. and Kotz, S. 2002. The standard two-sided power distribution and its properties: With applications in financial engineering. American Statistician. 56: 90–99.

Chapter 17

Monitoring the Blood Levels of Animals using Control Charts

Liaquat Ahmad[1], Sabqa Khan[2] and Muhammad Aslam*[3]

[1,2]Department of Statistics and Computer Science, University of Veterinary and Animal Sciences, Lahore 54000, Pakistan.

Emails: liaquatahmad@uvas.edu.pk; sabqakhan7@gmail.com

[3]Department of Statistics, Faculty of Science, King Abdulaziz University, Jeddah 21551, Saudi Arabia.

17.1 INTRODUCTION

Control chart is an important tool of statistical process control for the quick and early detection of out-of-control manufacturing processes. The efficient monitoring scheme helps the operators/attendants to avoid the losses of sudden disease attacks and deaths among the animals (Montgomery 2009). Basically, these control charts were developed for the monitoring of the production process of the manufacturing industry but during the last few decades, the technique of control chart is being used very commonly in many other disciplines (Abbasi and Miller 2013) as well. The application of control chart scheme in nuclear engineering, in the nuclear power plant control room, was presented by (Hwang et al. 2008). Likewise, Wang and Liang (2008) discussed the application of control chart techniques for monitoring university education. The application of the control chart scheme in the analytical laboratory was

*For Correspondence: Muhammad Aslam (aslam_ravian@hotmail.com)

presented by Masson (2007). Ahmad et al. (2014) studied the monitoring of coal quality through control charts' scheme which was studied by Woodall (2007). The application of quality control charts in the healthcare sector was presented by many researchers including Woodall (2007), who presented the control chart technique in healthcare monitoring and public health surveillance. Chen (1991) developed the monitoring system for the detection of increased rates of cancer. Morton (2005) presented the methods for hospital epidemiology and healthcare quality improvement. Webster, Pettitt and Cook (2005) presented the risk-adjusted CUCUM control chart for medical monitoring. The hematological profile of a group of 6046 normal dogs was explored by De Vries and Reneau (2010). Normal parturitic parameters of cats and dogs were addressed by von Heimendahl and Cariou (2009). The control chart for monitoring the animal production system was developed by De Vries and Reneau (2010).

In the Shewhart control chart technique, the judgment about the process control any time depends upon the recent measurements of the process and the trueness of control limits from the data. The customary X-bar charts are efficiently competent in identifying shifts in the statistical parameter due to their natural variations in the statistics. When studying the data that are variable, it is important to first monitor the mean of that character and then its variation or study both mean and variance at a time (Montgomery 2009).

This research is about the blood parameters of dogs and cats. The data were collected from labs of Pet Center and University Diagnostic Lab (UDL) of the University of Veterinary and Animal Sciences (UVAS), Lahore, Pakistan. We used 8 parameters consisting of white blood cells (WBC), hemoglobin, red blood cell (RBC) count, packed cell volume (PCV), mean cell volume (MCV), mean corpuscular hemoglobin (MCH), mean corpuscular hemoglobin concentration (MCHC), and platelets counts of blood samples of dogs and cats approached daily to these labs during January–April 2018. The control charts are the graphical presentation of the process data to monitor changes over the time as data plotted in time order. Control charts contain a central line (CL) for average value with two limits as the upper line for upper control limit (UCL) and a lower line for lower control limit (LCL); all of these limits are determined from the historical data. Control charts determine if the recent set of readings represents whether the parametric values have changed or are within the stable average by providing on-going statistical process tests. The test also checks the variation within the samples to determine whether the variation is within the setup stable average or not. A control chart provides the evidence that a parametric value is stable or not and a stable process can be predictable. Control charts, also known as Shewhart charts (after Walter A. Shewhart) or process-behavior charts, are control tools of statistics to decide if a business or manufacturing process or a veterinary statistical parameter is stable or not. Many control charts such as X-bar and R charts, EWMA and CUSUM charts of variable data and *p*-chart, *μ*-chart and *c*-chart for attribute data are commonly available/ used in quality improvement scenarios. For monitoring of parametric mean value, the traditional X-bar chart is examined widely.

Due to the efficient monitoring behavior of the X-bar chart scheme, this paper has been developed for the monitoring of blood parameters of the dogs and cats' data using the X-bar chart scheme. To the best of the knowledge of authors and by exploring the literature, it has been observed that no attention has been given to study the monitoring of the variation pattern in the blood parameters of the dogs and cats' data. The rest of the paper is organized as: the material and methods of the formulation of the proposed X-bar control chart has been discussed in Section 17.2. In Section 17.3, the results and discussion have been elaborated. The advantages of the proposed schemes have been described in Section 17.4. At the end the conclusion is written in Section 17.5.

17.2 MATERIAL AND METHODS OF THE FORMULATION OF THE X-BAR CONTROL CHART

In this section the control limits using X-bar charts have been constructed for the different blood parameters of the dogs and cats' data collected from January 2018 to April 2018 from the Pet Center and UDL of the UVAS, Lahore, Pakistan. It is one of the oldest teaching hospitals established in 1882 in Asia as Dog Hospital which was upgraded to Pet Center in 2002 with the upgradation of the college of veterinary sciences to the University level.

Design and Formulation of the X-bar Chart

In this section, we describe the design of the X-bar chart and the construction of the control limits.

Step I: Select the subgroup of size $n = 5$. Rational subgrouping is considered to minimize the variation of data within the subgroup.

Step-II: Data should be collected in the order that it is generated. After that, set the frequency at which the data will be collected.

Step-III: Before calculating the control limits, first calculate the number of subgroups k.

Step-IV: Calculate the subgroups average, for each subgroup as

$$\underline{X} = \frac{X_1 + X_2 + \cdots + X_n}{n} \quad (1)$$

where n = size of the each subgroup

Step-V: Calculate the subgroups range, for each subgroup as

$$R = X_{MAX} - X_{MIN} \quad (2)$$

where X_{MAX} and X_{MIN} are the maximum and the minimum value of individual samples in the subgroup, respectively.

Step-VI: Calculate the average range ($\underline{R}$) as

$$\underline{R} = \frac{R_1 + R_2 + \cdots + R_k}{K} \quad (3)$$

where K is the number of subgroups.

Step-VII: Calculate the overall process average $\overline{X}$ as

$$\overline{X} = \frac{\sum \underline{X}_i}{K} = \frac{\underline{X}_1 + \underline{X}_2 + \cdots + \underline{X}_K}{K} \quad (4)$$

Step-VIII: Calculate the control limits for X-bar charts. The upper control limit is represented by $UCL_{\underline{x}}$ and the lower control limit is represented $LCL_{\underline{x}}$.

$$UCL_{\underline{x}} = \overline{X} + A_2\underline{R} \quad (5)$$

$$LCL_{\underline{x}} = \overline{X} - A_2\underline{R} \quad (6)$$

where A_2 is the constant value of the control chart that depends upon the subgroup.

17.3 RESULTS AND DISCUSSION

In this section, the results and the discussions have been given. Control charts for the collected data have been constructed for four months for different parameters using the software SPSS version 20. The performance of the proposed X-bar control chart is evaluated under the usual normality assumption.

Using Eqs. (5) and (6), the X-bar chart for the collected data of eight blood parameters of WBC, hemoglobin levels, RBC, PCV, MCV, MCH, MCHC, and platelets counts of blood samples of dogs and cats are given as under (Figures 17.1 through 17.64).

17.3.1 Analysis of Dogs Blood Parameters

17.3.1.1 Analysis of WBC

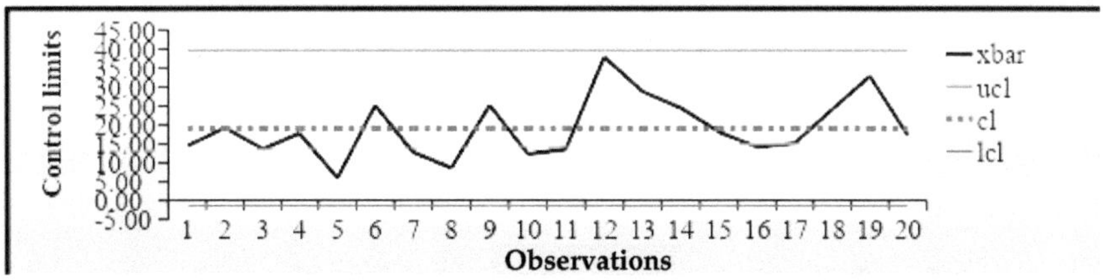

Figure 17.1 The X-bar chart for dog data for January.

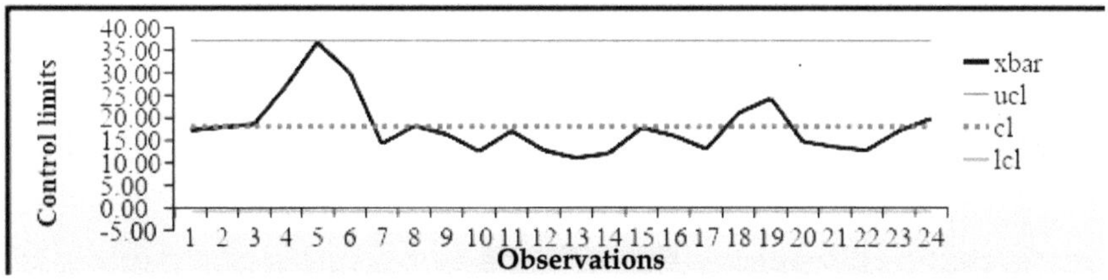

Figure 17.2 The X-chart for dog data for February.

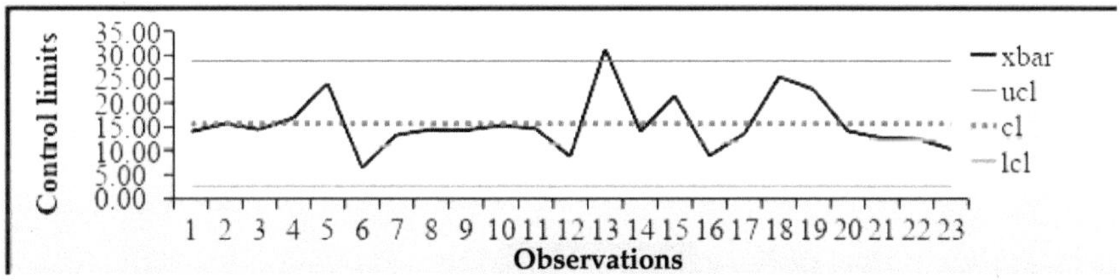

Figure 17.3 The X-chart for dog data for March.

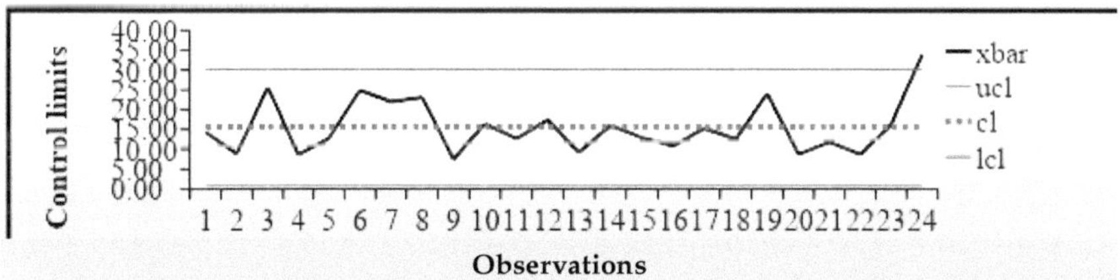

Figure 17.4 The X chart for dog data for April.

17.3.1.2 Analysis of Hemoglobin

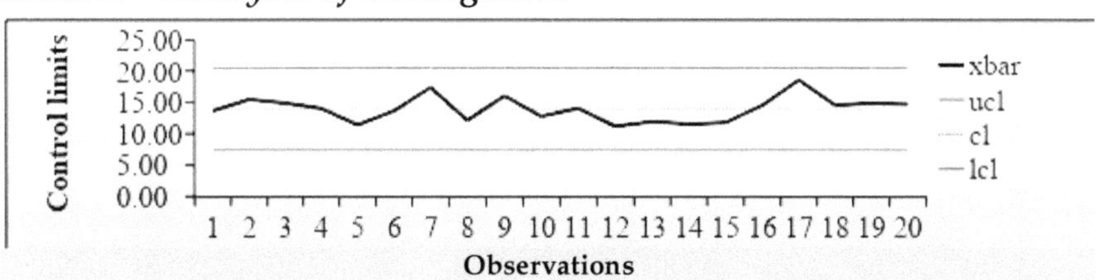

Figure 17.5 The X-bar chart for dog data for January.

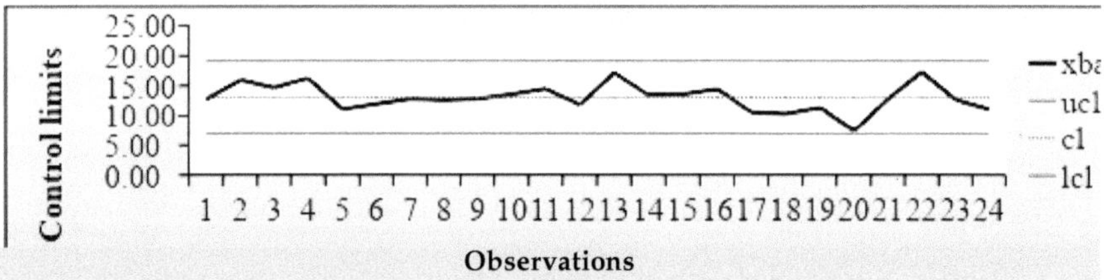

Figure 17.6 The X-bar chart for dog data for February.

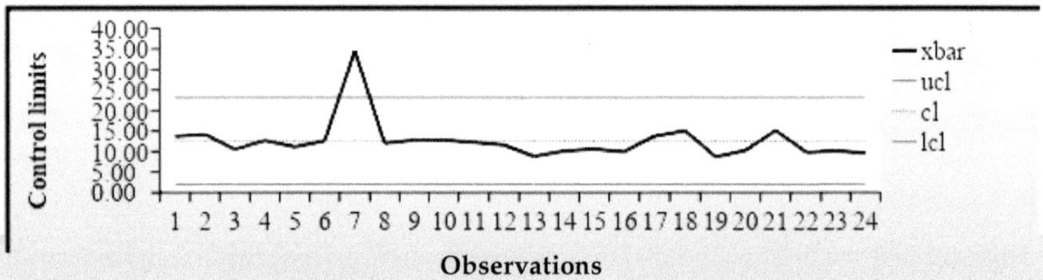

Figure 17.7 The X-bar chart for dog data for March.

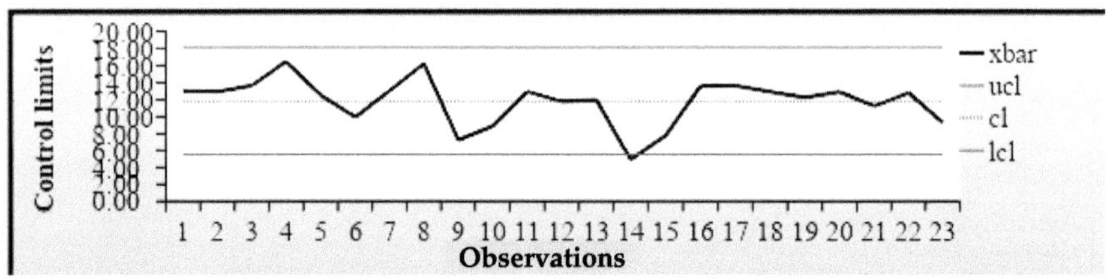

Figure 17.8 The X-bar chart for dog data for April.

17.3.1.3 Analysis of RBC count

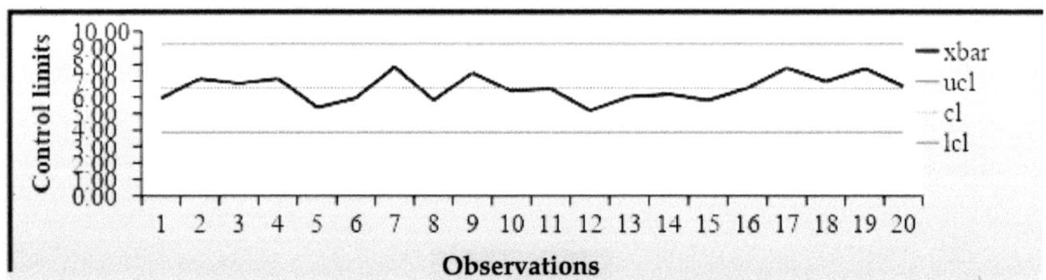

Figure 17.9 The X-bar chart for dog data for January.

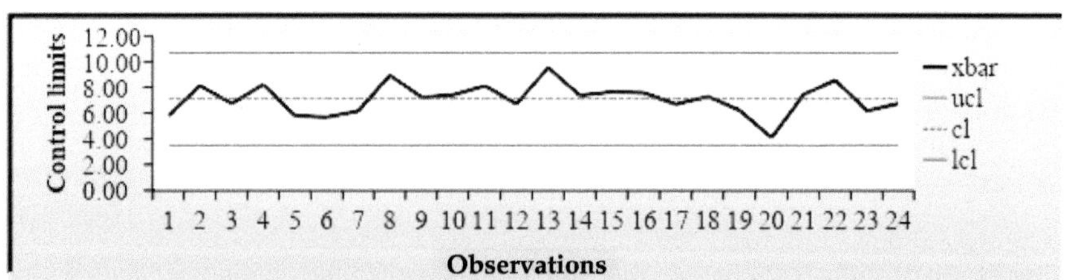

Figure 17.10 The X-bar chart for dog data for February.

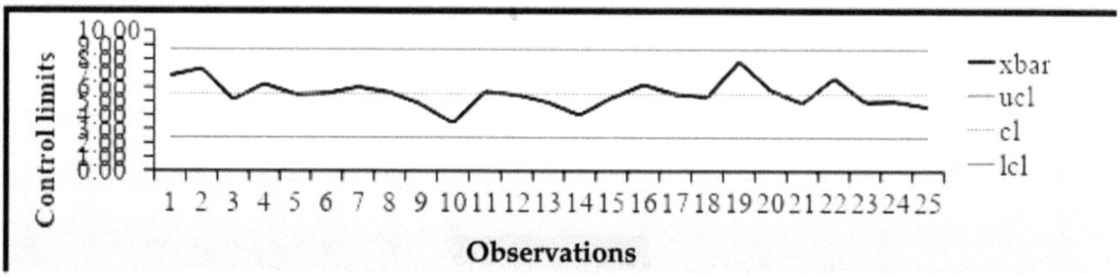

Figure 17.11 The X-bar chart for dog data for March.

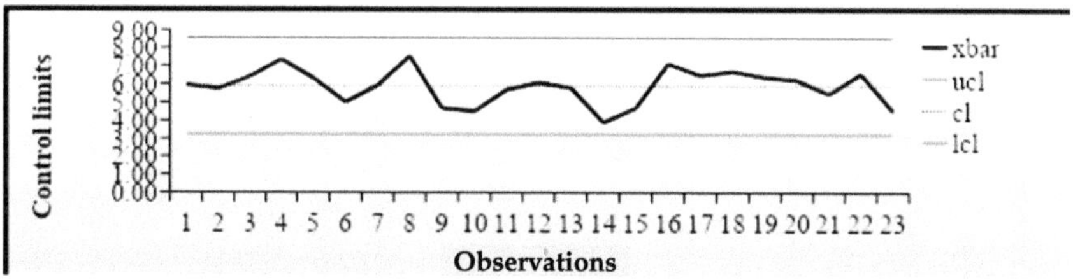

Figure 17.12 The X-bar chart for dog data for April.

17.3.1.4 *Analysis of PCV*

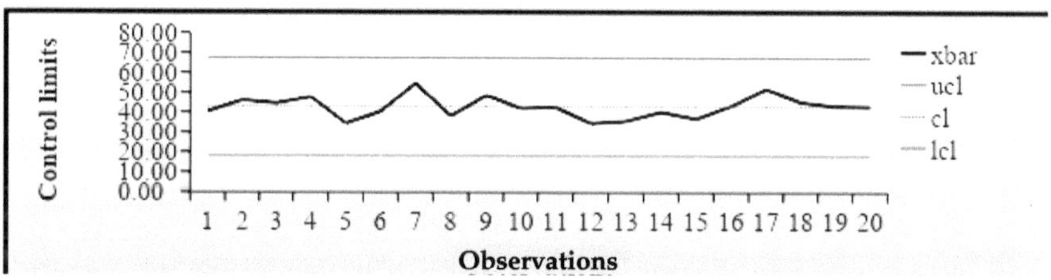

Figure 17.13 The X-bar chart for dog data for January.

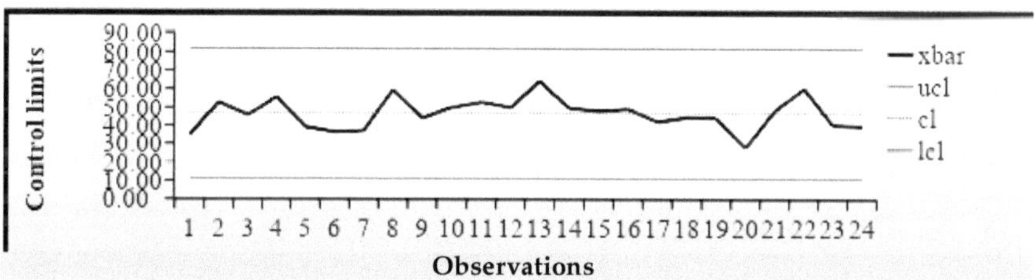

Figure 17.14 The X-bar chart for dog data for February.

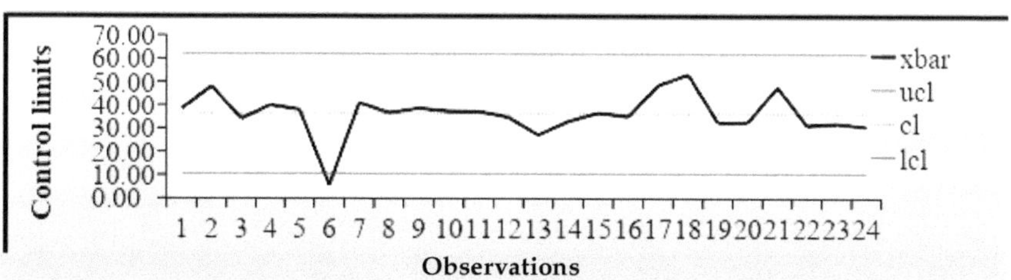

Figure 17.15 The X-bar chart for dog data for March.

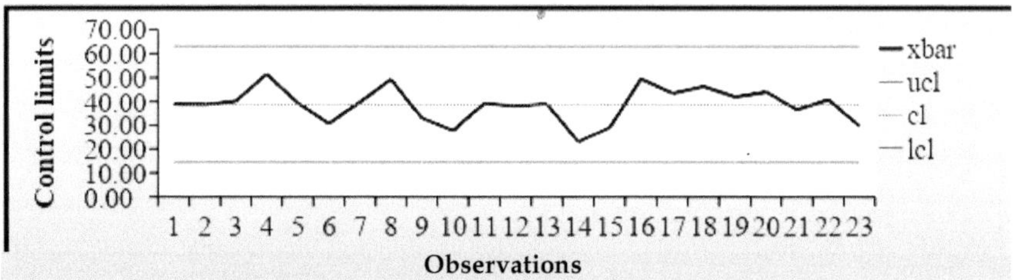

Figure 17.16 The X-bar chart for dog data for April.

17.3.1.5 *Analysis of MCV*

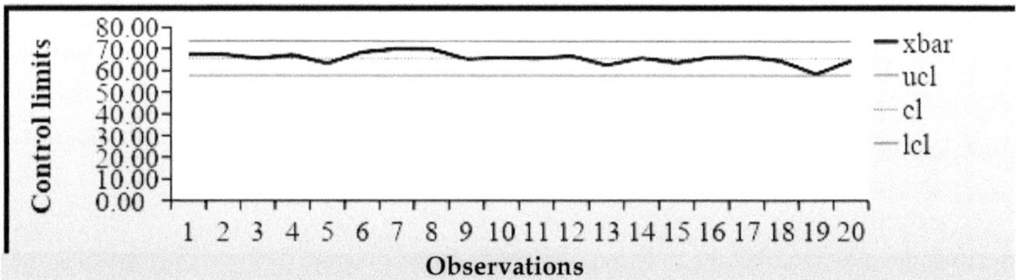

Figure 17.17 The X-bar chart for dog data for January.

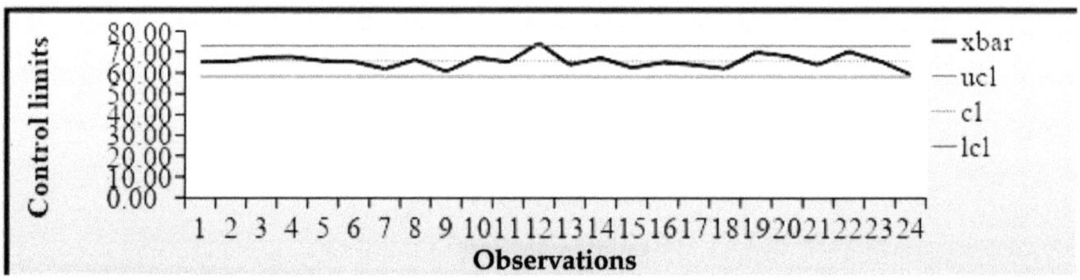

Figure 17.18 The X-bar chart for dog data for February.

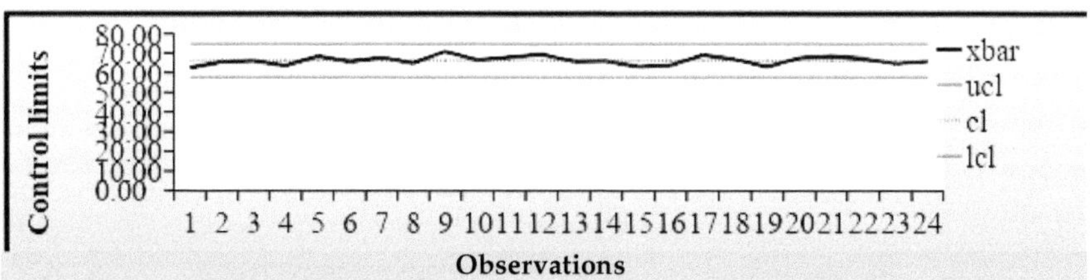

Figure 17.19 The X-bar chart for dog data for March.

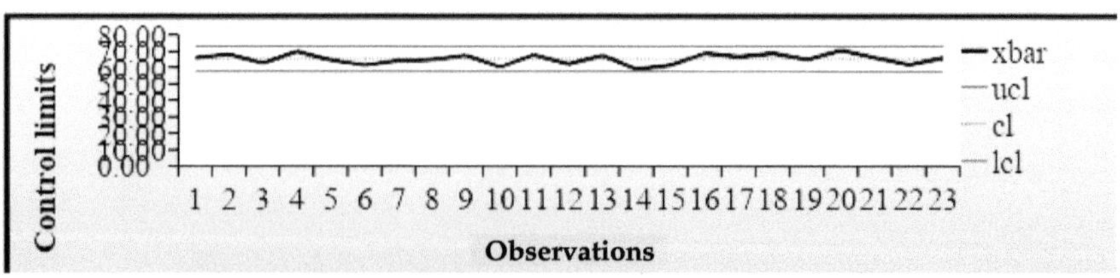

Figure 17.20 The X-bar chart for dog data for April.

17.3.1.6 Analysis of MCH

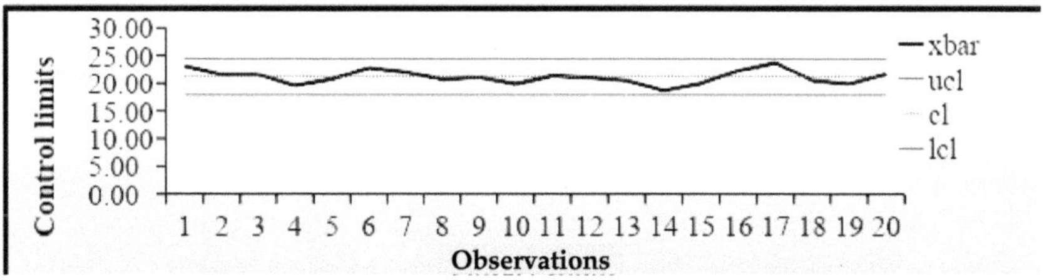

Figure 17.21 The X-bar chart for dog data for January.

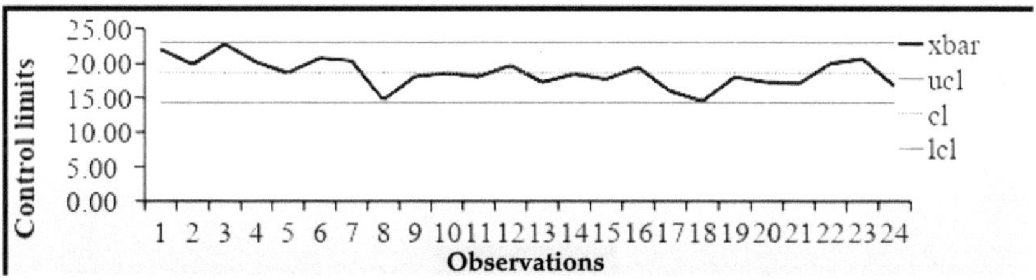

Figure 17.22 The X-bar chart for dog data for February.

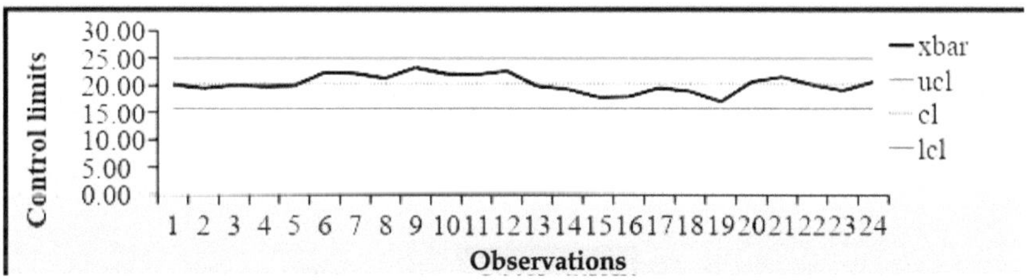

Figure 17.23 The X-bar chart for dog data for March.

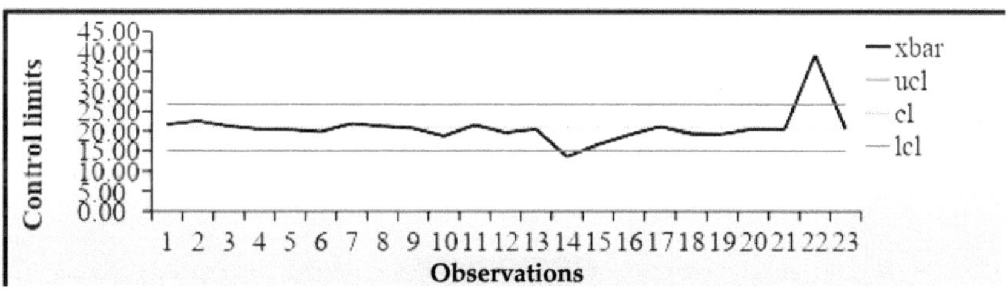

Figure 17.24 The X-bar chart for dog data for April.

17.3.1.7 Analysis of MCHC

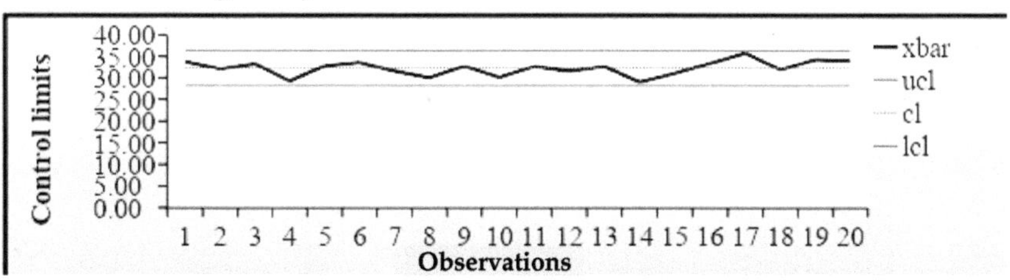

Figure 17.25 The X-bar chart for dog data for January.

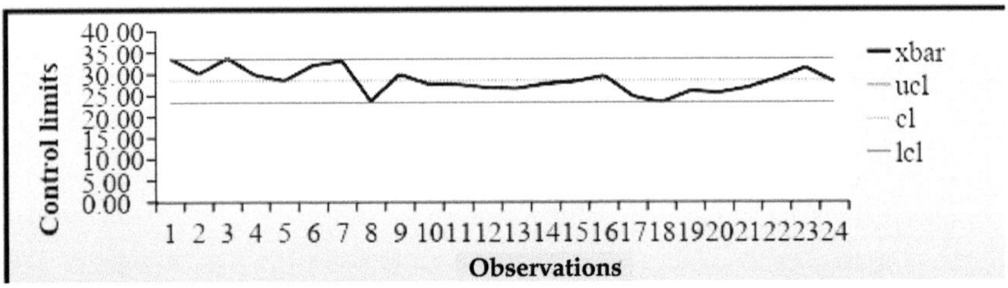

Figure 17.26 The X-bar chart for dog data for February.

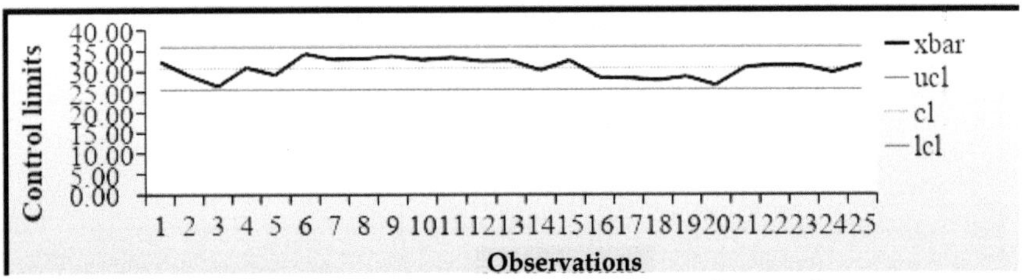

Figure 17.27 The X-bar chart for dog data for March.

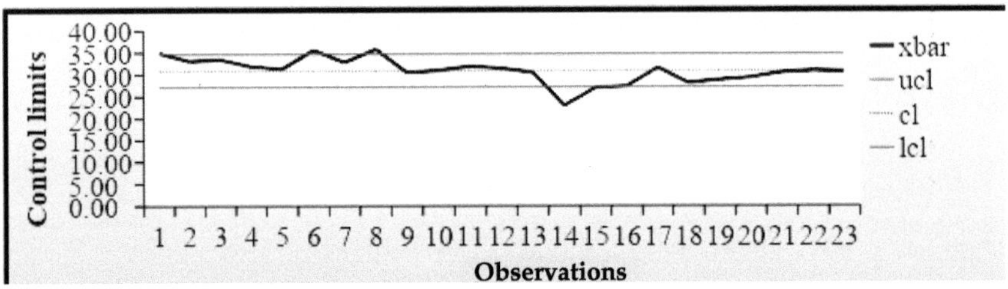

Figure 17.28 The X-bar chart for dog data for April.

17.3.1.8 *Analysis of Platelets*

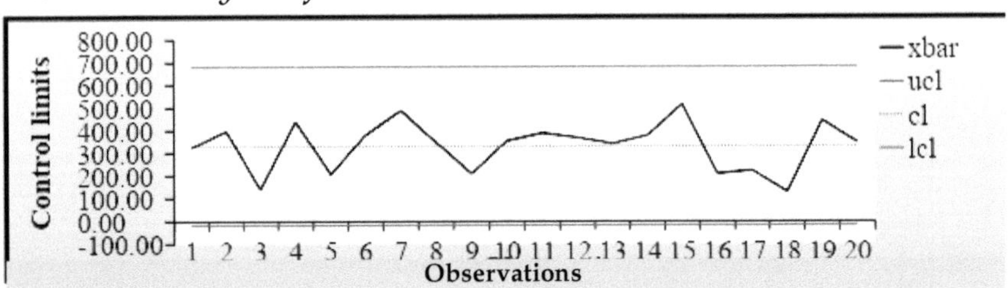

Figure 17.29 The X-bar chart for dog data for January.

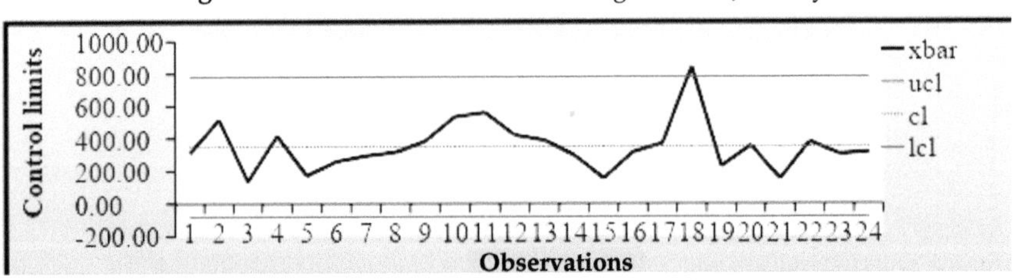

Figure 17.30 The X-bar chart for dog data for February.

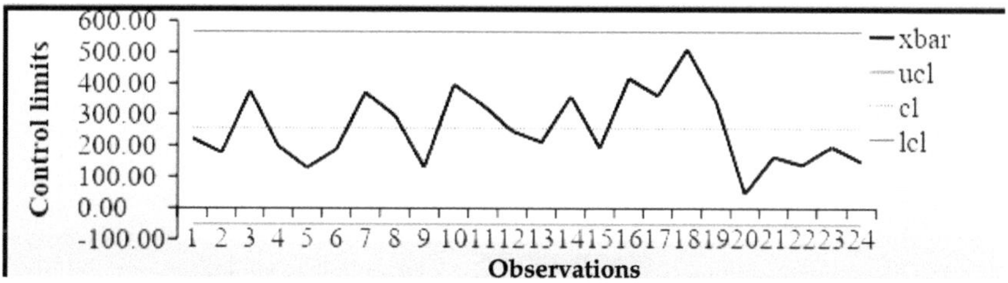

Figure 17.31 The X-bar chart for dog data for March.

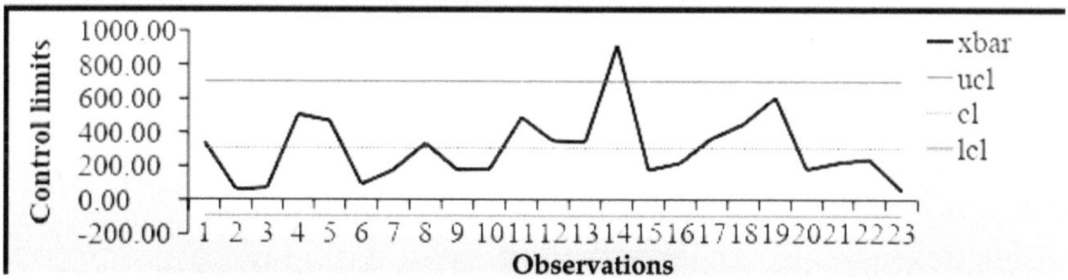

Figure 17.32 The X-bar chart for dog data for April.

17.3.2 Analysis of Cats Blood Parameters

17.3.2.1 Analysis of WBC

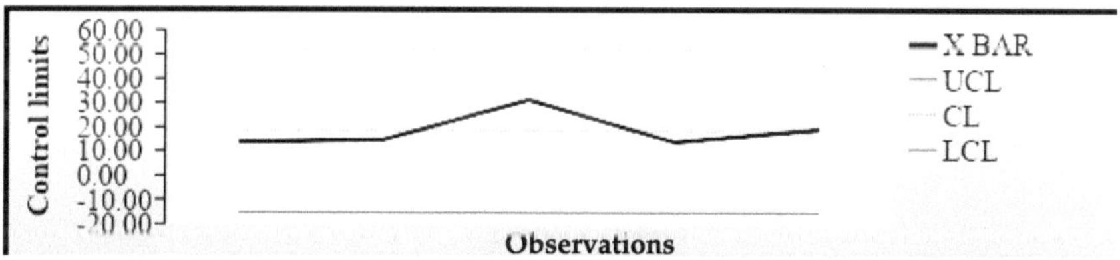

Figure 17.33 The X-bar chart for cat data for January.

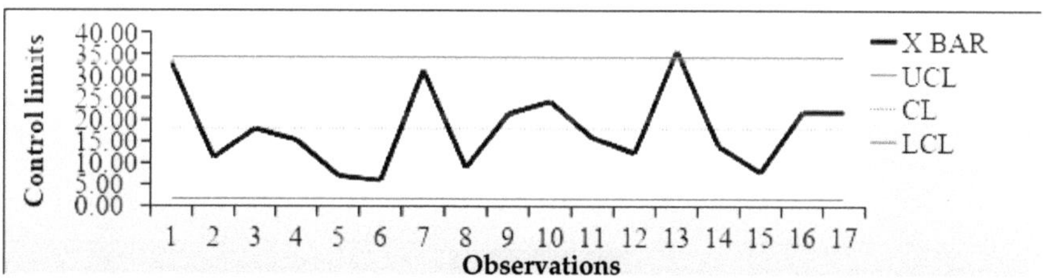

Figure 17.34 The X-bar chart for cat data for February.

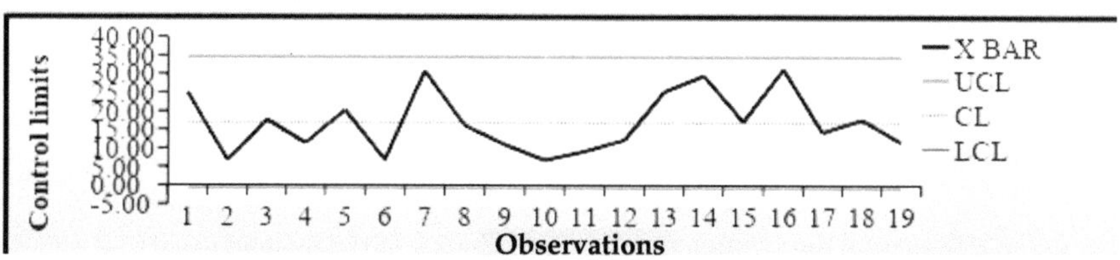

Figure 17.35 The X-bar chart for cat data for March.

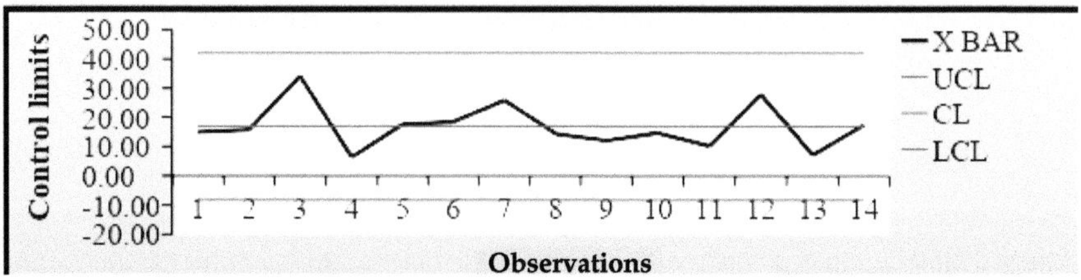

Figure 17.36 The X-bar chart for cat data for April.

17.3.2.2 *Analysis of Hemoglobin*

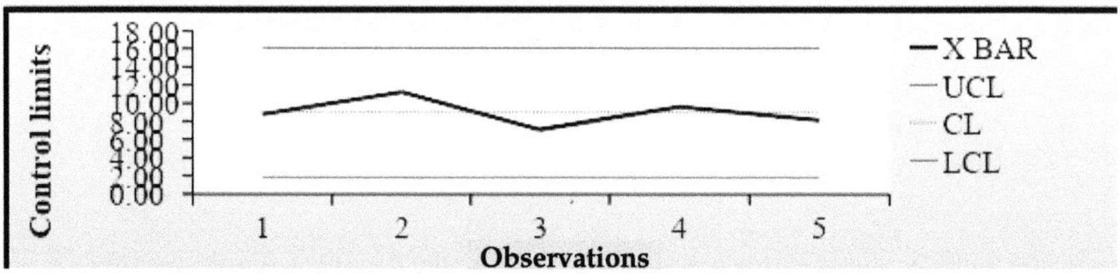

Figure 17.37 The X-bar chart for cat data for January.

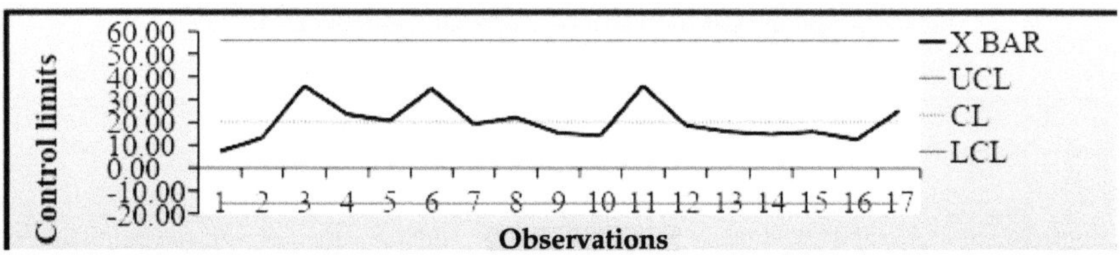

Figure 17.38 The X-bar chart for cat data for February.

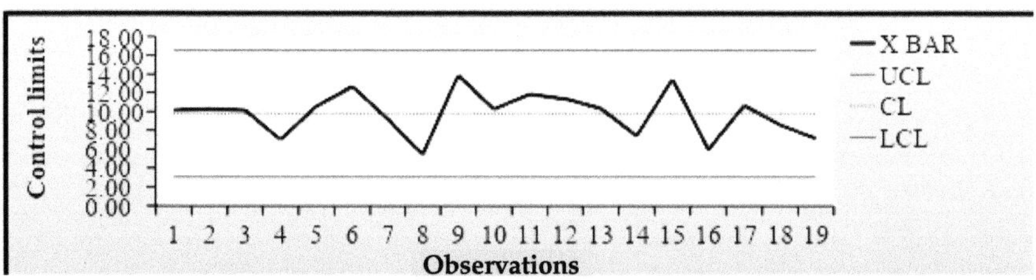

Figure 17.39 The X-bar chart for cat data for March.

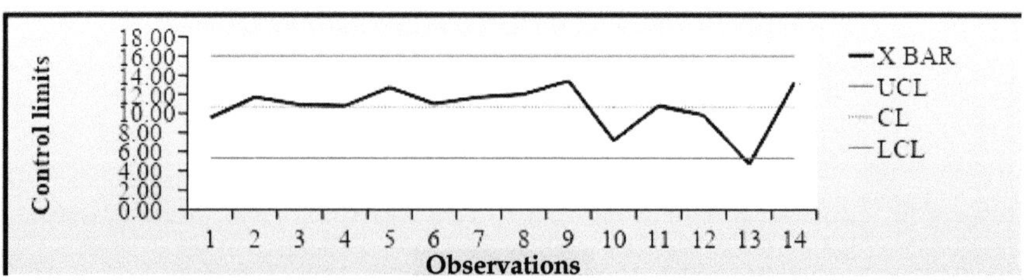

Figure 17.40 The X-bar chart for cat data for April.

17.3.2.3 Analysis of RCB Count

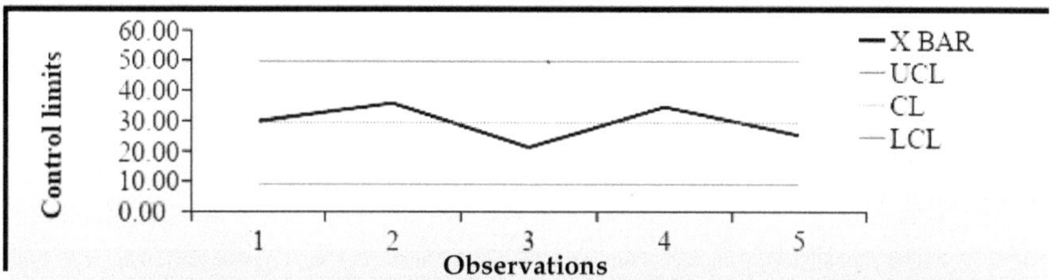

Figure 17.41 The X-bar chart for cat data for January.

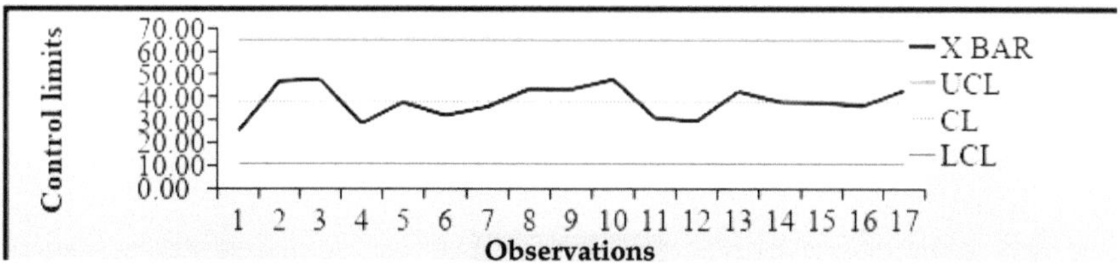

Figure 17.42 The X-bar chart for cat data for February.

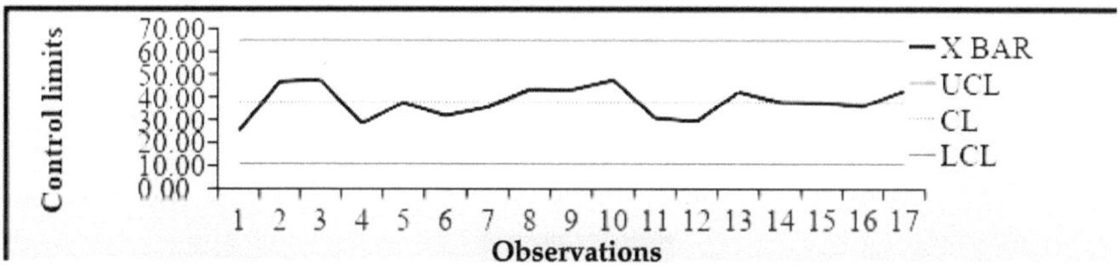

Figure 17.43 The X-bar chart for cat data for March.

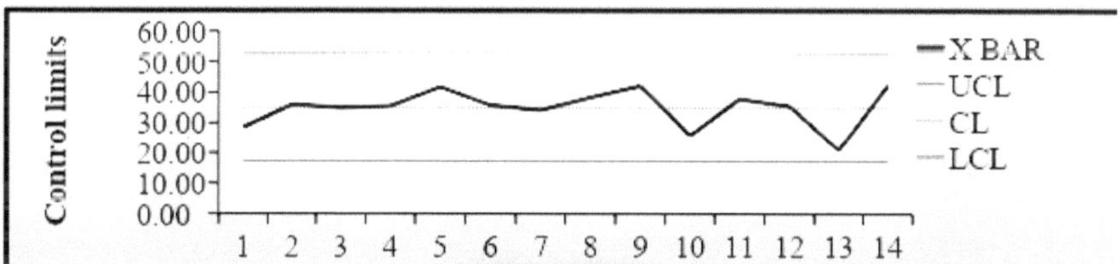

Figure 17.44 The X-bar chart for cat data for April.

17.3.2.4 Analysis of PCV

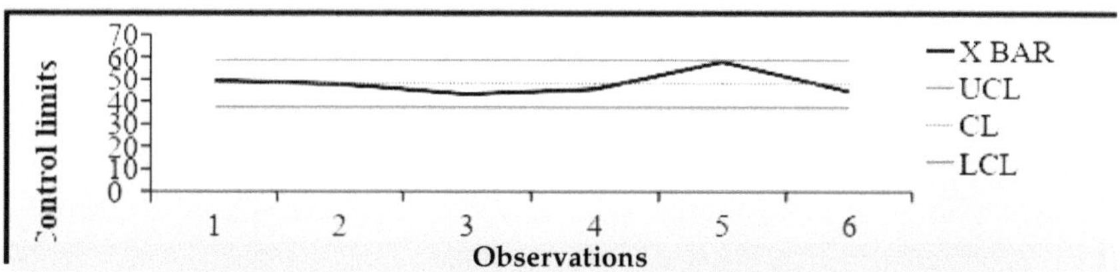

Figure 17.45 The X-bar chart for cat data for January.

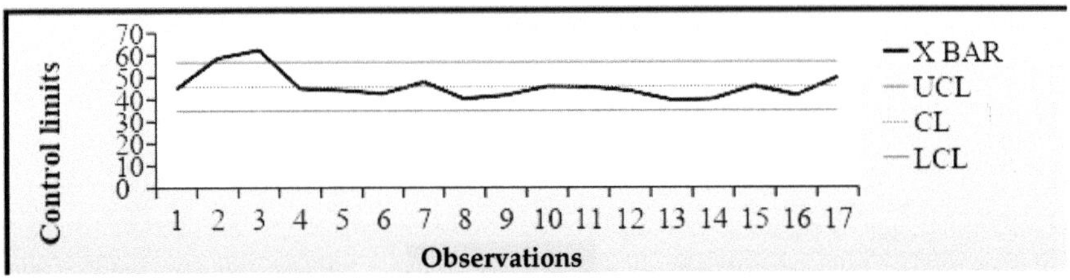

Figure 17.46 The X-bar chart for cat data for February.

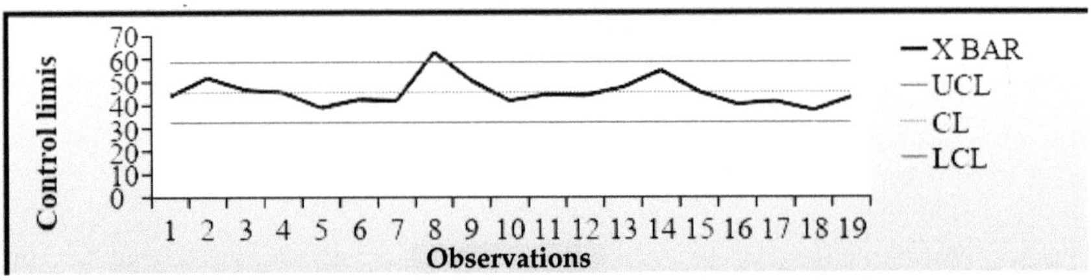

Figure 17.47 The X-bar chart for cat data for March.

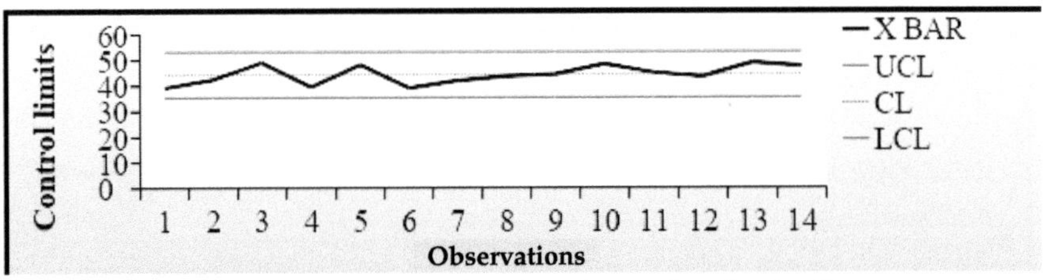

Figure 17.48 The X-bar chart for cat data for April.

17.3.2.5 *Analysis of MCV*

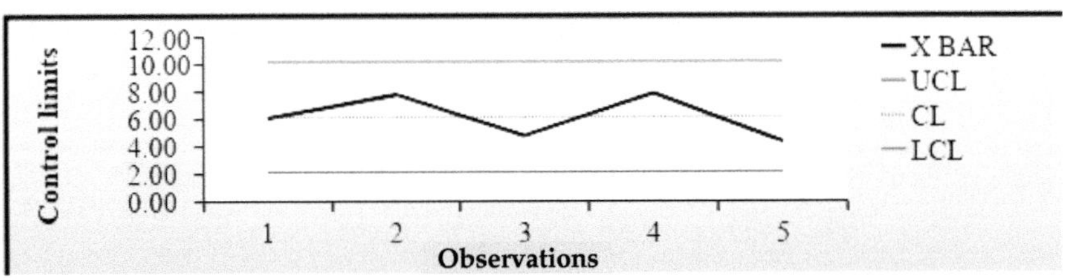

Figure 17.49 The X-bar chart for cat data for January.

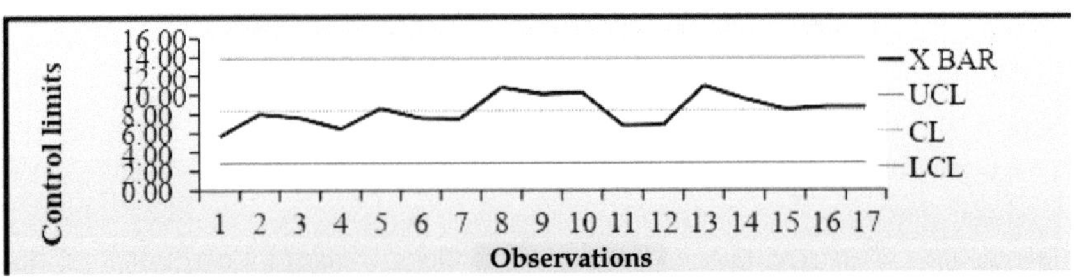

Figure 17.50 The X-bar chart for cat data for February.

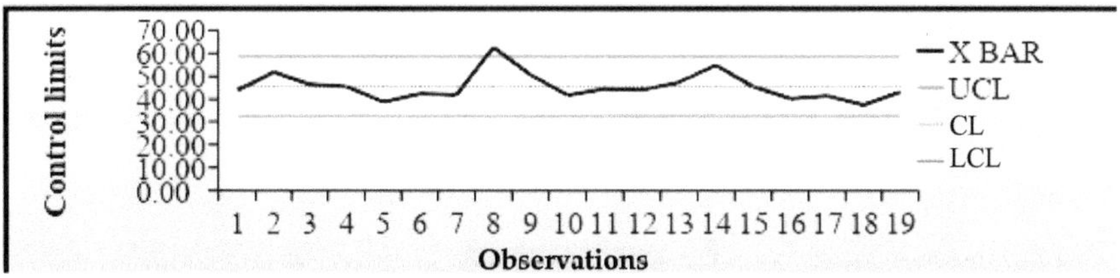

Figure 17.51 The X-bar chart for cat data for March.

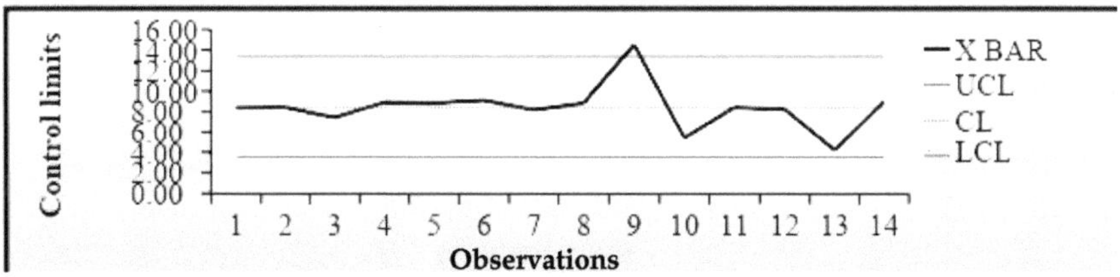

Figure 17.52 The X-bar chart for cat data for April.

17.3.2.6 *Analysis of MCH*

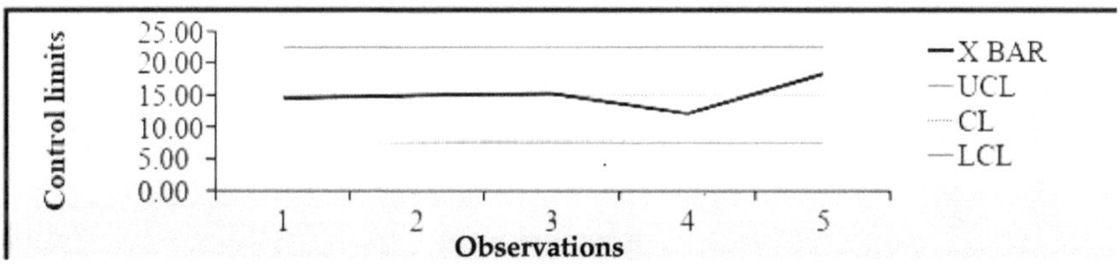

Figure 17.53 The X-bar chart for cat data for January.

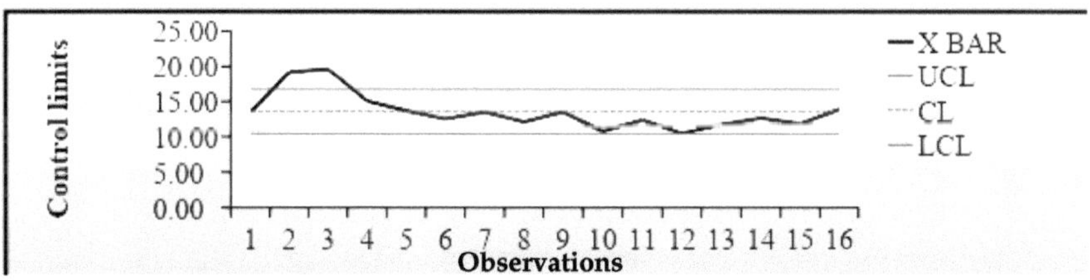

Figure 17.54 The X bar chart for cat data for February.

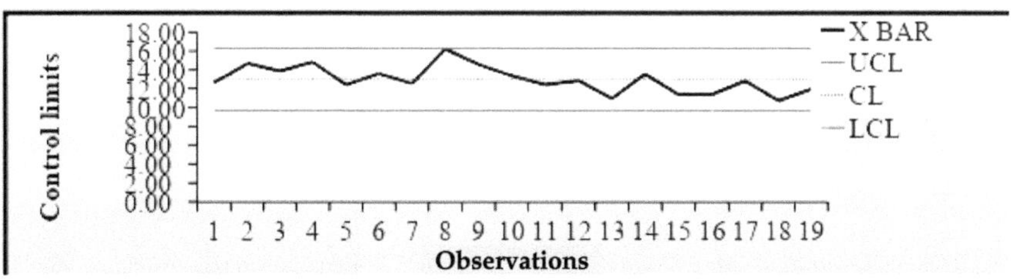

Figure 17.55 The X-bar chart for cat data March.

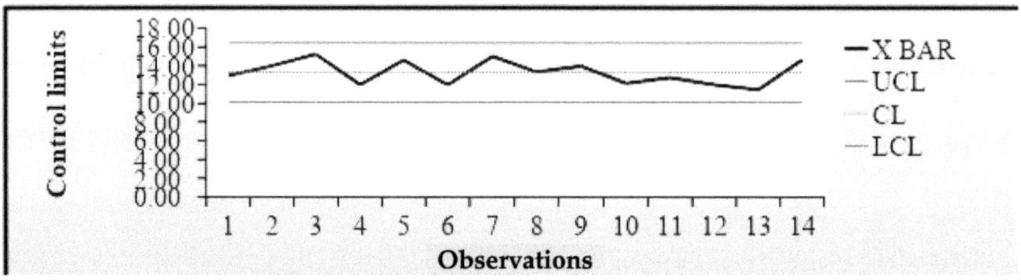

Figure 17.56 The X-bar chart for cat data for April.

17.3.2.7 Analysis of MCHC

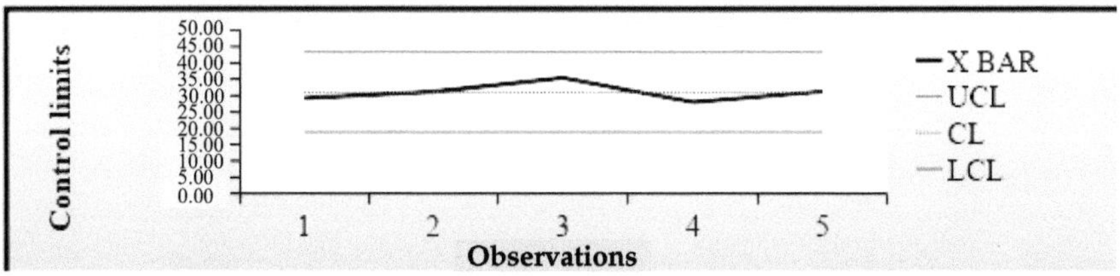

Figure 17.57 The X-bar chart for cat data for January.

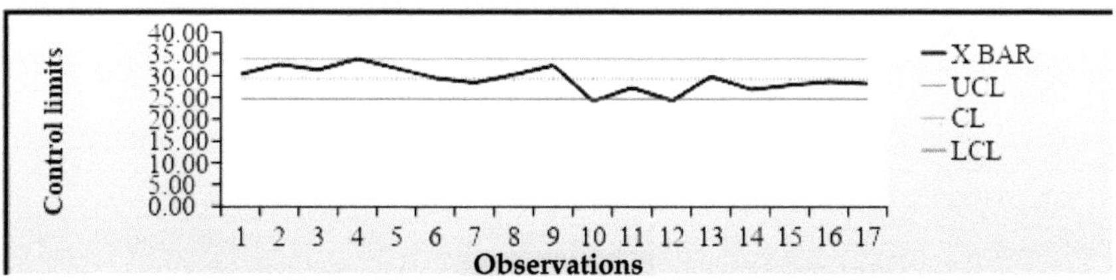

Figure 17.58 The X-bar chart for cat data for February.

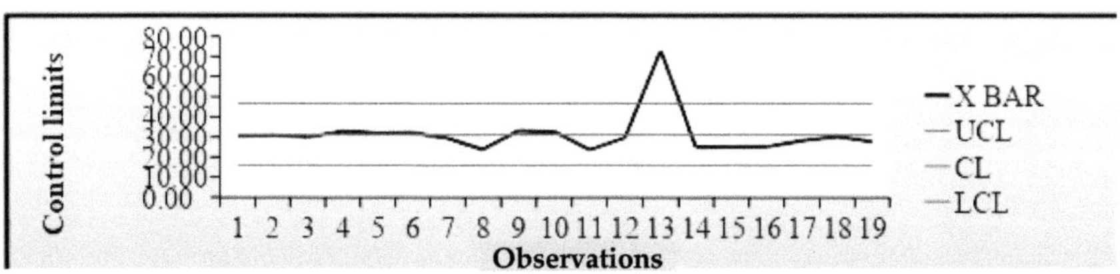

Figure 17.59 The X-bar chart for cat data for March.

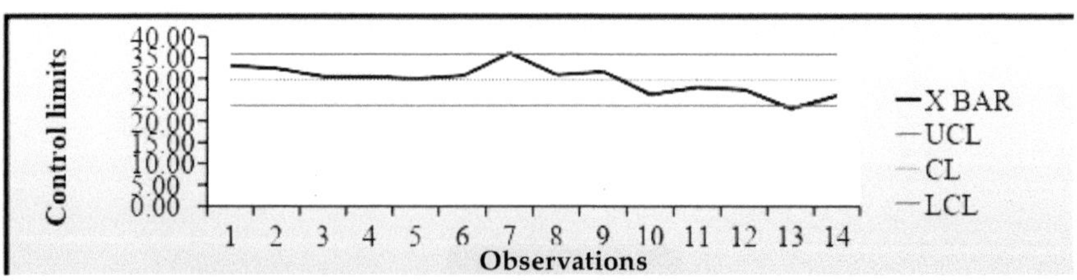

Figure 17.60 The X-bar chart for cat data for April.

17.3.2.8 Analysis of Platelets

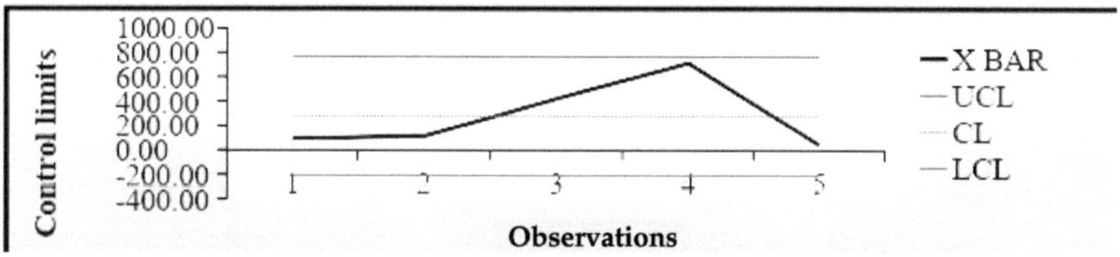

Figure 17.61 The X-bar chart for cat data for January.

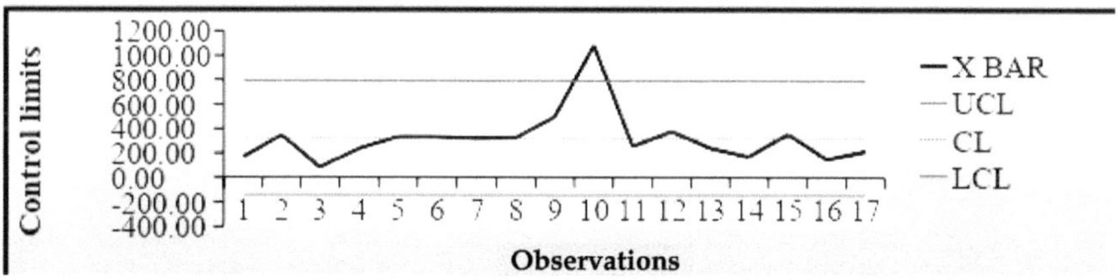

Figure 17.62 The X-bar chart for cat data for February.

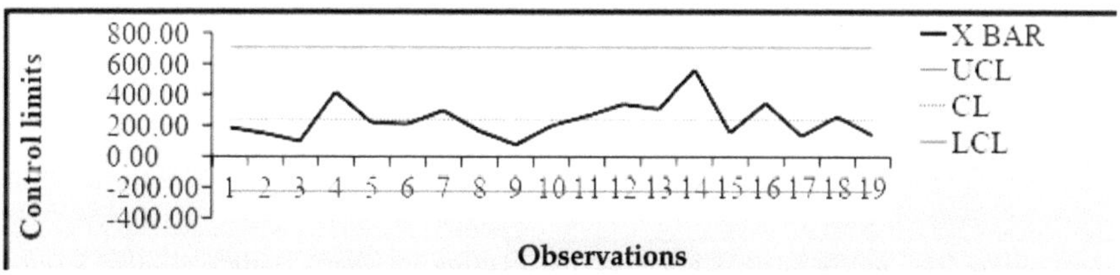

Figure 17.63 The X-bar chart for cat data for March.

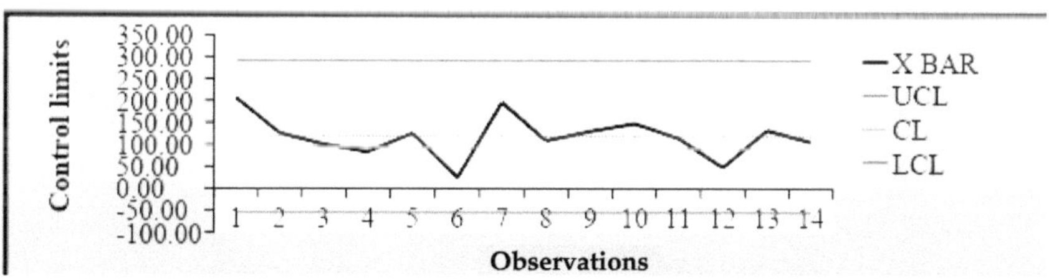

Figure 17.64 The X-bar chart for cat data for April.

17.4 THE ADVANTAGES OF THE PROPOSED SCHEMES

In this section, the advantages of the proposed X-bar control chart have been discussed to know the better performance in the form of early detection of the out-of-control situation by using the blood parameters of the dogs' and cats' data. Figures 17.1 through 17.32

are given for the dog data while Figures 17.33 through 17.64 are constructed for the cat data. It can be observed from the above that Figures 17.2–17.4, 17.7, 17.15, 17.18, 17.22, 17.24, 17.26, 17.28, 17.30, and 17.32 are showing the out-of-control observations in the dogs' parameters while Figures 17.34, 17.36, 17.40, 17.43, 17.46, 17.47, 17.51, 17.52, 17.54, 17.59, 17.60, and 17.62 are showing the out-of-control observations in the cats' parameters.

17.5 CONCLUSIONS

In this research, the control chart technique has been used for monitoring the veterinary related data of dogs and cats. Most commonly used technique, the X-bar chart, has been applied to study the variation of the data. We constructed the X-bar chart of the four months from January to April. So, the technique has also been applied for the data of dogs and cats. The application of these techniques has been introduced in the veterinary related data, which shows that the X-bar charts are simple to use as compared to any other technique which are not so simple to understand and apply on such data.

To check the variation of data of dogs and cats collected from the Pet Center and UDL of UVAS, Lahore, we applied ($\underline{X}$) control charts on their different blood parameters according to the sequence of dates on a monthly basis (January, February, March, April). Analysis shows that there is a lot of variation in the parameters of Dogs and Cats according to the status of each dog and cat. Each dog and cat contains a separate volume of blood parameters as per their health.

REFERENCES

Abbasi, S.A. and Miller, A. 2013. MDEWMA chart: An efficient and robust alternative to monitor process dispersion. Journal of Statistical Computation and Simulation. 83(2): 247–268.

Ahmad, L., Aslam, M. and Jun, C.-H. 2014. Coal quality monitoring with improved control charts. European Journal of Scientific Research. 125(2): 427–434.

Chen, R. 1991. Letter "Re: A monitoring system to detect increased rates of cancer incidence" by Sitter et al. (with reply). *American Journal of Epidemiology*. 134: 913–119.

De Vries, A. and Reneau, J. 2010. Application of statistical process control charts to monitor changes in animal production systems. Journal of animal science. 88(suppl._13): E11–E24.

Hwang, S.-L., Lin, J.-T., Liang, G.-F., Yau, Y.-J., Yenn, T.-C. and Hsu, C.-C. 2008. Application control chart concepts of designing a pre-alarm system in the nuclear power plant control room. Nuclear Engineering and Design. 238(12): 3522–3527.

Masson, P. 2007. Quality control techniques for routine analysis with liquid chromatography in laboratories. Journal of Chromatography A. 1158(1–2): 168–173.

Montgomery, C.D. 2009. Introduction to Statistical Quality Control, 6th Edn. John Wiley & Sons, Inc., New York.

Morton, A. 2005. Methods for hospital epidemiology and healthcare quality improvement. eICAT.

von Heimendahl, A. and Cariou, M. 2009. Normal parturition and management of dystocia in dogs and cats. Practice. 31(6): 254–261.

Wang, Z. and Liang, R. 2008. Discuss on applying SPC to quality management in university education. Paper presented at The 9th International Conference for Young Computer Scientists. (ICYCS 2008).

Webster, R., Pettitt, A. and Cook, D. 2005. Risk-adjusted CUSUM charts for medical monitoring: Some sources of uncertainty. *In*: *Unpublished Manuscript*.

Woodall, W.H. 2007. The use of control charts in health-care and public-health surveillance. Quality Control and Applied Statistics. 52(3): 253–256.

Chapter **18**

On Classical and Bayesian Inference for Bivariate Binomial Conditionals Distributions: Properties and Applications

Indranil Ghosh
University of North Carolina,
Wilmington, USA.

18.1 INTRODUCTION

Bivariate count data arise in many real-life scenarios. For example, in medicine, we may have pre-treatment and post-treatment measurements of the same individuals, or we may consider the incidence of two diseases in certain sites. Paired count data also arise in various other domains affecting our daily lives, such as economics, medical science, sports medicine, reliability of a production process, etc. However, there has not been much discussion and study of a bivariate discrete distribution that exhibits both positive and negative correlation. Consequently, not much work has been done regarding Bayesian estimating of the model parameters. This serves as one of the major motivations to carry out the present research work. This article is an extension of the earlier work by Ghosh et al. (2024) in which the authors discussed the properties of the said bivariate distribution and discussed the maximum likelihood estimation (MLE) using a copula-based approach. For a Bayesian analysis for a bivariate binomial distribution, see Crowder and Sweeting (1989) and the references cited therein. In this article, we discuss the estimation (under both the frequentist and the Bayesian paradigm) of the model parameters of a bivariate binomial conditionals distribution that appeared in Arnold et al. (1999). The rest of this paper is organized as follows. In Section 18.2, we introduce the bivariate binomial conditionals distribution due to Arnold et al. (1999). In Section 18.3, we discuss the maximum likelihood method of estimating the model parameters via an iterative process that is different from that used by Ghosh et al. (2024). Section 18.4 outlines Bayesian inference for the bivariate binomial conditional type distributions using informative and non-informative priors. The simulated results are presented in Section 18.5. Section 18.6 discusses the Bayesian estimation using the posterior mode(s) as the posterior summary for the model parameters. For illustrative purposes, a real-life dataset has been reanalyzed to exhibit the efficacy of the proposed two methods of estimation under the frequentist and under the Bayesian framework in Section 18.7. Finally, some concluding remarks are provided in Section 18.8.

18.2 BIVARIATE BINOMIAL CONDITIONALS DISTRIBUTIONS

We begin our discussion in this section by introducing a bivariate discrete distribution for which both sets of conditionals are univariate binomial according to Arnold et al. (1999).

Let us assume the following:

- $X|Y = y \sim$ binomial $(n_1, p_1(y))$, for each fixed $Y = y$.
- $Y|X = x \sim$ binomial $(n_2, p_2(y))$, for each fixed $X = x$. According to Eq. (1), the associated joint p.m.f. will be

$$P(X=x,Y=y)=K_B(n_1,n_2,p_1,p_2,\theta)\binom{n_1}{x}\binom{n_2}{y}p_1^x p_2^y(1-p_1)^{n_1-x}(1-p_2)^{n_2-y}\theta^{xy}, \tag{1}$$

where $p_1 \in (0, 1)$, and $p_2 \in (0, 1)$, and $\theta > 0$, and $x = 0, 1, 2, \ldots, n_1$; $y = 0, 1, 2, \ldots, n_2$; and $K_B(n_1, n_2, p_1, p_2, \theta)$ is the normalizing constant and

$$K_B^{-1} = K_B^{-1}(n_1,n_2,p_1,p_2,\theta) = \sum_{x=0}^{n_1}\sum_{y=0}^{n_2}\binom{n_1}{x}\binom{n_2}{y}p_1^x p_2^y(1-p_1)^{n_1-x}(1-p_2)^{n_2-y}\theta^{xy}.$$

Notice that if $\theta = 1$, then X and Y are independently distributed with Bin (n_1, p_1) and (n_2, p_2), respectively.

We will denote (henceforth, in short) the bivariate binomial conditionals distribution of the pair (X, Y) with the p.m.f. in (1) as *BBCD* $(n_1, n_2, p_1, p_2, \theta)$. Several useful structural properties of the joint p.m.f. in Eq. (1) have already been discussed in Ghosh et al. (2024).

In the next section we will focus our attention on the maximum likelihood estimation of the model parameters for the BBC distribution in Eq. (1) via a simple iterative strategy. The adopted strategy is different from the approach used in Ghosh et al. (2024). In that paper, the authors discussed maximum likelihood estimation using a copula-based approach.

18.3 ITERATIVE MAXIMUM LIKELIHOOD ESTIMATION STRATEGY FOR THE BBC DISTRIBUTION

At first, we observe the following:

$$\frac{\partial}{\partial p_1} K_B(n_1,n_2,p_1,p_2,\theta) = \theta K_B(n_1-1,n_2,p_1,p_2,\theta) \\ + n_1(1-p_1)K_B(n_1-1,n_2,p_1,p_2,\theta).$$

$$\frac{\partial}{\partial p_2} K_B(n_1,n_2,p_1,p_2,\theta) = \theta K_B(n_1,n_2-1,p_1,p_2,\theta) \\ + n_2(1-p_2)K_B(n_1,n_2-1,p_1,p_2,\theta).$$

$$\frac{\partial}{\partial \theta} K_B\left(n_1,n_2,p_1,p_2,\theta\right) = \theta K_B\left(n_1-1,n_2-1,p_1,p_2,\theta\right).$$

Next, the associated MLEs for a random sample of size m, $\{x_i, y_i\}$ drawn from the joint distribution, the associated log-likelihood will be

$$\ell = m\log(K_B(n_1,n_2,p_1,p_2,\theta)) + \sum_{i=1}^{m}\log\binom{n_1}{x_i} + \sum_{i=1}^{m}\log\binom{n_2}{y_i} \\ + \log p_1\sum_{i=1}^{m}x_i + \log p_2\sum_{i=1}^{m}y_i + \sum_{i=1}^{m}(n_1-x_i)\log(1-p_1) \\ + \sum_{i=1}^{m}(n_2-y_i)\log(1-p_2) + \log(\theta)\sum_{i=1}^{m}x_iy_i. \tag{2}$$

The associated likelihood equations will be

$$\frac{\partial \ell}{\partial p_1} = \frac{m(\theta K_B(n_1-1,n_2,p_1,p_2,\theta)+n_1(1-p_1)K_B(n_1-1,n_2,p_1,p_2,\theta))}{K_B(n_1,n_2,p_1,p_2,\theta)} - \frac{mn_1}{1-p_1} + \frac{1}{p_1(1-p_1)}\sum_{i=1}^{m} x_i. \tag{3}$$

$$\frac{\partial \ell}{\partial p_2} = \frac{m(\theta K_B(n_1,n_2-1,p_1,p_2,\theta)+n_2(1-p_2)K_B(n_1,n_2-1,p_1,p_2,\theta))}{K_B(n_1,n_2,p_1,p_2,\theta)} - \frac{mn_2}{1-p_2} + \frac{1}{p_2(1-p_2)}\sum_{i=1}^{m} y_i. \tag{4}$$

$$\frac{\partial \ell}{\partial \theta} = \frac{mK_B(n_1-1,n_2-1,p_1,p_2,\theta)}{K_B(n_1,n_2,p_1,p_2,\theta)} + \frac{\sum_{i=1}^{m} x_i y_i}{\theta}. \tag{5}$$

Therefore, we obtain the approximate asymptotic variance-covariance matrix for the MLEs by getting the inverse of the observed FIM, which is as follows:

$$I(\widehat{p_1},\widehat{p_2},\hat{\theta}) = \begin{bmatrix} -\frac{\partial^2 \ell}{\partial p_1^2} & -\frac{\partial^2 \ell}{\partial p_1 \partial p_2} & -\frac{\partial^2 \ell}{\partial p_1 \partial p_2} \\ \frac{\partial^2 \ell}{\partial p_2 \partial p_1} & -\frac{\partial^2 \ell}{\partial p_2^2} & -\frac{\partial^2 \ell}{\partial p_2 \partial \theta} \\ -\frac{\partial^2 \ell}{\partial \theta \partial p_1} & -\frac{\partial^2 \ell}{\partial \theta \partial p_2} & -\frac{\partial^2 J(\underline{\lambda})}{\partial \theta^3} \end{bmatrix} = \begin{bmatrix} Var(\widehat{p_1}) & & \\ & Var(\widehat{p_2}) & \\ & & Var(\hat{\theta}) \end{bmatrix} \tag{6}$$

The asymptotic variance-covariance matrix of the MLE $\underline{\Omega} = (\widehat{p_1},\widehat{p_2},\hat{\theta}_3)$ can be obtained from the inverse of the observed Fisher information matrix as

$$\mathbf{V} = \mathbf{I}^{-1}(\underline{\Omega}) \overset{def.}{=} \begin{bmatrix} v_{11} & v_{12} & v_{13} \\ & v_{22} & v_{23} \\ & & v_{33} \end{bmatrix}.$$

Under mild regularity conditions,

$$(\widehat{p_1}, \widehat{p_2}, \hat{\theta}) \sim N_3((p_1, p_2, \theta), \mathbf{V}).$$

Therefore, a 100 (1 – τ)% approximate confidence intervals of the parameters $\widehat{\lambda}_i$ will be

$$\widehat{\Omega}_i \pm Z(1-\tau/2) \times \sqrt{v_{ii}},$$

i = 1, 2, 3, where Z_q is the 100q-th upper percentile of the standard normal distribution.

Next, in this case, the elements of the observed FIM are

- $$\frac{\partial^2 \ell}{\partial p_1^2} = (t_1(2p_1 - 1)(p_1 \times (1-p_1))^{-2}) + \frac{mn_1}{(1-p_1)^2} - (K_B(n_1 - 1, n_2, p_1, p_2, \theta))^{-2}$$
$$\times (K_B(n_1, n_2 - 1, p_1, p_2, \theta) \times \{(\theta + n_1(1-p_1))$$
$$- m(\theta K_B(n_1 - 1, n_2, p_1, p_2, \theta) + n_1(1-p_1) K_B(n_1 - 1, n_2, p_1, p_2, \theta))^2\}.$$

- $$\frac{\partial^2 \ell}{\partial p_2^2} = \left(t_2(2p_2 - 1)(p_2 \times (1-p_2))^{-2}\right) + \frac{mn_2}{(1-p_2)^2} - (K_B(n_1, n_2 - 1, p_1, p_2, \theta))^{-2}$$
$$\times \left(K_B(n_1, n_2 - 1, p_1, p_2, \theta) \times \left\{(\theta + n_1(1-p_1))\right.\right.$$
$$\left.\left.- m\left(\theta K_B(n_1, n_2 - 1, p_1, p_2, \theta) + n_1(1-p_1) K_B(n_1, n_2 - 1, p_1, p_2, \theta)\right)^2\right\}\right).$$

- $$\frac{\partial^2 \ell}{\partial \theta \partial p_1} = \left(K_B(n_1, n_2, p_1, p_2, \theta)\right)^{-2}$$
$$\times \left(mK_B(n_1, n_2, p_1, p_2, \theta) \times \left\{\theta^2 K_B(n_1 - 2, n_2 - 1, p_1, p_2, \theta)\right.\right.$$
$$\left. + \theta n_1(1-p_1) K_B(n_1 - 2, n_2 - 1, p_1, p_2, \theta)\right\}$$
$$- \theta K_B(n_1 - 1, n_2 - 1, p_1, p_2, \theta)\left\{\theta K_B(n_1 - 1, n_2, p_1, p_2, \theta)\right.$$
$$\left.\left. + n_1(1-p_1) K_B(n_1 - 1, n_2 - 1, p_1, p_2, \theta)\right\}\right).$$

- Again,

$$\frac{\partial^2 \ell}{\partial\theta\partial p_2} = \left(K_B\left(n_1,n_2,p_1,p_2,\theta\right)\right)^{-2}$$
$$\times\Big(mK_B\left(n_1,n_2,p_1,p_2,\theta\right)\times\left\{\theta^2 K_B\left(n_1-1,n_2-2,p_1,p_2,\theta\right)\right.$$
$$\left.+\,\theta n_2\left(1-p_2\right)K_B\left(n_1-1,n_2-2,p_1,p_2,\theta\right)\right\}$$
$$-\,\theta K_B\left(n_1-1,n_2-1,p_1,p_2,\theta\right)\left\{\theta K_B\left(n_1,n_2-1,p_1,p_2,\theta\right)\right.$$
$$\left.+\,n_2\left(1-p_2\right)K_B\left(n_1,n_2-1,p_1,p_2,\theta\right)\right\}\Big).$$

- Again,

$$\frac{\partial^2 \ell}{\partial p_2\partial p_1}$$
$$= m\left(K_B\left(n_1,n_2,p_1,p_2,\theta\right)\right)^{-2}$$
$$\times\Big(\left(\theta+n_1\left(1-p_1\right)\right)K_B\left(n_1,n_2,p_1,p_2,\theta\right)\left\{\theta K_B\left(n_1-1,n_2-1,p_1,p_2,\theta\right)\right.$$
$$\left.+\,n_2\left(1-p_2\right)K_B\left(n_1-1,n_2-1,p_1,p_2,\theta\right)\right\}$$
$$-\left(\theta+n_1\left(1-p_1\right)\right)\left(\theta+n_2\left(1-p_2\right)\right)K_B\left(n_1-1,n_2,p_1,p_2,\theta\right)K_B\left(n_1,n_2-1,p_1,p_2,\theta\right)\Big).$$

Instead of using any optimization program/subroutine, we consider the following approach to obtain the MLEs of p_1, p_2, θ. This process was originally proposed and discussed in Arnold and Ghosh (2023). Observe that, the above likelihood equations can be rewritten as

$$p_1 = ((1-p_1)\{\theta K_B(n_1-1,n_2-1,p_1,p_2,\theta)+n_1(1-p_1)K_B(n_1-1,n_2,p_1,p_2,\theta)\})^{-1}$$
$$\times(mn_1p_1-t_1)K_B(n_1,n_2,p_1,p_2,\theta) \tag{7}$$

$$p_2 = ((1-p_2)\{\theta K_B(n_1-1,n_2-1,p_1,p_2,\theta)+n_2(1-p_2)K_B(n_1,n_2-1,p_1,p_2,\theta)\})^{-1}$$
$$\times(mn_2p_2-t_2)K_B(n_1,n_2,p_1,p_2,\theta) \tag{8}$$

$$\theta = -\frac{t_3K_B(n_1,n_2,p_1,p_2,\theta)}{K_B(n_1-1,n_2-1,p_1,p_2,\theta)}. \tag{9}$$

Next, we adopt the following simple (repetitive) process:

- First, we pick initial values for the $\Omega = (p_1, p_2, \theta)$.
- Next, use Eq. (7) with the three current values for the Ω's on the right side to update p_1.

- Next, use Eq. (8) with the three current values for the Ω's on the right side to update p_2.
- Finally, use Eq. (9) with the three current values for the Ω's on the right side to update θ.
- We continue this process until the process converges in the sense that we stop at stage m if $|p_i^m - p_i^{m+1}| < \varepsilon$, for i = 1, 2, and $|\theta_i^m - \theta_i^{m+1}| < \varepsilon$, where ε is a very small quantity < 0.0005.

18.4 SIMULATION STUDY

Let us assume that a random sample of size n is drawn from the joint p.m.f. in Eq. (1). In particular, we consider the sample sizes n = 60, 85, and 200 with the following 4 sets of choices of the model parameters:

(a) Choice 1: $p_1 = 0.25$, $p_2 = 0.35$ and $\theta = 0.17$.
(b) Choice 2: $p_1 = 0.35$, $p_2 = 0.43$ and $\theta = 0.29$.
(c) Choice 3: $p_1 = 0.56$, $p_2 = 0.48$ and $\theta = -0.35$.
(d) Choice 4: $p_1 = 0.64$, $p_2 = 0.52$ and $\theta = 0.45$.

Random samples from the BBC distribution are generated using the techniques discussed in Shin and Pasupathy (2010). The MLEs of p_1, p_2, and p_3 are obtained by adopting the strategy described in the previous section.

For each of these choices above, the following initial values of the parameters are considered

(a) Initial values for Choice 1: $p_1 = 0.11$, $p_2 = 0.13$ and $\theta = 0.15$.
(b) Initial values for Choice 2: $p_1 = 0.17$, $p_2 = 0.24$ and $\theta = 0.18$.
(c) Initial values for Choice 3: $p_1 = 0.27$, $p_2 = -0.17$ and $\theta = -0.12$.
(d) Initial values for Choice 4: $p_1 = 0.31$, $p_2 = 0.28$ and $\theta = 0.29$.

One may observe from Table 18.1, that the estimated MSEs for the three parameters p_1, p_2, and θ decrease as the sample size increases. However, for the estimated biases, there is not a steady decreasing pattern with the increase of sample sizes, and on the contrary, in some cases, it appears that there is a negligible amount (by 0.01–0.05) of increase. We observe that the direction of the estimated biases of the MLE of θ is the same as the sign of the true value of the parameter θ. Moreover, the estimated MSEs of θ are larger than the MSEs of p_1 and p_2. However, whether this is an anomaly or the general behavior can only be ascertained by an exhaustive simulation study with varying parameter choices and also with varying sample sizes will be considered in a separate article.

Table 18.1 Simulated coverage probabilities (CP) and average widths (AW) of the MLEs of the parameters in the BBCN distribution for various choices of p_1, p_2 and θ

Parameter choice		Based on asymptotic variances from inverting $I(\Omega)$							Based on bootstrap variances					
		p_1		p_2		θ		% of negative	p_1		p_2		θ	
	n	CP	AW	CP	AW	CP	AW	variances	CP	AW	CP	AW	CP	AW
Choice 1	60	0.945	1.216	0.951	0.563	0.992	2.352	0.0526	0.913	1.388	0.938	0.535	0.922	1.723
	85	0.939	1.234	0.952	0.487	0.986	1.261	0.082	0.814	1.344	0.938	0.485	0.917	1.543
	200	0.935	1.172	0.954	0.432	0.982	0.584	0.075	0.923	1.299	0.935	0.452	0.927	1.353
Choice 2	60	0.905	1.412	0.940	0.577	0.997	2.569	0.130	0.912	1.508	0.950	0.570	0.958	1.173
	85	0.882	1.183	0.943	0.498	0.993	1.139	0.170	0.940	1.422	0.945	0.508	0.959	1.023
	200	0.853	1.167	0.943	0.448	0.988	1.032	0.090	0.925	1.362	0.944	0.469	0.954	0.821
Choice 3	60	0.950	1.375	0.949	0.561	0.993	2.481	0.110	0.918	1.392	0.943	0.535	0.926	1.634
	85	0.941	1.292	0.956	0.484	0.986	1.254	0.170	0.904	1.345	0.935	0.486	0.914	1.354
	200	0.930	1.216	0.955	0.428	0.978	1.172	0.110	0.911	1.292	0.934	0.449	0.924	0.721
Choice 4	60	0.914	1.437	0.940	0.575	0.997	2.556	0.130	0.936	1.478	0.945	0.564	0.955	1.524
	85	0.865	1.289	0.943	0.496	0.993	1.218	0.090	0.935	1.406	0.944	0.504	0.953	1.023
	200	0.865	1.199	0.943	0.447	0.988	1.043	0.100	0.921	1.373	0.947	0.474	0.947	0.936

Furthermore, from Table 18.1, one may also observe the following:

- that the proportions of cases in which negative variance estimates are obtained is negligibly small.
- Additionally, the computed approximate confidence intervals based on bootstrap variances performs satisfactorily well. Note that these approximate confidence intervals can be used as an alternative when the asymptotic variances are negative (for pertinent details, see Ghosh and Ng (2019) and the references cited therein).

18.5 BAYESIAN INFERENCE

Quite often the classical methods of estimation for bivariate discrete probability models does not always yield satisfactory results due to several factors, such as likelihood involving ubiquitous normalizing constants, non-existence of efficient algorithms to obtain global maximums for the model parameter(s) as opposed to local maximums, etc., it is sensible to consider a Bayesian approach in this context. In this section, we begin our discussion on the Bayesian estimation by assuming the conjugate prior set-up at first for the joint p.m.f. as given in Eq. (1). In this case, we are dealing with a three parameter exponential family, so a conjugate prior will exist. The strategy adopted here is originally proposed and discussed in Arnold and Ghosh (2023) in the context of a Bayesian estimation for a bivariate Poisson conditionals distribution. We are implementing the procedure described in that paper. First, we reparametrize by defining new parameters as follows:

$$\eta_i = \log p_i,\ i = 1, 2, \quad \eta_3 = \log \theta.$$

Note that $\delta_1, \delta_2 \in (-\infty, 0)$ while $\delta_3 \in (-\infty, \infty)$.

The BBCD joint distribution in Eq. (1) can be re-written as

$$f(x,y;\vec{\eta}) = \widetilde{K}(\eta_1,\eta_2,\eta_3)\exp[\eta_1 x + \eta_2 y + \eta_3 xy + (n_1 - x)\log(1-\exp(\eta_1)) + (n_2 - y)\log(1-\exp(\eta_2))] \times \binom{n_1}{x}\binom{n_2}{y}, \tag{10}$$

where x and y are non-negative integers, and $\vec{\eta} = (\eta_1, \eta_2, \eta_3)$.

Therefore, the associated likelihood function corresponding to a sample of size m will then be

$$L(\underline{\eta}) = [\widetilde{K}(\eta_1,\eta_2,\eta_3)]^m \exp[\eta_1 t_1 + \eta_2 t_2 + \eta_3 t_3 + (mn_1 - t_1)\log(1-\exp(\eta_1)) + (mn_2 - t_2)\log(1-\exp(\eta_2))] \times \prod_{i=1}^{m}\binom{n_1}{x_i}\binom{n_2}{y_i}, \tag{11}$$

where $t_1 = \sum_{i=1}^{m} x_i,\ t_2 = \sum_{i=1}^{m} y_i,\ t_3 = \sum_{i=1}^{m} x_i y_i.$

As a conjugate prior, one may consider the following:

$$f(\underline{\xi}) \propto [\widetilde{K}(\xi_1, \xi_2, \xi_3)]^{\xi_0} \exp[\eta_1\xi_1 + \eta_2\xi_2 + \eta_3\xi_3]. \tag{12}$$

The corresponding posterior density will be of the same form, with adjusted hyperparameters, i.e.,

$$f(\underline{\eta} \mid \underline{t}) \propto [\widetilde{K}(\eta_1, \eta_2, \eta_3)]^{\xi_0 + m} \exp[(\eta_1 + t_1)\xi_1 + (\eta_2 + t_2)\xi_2 + (\eta_3 + t_3)\xi_3 \\ + (mn_1 - t_1)\log(1 - \exp(\eta_1) + (mn_2 - t_2)\log(1 - \exp(\eta_2)], \tag{13}$$

However, during this procedure of selecting the appropriate values of the hyperparameters, we observe that the posterior density is proportional to the likelihood of a sample of size $m^* = \xi_0 + m$ with sufficient statistics $t_{i,*} = \xi_i + t_i$, $i = 1, 2, 3$.

Next, for a sensible choice for the four hyperparameters, we will rely on the fact that our informed expert has had past experience with data very similar to the current dataset. This assumption is indeed quite strong but this works quite well in the present scenario of prior conjugacy. We ask the expert to provide for a typical value for observed X's which will be denoted by u_1, a typical value for the Y's to be denoted by u_2 and a typical value for the product of XY's to be denoted by u_3. Additionally, we ask for a number or index to indicate how confident he is about the three typical values. Denote this by m^{**}. This can alternatively be viewed as being a consequence of having observed an 'imaginary' sample of size m^{**} with corresponding sufficient statistics

$$\sum_{i-1}^{m^{**}} x_i = m^{**} u_1, \sum_{i=1}^{m^{**}} y_i = m^{**} u_2, \sum_{i=1}^{m^{**}} x_i y_i = m^{**} u_3.$$

Based on this information we choose as our 4 hyperparameters as follows: $\xi_0 = m^{**}$, $\xi_i = m^{**}u_i$, $i = 1, 2, 3$.

We make a note of the fact that one can rewrite the posterior density as a function of original p_i's and θ, if one wishes so. Consequently, the associated log-likelihood be the same as the log-likelihood given in Eq. (11) with suitably revised values for m, t_1, t_2 and t_3. Additionally, if we decide to use the posterior mode to estimate the p_i's and θ, one can apply the iterative scheme to find the location of the mode as discussed earlier. For the sake of brevity, this particular approach is not considered in the present paper.

18.6 A SIMULATION STUDY

Let us assume that the confidence index provided by our informed expert is $m^{**} = 10$, a small value, indicating that the expert is not at all sure about the values $u_1 = 4$, $u_2 = 6$, $u_3 = 7$, that are provided. Then, as per the suggestion made earlier, we have the following suggested values for the hyperparameters: $\eta_0 = 10$, $\eta_1 = 40$, $\eta_2 = 60$, $\eta_2 = 70$. Next, the marginal posteriors can be obtained as (proportional to)

- $$\Pi_1\left(\eta_1 \mid \vec{t^*}\right) \propto \exp\left[\eta_1(\xi_1 + t_1) + (mn_1 - t_1)\log(1 - \exp(\eta_1)\right]$$
$$\times \int_{-\infty}^{0}\int_{-\infty}^{\infty}\left[\widetilde{K}(\eta_1, \eta_2, \eta_3)\right]^{\xi_0 + m} \exp[(\eta_2 + t_2)\xi_2 + (\eta_3 + t_3)\xi_3$$
$$+ (mn_2 - t_2)\log(1 - \exp(\eta_2)\big]d\eta_2 d\eta_3.$$

- $$\Pi_2\left(\eta_2 \mid \vec{t^*}\right) \propto \exp\left[\eta_2(\xi_2 + t_2) + (mn_2 - t_2)\log(1 - \exp(\eta_2)\right]$$
$$\times \int_{-\infty}^{0}\int_{-\infty}^{\infty}\left[\widetilde{K}(\eta_1, \eta_2, \eta_3)\right]^{\xi_0 + m} \exp[(\eta_1 + t_1)\xi_1 + (\eta_3 + t_3)\xi_3$$
$$+ (mn_2 - t_2)\log(1 - \exp(\eta_2)]d\eta_1 d\eta_3.$$

- $$\Pi_3\left(\eta_3 \mid \vec{t^*}\right) \propto \exp\left[\eta_3(\xi_3 + t_3)\right]$$
$$\times \int_{-\infty}^{0}\int_{-\infty}^{0}[\widetilde{K}(\eta_1, \eta_2, \eta_3)]^{\xi_0 + m} \exp[(\eta_1 + t_1)\xi_1$$
$$+ (mn_1 - t_1)\log(1 - \exp(\eta_1) + (mn_2 - t_2)\log(1 - \exp(\eta_2)]d\eta_1 d\eta_2.$$

If instead we wish to use the posterior expectations of the λ_i's as our estimates, we will need to use numerical integration as follows:

- For p_1,

$$E(p_1 \mid \underline{t})$$
$$= \left(\int_{-\infty}^{0}\int_{-\infty}^{0}\int_{-\infty}^{\infty}[\widetilde{K}(\eta_1, \eta_2, \eta_3)]^{\xi_0 + m} \exp[(\eta_1 + t_1)\xi_1 + (\eta_2 + t_2)\xi_2 + (\eta_3 + t_3)\xi_3\right.$$
$$\left. + (mn_1 - t_1)\log(1 - \exp(\eta_1) + (mn_2 - t_2)\log(1 - \exp(\eta_2)]\right)^{-1}$$
$$\times \int_{-\infty}^{0}\int_{-\infty}^{0}\int_{-\infty}^{\infty}\exp(\eta_1)[\widetilde{K}(\eta_1, \eta_2, \eta_3)]^{\xi_0 + m} \exp[(\eta_1 + t_1)\xi_1 + (\eta_2 + t_2)\xi_2$$
$$+ (\eta_3 + t_3)\xi_3 + (mn_1 - t_1)\log(1 - \exp(\eta_1) + (mn_2 - t_2)\log(1 - \exp(\eta_2)].$$

- For p_2,

$$E(p_2 \mid \underline{t}) = \left(\int_{-\infty}^{0}\int_{-\infty}^{0}\int_{-\infty}^{\infty}[\widetilde{K}(\eta_1, \eta_2, \eta_3)]^{\xi_0 + m} \exp[(\eta_1 + t_1)\xi_1 + (\eta_2 + t_2)\xi_2\right.$$
$$\left. + (\eta_3 + t_3)\xi_3 + (mn_1 - t_1)\log(1 - \exp(\eta_1) + (mn_2 - t_2)\log(1 - \exp(\eta_2)]\right)^{-1}$$
$$\times \int_{-\infty}^{0}\int_{-\infty}^{0}\int_{-\infty}^{\infty}\exp(\eta_2)[\widetilde{K}(\eta_1, \eta_2, \eta_3)]^{\xi_0 + m} \exp[(\eta_1 + t_1)\xi_1 + (\eta_2 + t_2)\xi_2 + (\eta_3 + t_3)\xi_3$$
$$+ (mn_1 - t_1)\log(1 - \exp(\eta_1) + (mn_2 - t_2)\log(1 - \exp(\eta_2)].$$

- For θ,

$$E(\theta \mid \underline{t})$$

$$= \left(\int_{-\infty}^{0}\int_{-\infty}^{0}\int_{-\infty}^{\infty}[\widetilde{K}(\eta_1,\eta_2,\eta_3)]^{\xi_0+m}\exp[(\eta_1+t_1)\xi_1+(\eta_2+t_2)\xi_2+(\eta_3+t_3)\xi_3\right.$$

$$\left.+(mn_1-t_1)\log(1-\exp(\eta_1)+(mn_2-t_2)\log(1-\exp(\eta_2)]\right)^{-1}$$

$$\times\int_{-\infty}^{0}\int_{-\infty}^{0}\int_{-\infty}^{\infty}\exp(\eta_3)[\widetilde{K}(\eta_1,\eta_2,\eta_3)]^{\xi_0+m}\exp[(\eta_1+t_1)\xi_1+(\eta_2+t_2)\xi_2+(\eta_3+t_3)\xi_3$$

$$+(mn_1-t_1)\log(1-\exp(\eta_1)+(mn_2-t_2)\log(1-\exp(\eta_2)].$$

Note that higher-order moments can also be obtained (via numerical methods, of course). The choice of priors will have a significant impact on both bias and computational time. We consider the posterior mean as the Bayes estimates for the parameters. We also provide 95% credible intervals as a summary related to Bayesian estimation that are given in Table 18.2. We consider the following 4 different parameter settings:

- **Choice 1:** $\eta_1 = -2.5$; $\eta_2 = -1.3$; $\eta_3 = -0.25$.
- **Choice 2:** $\eta_1 = -1.8$; $\eta_2 = -0.98$; $\eta_3 = -0.45$.
- **Choice 3:** $\eta_1 = -1.58$; $\eta_2 = -1.87$; $\eta_3 = -0.55$.
- **Choice 4:** $\eta_1 = -0.46$; $\eta_2 = -0.92$; $\eta_3 = -0.65$.

18.6.1 Bayesian Analysis with Locally Uniform Priors

In this case, we consider the following locally uniform priors for the 3 parameters which are:

- $\Pi(\eta_1) \propto 1$, for $-\infty < \eta_1 < 0$.
- $\Pi(\eta_2) \propto 1$, for $-\infty < \eta_2 < 0$, and
- $\Pi(\eta_3) \propto 1$, for $\infty < \eta_3 < \infty$.

If we, before observing the imaginary sample, assume that the parameters had a flat joint prior (that are given above), then the posterior, after observing the imaginary sample would be just the likelihood of the imaginary sample, i.e.,

$$L^*(\underline{\delta}) \propto [\widetilde{K}(\eta_1,\eta_2,\eta_3)]^{m^{**}}\exp[\eta_1 m^{**}u_1+\eta_2 m^{**}u_2+\eta_3 m^{**}u_3]. \tag{14}$$

It is this posterior that we will use for a prior for the real dataset. Therefore, the resulting posterior combining the data likelihood given in Eq. (11) with the prior given in Eq. (15) will be

$$\Pi(\underline{\delta}\mid\vec{v},\vec{t}) \propto \left[\widetilde{K}(\eta_1,\eta_2,\eta_3)\right]^{m^{**}+m}$$

$$\exp\left[\eta_1(m^{**}u_1+t_1)+\eta_2(m^{**}u_2+t_2)+\eta_3(m^{**}u_3+t_3)\right]. \tag{15}$$

Table 18.2 Posterior summary for the BBC model under the conjugate prior assumption

	$\hat{p}_1$		$\hat{p}_2$		$\hat{\theta}$	
Parameter choices	**Posterior mean**	**95% HPD**	**Posterior mean**	**95% HPD**	**Posterior mean**	**95% HPD**
Choice 1	0.0793	(0.0495, 0.9523)	0.2613	(0.1745, 0.8739)	0.7415	(0.5203, 0.9735)
Choice 2	0.1935	(0.1248, 0.8495)	0.3612	(0.1388, 0.9231)	0.6256	(0.3974,0.7815)
Choice 3	0.1819	(0.0946, 0.5683)	0.2693	(0.1833, 0.8316)	0.5423	(0.3213, 0.7529)
Choice 4	0.6031	(0.1214, 0.9401)	0.3751	(0.1238, 0.8342)	0.5128	(0.3468, 0.7848)

Subsequently, the posterior means for the parameters are obtained which will be

- $E\left[p_1 \mid \underline{u}, \underline{t}\right]$

$$= \left(\int_{-\infty}^{0} \int_{-\infty}^{0} \int_{-\infty}^{\infty} \left[\widetilde{K}(\eta_1, \eta_2, \eta_3) \right]^{m^{**}+m} \exp\left[\eta_1(m^{**}u_1 + t_1 + 1) + \eta_2(m^{**}u_2 + t_2) \right] \right.$$
$$\left. + \eta_3(m^{**}u_3 + t_3) \right] d\eta_1 d\eta_2 d\eta_3 \bigg)$$
$$\times \left(\int_{-\infty}^{0} \int_{-\infty}^{0} \int_{-\infty}^{\infty} \left[\widetilde{K}(\eta_1, \eta_2, \eta_3) \right]^{m^{**}+m} \exp\left[\eta_1(m^{**}u_1 + t_1) + \eta_2(m^{**}u_2 + t_2) \right. \right.$$
$$\left. + \eta_3(m^{**}u_3 + t_3) \right] d\eta_1 d\eta_2 d\eta_3 \bigg)^{-1}.$$

- $E\left[p_2 \mid \underline{u}, \underline{t}\right]$

$$= \left(\int_{-\infty}^{0} \int_{-\infty}^{0} \int_{-\infty}^{\infty} \left[\widetilde{K}(\eta_1, \eta_2, \eta_3) \right]^{m^{**}+m} \exp\left[\eta_1(m^{**}u_1 + t_1) + \eta_2(m^{**}u_2 + t_2 + 1) \right. \right.$$
$$\left. + \eta_3(m^{**}u_3 + t_3) \right] d\eta_1 d\eta_2 d\eta_3 \bigg)$$
$$\times \left(\int_{-\infty}^{0} \int_{-\infty}^{0} \int_{-\infty}^{\infty} \left[\widetilde{K}(\eta_1, \eta_2, \eta_3) \right]^{m^{**}+m} \exp\left[\eta_1(m^{**}u_1 + t_1) + \eta_2(m^{**}u_2 + t_2) \right. \right.$$
$$\left. + \eta_3(m^{**}u_3 + t_3) \right] d\eta_1 d\eta_2 d\eta_3 \bigg)^{-1}.$$

- $E\left[\theta \mid \underline{u}, \underline{t}\right]$

$$= \left(\int_{-\infty}^{0} \int_{-\infty}^{0} \int_{-\infty}^{\infty} \left[\widetilde{K}(\eta_1, \eta_2, \eta_3) \right]^{m^{**}+m} \exp\left[\eta_1(m^{**}u_1 + t_1) + \eta_2(m^{**}u_2 + t_2) \right. \right.$$
$$\left. + \eta_3(m^{**}u_3 + t_3 + 1) \right] d\eta_1 d\eta_2 d\eta_3 bigg)$$
$$\times \left(\int_{-\infty}^{0} \int_{-\infty}^{0} \int_{-\infty}^{\infty} \left[\widetilde{K}(\eta_1, \eta_2, \eta_3) \right]^{m^{**}+m} \exp\left[\eta_1(m^{**}u_1 + t_1) + \eta_2(m^{**}u_2 + t_2) \right. \right.$$
$$\left. + \eta_3(m^{**}u_3 + t_3) \right] d\eta_1 d\eta_2 d\eta_3 \bigg)^{-1}.$$

In this case, we use the same set of 4 different choices of the model parameters listed earlier for the previous simulation study. The summary of the posterior simulation study is represented in Table 18.3.

18.7 REAL-DATA APPLICATION

To illustrate the feasibility of the proposed two Bayesian approaches in Section 18.6, we consider the data about seeds and plants grown in Rao (1990) and also in Table 18.1 in Lakshminarayana et al. (1999) and in Ghosh et al. (2024). The dataset reports the number of seeds and

Table 18.3 Posterior summary for the BBC model under the non-informative prior assumption

	$\hat{p}_1$		$\hat{p}_2$		$\hat{\theta}$	
Parameter choices	**Posterior mean**	**95% HPD**	**Posterior mean**	**95% HPD**	**Posterior mean**	**95% HPD**
Choice 1	0.0821	(0.0326, 0.9615)	0.2445	(0.0832, 0.8468)	0.7643	(0.5138, 0.8736)
Choice 2	0.1834	(0.1126, 0.8632)	0.3284	(0.1276, 0.9536)	0.6562	(0.2615, 0.6816)
Choice 3	0.1774	(0.0836, 0.5817)	0.2547	(0.1139, 0.8468)	0.5188	(0.4832, 0.9644)
Choice 4	0.5834	(0.1135, 0.9624)	0.3629	(0.1056, 0.8627)	0.5029	(0.3992, 0.7653)

Table 18.4 Goodness of fit summary for the Lens data under the BBCD model)

	$\hat{p}_1$		$\hat{p}_2$		$\hat{\theta}$	
Parameter choices	**Posterior mean**	**95% HPD**	**Posterior mean**	**95% HPD**	**Posterior mean**	**95% HPD**
Conjugate prior set-up	0.1439	(0.0826, 0.7384)	0.1538	(0.1138, 0.6534)	0.7326	(0.2645, 0.9643)
Locally uniform prior set-up	0.1861	(0.0326, 0.8132)	0.1647	(0.1065, 0.6836)	0.7419	(0.2248, 0.9841)

plants grown over a plot of size five square feet. For pertinent details on this particular dataset and the applicability of the bivariate binomial conditionals distribution as a reasonable fit for this dataset, see Ghosh et al. (2021). In this subsection, we reanalyze this dataset under the Bayesian paradigm.

Next, for the Bayesian analysis, we make a note of the following:

- For the conjugate prior set-up, we consider the following values of the hyperparameters: $\eta_0 = 0.435$, $\eta_1 = -1.237$, $\eta_2 = 0.432$, $\eta_3 = -1.465$.
- For the locally uniform prior set-up, we consider the joint prior as given in Eq. (14).

The parameter estimates (posterior mean, highest posterior density interval) under both the conjugate prior and the locally uniform priors are provided in Table 18.4.

Under the flat prior setup, the length of 95% HPD intervals are slightly wider as can be observed from Table 18.4, second row—third, fifth, and the seventh column values.

18.8 CONCLUSION

Modeling of bivariate paired count data is an open problem because of the inadequate class of bivariate discrete distributions, which if available, might explain the true dependence structure effectively. In this paper, we focus on the classical inference (using MLE via an iterative procedure) and Bayesian inference (using both conjugate priors and locally uniform priors) for a bivariate discrete probability distribution for which both the conditionals belong to a univariate binomial distribution with suitable parameters, according to Arnold et al. (1999) that exhibits both the positive and negative correlation. The iterative algorithm for the maximum likelihood method under the frequentist setup adopted in this paper has the noteworthy advantage that we don't need any optimizing root finding subroutines and this iterative procedure can be implemented in any programming environment via some user defined package(s) such as in R, SAS, Mathematica, etc. On the Bayesian inferential aspect, both the conjugate and the locally uniform prior setup are considered. It is to be noted that although prior conjugacy is quite natural for the joint p.m.f. of the form as given in Eq. (10), but from a real-life perspective it is hard to imagine that such a set of priors will exist. However, a routine Bayesian analysis can be done without any difficulty. An exhaustive study under both the classical and Bayesian paradigm for a bivariate binomial conditional distribution and in higher dimensions can be considered from a real-life perspective where such a model will be useful. However, that is beyond the scope of this current paper and will be discussed in a separate article.

REFERENCES

Arnold, B.C., Castillo, E. and Sarabia, J.M. 1999. Conditional Specification of Statistical Models. Springer, New York.

Arnold, B.C. and Ghosh, I. 2023. On classical and Bayesian inference for bivariate Poisson conditionals distributions: Theory, methods, and applications. https://doi.org/10.48550/arXiv.2301.04251.

Crowder, M. and Sweeting, T. 1989. Bayesian inference for a bivariate binomial distribution. Biometrika. 76(3): 599–603.

Ghosh, I. and Ng, H.K.T. 2019. A class of skewed distributions with applications in environmental data. Communications in Statistics: Case Studies, Data Analysis and Applications. 5(4): 346–365.

Ghosh, I., Marques, F. and Chakraborty, S. 2024. A form of bivariate binomial conditionals distributions. Communications in Statistics - Theory and Methods. 1–0. https://doi.org/10.1080/03610926.2024.2315294

Lakshminarayana, J., Pandit, S.N.N. and Srinivasa Rao, K. 1999. On a bivariate Poisson distribution. Communications in Statistics: Theory and Methods. 28(2): 267–276.

Rao, S. 1990. Experimental studies on the yield of groundnuts in coastal region. Technical Report. Andhra University, Visakhapatnam.

Shin, K. and Pasupathy, R. 2010. An algorithm for fast generation of bivariate Poisson random vectors. INFORMS Journal on Computing, 22(1): 81–92.

Chapter **19**

Applications of the Cos-Modified Lindley Distribution on Climate Data

Lishamol Tomy[1], Veena G.[2*] and Christophe Chesneau[3]

[1]Department of Statistics, Deva Matha College, Kuravilangad, Kerala-686633, India. Email: lishatomy@gmail.com

[2]Department of Statistics, St.Thomas College, Palai, Kerala-686574, India. Email: veenagpillai@hotmail.com

[3]Department of Mathematics, Université de Caen Basse-Normandie, UFR de Sciences, F-14032, Caen, France. Email: christophe.chesneau@unicaen.fr

19.1 INTRODUCTION

Recent years have seen a significant increase in the amount of research in climatology, environmental studies, and oceanology. Rainfall and precipitation studies are critical for understanding the local climate as well as analyzing the effects these events have on various aspects of society. The distribution pattern of the climatic data, which includes precipitation and rainfall data, is one of the key factors of such analysis. Establishing a (probability) distribution that offers a good fit to these datasets has long been a topic of interest in the science of meteorology. Lifetime distributions defined on $[0, +\infty]$ are commonly used in this

*For Correspondence: Veena G. (veenagpillai@hotmail.com)

regard. The impact of rainfall on the wheat output in Rothamsted was investigated by Fisher (1934). He demonstrated that agricultural productivity is affected more by the distribution of rainfall during a season than by its total amount.

After applying the approach to the distribution of sunlight, Tippet (1926) discovered that sunlight has a positive impact on wheat crops all year long. Kulandaivelu (1984) used an incomplete gamma distribution model to fit the daily precipitation data for Coimbatore over a period of 70 years to determine weekly totals. The figures show when the rainy season will likely start, how long the drought will last, and when it will finish.

Manning (1950) pioneered the introduction of yet another fruitful area of research in the study of rainfall dispersion. He changed the rainfall's skew frequency distribution to be as close to the theoretical normal distribution as possible. In this study, we delve into the application of a weighted trigonometric distribution established by Tomy et al. (2023) on datasets from climatology.

Numerous studies in the fields of biomedicine, reliability, ecology, and branching processes use weighted distributions. It is particularly important for lifetime data modeling and analysis. Fisher (1934) provided the idea of weighted distributions to model ascertainment bias. When standard distributions were not suitable to accurately capture observations with equal probabilities, Rao (1965) later developed this idea in a cohesive approach while modeling the statistical data. As a result, weighted models were made to be used in these circumstances to capture the data in accordance with some weighted function. Some of the continuous distributions that have a trigonometric weight function include the three-parameter new exponential trigonometric (NET) distribution introduced by Bakouch et al. (2018). A particular feature of this distribution is the presence of a periodic hazard rate function (hrf).

The cos-Lindley (CL) model developed by Chesneau et al. (2023) is also another remarkable distribution. Thanks to its oscillatory properties, this distribution won on some important aspects against its competitors, which include the Lindley, modified Lindley, and NET distributions.

The current research is intended to provide an application of the weighted cosmodified Lindley (Wc-ML) distribution, which usually gives the best fit against its counterpart CL and exponential distributions. This fit is generally evaluated using two goodness-of-fit (gof) tests. Subsequently, we chose the Wc-ML distribution with the best fit to the values of the meteorological parameters based on the monthly precipitation dataset and the monthly rainfall dataset at two different sites.

The structure of this chapter is as follows: Section 19.2 provides a preliminary study of the Wc-ML distribution. The applicability of the model to two real-world climatic datasets is studied in Section 19.3. Section 19.4 highlights the results and concludes the chapter.

19.2 PRELIMINARIES

Let us consider a random variable X following the Wc-ML distribution with parameters $\alpha \in [-1, 1]$, and β, $\theta > 0$. Then the corresponding probability density function (pdf) is given by

$$f_{Wc\text{-}ML}(x;\alpha,\beta,\theta) = \begin{cases} W_c^{-1} \dfrac{\theta}{1+\theta}[1-\alpha\cos(\beta x)]e^{-2\theta x}\left[(1+\theta)e^{\theta x}+2\theta x-1\right], & \text{if } x>0 \\ 0, & \text{if } x \le 0, \end{cases}$$

where $$W_c = \frac{\theta}{1+\theta}\left[\frac{\theta+1}{\theta} - \frac{\alpha\theta(1+\theta)}{\theta^2+\beta^2} + \frac{4\theta\alpha\beta^2}{(4\theta^2+\beta^2)^2}\right]. \tag{1}$$

The pdf of the Wc-ML distribution shows (i) oscillating shapes with decreasing and non-monotonic tendency, and (ii) unimodal shapes for various parameter values.

The cumulative density function (cdf) of the Wc-ML distribution is given by

$$F_{Wc\text{-}ML}(t;\alpha,\beta,\theta) = \begin{cases} W_c^{-1}\dfrac{\theta}{1+\theta}\left[\dfrac{1+\theta}{\theta}\left(1-e^{-\theta t}\right) - te^{-2\theta t}\right. \\ -\dfrac{\alpha(1+\theta)e^{-\theta t}}{\theta^2+\beta^2}\left\{\beta\sin(\beta t) - \theta\cos(\beta t) + \theta e^{\theta t}\right\} \\ -\dfrac{2\theta\alpha e^{-2\theta t}}{\left(4\theta^2+\beta^2\right)^2}\left\{\left(4\theta^2-\beta^2\right)e^{2\theta t} + \beta\left(\beta^2 t + 4\theta^2 t + 4\theta\right)\sin(\beta t)\right. \\ \left. +\left[\beta^2(1-2\theta t) - 4\theta^2(2\theta t+1)\right]\cos(\beta t)\right\} \\ \left. +\dfrac{\alpha e^{-2\theta t}}{4\theta^2+\beta^2}\left\{\beta\sin(\beta t) - 2\theta\cos(\beta t) + 2\theta e^{2\theta t}\right\}\right], & \text{if } t>0 \\ 0, & \text{if } t \le 0, \end{cases}$$

where W_c is indicated as in Eq. (1).

The hrf of the Wc-ML distribution illustrates an oscillating graph with bimodal and multimodal shapes.

Tomy et al. (2024) suggested that the Wc-ML distribution is a good candidate for the modeling the animal migration data and the failure data. The Wc-ML distribution was compared with other competitive distributions, which include the NET, exponential, and CL distributions.

19.3 DATA ANALYSIS

This section is relevant in proving the applicability of the Wc-ML distribution over other existing distributions. Here, the considered competitive distributions are the exponential (Expo) and CL distributions. The software used for statistical analysis is the R software Team (2005).

Table 19.1 displays the pdfs of the competitive distributions.

Table 19.1 Presentation of the pdfs of the competitive distributions

Model	pdf
Expo	$\theta e^{-\theta y}$
CL	$\left\{\frac{1+\theta}{\theta^2}-\frac{\alpha\theta}{\theta^2+\beta^2}-\frac{\alpha\left(\theta^2-\beta^2\right)}{\left(\theta^2+\beta^2\right)^2}\right\}^{-1}\left[1-\alpha\cos(\beta y)\right](1+y)e^{-\theta y}$

19.3.1 Methodology

- As an initial step, we study the characteristics of the datasets in terms of mean, standard deviation (SD), skewness, and kurtosis.
- A box plot and a total time on test (TTT) plot are shown to understand the datasets better. We refer the readers to Aarset (1987) for a more in-depth interpretation of the TTT plot.
- The estimates of α, β and θ, denoted as $\widehat{\alpha}$, $\widehat{\beta}$, and $\widehat{\theta}$, respectively, along with their standard errors (SEs), of the proposed and competitive distributions are obtained.
- The gof criteria, such as the Akaike information criterion (AIC) and Bayesian information criterion (BIC), are used to select the best fitted distribution.

  They are defined by

  $$\text{AIC} = 2k - 2\text{LL}, \qquad \text{BIC} = -2\text{LL} + k\log(n),$$

  where LL is the log-likelihood function taken at the estimates of the distribution, and k is the number of parameters.
- The adequacy measures used here are the Kolmogorov-Smirnov (KS), Anderson Darling (AD), and Cramér von-Mises (CVM)

statistics. The p-value corresponding to the KS statistic is also considered.

Suppose $y_1, y_2, \ldots, y_n$ represent the data and $y_{(1)}, y_{(2)}, \ldots, y_{(n)}$ be their ordered values. As an example, the CVM, AD, and KS statistics applied to the Wc-ML distribution are defined by

$$\mathrm{CVM} = -n - \sum_{i=1}^{n} \frac{2i-1}{n}\left[\log(F_{Wc\text{-}ML}(y_{(i)};\hat{\alpha},\hat{\beta},\hat{\theta})) + \log(S_{Wc\text{-}ML}(y_{(n+1-i)};\hat{\alpha},\hat{\beta},\hat{\theta}))\right],$$

where $S_{Wc\text{-}ML}(y;\alpha,\beta,\theta)$ is the survival function of the Wc-ML distribution,

$$AD = \frac{1}{12n} + \sum_{i=1}^{n}\left(F_{Wc\text{-}ML}(y_{(i)};\hat{\alpha},\hat{\beta},\hat{\theta}) - \frac{2i-1}{2n}\right)^2$$

and

$$\mathrm{KS} = \max_{i=1,2,\ldots,n}\left\{F_{Wc\text{-}ML}(y_{(i)};\hat{\alpha},\hat{\beta},\hat{\theta}) - \frac{i-1}{n}, \frac{i}{n} - F_{Wc\text{-}ML}(y_{(i)};\hat{\alpha},\hat{\beta},\hat{\theta})\right\},$$

respectively. For more information on these adequacy measures, we refer to Anderson and Darling (1954) and Darling (1957).

- The distribution having the highest p-value and the minimum values for the adequacy measures and the gof measures are considered as the best distribution in modeling the datasets.
- For a visual depiction of the best distribution and to support the numerical results obtained, the estimated pdfs (epdfs) and estimated cdfs (ecdfs) are plotted.

19.3.2 Datasets

In this subsection, we provide the datasets considered for the study.

19.3.2.1 Monthly Precipitation Dataset

The first dataset is made up of 30 consecutive March precipitation (in inches) values provided by Hinkley (1977). The data are: 0.77, 1.74, 0.81, 1.2, 1.95, 1.2, 0.47, 1.43, 3.37, 2.2, 3, 3.09, 1.51, 2.1, 0.52, 1.62, 1.31, 0.32, 0.59, 0.81, 2.81, 1.87, 1.18, 1.35, 4.75, 2.48, 0.96, 1.89, 0.9, 2.05.

19.3.2.2 Monthly Rainfall Dataset

In this chapter, we consider a dataset related to the total monthly rainfall during April. It is obtained from the Department of Water Resources and Power agency manager of water resources in the State of São Paulo including a period from 1960 to 2014. The data are: 59.00, 102.20, 17.30,

23.00, 50.60, 27.00, 203.00, 40.90, 53.00, 177.40, 94.60, 129.40, 76.00, 93.20, 22.80, 98.80, 77.70, 204.20, 16.90, 55.10, 103.90, 34.90, 39.70, 137.70, 104.20, 117.60, 17.10, 120.80, 164.90, 50.20, 172.80, 58.50, 112.40, 24.50, 32.80, 64.00, 72.10, 139.30, 0.50, 70.90, 0.80, 82.70, 108.60, 32.30, 13.60, 25.70, 135.80, 136.80, 89.70, 139.20, 102.80, 97.30, 60.60.

19.3.3 Statistical Analysis

Precipitation dataset

Table 19.2 gives the basic descriptive analysis of the dataset, which include the mean, median, SD, skewness, and kurtosis.

Table 19.2 Descriptive statistics of the precipitation dataset

Mean	Median	SD	Skewness	Kurtosis
1.675	1.470	1.000616	1.086682	4.206884

From Table 19.2, for the precipitation dataset, the skewness is 1.086682, which means that the data are positively skewed. The kurtosis of the dataset is 4.206884, which indicates the peak of the curve is leptokurtic in nature.

Figure 19.1 displays the box plot and TTT plot for the precipitation dataset.

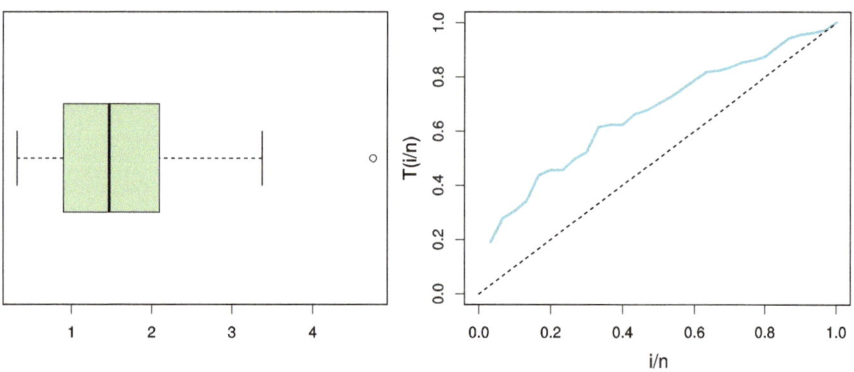

(a) Positively skewed data (b) Increasing hrf

Figure 19.1 (a) Box plot and (b) TTT plot of the precipitation dataset.

Observations from Figure 19.1:

1. The box plot shows that the dataset is almost symmetric, with the portion of the left side of the median being equal to the right side.

2. The colored TTT line in the TTT plot is concave, meaning that the corresponding empirical hrf of the data is increasing.

Tables 19.3 and 19.4 provide the estimates and SEs, adequacy measures, and gof measures for the precipitation dataset.

Table 19.3 Estimates and SEs of the precipitation dataset

Model	$\hat{\alpha}$	$\hat{\beta}$	$\hat{\theta}$
Expo	–	–	0.597
	–	–	(0.109)
CL	0.8377	1.5701	1.1116
	(0.2421)	(0.1322)	(0.2174)
Wc-ML	0.7443	1.5590	0.8071
	(0.2898)	(0.1587)	(0.1774)

Table 19.4 Adequacy measures and gof measures for the precipitation dataset

Model	AIC	BIC	AD	CVM	KS	p-value
Expo	92.9488	94.35	2.51387480	0.45389002	0.23518729	0.07239022
CL	84.8812	89.0848	0.26157860	0.03196342	0.08050751	0.99001851
Wc-ML	84.7425	88.946	0.19920660	0.02241366	0.07518279	0.99578514

From Table 19.4, it can be observed that the Wc-ML distribution has the highest p-value of 0.99578515, with the least values for adequacy measures and gof statistics.

Figure 19.2 shows a visual representation of the epdfs and ecdfs of the models to supplement Table 19.3.

Figure 19.2 supports the findings presented in Tables 19.3 and 19.4.

Rainfall dataset

Table 19.5 provides the descriptive statistics of the rainfall dataset.

Table 19.5 Descriptive statistics of the rainfall dataset

Mean	Median	SD	Skewness	Kurtosis
80.88	76	52.24407	0.4940028	2.511669

Some observations from Table 19.5:

- The skewness and kurtosis show that the data are positively skewed and are platykurtic in nature.

Figure 19.3 displays the box plot and TTT plot of the rainfall dataset.

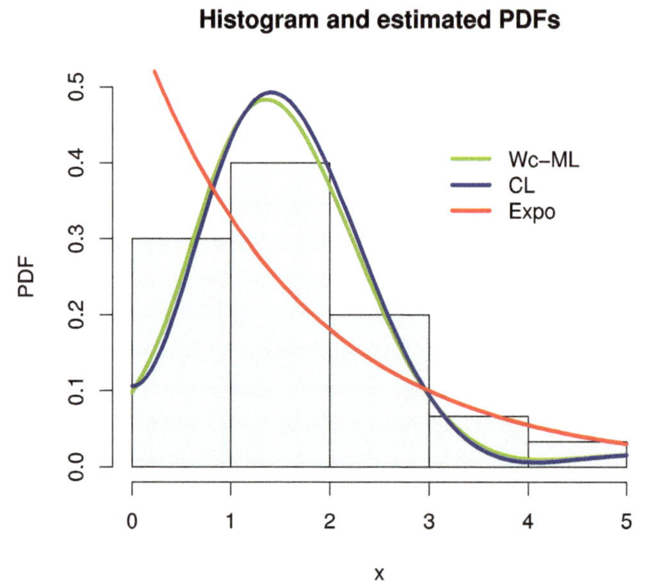

(a) The epdfs of the considered models over the corresponding the histogram of the data

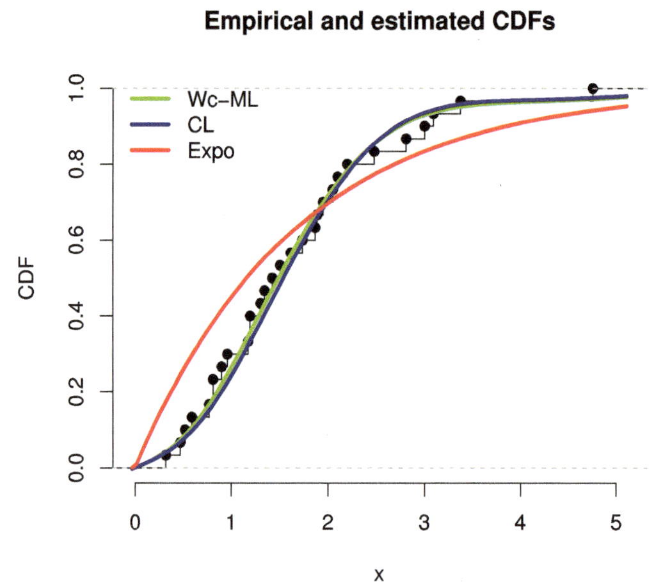

(b) The ecdf of the considered models over the corresponding empirical cdf of the data

Figure 19.2 Illustration of the (a) epdfs and (b) ecdfs of the considered models for the precipitation dataset.

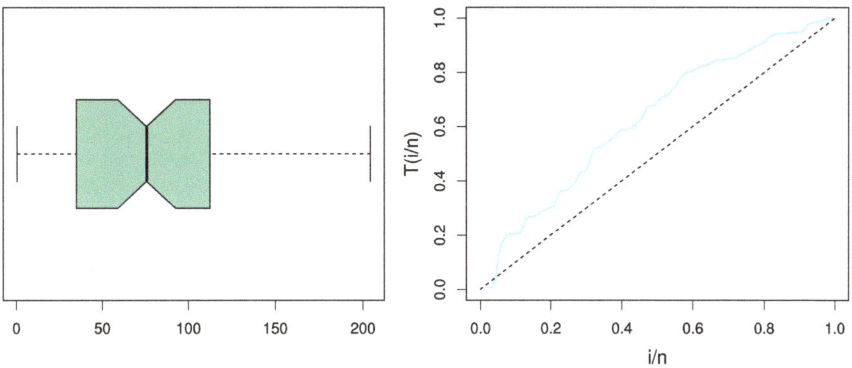

(a) Almost symmetric data (b) Increasing hrf

Figure 19.3 (a) Box plot and (b) TTT plot of the rainfall dataset.

Some observations on Figure 19.3 are:

1. The box plot supports the positive skewness of the data, as seen in Table 19.5. Additionally, the portion to the left of the median is shorter compared to the other portion in both the box plots, indicating that smaller values are more closely spaced out than the larger ones. The typical normal distribution is insufficient for modeling these data, as a quick observation will show.
2. The colored TTT line helps illustrate that the TTT plot is concave, depicting the increasing nature of the corresponding empirical hrf of the dataset.

Tables 19.6 and 19.7 provide the obtained estimates and SEs along with the gof metrics for the rainfall dataset.

Table 19.6 Estimates and SEs for the rainfall dataset

Model	$\hat{\alpha}$	$\hat{\beta}$	$\hat{\theta}$
Expo	–	–	0.012364
	–	–	(0.001687)
CL	0.820948	0.101017	0.023589
	(0.194310)	(0.003999)	(0.002344)
Wc-ML	0.225715	0.568790	0.015104
	(0.174964)	(0.006588)	(0.001580)

Table 19.7 Adequacy measures and gof measures for the rainfall dataset

Model	**AIC**	**BIC**	**AD**	**CVM**	**KS**	***p*-value**
Expo	573.6584	579.6287	2.33007654	0.42156822	0.16053058	0.11642126
CL	593.3795	599.2904	1.1974418	0.1391982	0.1268972	0.3322465
Wc-ML	573.105	579.0158	0.8347991	0.1020632	0.1055693	0.5600968

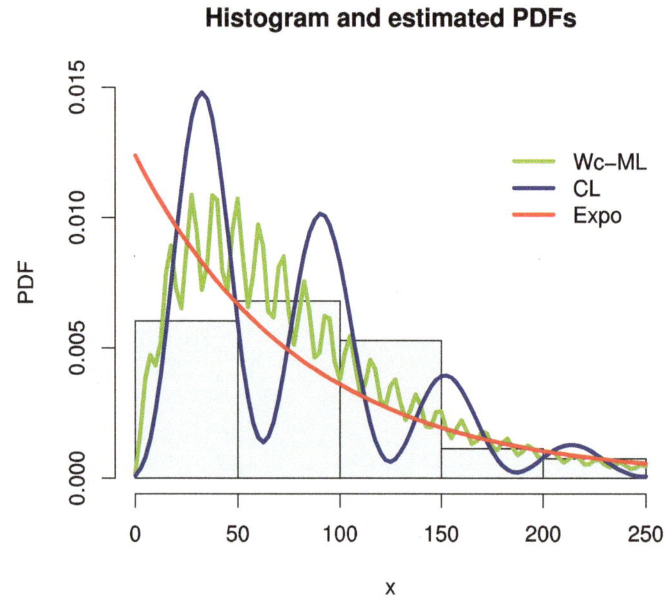

(a) The epdfs of the considered models over the corresponding the histogram of the data

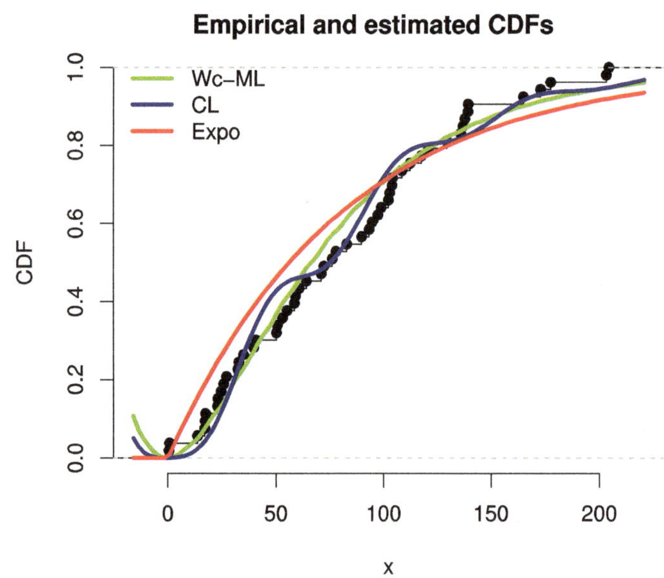

(b) The ecdf of the considered models over the corresponding empirical cdf of the data

Figure 19.4 Plots of the (a) epdf and (b) ecdf for the rainfall dataset.

From Table 19.7, the Wc-ML distribution has the highest p-value of 0.56000968, with the lowest values for the AIC, BIC, AD, and CVM statistics. From these results, we can conclude that the Wc-ML distribution provides the best fit to the rainfall dataset. The illustration of the epdfs and ecdfs in Figure 19.4 supports the numerical findings.

From Figure 19.4 and Table 19.7, we can observe that the Wc-ML distribution gives a better fit to both rainfall datasets compared to other distributions.

19.4 CONCLUSION

In this article, we studied the statistical applications of the Wc-ML distribution on two climatic datasets. The applicability of this distribution is determined by using the adequacy measures and the goodness-of-fit metrics. It is studied on two real-world datasets against the existing cos-Lindley and exponential distributions. It is found that the Wc-ML distribution can be used to model such data and can be used as an alternative to the competitive distributions.

REFERENCES

Aarset, M.V. 1987. How to identify a bathtub hazard rate. IEEE Transactions on Reliability. 106–108.

Anderson, T.W. and Darling, D.A. 1954. A test of goodness of fit. Journal of the American Statistical Association. 49(268): 765–769.

Bakouch, H.S., Chesneau, C. and Leao, J. 2018. A new lifetime model with a periodic hazard rate and an application. Journal of Statistical Computation and Simulation. 88(11): 2048–2065.

Chesneau, C., Bakouch, H. S., Tomy, L. and Veena, G. 2023. The Cos-Lindleymodel with Modelling to Service times and Failure times data. International Journal of Modelling and Simulation, 1–12.

Darling, D.A. 1957. The Kolmogorov-Smirnov, Cramér-von Mises tests. The Annals of Mathematical Statistics. 28(4): 823–838.

Fisher, R.A. 1934. The effect of methods of ascertainment upon the estimation of frequencies. Annals of Eugenics. 6(1): 13–25.

Hinkley, D. 1977. On quick choice of power transformations. Journal of Royal Statistical Society, Series C. 26: 67–69.

Kulandaivelu, R. 1984. Probability analysis of rainfall and evolving cropping system for Coimbatore. Mausam. 3: 257–258.

Manning, H.L. 1950. Confidence limits of expected monthly rainfall. The Journal of Agricultural Science. 40: 169–176.

Rao, C.R. 1965. On discrete distributions arising out of methods of ascertainment. Sankhyā: The Indian Journal of Statistics, Series A.

Team, R.D.C. 2005. R foundation for statistical computing. R: A Language and Environment for Statistical Computing.

Tippet, L.H.C. 1926. On the effect of sunshine on wheat yield at Rothamsted. The Journal of Agricultural Science. 2: 159–165.

Tomy, L., Veena, G. and Chesneau, C. 2024. The Cos-Modified Lindley distribution: Properties, Estimation and Application. Statistical Outliers and Related Topics. Boca Raton, Florida: CRC Press, Taylor & Francis Group.

Chapter **20**

Classification of Rainfall Regions in the UK Using Spatial Distances

A.H.M. Rahmatullah Imon*[1],
Noora N Saleh[2] and Sirajum Munira Khan[3]

[1]Department of Mathematical Sciences,
Ball State University, Muncie, Indiana, USA.
Email: rimon@bsu.edu

[2]Department of Mathematics, College of Education,
Al-Qadisiyah University, Iraq.
Email: noora.saleh@qu.edu.iq

[3]Department of Bioinformatics and Biostatistics,
School of Public Health and Information Sciences,
University of Louisville, Kentucky, USA.
Email: munirakhan09@gmail.com

20.1 INTRODUCTION

Spatial statistics includes formal techniques, which study entities using their topological, geometric, or geographic properties. The phrase is often used in a more restricted sense to describe techniques applied to the analysis of geographic data. Rainfall may be considered as a spatial data as there are reasons to believe that the amount of rainfall

*For Correspondence: A.H.M. Rahmatullah Imon (rimon@bsu.edu)

depends on the geographical location of a place. In this study, we would consider rainfall data of the U.K. First, we will check whether there is any difference in average rainfall among different weather stations. If so, we will employ spatial analysis to identify those different regions. In spatial statistics, classification of clusters and the identification of outliers grew up together. We usually develop a method to classify the major clusters and those observations that fail to match with any of the major clusters are identified as outliers. The identification of spatial outliers is important because it can reveal hidden but valuable knowledge in many applications. An excellent review of different aspects of spatial outliers is available in the literature by Haslett et al. (1991), Anselin (1996), Shekhar et al. (2002), Huang and Qin (2004), Lu et al. (2004), Hadi et al. (2009), Hadi and Imon (2019), and others. Traditional outlier detection methods mainly focus on "global comparisons" and identify observations which stand apart from the majority of the entire dataset. For small data this can be done with ease, but for big data or when data appears in clusters it is not easy to tell which portion of the data is major and which portion is minor. However, if we are able to find major clusters in the data, which altogether contain almost all of the observations, and if there exist few scatter data points which fail to accommodate with any of the major clusters, these isolated points are called outliers in the spatial statistics literature (Hadi and Imon 2019). Spatial outliers are those observations whose characteristics are markedly different from their spatial neighbors. In other words, spatial outliers are those observations which are responsible for neighborhood instabilities as defined by Shekhar et al. (2002). The same thing happens when observations belong to different clusters. The boundary point of one cluster shows a marked difference with the boundary point of the neighboring cluster. This motivates us to update a spatial outlier detection technique to a classification technique based on successive differencing of observations. The existing simple classification methods usually suffer from masking (false negative) and/or swamping (false negative). On the other hand, sophisticated outlier detection methods are computationally very intensive and hence are not very friendly for big data. For this reason we need a classification technique which is easy to execute and robust at the same time.

20.2 CLASSIFICATION BASED ON ROBUST SPATIAL DISTANCES

Spatial z test proposed by Shekhar et al. (2002) is a very simple, easily understood and popular technique for the identification of spatial outliers. However, it is now evident (Hadi and Imon 2019) that this test may often fail to identify multiple outliers. We begin this section with a motivational example. Table 20.1 contains 20 observations generated from five clusters. Observations 1–5 come from *Uniform* (15, 20), observations 6–9 come from *U* (0–5), observations 10–13 come from *U* (20, 25), observations 14–16 come from *U* (10, 15), and observations 17–20 come from *U* (25, 30). The index plot of this data as shown in Figure 20.1 clearly exhibits these five clusters.

Table 20.1 Artificial spatial data with five clusters

Index	Attributes	Index	Attributes	Index	Attributes	Index	Attributes
1	15.7501	6	1.9707	11	22.1987	16	12.6324
2	17.3298	7	1.7515	12	20.9845	17	28.5693
3	19.6610	8	4.1509	13	24.4976	18	26.0679
4	15.2828	9	1.8965	14	10.0311	19	29.7140
5	18.7305	10	23.5442	15	11.6561	20	25.3638

Figure 20.1 Index plot of artificial spatial data.

In order to locate points, which are responsible for neighborhood instabilities, we consider the first difference of observations. For a set of observations $X_1, X_2, \ldots, X_n$, we define a set of differenced attributes as

$$D_i = X_i - X_{i-1}, \quad i = 2, 3, \ldots, n \tag{1}$$

We ompute D_i values for the artificial data and the results are presented in Table 20.2.

Table 20.2 First order difference in attributes for the artificial spatial data

Index	D_i	Index	D_i	Index	D_i	Index	D_i
1	*	6	−16.7598	11	−1.3455	16	0.9763
2	1.5797	7	−0.2191	12	−1.2143	17	15.9369
3	2.3312	8	2.3993	13	3.5131	18	−2.5014
4	−4.3782	9	−2.2543	14	−14.4665	19	3.6461
5	3.4477	10	21.6477	15	1.6250	20	−4.3502

Figure 20.2 shows the difference of each observation from its neighbors. The global version of statistics will essentially look at the attribute values in the y-axis and the maximum and the minimum observations such as 7 and/or 19 are candidate outliers. Nevertheless, we see no unusual behavior of these two observations at all. In contrast to that, the spatial version searches for unusual spikes such as the pairwise differences of observations (5, 6), (9, 10), (13, 14), and (16, 17) as shown in Figure 20.2 which are very different from their neighbors. They look like spatial outliers because they violate the law of geography that the nearby observations should be very close to one another. However, graphical methods are very subjective in nature. For this reason, we need to employ a quantitative test to confirm our suspicion.

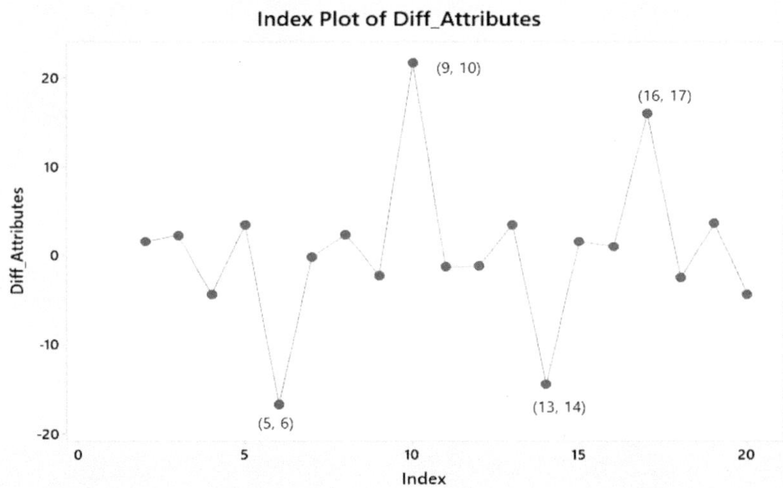

Figure 20.2 Index plot of differences of attributes for the artificial spatial data.

At this point, we would like to employ an analytical test. Shekhar et al. (2002) proposed a standard normal z test on the difference of attributes that we call a spatial z test. Under a normal assumption, a single value D_i as defined in (1) may be considered as an outlier if it falls outside a certain range of the standard deviation. A traditional measure of the 'outlyingness' of an observation D_i with respect to a sample is the ratio between its distance to the sample mean and the sample standard deviation

$$z_i = \frac{D_i - \underline{D}}{s_{D_i}} \quad \text{for } i = 2, 3, \ldots, n \tag{2}$$

z-values defined in Eq. (2) will be called spatial z scores since they are calculated on spatial differences. Observations with $|z_i| > 3$ are traditionally deemed as suspicious (the three-sigma rule), based on the fact that they would be very unlikely under normality, since $P(|z| > 3) = 0.003$ for a random variable z with a standard normal distribution.

For this motivational example, we calculate the spatial z scores for the difference of attributes and the scores are presented in Table 20.3 and also in Figure 20.3. Since z scores are computed on the differences of attribute values, each point appears in pairs and if it is a genuine outlier the pairs should exhibit unusual patterns.

Table 20.3 Spatial z scores for the artificial spatial data

Index	z_Diff	Index	z_Diff	Index	z_Diff	Index	z_Diff
1	*	6	−2.01967	11	−0.21657	16	0.05501
2	0.12560	7	−0.08482	12	−0.20123	17	1.80504
3	0.21350	8	0.22147	13	0.35176	18	−0.35179
4	−0.57132	9	−0.32289	14	−1.75141	19	0.36732
5	0.34410	10	2.47305	15	0.13090	20	−0.56805

Table 20.3 shows that there is no pair of observations that corresponds to z-value higher than 3 in absolute value. Figure 20.3 shows that although there exists four z values which are relatively higher or lower than most of the cases, but none of them exceeds the cut-off margin ±3. Thus, the spatial z score does not identify even a single observation as unusual.

At this point the biggest question is why the spatial z test fails to identify even a single observation to be unusual while Figure 20.3 shows the possible existence of four of them. When we look at the spatial z statistic as defined in Eq. (2) more carefully, we observe that it contains two components, mean, and standard deviation both of which may be severely affected in the presence of unusual observations. It motivates us to develop a new spatial z test which should be very robust in the presence of unusual observations. Here we replace mean and standard deviation by their robust counterparts. Sample median is a very effective robust measure of location. For the measure of dispersion, we can use the median absolute deviation (MAD). For the set of differenced attributes, D_i, we compute the median absolute deviation (MAD) defined as

$$MAD\ (D_i) = Median\ \{|D_i - Median\ (D_i)|\}/0.6745 \tag{3}$$

Figure 20.3 Index plot of spatial *z* scores of differences of attributes for the artificial data.

Here the correction factor 0.6745 is used to make the efficiency of MAD comparable with the standard deviation for a normal sample. Hence replacing the mean by the median and the standard deviation by the median absolute deviation, we define a robust spatial *z* score as

$$z_i^l = \frac{D_i - Median(D_i)}{MAD(D_i)} \tag{4}$$

Observations with $|Z_i^l| > 3$ will be identified as spatial outliers. It is worth mentioning that the robust spatial z scores defined in Eq. (4) are analogous to Hadi and Imon (2019) scores but here *D* statistics are different. We compute the robust spatial *z* scores for the motivational example and they are presented in Table 20.4 and in Figure 20.4.

Table 20.4 Robust spatial *z* scores for the artificial spatial data

Index	Robust_z	Index	Robust_z	Index	Robust_z	Index	Robust_z
1	*	6	**−7.18211**	11	−0.94148	16	−0.00151
2	0.24280	7	−0.48548	12	−0.88836	17	**6.51888**
3	0.54704	8	0.57462	13	1.02554	18	−2.15797
4	−2.16929	9	−1.30944	14	**−6.25364**	19	0.90101
5	0.99905	10	**8.36747**	15	0.26113	20	−1.40949

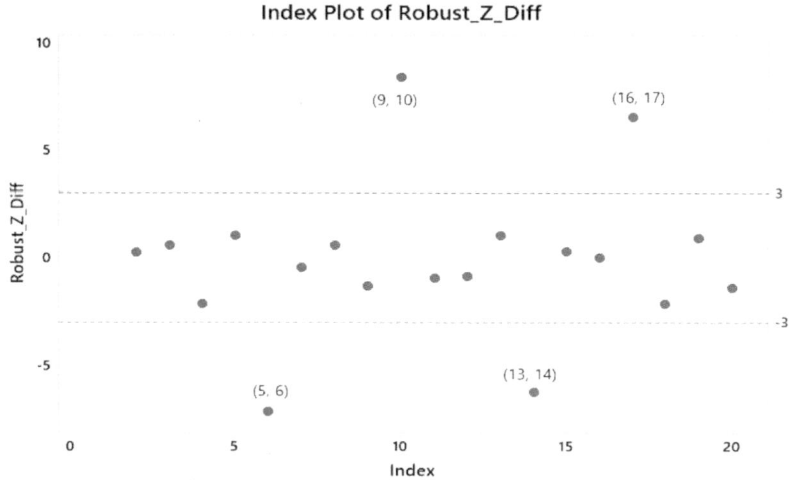

Figure 20.4 Index plot of robust spatial z scores of differences of attributes for the artificial data.

Results presented in Table 20.4 shows that there are four pairs of observations which have robust z-scores higher than 20.3 in absolute value. We get a clear view from Figure 20.4 that the pairwise differences of observations (5, 6), (9, 10), (13, 14), and (16, 17) are definitely different from other pairs and their corresponding robust z scores fall outside the acceptance zone (–3, 3). These four boundaries determine the existence of five clusters in this data and they are as follows:

Cluster 1: {1–5}, Cluster 2: {6–9}, Cluster 3: {10–13}, Cluster 4: {14–16}, Cluster 5: {17–20}.

20.3 CLASSIFICATION OF BRITISH RAINFALL REGIONS BASED ON ROBUST SPATIAL Z SCORES

The British Isles have variable weather that changes from day to day between different regions and within different regions. Consequently, it is very difficult to predict or forecast the weather. It is well known that different parts of the U.K. experience different regional climates. The variability of weather and climate in different regions of the British Isles is due to the different air masses that meet over the Isles. We anticipate that the distances of regions from the sea may be a very crucial factor for the variability of rainfall in different regions of the U.K.

We have taken climate data from Alanazi (2017) who took it from the British weather website recorded by all 26 weather stations in the U.K. These stations are Aberporth, Armagh, Bradford, Braemar, Camborne, Chivenor, Cwmystwyth, Dunstaffnage, Durham, Eastbourne, Eskdalemuir, Hurn, Lerwick, Leuchars, Lowestoft, Manston, Nairn, Oxford, Paisley, Ringway, Ross-on-Wye, Shawbury, Sheffield, Southampton, Sutton Bonington, and Tiree. This website contains monthly rainfall data together with some other climate variables up to 2016. All weather stations have data from more than hundred years back, the oldest record we have is data from 1886. We have ordered the weather stations according to their distances from the sea. The average monthly rainfall for all 26 weather stations (in mm) are presented in Table 20.5 and also in Figure 20.5.

Table 20.5 Average monthly rainfall of 26 weather stations in the UK

Station	Average_Rainfall (in mm)	Station	Average_Rainfall (in mm)	Station	Average_Rainfall (in mm)
10	143.350	2	75.380	18	54.801
7	145.850	3	76.050	8	54.790
6	149.630	17	78.810	16	52.724
12	96.440	24	64.690	4	48.920
25	98.560	9	67.160	15	49.250
19	98.290	20	68.890	11	50.580
5	96.880	13	57.130	14	50.206
1	95.012	21	60.040	26	51.000
23	72.020	22	56.520		

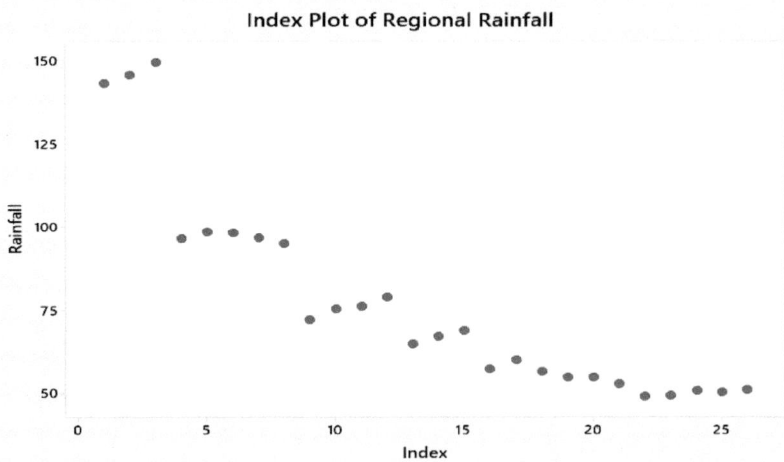

Figure 20.5 Average monthly rainfall of 26 weather stations in the UK.

Figure 20.5 shows the existence of several clusters in the dataset. The graphical inspection is always subjective. One possible classification is as follows:

Cluster 1: {10, 7, 6}, Cluster 2: {12, 25, 19, 5, 1}, Cluster 3: {23, 2, 3, 17}, Cluster 4: {24, 9, 20}, Cluster 5: {13, 21, 22, 18, 8, 16, 4, 15, 11, 14, 26}.

Of course, this classification is not unique. Looking at the same graph different experimenters may come with different classifications. Now we classify the rainfall regions using our proposed method and the resulting spatial *z* and robust spatial *z* values are presented in Table 20.6 and also in Figure 20.6.

Table 20.6 Spatial *z* and robust spatial *z* score for the U.K. average monthly rainfall data

Station	*Spatial z*	*Robust spatial z*	*Station*	*Spatial z*	*Robust spatial z*
10	*	*	9	0.513476	1.71924
7	0.51597	1.7401	20	0.451832	1.20416
6	0.62260	2.6311	**13**	−0.671917	**−8.18551**
12	**−4.12313**	**−37.0227**	21	0.550129	2.02550
25	0.48432	1.4756	22	0.014495	−2.45008
19	0.28523	−0.1879	18	0.164522	−1.19650
5	0.19026	−0.9814	8	0.306802	−0.00766
1	0.15211	−1.3002	16	0.135616	−1.43803
23	−1.60757	**−16.0035**	4	−0.009163	−2.64776
2	0.58761	2.3387	15	0.335209	0.22970
3	0.36353	0.4664	11	0.418511	0.92574
17	0.53763	1.9211	14	0.276564	−0.26032
24	−0.86851	**−9.8282**	26	0.373861	0.55266

We observe from the spatial *z* values that only one pair of observations (6, 12) exceeds 3 in absolute value. Hence there are only two clusters in the data and they are:

Cluster 1: {10, 7, 6}

Cluster 2: {12, 25, 19, 5, 1, 23, 2, 3, 17, 24, 9, 20, 13, 21, 22, 18, 8, 16, 4, 15, 11, 14, 26}.

The above classification does not look satisfactory.

When we employ the proposed classification technique based on robust spatial *z* we find four pairs of observations (6, 12), (1, 23), (17, 24), and (13, 20) have robust spatial *z* values exceeding 3. These four boundaries provide five clusters as shown below. A graphical display of this classification is given in Figure 20.6 as well.

Cluster 1: {6, 7, 10}, Cluster 2: {12, 25, 19, 5, 1}, Cluster 3: {23, 2, 3, 17}, Cluster 4: {24, 9, 20}, Cluster 5: {13, 21, 22, 18, 8, 16, 4, 15, 11, 14, 26}.

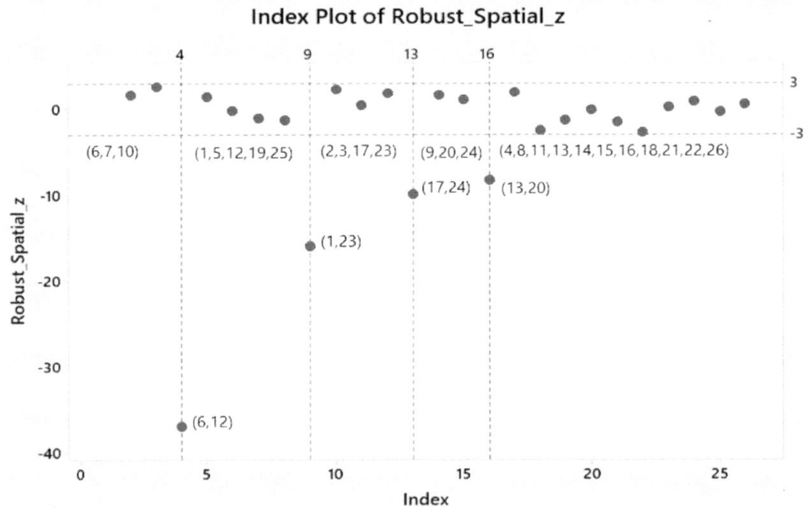

Figure 20.6 Robust spatial *z* scores for the U.K. average monthly rainfall data.

Based on this we can classify the rainfall regions in this way.

Highest Rainfall Region: Chivenor, Cwmystwyth, Eastbourne

High Rainfall Region: Aberporth, Camborne, Hurn, Paisley, Sutton Bonington

Moderate Rainfall Region: Armagh, Bradford, Nairn, Sheffield

Moderate to Low Rainfall Region: Durham, Ringway, Southampton

Low Rainfall Region: The rest of the country

It is worth mentioning that the several existing classification techniques may produce the same set of results but they are computationally very extensive. They require variance covariance matrices of all possible subsets of observations and which could be very complicated in big data. When the computation is huge it does not only take longer time, the results may have lower accuracy as well. Our proposed method requires only the median of the observations. It does not even require a variance or a standard deviation. It must be very friendly to big data.

20.4 CONCLUSIONS

In this book chapter, our main objective was to classify different rainfall regions in the U.K. based on the average monthly rainfall. In order to do that we develop a new classification technique which is designed exclusively for spatial data. Unlike the other classification techniques, it does not involve any complicated variance covariance structures of data, it requires only the median of the observations and also is resistant to outliers. A numerical example shows that the proposed robust spatial z method can classify clusters successfully for an artificial data. When we employ this method to the U.K. rainfall data, it classifies the entire U.K. in to five different rainfall regions according to the average monthly rainfall.

REFERENCES

Alanazi, N.N. 2017. Modeling and Prediction of Rainfall Data. Unpublished M.S. Thesis, Department of Mathematical Sciences, Ball State University, USA.

Anselin, L. 1996. The moran scatterplot as an ESDA tool to assess local instability in spatial association. In: Spatial Analytical Perspectives on GIS. M. Fischer, H. Scholten, and D. Unwin (Eds.). Taylor and Francis, London, 111–125.

Hadi, A.S., Imon, A.H.M.R. and Werner, M. 2009. Detection of outliers. Wiley Interdisciplinary Reviews: Computational Statistics. 1: 57–70.

Hadi, A.S. and Imon, A.H.M.R. 2019. Identification of multiple outliers in spatial data. International Journal of Statistical Sciences. 16: 87–96.

Haslett, J., Brandley, R., Craig, P., Unwin, A. and Wills, G. 1991. Dynamic graphics for exploring spatial data with applications to locating global and local anomalies. The American Statisticians. 45: 234–242.

Huang, T. and Qin, X. 2004. Detecting outliers in spatial database. In: Proceedings of the Third International Conference on Image and Graphics. Hong Kong. 556–559.

Lu, C., Chen, D. and Kou, Y. 2004. Multivariate spatial outlier detection. International Journal of Artificial Intelligence Tools. 13: 801–812.

Shekhar, S., Lu, C. and Zhang, P. 2002. Detecting graph-based spatial outlier. Intelligent Data Analysis. 6: 451–468.

Chapter **21**

Identification of High Leverage Points in High Dimensional Sparse and Non-Sparse Data

Siti Zahariah[1] and Habshah Midi*[2]

[1]Applied Statistics and Data Science Cluster, Universiti Kuala Lumpur Malaysian Institute of Information Technology (UniKL MIIT), Kuala Lumpur, Malaysia. Email: sitizahariah@unikl.edu.my, +60173788681.

Institute for Mathematical Research, Universiti Putra Malaysia, 43400 UPM, Serdang, Selangor, Malaysia.

[2]Institute for Mathematical Research, Universiti Putra Malaysia, 43400 UPM Serdang, Selangor, Malaysia.

Department of Mathematics and Statistics, Universiti Putra Malaysia, 43400 UPM Serdang, Selangor, Malaysia.
Email: habshah@upm.edu.my, +60133925520.

21.1 INTRODUCTION

The analysis of high dimensional data has become increasingly important in many fields such as applied sciences, engineering, and medicines. For instance, there are tens of thousands of gene expression values available in tumor classification utilizing genomic data; however, the number of arrays is only at the order of ten. Other applications in high dimensional data (HDD) are in chemometrics, fraud detection, climate studies, and satellite processing. HDD that refer to a situation when the number of

*For Correspondence: Habshah Midi (habshah@upm.edu.my, habshahmidi@gmail.com)

predictor variables (p) is much larger than the sample size (n), forms a major statistical challenge in terms of data classification and other statistical analyses. In HDD, a matrix related to some algorithms may become singular. This characteristic leads to a sparsity problem known as the curse of dimensionality phenomena (Lee et al. 2011). In the curse of high dimensionality, conventional statistical methods do not work well, and most of them fail to perform especially when dealing with contaminated data or outliers.

Several versions of outliers exist in regression problems, such as vertical outliers, residual outliers, and high leverage points (HLPs). Residual outliers refer to any observations that have a large residual while vertical outliers refer to observations that are extreme or are outlying in the Y-space. On the other hand, HLPs refer to observations which fall far from the majority of the explanatory variables or are outlying in the X-space. Among the three types of outliers, leverage point has the most detrimental effect on the computed values of various estimates which leads to misleading conclusion about the fitted regression model (Habshah et al. 2009).

Therefore, a HLP detection in HDD has become an issue of great importance to study although it has not received significant attention from statisticians. Accurate identification of HLPs play an important role in statistical analysis as incorrect detection of high leverage points will substantially affect the standard error of estimates and cause multicollinearity problems, masking and/or swamping of outliers, overfitting and/or underfitting of a model which will lead to a false prediction. Most existing algorithms in HDD treat abnormal observations as outliers. Not many works have focused on dealing with HLPs under the curse of dimensionality. It is not easy to correctly capture the HLPs in HDD since the data have a large dimensionality. It is very complicated, time-consuming, and computationally intensive. With the increase of dimensionality, the efforts needed to identify the HLPs become more challenging and trickier.

Most of the conventional outlier detection methods fail to correctly identify outliers (Serfling and Wang 2012) mainly because of the masking and or swamping effects. 'Swamping' refers to 'good' observations incorrectly declared as outliers, while 'masking' refers to a case where outliers are incorrectly declared as inliers (Rousseeuw and Leroy 1987). Daniel Peña and Yohai (1991) noted that the HLPs are mainly responsible for the masking and swamping of outliers in a linear regression. Robustness and resistance against both masking and swamping are vital to avoid misleading conclusions.

Thus, the detection of HLPs in HDD needs strict attention as it has a massive number of predictor variables. The existing classical method based on the Mahalanobis distance is not applicable in HDD since the covariance matrix is not invertible (Rousseeuw 1985). There are many HLPs detection algorithms available in the literature; however, most of the methods are not applicable for HDD because the covariance matrix is not invertible. Moreover, not many works have focused on HLPs detection in HDD, specifically for high dimensional sparse data where most of the features have values close to zero. As a remedy to this issue, Hubert et al. (2005) proposed the use of the Robust Principal Component Analysis (ROBPCA) to diagnose bad and good leverage points in HDD. However, the ROBPCA procedure suffers from masking and swamping effects. Boudt et al. (2018) developed the Minimum Regularized Covariance Determinant (MRCD) technique and used it to compute RMD to detect outliers (hereinafter referred to as HLPs) for HDD. The method is very successful for the detection of HLPs in high dimensional. Nevertheless, our investigation shows that the performance of the RMD-MRCD method deteriorates as the number of independent variables increases.

In this chapter, a new method called the MRCD based on the principal component analysis (MRCD-PCA) is developed. The RMD-MRCD-PCA is developed by incorporating the Principal Component Analysis (PCA) in the MRCD algorithm whereby this robust approach shrinks the covariance matrix to make it invertible and thus, can be employed to compute the RMD for HDD. The performance of our proposed method is investigated by a Monte Carlo simulation study and real examples.

This chapter is organized as follows: The overview of a few existing HLP detection methods is presented in Section 21.2; Section 21.3 describes the proposed RMD-MRCD-PCA method for the detection of HLPs in high dimensional sparse data, which involves the combined use of the MRCD to identify suspicious observations and PCA as a reduction method. The simulation study is presented in Section 21.4 and application on real datasets is discussed in detail in Section 21.5. Section 21.6 presents the concluding remarks.

21.2 HIGH LEVERAGE POINT DETECTION MEASURES IN A LINEAR REGRESSION

In this section, some of the commonly used methods for the identification of HLPs in a linear regression model is summarized.

21.2.1 Mahalanobis Distance (MD)

Mahalanobis (1936) defined Mahalanobis Distance (MD) as a measure of deviation of a data point from its center. Let write the ith vector of predictor variables as

$$x_i^T = (1, x_1, x_2, \ldots, x_p) = (1, t_i)$$

where t_i is a p-dimensional row vector. The mean vector and the variance covariance matrix are calculated as

$$\bar{t} = \frac{1}{n}\sum_{i=1}^{n} t_i \text{ and } C = \left(\frac{1}{n-1}\right)\sum_{i=1}^{n}(t_i - \bar{t})^T (t_i - \bar{t}) \text{ respectively.}$$

Subsequently, the (*MD*) for each observation is written as follows:

$$MD_i = \sqrt{(t_i - T(X))C(X)^{-1}(t_i - T(X))^T} \qquad i = 1, 2, 3, \ldots, n \tag{1}$$

where $T(X)$ is the mean vector $(\bar{t})$ and $C(X)$ is the variance covariance matrix (C). Any observation is considered as HLP if its corresponding MD_i value exceeds the cutoff points $\sqrt{\chi^2_{p,0.95}}$.

21.2.2 Minimum Covariance Determinant (MCD)

Rousseeuw (1985) introduced the Minimum Covariance Determinant (MCD) estimator, which is a highly robust estimator of multivariate location and scatter. The MCD estimator is widely used by statistical practitioners, but it is not computationally efficient. To overcome this drawback, Rousseeuw and Driessen (1999) developed the Fast-Minimum Covariance Determinant algorithm. The main purpose of MCD is to find a subset of h data points with the smallest determinant of the covariance matrix, where $\frac{n}{2} \le h \le n$. A robust location estimator $\hat{\mu}_{mcd}$ is an average of these h data points, while a robust scatter estimator, C_{mcd}, is a variance covariance matrix of h data points multiplied by consistency factor, c_α. Croux and Haesbroeck (1999) noted that c_α equals $\frac{\alpha}{F_{\chi^2_{p2}}(x^2_{p,2})}$ where $\alpha = \frac{(n-h)}{n}$; n is a dimension of dataset; and $h = 0.75n$ or $h = (n + p + 1)/2$. The asymptotic efficiency of MCD is almost 50% when $h = 0.75n$ or $h = (n + p + 1)/2$. Hubert et al. (2018) stated that the MCD location and scatter estimator have the highest breakdown values and thus, become more robust.

MCD estimators can be determined when $h > p$; otherwise, the covariance matrix will be singular, and the determinant of the covariance matrix will be zero. The principal rule of thumb suggested by Rousseeuw and Zomeren (1990) and Hubert et al. (2012) is to avoid the curse of dimensionality, and it requires $n > 5p$.

21.2.3 Robust Mahalanobis Distance (RMD)

Habshah et al. (2009) noted that observations with a huge value of MD are not necessarily HLPs due to swamping problem, while HLPs need not really have high MD values due to masking problem. These problems occur because the HLPs affect the classical location and scatter estimates as the mean and sample covariance matrix of MD are not robust. Therefore, the estimated distance will not be reliable. The alternative to this drawback of the classical MD is to find a robust location and scatter estimates that are resistant to HLPs or outlying observations. Rousseeuw and Leroy (1987) suggested using Robust Mahalanobis Distance (RMD) as a diagnostic tool for detection of HLPs by replacing the classical mean vector, $T(X)$ and classical covariance matrix, $C(X)$ of MD_i in Eq. (1) by robust estimators such as Minimum Volume Ellipsoid (MVE) or MCD.

The robust Mahalanobis Distance (RMD) is defined as

$$RMD_i = \sqrt{(t_i - T(X))C(X)^{-1}(t_i - T(X))^T} \quad i = 1, 2, 3, \ldots, n \tag{2}$$

where $T(X)$ is robust mean vector and $C(X)$ is robust variance covariance matrix.

An observation is considered as HLPs if its corresponding (RMD value) exceeds the cutoff point, i.e., $\sqrt{\chi^2_{p,0.5}}$ as proposed by Rousseeuw and Leroy (1987). Nonetheless, this cutoff point poses several shortcomings when applied to real applications, as stated below:

- This cutoff point is constructed based on the assumption that the p-dimensional variables follow a multivariate normal distribution. However, in a real-life problem, there is no guarantee that data would come from a multivariate normal distribution.
- This cutoff point depends only on the dimension of the regressors but does not take into consideration the number of observations.

Hence, to remedy these shortcomings, Rahmatullah Imon (2005) recommended a confidence-bound type of cutoff point for the robust Mahalanobis distances as

$$Median(RMD_i) + 3MAD(RMD_i) \tag{3}$$

where

$$MAD = \frac{med\{|RMD_i - med(RMD_i)|\}}{0.6745}$$

However, the RMD approach as stated in Eq. (2) is not relevant to HDD because the covariance matrix, $C(X)$ is not invertible. In this regard, calls for new improved MCD estimator, which is a highly robust estimator of multivariate location and scatter to handle these issues have become an urgent necessity.

21.2.4 Minimum Regularized Covariance Determinant (MRCD)

There is a primary constraint in the MCD system to be applied to HDD. For the MCD, the criterion of dimension p must satisfy $p < h$ for any h-subset. This requirement must be fulfilled in order to obtain a non-singular covariance matrix. An improvement to the MCD is of high importance to make it work for HDD. Boudt et al. (2018) developed a new modification of the MCD, the so-called MRCD. The fundamental objective of the MRCD is to substitute a regularized covariance estimate for the MCD subset-based covariance. H-subset of MRCD that minimizes the determinant of regularized covariance of MRCD, $K(H)$ is as shown below:

$$H_{mrcd} = \arg \min_{H \in \mathrm{H}_h} (\det K(H))^{1/p} \tag{4}$$

where H_h is set of all H subsets. $K(H)$ represents a regularized covariance matrix in MRCD and can be written as

$$K(H) = \rho T + (1 - \rho) c_\alpha S(H) \tag{5}$$

where $S(H) = h^{-1}(X_H - m(H))^T(X_H - m(H))$ is a sample covariance estimate based on subset H with $h \times p$ submatrix of X and can be denoted as X_H. The mean estimate subset H is $m(H) = h^{-1}X_H^T$ with $h = \frac{n}{2}$ where h is a number of observations of subset H.

T is a predetermined, symmetric, and positive definite target matrix. In other words, we assume $T = I_p$ where I_p is an identity matrix with dimension $p \times p$. The identity matrix is used as a target matrix because it has very good statistical properties and is well-conditioned in high dimensional datasets. $\rho = (0, 1]$ is a regularization intensity parameter and the consistency factor, c_α is to obtain the consistency at the normal multivariate distribution and unbiased at small sample. The consistency factor used in the MRCD method is the same as the one used in the MCD procedure. The value of ρ is set such that $K(H)$ is well-conditioned such that $\frac{\lambda_{\max}}{\lambda_{\max}} \leq 1000$, where $\lambda_{\max}$ and $\lambda_{\min}$ are the largest and the smallest eigenvalues of $c_\alpha S(H)$, respectively. The eigenvalue of MRCD covariance is equals to $\rho + (1 - \rho)\lambda$, and the regularization is used when needed. Then C-steps are applied until the estimated MRCD covariance converges. The C-step of Boudt et al. (2018) of MCD has been generalized to regularized covariance matrices in the MRCD method. In MCD, The C-step theorem computes the mean and the covariance matrix of h-subset H_1, and then puts the observations with smallest Mahalanobis distance in a new subset H_2. The C-step theorem proves that the covariance determinant of H_2 is less than or equal to H_1. The generalized C-step theorem discussed at length in this section.

The generalized C-steps theorem is summarized as follows:

Starting from an h-subset H_1 compute

$$m_1 = \frac{1}{h}\sum_{i \in H_1} x_i \quad \text{and} \quad S_1 = \frac{1}{h}\sum_{i \in H_1} (x_i - m_i)^T (x_i - m_i).$$

The matrix $K_1 = \rho T + (1 - \rho)c_\alpha S_1$ is positive definite, hence invertible. Then compute

$$d_1(i) = (x_i = m_1)K_1^{-1}(x_i - m_1)^T \quad i = 1, \ldots, n.$$

Let H_2 be an h-subset for which

$$\sum_{i \in H_2} d_1(i) \le \sum_{i \in H_1} d_1(i)$$

and compute

$$m_2 = \frac{1}{h}\sum\nolimits_{i \in H_2} x_i, S_2 = \frac{1}{h}\sum\nolimits_{i \in H_2} (x_i - m_2)^T (x_i - m_2) \text{ and } K_2 = \rho T + (1 - \rho)S_2.$$

The $\det(K_2) \le \det(K_1)$.

With equality if and only if $m_1 = m_2$ and $K_1 = K_2$. The iteration process in the C-steps theorem stop if $\det(K_2) = 0$ or $\det(K_2) = \det(K_1)$; otherwise the iteration process continues until it converges.

By using this generalized C-steps, the MRCD regularized covariance matrix is constructed to find the MRCD subset.

Below is the algorithm for the MRCD:

Step 1: Compute the median of each variable and put them in location vector, V_x.

Step 2: Compute the scale of each variable by using Q_n estimator and put the scale in the diagonal matrix, D_x. Rosseeuw and Croux (1993) defined Q_n as follows:

$$Q_n = c\{|x_i - x_j|;\ 1 \le i < j \le n\}_{(k)}, \quad c = 2.2219 \tag{6}$$

where $k = \binom{h}{2} \approx \frac{\binom{n}{2}}{4}$ and $h = \left[\frac{n}{2}\right] + 1$. We take the kth order statistic of the $\binom{n}{2} = \frac{n(n+1)}{2}$ pairwise distances $|x_i - x_j|$. The constant 2.2219 ensures that the Q_n estimator is asymptotically consistent at normal distribution.

Step 3: Standardize each of the observations, $u_i = D_x^{-1}(x_i - v_x)$

Step 4: Perform Singular Value Decomposition (SVD) of the target matrix, $T = Q\Lambda Q^T$. T is an identity matrix, where Λ is the diagonal matrix holding the eigenvalues of T, and Q is the orthogonal matrix containing the eigenvectors of the corresponding eigenvalues of T.

Step 5: Then compute the transformed standardized observation, $w_i = \Lambda^{-\frac{1}{2}} Q^T u_i = \frac{Q^T u_i}{\sqrt{\Lambda}}$

Step 6: Compute 6 initial location estimates, m^j and scatter estimates, s^j of W by following Hubert et al. (2012).

(a) The hyperbolic tangent of standardized data
(b) Spearman Correlation matrix
(c) Tukey normal scores
(d) The spatial sign covariance matrix
(e) BACON
(f) Raw OGK estimates of scatter.

Step 7: Determine the subset of H_0^j of W with the lowest Mahalanobis Distance in terms of m^j and s^j where m^j is the location estimate, and s^j is the scatter estimate of W in Step 6 for $j = 1, 2, 3, 4, 5, 6$ and $i = 1, 2, 3, \ldots, n$

$$d_i = \sqrt{(w_i - m^j)(s^j)^{\ 1}(w_i - m^j)^T} \tag{7}$$

Step 8: Calibrate intensity parameter by setting the value of ρ at the lowest non-negative value for which $\rho^j 1 + (1 - \rho^j)C_\alpha S(H_0^j)$ is well-conditioned. For each subset H_0^j, determine the smallest value of ρ such that $0 \leq \rho \leq 1$. Denote the value as ρ_0^j. Intensity parameter ρ is a scalar weight of the regularized covariance matrix of h-subsets H_0^j of the transformed standardized data, W. Then the smallest value of $0 \leq \rho^j \leq 1$ is determined for which $\rho^j I + (1 - \rho^j)C_\alpha S(H_0^j)$ is well-conditioned. c_α is a consistency factor, and $\rho^j \in (0, 1]$.

Step 9: If $\max_j \rho_0^j \leq 0.1$, set $\rho = \max_j \rho_0^j$, or else set $\rho = \max\{0.1, \textit{meidan}\ \rho_0^j\}$.

Step 10: For the initial subset H_0^j for which $\rho_0^j \leq \rho$, repeat the generalized C-steps using $\rho^j I + (1 - \rho^j)C_\alpha S(H_0^j)$ until convergence. Denote resulting subsets as H^j.

Step 11: Choose the final subset, H_{MRCD}, that has the lowest determinant among the ones obtained from the six initial subsets.

Step 12: Determine the final MRCD location and scatter estimates from the chosen H_{MRCD} in Step 11.

MRCD location estimates, $M_{MRCD} = V_x + D_x m_w(H_{MRCD})$

MRCD scatter estimates,

$$K_{MRCD} = D_x Q \Lambda^{-\frac{1}{2}} \left[\rho I + (1-\rho) S_w (H_{MRCD})\right] \Lambda^{-\frac{1}{2}} Q^T D_x$$

where $m_w(H_{MRCD})$ is the location estimate based on subset H_{MRCD}. W consists of the transformed standardized observations in Step 5.

Then RMD can be computed based on the MRCD location and scatter estimates. According to Boudt et al. (2018), data points having RMD-MRCD values larger than $\sqrt{\chi^2_{p,0.99}}$ are considered as HLPs.

21.2.5 Principal Component Analysis (PCA)

Principal Component Analysis (PCA) is an unsupervised dimension reduction procedure of Hubert et al. (2005), and it is commonly used in multiple linear regression analyses. It was first established by Hotelling (1933) before it became a prevalent method in many fields such as chemometrics, engineering, computer vision, face recognition, and other domains such as in gene selection and tumor classification as illustrated by Feng et al. (2019). PCA analysis aims to find a small number of linear combinations of the predictors that can be used to summarize data without losing too much information. This statistical method transforms a broad set of correlated variables into a smaller number of uncorrelated factors called principal components. These orthogonal principal components solve the multi-collinearity problem.

21.2.5.1 *Data Matrix*

Let X be $n \times p$ matrix where n is a number of observations, and p is a set of predictive variables $x_1, x_2, \ldots, x_p$. The set of X represent a random observation of $x_1, x_2 \ldots, x_p$. The issue is selecting a subset of the predictive variables which contains most information of the dataset.

21.2.5.2 *Variance-Covariance Matrix*

The PCA procedure involves the covariance structure of data to obtain the values of eigenvalue and eigenvectors. Let σ_{ij} denotes the covariance between X_i and X_j in the data matrix. The matrix σ_{ij} is written as Σ and known as a variance-covariance matrix. The values of the diagonal element in matrix Σ are the variances of X_i. This covariance matrix Σ must be a symmetric and a square matrix.

21.2.5.3 *Linear Combination*

The principal components (PCs) of PCA are the linear combination of the original variables, $x_{i1}, x_{i2}, \ldots, x_{ip}$, that can be expressed as $\sum a_i x_i$, where a_i

is the scalar. A principal component is a normalized linear combination of the original predictors in a dataset if $\sum|a_i| = 1$.

21.2.5.4 Linear Independence

The PCA method successfully solves the problem of correlated variables by choosing a set of orthogonal PCs. This is a potent mathematical property. All the principal components are independent of each other. The first principal component, PC1, is uncorrelated to PC2, and PC2 and PC3 are uncorrelated and so on. It can be proven by computing the product of $b_i^T b_j = 0$, where $i, j = 1, 2, \ldots, p$. The linear combination of variables for PCs is formulated as $t_i = x_{i1}b_1 + x_{i2}b_2 + \cdots + x_{ip}b_p$, where $i = 1, 2, \ldots, n$, and $b = b_1, b_2 \ldots, b_p$ are the loading vectors, and p is the number of predictor variables.

21.2.6 Robust Principal Component Analysis (ROBPCA)

Hubert et al. (2005) developed a new approach of robust PCA known as ROBPCA. The procedure is a combination of projection pursuit and robust covariance estimation. Projection pursuit is used to reduce the dimension of HDD into the low dimensional dataset. Robust covariance estimator based on MCD is then applied to this low dimensional data set. There are two distances used in the ROBPCA approach to determine outliers in PCA: robust score distance (SD) and orthogonal distance (OD). The score distance (SD) measures the Principal Component (PC) score distance within the PCA subspace. The distance can be defined as

$$SD_i = \sqrt{\sum_{j=1}^{k} \frac{t_{ij}^2}{l_j}}, \tag{8}$$

with j as the number of principal components, and a maximum number of PCs is k; t_{ij} is the PC score; and l_j is the robust eigenvalues corresponding to the PCs.

The orthogonal distance (OD) measures the distance between an observation, x_i and its projection, $\hat{x}_i = \hat{\mu} + p_{p,k}t_i$ in the k-dimensional PCA subspace. The OD can be written as

$$OD_i = \| x_i - \hat{\mu} - P_{p,k}t_i \| \tag{9}$$

The cutoff point for SD is $\sqrt{\chi^2_{k,0.975}}$ with the assumption that the PC scores are normally distributed, and the cutoff point for OD is $(\hat{\mu}_{mcd} + \hat{\sigma}_{mcd}\, z_{0.975})^{3/2}$, where $z_{0.975}$ is the 97.5% quantile of the Gaussian distribution.

The PCA has a broad range of applications across genetics, facial recognition, and classification areas (Jolliffe 2002). Although PCA has been widely used in many applications, this classical approach is not robust. As such, many robust and sparse PCA methods have then been developed in the literature to overcome the shortcomings of the classical PCA. The robust and sparse PCA has produced great results in many areas, such as in gene selection and tumor classification as shown by Feng et al. (2019).

21.3 ROBUST MAHALANOBIS DISTANCE BASED ON THE COMBINED METHOD OF MINIMUM REGULARIZED COVARIANCE DETERMINANT AND PRINCIPAL COMPONENT ANALYSIS (MRCD-PCA)

Boudt et al. (2018) proved that the MRCD successfully provides the final robust estimator of location and scatter that can be used in the computation of RMD for HDD. However, through our investigation, for high dimensional sparse data, its performance deteriorates as the number of p increases. The shortcoming of the RMD-MRCD method has motivated us to improvise the RMD-MRCD technique by incorporating the PCA method in the algorithm of the MRCD. The proposed method has shown to reduce the dataset by using the PCA at the outset. Subsequently, the fitted $\widehat{X}$ was obtained. Since the fitted $\widehat{X}$ matrix is transformed based on the chosen number of principal components such that its cumulative variance is at least 80%, this results in fitted $\widehat{X}$ matrix with reduced noise (Varmuza and Filzmoser 2008). Hence, the RMD-MRCD-PCA involves two steps. In the first step, the PCA reduces the dimension of dataset and generates a fitted $\widehat{X}$ matrix in the original dimension p, and in the second step, the fitted $\widehat{X}$ matrix will be shrunk and to yield an invertible covariance matrix for HDD. The MRCD was then performed on these fitted $\widehat{X}$ to determine the robust mean and robust covariance of HDD. By using these robust estimators, the distance of each observation was computed by employing the RMD based on MRCD-PCA:

$$RMD_i(MRCD-PCA)=\sqrt{(\hat{x}_i-\widehat{\mu}_{mrcd-pca})\widehat{\Sigma}^{-1}_{mrcd-pca}(\hat{x}_i-\widehat{\mu}_{mrcd-pca})^T} \tag{10}$$

where the $\widehat{\mu}_{mrcd-pca}$ and $\widehat{\Sigma}^{-1}_{mrcd-pca}$ are the robust mean and location estimator of MRCD-PCA, respectively. Finally, the robust cutoff point was used to verify the correct HLPs.

The MRCD-PCA technique can be summarized as follows:

Step 1: For each observation x_{ij}, compute the centered data matrix by subtracting the median of each column x_j.

$$x_{ij} - median\,(x_j) \tag{11}$$

Median is used instead of sample mean because median is not easily affected by outliers.

Step 2: In the second step, the centered data matrix will be transformed and reduced by using the PCA method. The number of principal components k is chosen based on the cumulative variance, which is the first k loadings $\geq 80\%$ (Rousseeuw and Croux 1993). The $n \times k$ matrix of PCA projections (scores) can be written as $Z = XV$ where V is the $p \times k$ matrix (eigenvector matrix).

Step 3: We reconstruct the original data based on these k principal components, and we map them back to p dimensions.

$$\widehat{X} = ZV^T \tag{12}$$

Step 4: Then the minimum regularized covariance determinant is performed on the fitted data, $\widehat{X}$, to determine the robust mean and robust covariance estimator for HDD.

Step 5: Finally, based on the estimated mean and the estimated covariance matrix of MRCD-PCA, the RMD for each observation is computed as defined in Eq. (10).

Step 6: Any observations that exceed the cutoff point are declared as HLPs. Since the distribution of the RMD-MRCD-PCA is intractable, as per Hadi (1992), the following cutoff point is suggested:

$$median(RMD_{mrcd\,-\,pca}) + 3MAD(RMD_{mrcd\,-\,pca}), \tag{13}$$

where $MAD(RMD_{mrcd-pca})$ is a median absolute deviation of $RMD_{mrcd-pca}$. It is defined as

$$MAD(RMD_{mrcd-pca}) = \frac{median\left|RMD_{mrcd-pca_i} - median\,(RMD_{mrcd-pca})\right|}{0.6745}$$

for $i = 1, 2, 3, ..., n$.

21.4 MONTE CARLO SIMULATION STUDY

The simulation study is conducted similar to that of Boudt et al.'s (2018) simulation design. Two different sample sizes of $n = 50$ and $n = 100$ with four different dimensions of dataset, $p = 200, 300, 700,$

and 1000 throughout the 500 simulations are generated. Three principal components were employed as in Hubert et al. (2005). Since the MRCD estimators are location and scale equivariant, following Agostinelli et al. (2015), we assume, without loss of generality, the mean $\mu = 0$, and the variances in diagonal elements of Σ are all equal to 1 where Σ is a correlation matrix. The behavior of the RMD-MRCD-PCA method relies on the covariance structure to identify HLPs. Hence, to account for the lack of affine equivariance of the RMD-MRCD-PCA estimator, the method by Agostinelli et al. (2015) is adapted by generating a correlation matrix randomly with condition number fix at 100 for each replication of the simulated data. Therefore, the performance of the estimator was not bound to a particular choice of a correlation matrix. A clean observation was generated from $x_i \sim N_p(0, I)$ for $i = 1, 2, 3, \ldots, n - m$. To contaminate the dataset with HLPs, similar data as in Maronna and Zamar's (2002) is generated. The smallest eigenvalue along the eigenvector direction of Σ and denoted it as a_0 is determined. This is the direction where the contamination is the hardest to detect. For the contamination model is generated as $x_i \sim N_p(y_0, \delta^2 I)$ for $i > n - m$, where $y_0 = ka_0$ and $\delta = 0.1$. Following Boudt et al. (2018), the distance between the outliers and clean data is set as $k = 50$. To identify the HLPs, four different contamination levels are considered, at 4%, 10%, 20%, and 30%. Boudt et al. (2018), on the other hand, in their simulation study, considered only a fraction of 20% and 40% of contamination and presented the results in terms of the mean squared error (MSE) of the scatter estimates to show that their scatter estimates are more efficient than the Orthogonalized Gnanadesikan-Kettenring (OGK) method. Their work did not place an emphasis on the identification of HLPs. Since the main aim of our study is to illustrate the merit of our proposed RMD-MRCD-PCA in detecting HLPs for huge $p \gg 700$, the percentage of correct detection of HLPs, masking and swamping effects will be presented and compared with the MRCD and ROBPCA methods.

The results of the percentage of correct identification of HLPs, masking and swamping are exhibited in Tables 21.1 and 21.2. The results clearly show that our newly proposed RMD-MRCD-PCA method has correctly identified all the HLPs with no swamping effect and with a very small percentage of masking, irrespective of the level of contamination and different dimensions. It can be observed that the performance of the RMD-MRCD and the RMD-MRCD-PCA is fairly close to each other for $p = 200$, in which both methods can successfully detect all the HLPs with no masking and swamping effect. Nonetheless, as the number of p increases, the performance of the RMD-MRCD worsens. The results in Tables 21.1 and 21.2 clearly show the poor performance of the RMD-MRCD method. For example, for $n = 50$, $p = 300$ and 10% HLPs, the RMD-MRCD can only detect 42.08% outliers with a high percentage of

Table 21.1 Percentage of correct detection of HLPs, masking and swamping by RMD-MRCD, ROBPCA and RMD-MRCD-PCA, n = 50

				4% contamination (2 HLPs)					
	% of correct detection			**% of masking**			**% of swamping**		
p	RMD-MRCD	ROBPCA	RMD-MRCD-PCA	RMD-MRCD	ROBPCA	RMD-MRCD-PCA	RMD-MRCD	ROBPCA	RMD-MRCD-PCA
200	100	100	100	0	0	0	0	0.46	0
300	100	100	100	0	0	0	0	0.46	0
700	0	100	100	100	0	0	0	0.47	0
1000	0	100	100	100	0	0	0	0.49	0
				10% contamination (5 HLPs)					
	% of correct detection			**% of masking**			**% of swamping**		
p	RMD-MRCD	ROBPCA	RMD-MRCD-PCA	RMD-MRCD	ROBPCA	RMD-MRCD-PCA	RMD-MRCD	ROBPCA	RMD-MRCD-PCA
200	100	100	100	0	0	0	0	6.62	0
300	42.08	100	100	57.92	0	0	0	7.00	0.4
700	0.16	100	100	99.84	0	0	0	7.77	0.4
1000	0	100	100	100	0	0	0	8.06	0.63
				20% contamination (10 HLPs)					
	% of correct detection			**% of masking**			**% of swamping**		
p	RMD-MRCD	ROBPCA	RMD-MRCD-PCA	RMD-MRCD	ROBPCA	RMD-MRCD-PCA	RMD-MRCD	ROBPCA	RMD-MRCD-PCA
200	100	100	100	0	0	0	0	3.36	0
300	39.94	100	100	60.06	0	0	0	4.68	0.14
700	0	100	100	100	0	0.4	0	6.75	0.15
1000	0	100	100	100	0	0.62	0	7.90	0.15
				30% contamination (15 HLPs)					
	% of correct detection			**% of masking**			**% of swamping**		
p	RMD-MRCD	ROBPCA	RMD-MRCD-PCA	RMD-MRCD	ROBPCA	RMD-MRCD-PCA	RMD-MRCD	ROBPCA	RMD-MRCD-PCA
200	99.96	100	100	0.04	0	0	0	0.12	0.064
300	37.6	100	100	62.4	0	0	0	0.05	0.036
700	0	100	100	100	0	0	0	0.24	0.06
1000	0	100	100	100	0	1.6	0	0.08	0.012

masking. The percentage of correct detection of HLPs becomes smaller as the number of HLPs increases. The RMD-MRCD method utterly fails to detect HLPs when the dimensions of p become huge, specifically when p is 700 and 1000. For instance, for n = 50, p = 700 and at 4% HLPs, the RMD-MRCD fully suffers (100%) from masking effect. The RMD-MRCD-PCA becomes sturdy for both sample sizes, and as we can see, the percentage of swamping is getting smaller with the increased number of p. The ROBPCA method, in contrast, manages to correctly

Table 21.2 Percentage of correct detection of HLPs, masking and swamping by RMD-MRCD, ROBPCA and RMD-MRCD-PCA, $n = 100$

	4% contamination (2 HLPs)								
	% of correct detection			**% of masking**			**% of swamping**		
p	RMD-MRCD	ROBPCA	RMD-MRCD-PCA	RMD-MRCD	ROBPCA	RMD-MRCD-PCA	RMD-MRCD	ROBPCA	RMD-MRCD-PCA
200	100	100	100	0	0	0	0	0.27	0.016
300	100	100	100	0	0	0	0	0.28	0.028
700	0	100	100	100	0	0	0	0.31	0.04
1000	0	100	100	100	0	0	0	0.32	0.02
	10% contamination (5 HLPs)								
	% of correct detection			**% of masking**			**% of swamping**		
p	RMD-MRCD	ROBPCA	RMD-MRCD-PCA	RMD-MRCD	ROBPCA	RMD-MRCD-PCA	RMD-MRCD	ROBPCA	RMD-MRCD-PCA
200	97.43	100	100	2.57	0	0	0	4.97	0.16
300	99.16	100	100	0.84	0	0	0	5.68	0.106
700	2.9	100	100	97.1	0	0	0	6.94	0.084
1000	0	100	100	100	0	0.4	0	2.85	0.002
	20% contamination (10 HLPs)								
	% of correct detection			**% of masking**			**% of swamping**		
p	RMD-MRCD	ROBPCA	RMD-MRCD-PCA	RMD-MRCD	ROBPCA	RMD-MRCD-PCA	RMD-MRCD	ROBPCA	RMD-MRCD-PCA
200	97.51	100	100	2.49	0	0	0	2.38	0.018
300	95	100	100	4.9	0	0	0	3.02	0.01
700	12.03	100	100	89.79	0	0	0	5.06	0.038
1000	0	100	100	100	0	0.72	0	13.52	0.056
	30% contamination (15 HLPs)								
	% of correct detection			**% of masking**			**% of swamping**		
p	RMD-MRCD	ROBPCA	RMD-MRCD-PCA	RMD-MRCD	ROBPCA	RMD-MRCD-PCA	RMD-MRCD	ROBPCA	RMD-MRCD-PCA
200	99.6	100	100	0.4	0	0	0	0.02	0.006
300	88.53	100	100	11.47	0	0	0	0.02	0.01
700	4.766	100	100	95.23	0	0	0	0.02	0.004
1000	0	100	100	100	0	0.4	0	0.04	0.002

detect all the HLPs; however, the number of swamping of outliers is relatively high. As can be seen in Tables 21.1 and 21.2, the percentage of swamping of ROBPCA is higher than the RMD-MRCD and our proposed RMD-MRCD-PCA methods for 4%, 10%, and 20% of contaminations. Nonetheless, the percentage of swamping for the ROBPCA method can be considered small for 30% contamination. The ROBPCA method suffers from a swamping problem when the percentage of HLPs is less than 30%. On the other hand, the performance of the ROBPCA is fairly close to the RMD-MRCD-PCA when the number of HLPs in the dataset is more than 30%. However, its percentage of swamping is slightly more than that when using the RMD-MRCD-PCA method.

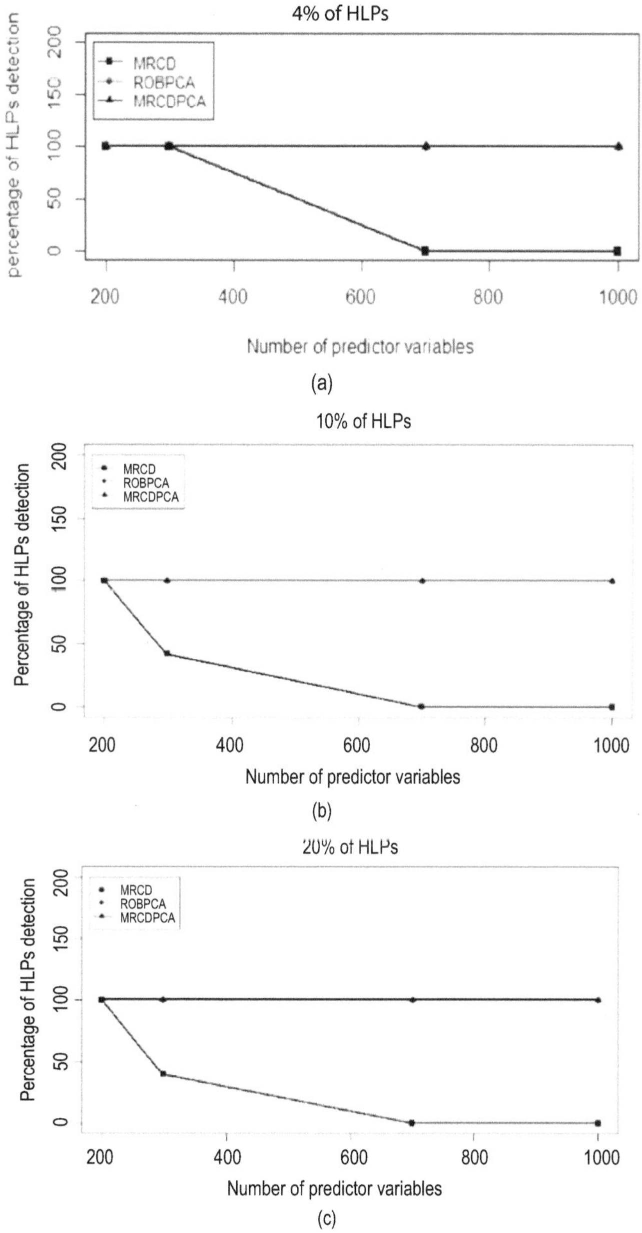

Figure 21.1 *(Contd.)*

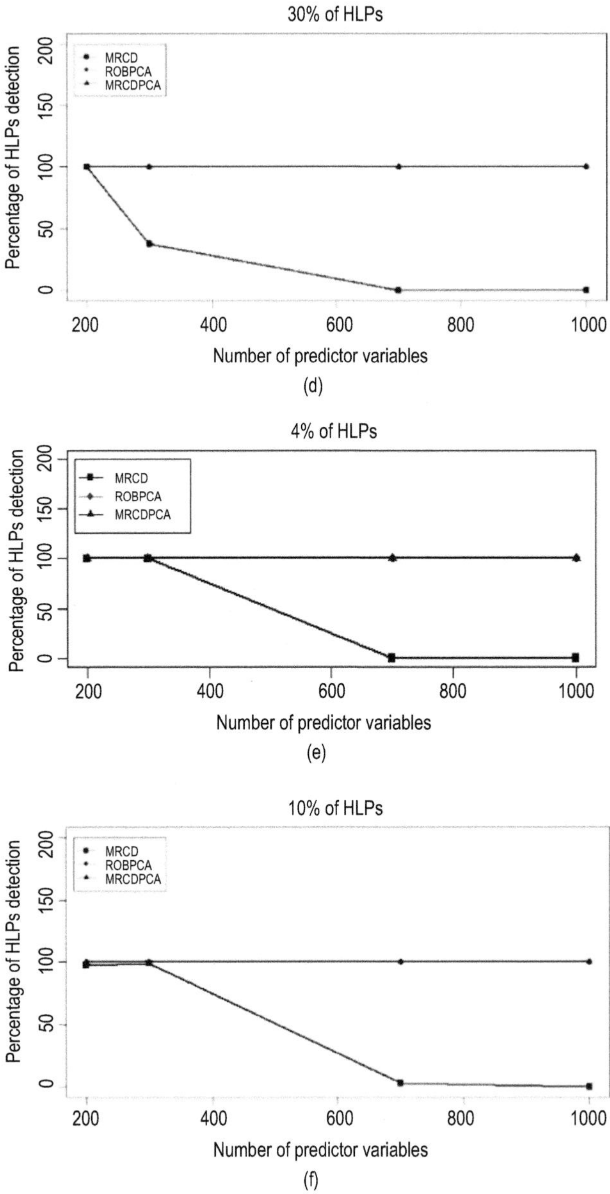

Figure 21.1 *(Contd.)*

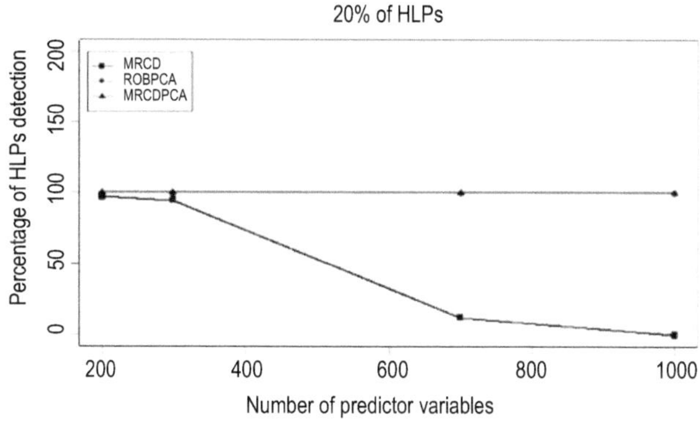

(g)

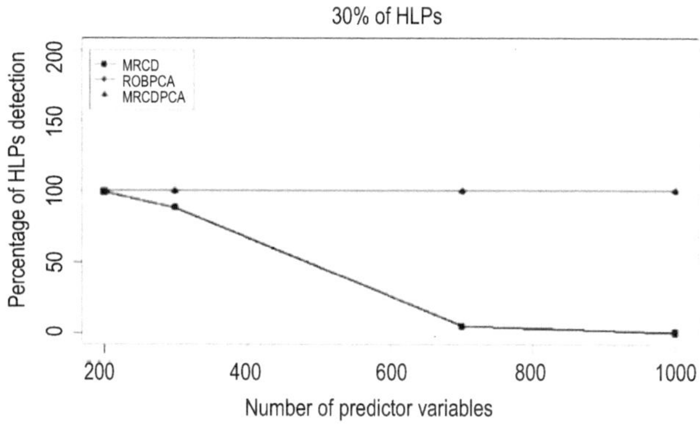

(h)

Figure 21.1 $n = 50$ for graph (a)–(d) and $n = 100$ for graph (e)–(h).

Figure 21.1 displays a clear picture of the three methods in the detection of HLPs. We can observe from Figure 21.1 that the plot for the RMD-MRCD-PCA is very stable for all the percentages of contamination and all sizes of n. Similarly, the ROBPCA can detect 100% of HLPs for all cases. As a result, the line of the ROBPCA plot overlaps with the RMD-MRCD-PCA plot. Nonetheless, as already discussed, the ROBPCA has a high swamping effect for contaminations less than 30%, which is not desirable. On the other hand, the RMD-MRCD plot decreases as the number of p increases, which indicates that the RMD-MRCD method fails to detect HLPs for p greater than 700. It is very interesting to note that for $p > 700$, the RMD-MRCD-PCA is still able to successfully detect the HLPs.

21.5 NUMERICAL EXAMPLE

In this section, three real-world experimental data are used to demonstrate the merit of our proposed technique compared to other methods.

21.5.1 Octane Dataset

Our first example is an octane dataset. This dataset was used by Hubert et al. (2005) and Boudt et al. (2018). It consists of near-infrared (NIR) absorbance spectra with $p = 226$ wavelengths and $n = 39$ gasoline samples. Hubert et al. (2005) declared that there were six HLPs in this dataset, i.e., observations 25, 26, 36, 37, 38, and 39, which contained added ethanol. The RMD-MRCD-PCA was applied to Octane data, and the performance was investigated. The PCA technique reduced the dimension of the Octane dataset into two principal components, which are the cumulative variance and the Octane dataset as 98%. Based on these 2 PCs, the dataset will be reconstructed to map back to its original p dimension, $p = 226$. From this fitted dataset, the MRCD procedure was performed to determine its robust location and robust scatter. Next, the RMD was computed to identify those high leverage points. It is interesting to note that our method had successfully spotted the six samples with added alcohol in observations 25, 36, 37, 39, 38, and 26. Similar results were also obtained using the ROBPCA and the RMD-MRCD methods as these methods were also able to spot the same six observations. The six detected observations can be seen clearly from the RMD-MRCD index plot in Figure 21.2, ROBPCA index plot in Figure 21.3, and the RMD-MRCD-PCA index plot in Figure 21.4. These results suggest that the RMD-MRCD and the ROBPCA are able to detect the same number of HLPs as detected by the RMD-MRCD-PCA. It is very interesting to note that the results of the RMD-MRCD and the RMD-MRCD-PCA methods for both real datasets are consistent with the results of the simulation study, where for p reasonably close to 200, (for octane data, p is 226), both methods have correctly identified the genuine HLPs.

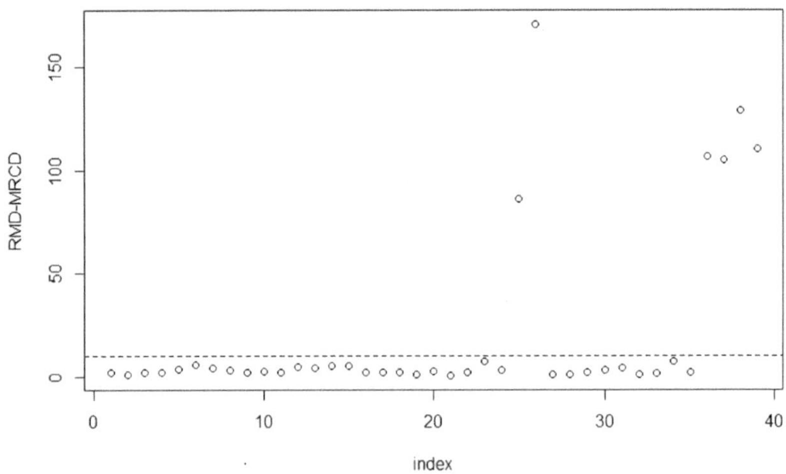

Figure 21.2 Index plot of RMD-MRCD for Octane.

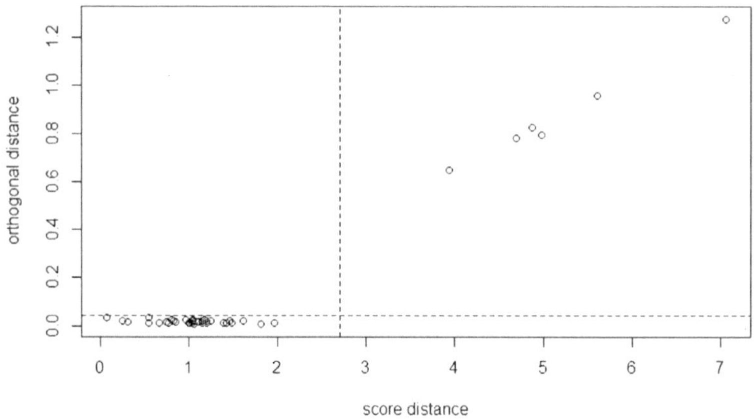

Figure 21.3 Index plot of ROBPCA for octane.

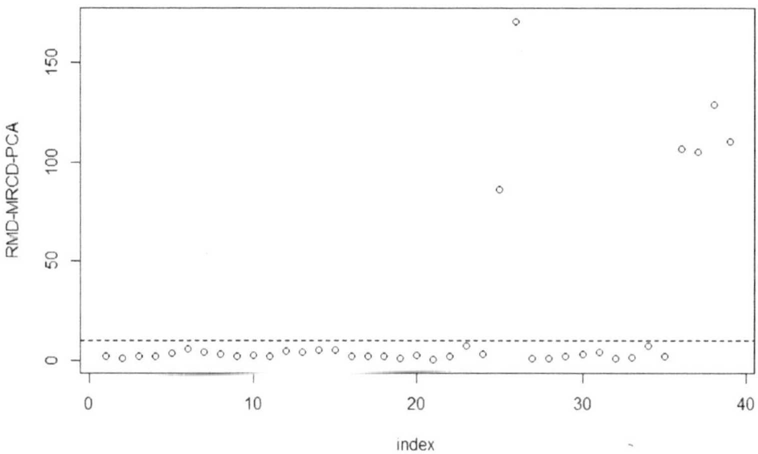

Figure 21.4 Index plot of RMD-MRCD PCA for octane.

The RMD-MRCD-PCA for Octane data can be applied within R-program as follows:

The Octane dataset is available in R package.

library (robustbase)
library (rrcov)
octane = data(octane)
data = octane [,2: dim(data) [2]] # data contains only x-variables
mx = apply (data, 2, "median")
x = sapply (1: p, function(i) (data [, i]-mx[i]))

The Octane data is reduced by using PCA.

xpca = prcomp(x)
nComp = 2

Convert the Octane data into its original dimension based on 2 principal components,

```
xhat = xpca$x [,1: nComp] %*% t (xpca$rotation [,1: nComp])
x1 = sapply (1: p, function(i) (xhat [, i] +mx[i]))
```

Robust Mahalanobis distance based on MRCD is executed to compute the robust distance for each observation.

```
rd = sqrt (mahalanobis (x1, CovMrcd(x1)$center, CovMrcd(x1)$cov))
```

output:

[1,] 1.3685 [2,] 0.2799 [3,] 2.4900 [4,] 1.8555 [5,] 5.0857 [6,] 6.5955 [7,] 3.1839 [8,] 2.2027 [9,] 2.2029 [10,] 4.3664 [11,] 1.6136 [12,] 5.6570 [13,] 2.8585 [14,] 5.9952 [15,] 5.9480 [16,] 1.4820 [17,] 1.9494 [18,] 2.7335 [19,] 0.9223 [20,] 3.8804 [21,] 0.5088 [22,] 4.0336 [23,] 7.9197 [24,] 1.9179 [25,] 65.0111 [26,] 128.5724 [27,] 1.3015 [28,] 1.4211 [29,] 3.8964 [30,] 4.5721 [31,] 3.0178 [32,] 2.3283 [33,] 1.8909 [34,] 7.8747 [35,] 2.4099 [36,] 80.8705 [37,] 79.2379 [38,] 97.1860 [39,] 83.3976

Command for the robust cutoff point is presented as follows:

```
mad.rd = median (abs (rd-median (rd)))/0.6745
cutoff = median (rd) + 3*mad.rd
```

output:

cutoff = 9.931218

Detection of high leverage points,

```
rd > cutoff
d = 0
for (i in 1: n) {
   if (rd[i] > cutoff)
      d = c (d, i)
}
d = d [-1]
list (rd = d, cutoff = cutoff, rd = rd)
```

output:

observations that exceed cut-off points are observations 25, 26, 36, 37, 38, and 39.

The graph plot,

```
plot (rd, xlab = "index", ylab = "RMD-MRCD-PCA")
abline (h = cutoff, lty = 2)
```

The presented command can be applied to Biscuit dough and Glass spectra dataset.

21.5.2 Biscuit Dough Dataset

The second real example in this chapter is the Biscuit dough dataset. The data consist of 40 samples and 700 of NIR spectra 1200 nm to 2400 nm dimension. Brown et al. (2001) and Hubert and Branden (2003) declared that there is one HLP on observation 23.

The RMD-MRCD-PCA was applied to Biscuit dough data, and the performance was investigated. The PCA technique reduced the dimension of the Biscuit data set into three principal components. Based on these 3 PCs, the dataset will be reconstructed to map back to its original p dimension, p = 700. From this fitted dataset, the MRCD procedure was performed to determine its robust location and robust scatter. Next, the RMD based on MRCD-PCA was computed to identify those HLPs.

It is interesting to note that our method had successfully spotted one HLP in observation 23. Figure 21.5 clearly indicates that our suggested technique, RMD-MRCD-PCA correctly identifies observation 23 as an outlying point, as its value exceeds the proposed robust cutoff point line. However, as shown in Figure 21.6, the RMD-MRCD has a masking problem since it does not detect any HLPs in the dataset. It is important to note that, despite the fact that there is only one outlier in the dataset, the performance of RMD-MRCD diminishes as the number of dimensions exceeds 700. Similar to this, the ROBPCA approach is also affected by the swamping effect as it detects numerous HLPs on observations 7, 8, 20, 21, 23, and 24. Figure 21.7 clearly displays that there are many observations scattered above the line of cutoff point.

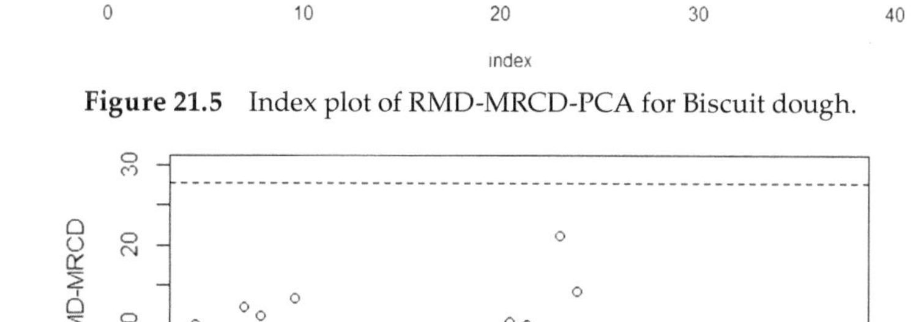

Figure 21.5 Index plot of RMD-MRCD-PCA for Biscuit dough.

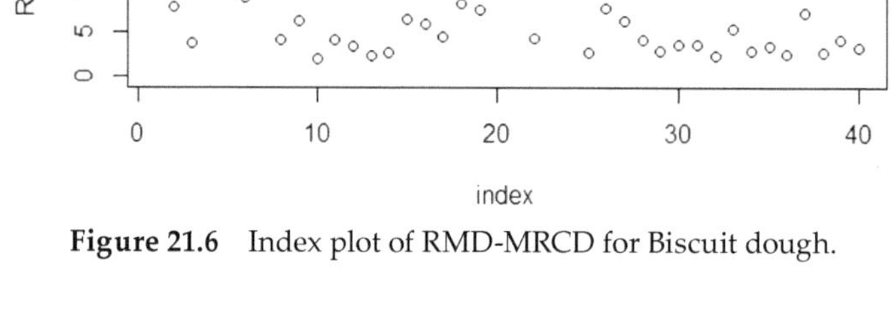

Figure 21.6 Index plot of RMD-MRCD for Biscuit dough.

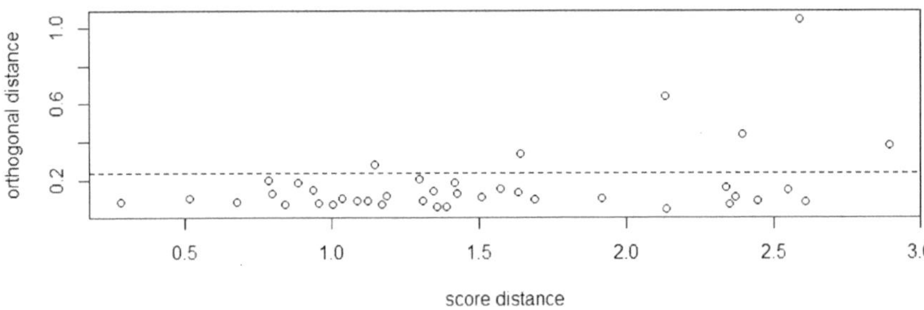

Figure 21.7 Index plot of ROBPCA for Biscuit dough.

21.5.3 Glass Spectra Dataset

Our third example involves EPXMA spectra data, where $p = 750$ wavelengths from 180 different samples of glass (Lemberge et al. 2000). A Joel JSM 6300 scanning electron microscope with an energy-dispersive Si (Li) X-ray detection (SEM-EDX) device was used to carry out the chemical analysis. This dataset has been used by Hubert et al. (2005) to classify the outliers by using the ROBPCA method. As presented in Figure 21.8, ROBPCA located three groups of observations as HLPs (cases 143–179; cases 57–63; and cases 74–76). According to Hubert et al. (2005), observation (180) is an isolated case because it is not in the two groups of detected outliers. We applied the RMD-MRCD-PCA technique to identify the HLPs in the dataset. As displayed in the index plot of RMD-MRCD-PCA in Figure 21.9, this dataset has two groups of HLPs. One major group of HLPs is from points (143–180), which are the last 38 samples in the Glass dataset, while the other group of HLPs was from points (58–63) as well as points (74 and 76). There is a minor difference in the detected high leverage points of RMD-MRCD-PCA as compared to the ROBPCA method based on the MCD. In the ROBPCA method, cases 57 and 75 are declared as HLPs by Hubert et al. (2005). However, the RMD-MRCD-PCA does not spot those two cases as HLPs. Hubert et al. (2015) stated that ROBPCA is suffering from bias when there is a large number of contaminated data, and it cannot handle data with a sparse structure. Since the ROBPCA method is incorporated with the MCD procedure, this MCD system is suffering from swamping problems as noted by Rahmatullah Imon (2005). For this spectra data, the RMD-MRCD completely fails to identify even a single HLP. These results seem to suggest that the RMD-MRCD is unable to identify HLPs for a sparse feature dataset with massive dimension. In fact, most of the observations for this Glass Spectra dataset are very small and close to zero, which can be classified as sparse data. The RMD-MRCD-PCA, on the other hand, is able to detect them.

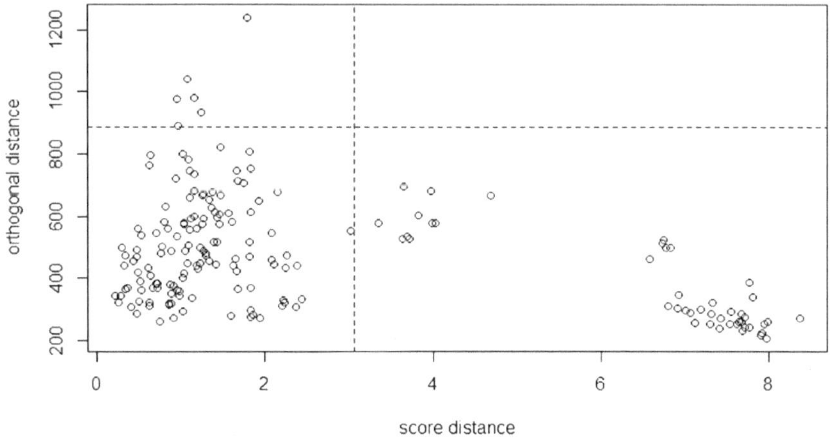

Figure 21.8 Index plot of the ROBPCA for Glass spectra.

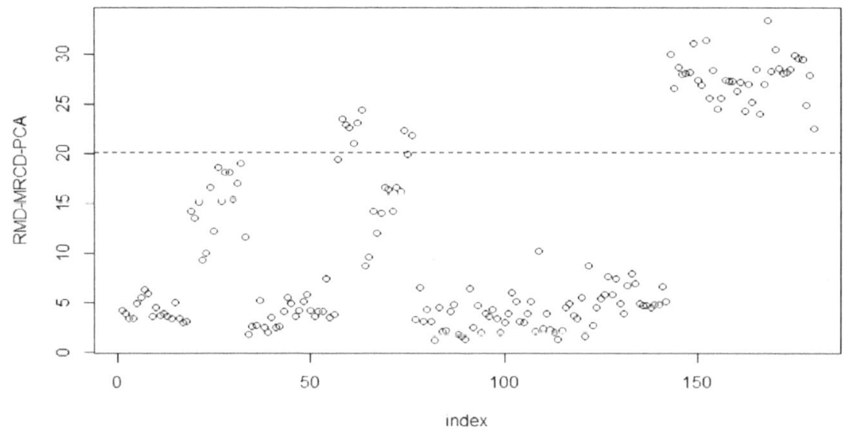

Figure 21.9 Index plot of the RMD-MRCD-PCA for Glass spectra.

21.6 CONCLUSION

The main aim of this chapter was to develop a novel diagnostic method to identify HLPs in HDD. In this study, we propose the RMD-MRCD-PCA method by integrating the PCA into the MRCD algorithm. We have made comparisons between the RMD-MRCD-PCA and the existing RMD-MRCD and ROBPCA methods. The results acquired from the simulation study and real examples suggest that for p near 200, the performance of the RMD-MRCD is relatively close to the performance of the RMD-MRCD-PCA. However, the performance of the RMD-MRCD rapidly declines as the number of independent variables increases, i.e., for p equals 700 and higher. On the other hand, the performance of the ROBPCA is poor with high swamping effect for contaminations less than 30%. Nevertheless, its performance is reasonably close to the RMD-MRCD-PCA when the number of HLPs in the dataset exceeds 30%. Our proposed RMD-MRCD-PCA method has successfully identified the HLPs with negligible swamping effects, irrespective of the number of independent variables, sample sizes, and percentage of contaminations.

REFERENCES

Agostinelli, C., Leung, A., Yohai, V.J. and Zamar, R.H. 2015. Robust estimation of multivariate location and scatter in the presence of cellwise and casewise contamination. *Test*. 24(3): 441–461. https://doi.org/10.1007/s11749-015-0450-6.

Boudt, K., Rousseeuw, P.J., Vanduffel, S. and Verdonck, T. 2018. The minimum regularized covariance determinant estimator. Statistics and Computing. https://doi.org/10.1007/s11222-019-09869-x.

Brown, P.J., Fearn, T. and Vannucci, M. 2001. Bayesian wavelet regression on curves with application to a spectroscopic calibration problem. *Journal of the American Statistical Association*. 96(454): 398–408. https://doi.org/10.1198/016214501753168118.

Croux, C. and Haesbroeck, G. 1999. Influence function and efficiency of the minimum covariance determinant scatter matrix estimator. Journal of Multivariate Analysis. 190(2): 161–190. https://doi.org/https://doi.org/10.1006/jmva.1999.1839.

Daniel Peña and Victor J. Yohai*. 1991. The detection of influential subsets in linear regression using an influence matrix. Journal of Chemical Information and Modeling. 53(9): 1689–1699. https://doi.org/10.1017/CBO9781107415324.004.

Feng, C.M., Xu, Y., Liu, J.X., Gao, Y.L. and Zheng, C.H. 2019. Supervised discriminative sparse PCA for com-characteristic gene selection and tumor classification on multiview biological data. IEEE Transactions on Neural Networks and Learning Systems. 1–12.

Habshah, M., Norazan, M.R. and Imon, A.H.M.R. 2009. The performance of diagnostic-robust generalized potentials for the identification of multiple high leverage points in linear regression. Journal of Applied Statistics. 36(5): 507–520. https://doi.org/10.1080/02664760802553463.

Hadi, A.S. 1992. Identifying Multiple Outliers in Multivariate Data. Journal of the Royal Statistical Society: Series B (Methodological). 54(3): 761–771. https://doi.org/10.1111/j.2517-6161.1992.tb01449.x.

Hubert, M. and K. Vanden, Branden. 2003. Robust methods for partial least squares regression. Journal of Chemometrics. 17(10): 537–549. https://doi.org/10.1002/cem.822.

Hubert, Mia, Rousseeuw, P.J. and K. Vanden Branden. 2005. ROBPCA: A new approach to robust principal component analysis. Technometrics. (47). https://doi.org/10.1198/004017004000000563.

Hubert, Mia, Rousseeuw, P.J. and Verdonck, T. 2012. A deterministic algorithm for robust location and scatter. Journal of Computational and Graphical Statistics. 21(3): 618–637. https://doi.org/10.1080/10618600.2012.672100.

Hubert, Mia, Reynkens, T., Schmitt, E. and Verdonck, T. 2015. Sparse PCA for high-dimensional data with outliers. Technometrics. 58(4): 424–434. https://doi.org/10.1080/00401706.2015.1093962.

Hubert, M., Debruyne, M. and Rousseeuw, P.J. 2018. Minimum covariance determinant and extensions. Wiley Interdisciplinary Reviews: Computational Statistics. 10(3): 1–11. https://doi.org/10.1002/wics.1421.

Jolliffe, I.T. 2002. Principal Component Analysis, 2nd Edn. Encyclopedia of Statistics in Behavioral Science. 30(3): 487. https://doi.org/10.2307/1270093.

Lee, S., Schowe, B. and V. Sivakumar. 2011. Feature Selection for High-Dimensional Data with RapidMiner. doi: 10.17877/DE290R-14289.

Lemberge, P., De Raedt, I., Janssens, K.H., Wei, F. and Van Espen, P.J. 2000. Quantitative analysis of 16-17th century archaeological glass vessels using PLS regression of EPXMA and μ-XRF data. *Journal of Chemometrics.* 14(5–6): 751–763. https://doi.org/10.1002/1099-128X(200009/12)14:5/6<751::AID-CEM622>3.0.CO;2-D.

Mahalanobis, P.C. 1936. On the Generalised Distance in Statistics. Sankhya A 80 (Suppl 1), 1–7 (2018). https://doi.org/10.1007/s13171-019-00164-5.

Maronna, R.A. and Zamar, R.H. 2002. Robust estimates of location and dispersion for high-dimensional datasets. Technometrics. 44(4): 307–317. https://doi.org/10.1198/004017002188618509.

Rahmatullah Imon, A.H.M. 2005. Identifying multiple influential observations in linear regression. Journal of Applied Statistics. 32(9): 929–946. https://doi.org/10.1080/02664760500163599.

Rousseeuw, P. 1985. Multivariate Estimation With High Breakdown Point. Mathematical Statistics and Applications, (June). https://doi.org/10.1007/978 91 009 5438-0.

Rousseeuw, P.J. and Leroy, A.M. 1987. Robust Regression and Outlier Detection. (Vol. 152). Bristol-Myers-Squibb B-1170 Brussels, Belgium: John Wiley & Sons,. https://doi.org/10.2307/2982847.

Rousseeuw, P.J. and van Zomeren, B.C. 1990. Unmasking multivariate outliers and leverage points. Journal of the American Statistical Association. https://doi.org/10.1080/01621459.1990.10474920.

Rousseeuw, P.J. and Croux, C. 1993. Alternatives to the median absolute deviation. Journal of the American Statistical Association. 88(424): 1273–1283. https://doi.org/10.1080/01621459.1993.10476408.

Rousseeuw, P.J. and Driessen, K. Van. 1999. A Fast Algorithm for the Minimum Covariance Determinant Estimator. Technometrics. 41.

Serfling, R. and Wang, S. 2012. General foundations for studying masking and swamping robustness of outlier identifiers. Statistical Methodology. 20 (May), 79–90. https://doi.org/10.1016/j.stamet.2013.08.004.

Varmuza, K. and Filzmoser, P. 2008. Introduction to Multivariate Statistical Analysis in Chemometrics. Applied Spectroscopy (Vol. 64). CRC Press. https://doi.org/10.1201/9781420059496.

Chapter **22**

Design of Control Chart Using Gini-Process Capability Index and Resampling

Gadde Srinivasa Rao*[1], Liaquat Ahmad[2], Muhammad Azam[3], Ishmal Shahzadi[4] and Muhammad Aslam[5]

[1*]Department of Mathematics and Statistics, The University of Dodoma, Dodoma, P.O. Box: 259, Tanzania. Email: gaddesrao@gmail.com

[2,3]Department of Statistics, University of Veterinary & Animal Sciences, Lahore, Pakistan. Email: liaquatahmad@uvas.edu.pk; mazam@uvas.edu.pk

[4]Department of Statistics, Forman Christian College (A Chartered University) Lahore, 54600, Pakistan. Email: ishahzadiaslam@gmail.com

[5]Department of Statistics, Faculty of Science, King Abdulaziz University, Jeddah 21551, Saudi Arabia. Email: aslam_ravian@hotmail.com

22.1 INTRODUCTION

To monitor and mend the production process, the control chart technique is the main principle in the Statistical Process Control (SPC). The common practise in most of the industries is to maximize the profit; the main objective is to detect an uncommon cause or causes early during the

*For Correspondence: G. Srinivasa Rao (gaddesrao@gmail.com)

process monitoring techniques to minimize the defective number of items or rework items (Montgomery 2009). There are many flourishing reports of rendering these techniques in famous organizations or industries such as General Motors, IBM, LG, Philips, Google, Facebook, and so on. Control charts play an important role in statistical monitoring authorities.

Process capability index (PCI) refers to the study of potentials and capabilities of the production process with some specific specification limits. The PCI is used to study both the average, and dispersion of the production process. The methodology and process of PCI is an important tool for examining the process capability which is easy to understand and comprehend. A process having larger values of these indices is considered as having higher process capability. The process is declared as inadequate if the index value is less than 1 and an excellent process if the value is greater than 2 (Sennaroglu and Senvar 2015). Several process capability indices are available in the literature but the most common and important ones are C_p, C_{pk}, C_{pm}, C_{pmk} which account the location and the variability persisting in the process in accordance with the specification limits (Kane 1986). Among these PCIs the C_p and C_{pk} are simply distinct traditions to estimate the permissible dispersion and the inherent variability in the process. The development of the control chart based on the PCIs can be seen in Ahmad et al. (2014a, b), Chen, Huang and Li (2001), Palmer and Tsui (1999) Pearn (1998), Sullivan (1984) and Aslam, Mohsin and Jun (2017).

Most of the industrial data are rarely realized of large sample size due to two constraints such as time and cost, hence in these situations we get incomplete or limited data that should be used by the professionals or engineers. Variability is the pivotal quantitative measure to study the PCIs. If the data follows normal distribution, we can use population standard deviation or its estimated sample standard deviation. Gini (1912) developed a Gini's mean difference (GMD) measure to obtain the standard deviation if data belongs to some non-normal. The GMD measure for the non-normal distributions has been studied by many quality control researchers. This measure is extremely useful when applied under the appropriate condition of highly skewed distributions (Gerstenberger and Vogel 2015, Riaz and Abbasi 2009, Riaz and Does 2008, Riaz and Saghirr 2007, Yitzhaki 2003). The median absolute deviation and interquartile range as important measures of variability have been suggested by Nanthakumar and Vijayalakshmi (2015) and Yitzhaki (2003).

It is vital to note that the thought of repetitive sampling is distinct from the double sampling plans. The repetitive sampling scheme in process control is repeated in anticipation of a decision that can be made on the basis of the present sample if the investigator could not reach a conclusion from the earlier samples. The principle of control charts based on repetitive sampling consists of inner control limits and outer control limits. The process is stated as in-control if a depicted statistic comes down within the inner control limits whereas it is stated out-of-control if the statistic is depicted away from the outer control limits. If the statistic is in between the inner and outer control limits, a new sample should be collected and tested. More details on repetitive sampling with respect to control charts can be seen in Sherman (1965), Balamurali and Jun (2006), Ahmad et al. (2014a, b), Huang et al. (2019).

The ARL is used to evaluate the efficiency of any chart (Montgomery 2009). The prime purpose of ARL is to avoid a huge number of defects in the production process (Cheng and Chen 2011, Garcia-Diaz and Aparisi 2005; Knoth 2007). This technique has been utilized by several researchers of quality control including (Ahmad et al. 2016a, b, Ahmad et al. 2014a, b, Khan et al. 2017, Maravelakis 2005). Thus, in this article a control chart based on repetitive group sampling for C_{pk} index for GMD measure of dispersion for Weibull distribution (WD) has been presented and assessed using ARLs. The rest of the article is ordered as: the concept of GMD calculation has been given in Section 22.2. Section 22.3 deals with the design of the proposed control chart and Section 22.4 with calculations of ARLs. In Section 22.5, practical examples have been given and explained for the proposed methodology. Closing remarks are included at the end.

22.2 GINI'S MEAN DIFFERENCE

The Gini's mean difference for a random sample of n arranged values can be calculated as

$$G_n = \frac{2}{n(n-1)} \sum_{j=1}^{n} \sum_{i=1}^{n} |x_i - x_j| \tag{1}$$

$$G_n = \frac{2}{n(n-1)} \sum_{i=1}^{n} [(x_i - x_1) + (x_i - x_2) + \cdots + (x_i - x_{i-1})] \tag{2}$$

$$G_n = \frac{2}{n(n-1)} \sum_{i=1}^{n} (2i - n - 1)x_{(i)}. \tag{3}$$

Downton (1966) developed an unbiased estimator of standard deviation under the normal distribution as

$$\sigma^* = \frac{c \sum_{i=1}^{n} (2i - n - 1)x_{(i)}}{n(n-1)} \tag{4}$$

here $c=\sqrt{\pi}=1.77245$. Finally, David (1968) suggested that $\sigma^*=0.8862*G_n$ can be used as a measure of dispersion, then GMD is explained as

$$G_n=\frac{2}{n(n-1)}\sum_{i=1}^{n}(2i-n-1)x_{(i)} \tag{5}$$

It may be written as

$$G_n=\frac{2}{n(n-1)}\sum_{i=1}^{n}((i-1)-(n-i))x_{(i)} \tag{6}$$

$$G_n=\frac{2}{n(n-1)}\left[\sum_{i=1}^{n}(i-1)x_{(i)}-\sum_{i=1}^{n}(n-i)x_{(i)}\right] \tag{7}$$

$$G_n=\frac{2}{n(n-1)}[U-V] \tag{8}$$

where $U=\sum_{i=1}^{n}(i-1)x_{(i)}$ and $V=\sum_{i=1}^{n}(n-i)x_{(i)}$. Expending the same method as compared to Nair (1936) and Lomnicki (1952), the GMD for WD (Huang and Hwang 2005) is given by

$$E(G_n)=\left(2-2^{1-\frac{1}{\beta}}\right)\frac{\Gamma\left(1+\frac{1}{\beta}\right)}{\lambda}=\sigma_{gw} \tag{9}$$

where p.d.f of WD is

$$f(x)=\lambda\beta(\lambda x)^{\beta-1}e^{-(\lambda x)^{\beta}}, \tag{10}$$

here λ and β are the scale and the shape parameters, respectively.

PCIs based on GMD

The PCI is a very commonly used tool of production process for evaluating its capability of producing its articles according to predefined specifications. Many indices have been developed for analyzing its mean value and the pattern of variability for the normal and non-normal distributions.

The C_{pk} is the commonly used PCI which is expressed as

$$C_{pk}=\text{Min}\left\{\frac{USL-\mu}{3\sigma},\frac{\mu-LSL}{3\sigma}\right\} \tag{11}$$

here USL and LSL are the upper and lower specification limits, μ is the process mean, and σ is the process standard deviation.

Clements (1989) developed C_{pk} index for the non-normal data as

$$C_{pk} = \text{Min}\left\{\frac{USL - m}{\xi_{0.99865} - m}, \frac{m - LSL}{m - \xi_{0.00135}}\right\} \tag{12}$$

where, $\xi_{0.99865}$ and $\xi_{0.00135}$ are the values of the 99.865th and 0.135th quantiles. Pearn and Chen (1997) have given the improved formula to Clements (1989) by substituting 3σ in (11) with $(\xi_{0.99865} + \xi_{0.00135})/2$ and can be defined as

$$C_{Npk} = 2\ \text{Min}\left\{\frac{USL - m}{\xi_{0.99865} - m}, \frac{m - LSL}{m - \xi_{0.00135}}\right\}. \tag{13}$$

To compute C_{pk} under GMD to measure the dispersion when the observations under study follow a Weibull distribution, the following adjustment in the above non-normal PCI_S formulas.

$$C_{Npkg} = \text{Min}\left\{\frac{USL - m}{2.6586\sigma_{gw}}, \frac{m - LSL}{2.6586\sigma_{gw}}\right\}. \tag{14}$$

For incorporating the waiting time in the above mentioned Eq. (14), the following C_{Npkg} with LSL which is equal to zero can be developed as

$$C_{Npkg} = \text{Min}\left\{\frac{USL - m}{2.6586\ \sigma_{gw}}, \frac{m}{2.6586\sigma_{gw}}\right\}. \tag{15}$$

22.3 DESIGN OF PROPOSED CONTROL CHART

Assume that X be the random variable following the Weibull distribution with pdf described in Eq. (10). Further assume that the shape parameter is β_0 and the scale parameter be denoted by λ. The statistics C_{Npkg} of the proposed GMD has been plotted on the control chart and the process is declared as out-of-control if it falls outside the upper or lower control limit. Following steps have been suggested for the construction of the proposed control chart:

Step 1: Select a random sample of size n from the manufacturing process to find the quality of the variable X.

Step 2: Determine the PCI $\widehat{C}_{Npkg}$ using the formula specified below:

$$\widehat{C}_{Npkg} = \text{Min}\left\{\frac{USL - \widehat{m}}{2.6586\sigma_{gw}}, \frac{\widehat{m}}{2.6586\sigma_{gw}}\right\} \tag{16}$$

where $\widehat{m}$ is the sample median of the sample data.

Step 3: The process is declared as out-of-control if $\widehat{C}_{Npkg} \geq UCL_1$ or $\widehat{C}_{Npkg} \leq LCL_1$ and in-control if $LCL_2 \leq \widehat{C}_{Npkg} \leq UCL_2$; otherwise, repeat the process from Step 1.

Let $\underline{C}_{Npkg}$ be the mean and S_{Npkg} be the standard deviation of the in-control process then the two pairs of the proposed chart can be written as

$$LCL_1 = \bar{C}_{Npkg} - K_1 S_{C_{Npkg}} \tag{17a}$$

$$UCL_1 = \bar{C}_{Npkg} + K_1 S_{C_{Npkg}} \tag{17b}$$

$$LCL_2 = \bar{C}_{Npkg} - K_2 S_{C_{Npkg}} \tag{17c}$$

$$UCL_2 = \bar{C}_{Npkg} + K_2 S_{C_{Npkg}} \tag{17d}$$

where k_1 and k_2 are control chart coefficients to be estimated. The mean and variance of the statistic C_{Npkg} are calculated as

$$\bar{C}_{Npkg} = \frac{1}{N}\sum_{i=1}^{N} \hat{C}_{Npkgi}$$

$$S^2_{C_{Npkg}} = \frac{1}{N}\sum_{i=1}^{N}\left(\hat{C}_{Npkgi} - \bar{C}_{Npkg}\right)^2$$

The probability of the out-of-control process can be calculated as

$$P_{out.0} = P\left(\hat{C}_{Npkg} > UCL_1 \mid \beta = \beta_0\right) + P\left(\hat{C}_{Npkg} < LCL_1 \mid \beta = \beta_0\right). \tag{18}$$

The probability that the repetition of the process ($P_{rep.0}$) when process shape parameter as in-control (i.e. $\beta = \beta_0$) is obtained as follows:

$$\begin{aligned} P_{rep.0} &= P\left(UCL_2 \le \hat{C}_{Npkg} \le UCL_1 \middle| \beta = \beta_0\right) \\ &\quad + P\left(LCL_1 \le \hat{C}_{Npkg} \le LCL_2 \middle| \beta = \beta_0\right). \end{aligned} \tag{19}$$

Thus, the probability of the out-of-control process (P^0_{out}) for the repetitive sampling is calculated as

$$P^0_{out} = \frac{P^0_{out.0}}{1 - P_{rep.0}}. \tag{20}$$

The ARL of the in-control process denoted by ARL_0 is calculated as

$$ARL_0 = \frac{1}{P^0_{out}}. \tag{21}$$

Let us assume that the shape parameter of the Weibull distribution is shifted from $\beta = \beta_0$ to $\beta = \beta_1 = c\beta_0$ while the scale parameter remains unchanged then the probability of the out-of-control process denoted by ($P_{out.1}$) is calculated as

$$P_{out.1} = P\left(\hat{C}_{Npkg} > UCL_1 \middle| \beta = \beta_1\right) + P\left(\hat{C}_{Npkg} < LCL_1 \middle| \beta = \beta_1\right). \tag{22}$$

The probability that the repetition of the process ($P_{rep.1}$) when the process shape parameter shifted can be obtained as follows:

$$P_{rep.1} = P\left(UCL_2 \leq \widehat{C}_{Npkg} \leq UCL_1 \middle| \beta = \beta_1\right) + P\left(LCL_1 \leq \widehat{C}_{Npkg} \leq LCL_2 \middle| \beta = \beta_1\right). \quad (23)$$

Thus, the probability of the process being declared to be out-of-control (P^1_{out}) for the proposed control chart based on repetitive sampling when the process is shifted can be obtained as follows:

$$P^1_{out} = \frac{P_{out.1}}{1 - P_{rep.1}}. \quad (24)$$

Therefore, the out-of-control ARL (denoted by ARL_1) is computed as follows:

$$ARL_1 = \frac{1}{P^1_{out}}. \quad (25)$$

22.4 CALCULATION OF ARL

The performance evaluation of the proposed chart is supported by using the Monte Carlo simulation. The simulation technique is used to get the values of control chart coefficients k_1 and k_2, out-of-control (ARL_1) for several shift values of the shape parameter β. If we consider the r_0 act as the pre-defined in-control ARL_0, then the value of control chart coefficients k_1 and k_2 are obtained such that $ARL_0 \geq r_0$. Using Eqs. (22) to (25) and control chart coefficients k_1 and k_2, we can obtain the out-of-control ARLs and standard deviation of run lengths (SDRLs) for several shifts' values of c. The outline of algorithm is given below:

Step 1: Generate a random sample of subgroup size n from WD with predefined in-control process parameters. Generate N such subgroups.

Step 2: Calculate mean and standard deviation of $\widehat{C}_{Npkg}$.

Step 3: Prefix in-control ARL value as r_0 (say).

Step 4: Compute chart coefficients k_1 and k_2 in such a way that $ARL_0 \geq r_0$.

Step 5: Find ARLs of out-of-control and SDRLs using control chart coefficients k_1 and k_2 for different shifts' values c.

The ARLs of the suggested chart for $\lambda = 2$, $\beta = 3.6$ and $n = 5, 7$, and 10 are studied and displayed in Tables 22.1 and 22.2 at $USL = 1.805$, where USL is the upper specification limit. Likewise, the values of ARL for $\lambda = 3$, $\beta = 4$ and $n = 5$ and 10 are given in Tables 22.3 and 22.4. The following are the interpretations noticed from Tables 22.1–22.4 for the suggested chart.

1. The ARL and SDRL values decrease as n increases from 5 to 10 with other fixed parameters.
2. The ARL and SDRLs decrease as (λ, β) increases from (2.0, 3.6) to (3.0, 4.0) when other parameters remain the same.
3. The ARL and SDRLs decrease as c increases when other parameters remain the same.

Table 22.1 The ARLs and SDRLs for the suggested chart when $\lambda = 2$, $\beta = 3.6$ and $ARL_0 = 370$

c	$n = 5$, $k_1 = 2.8015$, $k_2 = 2.4737$		$n = 7$, $k_1 = 2.6951$, $k_2 = 2.4129$		$n = 10$, $k_1 = 2.7351$, $k_2 = 2.4488$	
	ARL1	**SDRL**	**ARL1**	**SDRL**	**ARL1**	**SDRL**
1.00	370.29	369.79	370.48	369.98	370.05	369.55
1.02	343.62	343.12	338.10	337.60	328.26	327.76
1.04	320.18	319.68	295.41	294.91	294.48	293.98
1.06	292.16	291.66	267.77	267.27	266.32	265.82
1.08	260.46	259.96	250.26	249.76	248.91	246.40
1.10	231.07	230.57	227.85	226.35	225.67	225.17
1.12	205.14	204.64	202.73	202.23	201.54	201.04
1.14	190.42	189.92	188.01	187.51	186.30	185.80
1.16	168.94	168.44	167.56	167.06	163.11	162.61
1.18	149.70	149.20	147.77	147.27	146.01	145.51
1.20	137.11	136.61	136.59	136.09	134.77	134.27
1.22	124.39	123.89	121.64	121.14	120.81	120.31
1.24	107.76	107.26	106.98	106.48	104.72	104.22
1.26	95.79	95.28	93.10	92.60	88.90	88.40
1.28	82.58	82.07	80.77	80.27	79.19	78.69
1.30	76.42	75.92	74.94	74.44	73.32	72.82
1.32	66.01	65.51	64.39	63.89	63.08	62.58
1.34	55.10	54.60	54.03	53.53	52.48	51.97
1.36	50.64	50.14	48.57	48.07	45.18	44.67
1.38	43.68	43.17	43.28	42.78	38.48	37.98
1.40	37.44	36.94	35.59	35.09	51.45	50.95
1.42	34.84	34.34	42.20	41.70	35.48	34.98
1.44	30.91	30.40	27.31	26.80	25.96	25.46
1.46	27.12	26.62	25.41	24.91	23.79	23.29
1.48	23.14	22.64	22.71	22.21	21.03	20.52
1.50	20.60	20.10	20.51	20.01	19.23	18.73

Table 22.2 The ARLs and SDRLs for the suggested chart when $\lambda = 2$, $\beta = 3.6$ and $ARL_0 = 300$

c	$n = 5$		$n = 7$		$n = 10$	
	$k_1 = 2.7498$		$k_1 = 2.9645$		$k_1 = 2.9273$	
	$k_2 = 2.4776$		$k_2 = 2.5871$		$k_2 = 2.5899$	
	ARL1	**SDRL**	**ARL1**	**SDRL**	**ARL1**	**SDRL**
1.00	300.24	299.74	300.33	299.83	300.13	299.63
1.02	283.65	283.15	274.76	274.26	273.63	273.13
1.04	260.07	259.57	253.18	252.68	252.75	252.25
1.06	231.55	231.05	228.78	228.28	226.99	226.49
1.08	207.54	207.04	206.00	205.50	203.53	203.03
1.10	182.45	181.95	180.13	179.63	177.61	176.11
1.12	166.60	166.10	165.64	164.14	161.32	160.82
1.14	150.34	149.84	149.18	148.68	144.95	144.45
1.16	133.75	133.25	131.18	130.68	130.39	129.89
1.18	122.65	122.15	121.13	120.63	120.11	119.61
1.20	106.62	106.12	105.28	104.78	101.96	101.45
1.22	103.51	103.00	102.82	102.32	101.97	101.47
1.24	84.43	83.93	81.95	81.45	80.37	79.87
1.26	78.97	78.47	77.34	76.83	74.36	73.86
1.28	74.68	74.18	73.84	73.34	69.95	69.44
1.30	66.49	65.99	64.76	64.25	62.16	61.66
1.32	59.39	58.88	56.49	55.99	54.76	54.25
1.34	53.18	52.68	49.34	48.84	47.95	47.45
1.36	48.72	48.22	42.62	42.11	41.16	40.66
1.38	38.55	38.05	36.64	36.14	35.25	34.75
1.40	32.31	31.80	31.60	31.09	30.19	29.69
1.42	30.99	30.48	29.65	29.15	28.31	27.80
1.44	28.51	28.00	27.61	27.11	26.65	26.15
1.46	25.37	24.87	24.19	23.68	22.97	22.46
1.48	17.53	17.02	15.06	14.55	13.25	12.75
1.50	15.28	14.77	14.54	14.02	12.58	12.08

Table 22.3 The ARLs and SDRLs for the suggested chart when $\lambda = 3$, $\beta = 4$ and $ARL_0 = 370$

c	$n = 5$		$n = 7$		$n = 10$	
	$k_1 = 3.1603$		$k_1 = 2.8277$		$k_1 = 2.7475$	
	$k_2 = 2.6712$		$k_2 = 2.4798$		$k_2 = 2.4554$	
	ARL1	**SDRL**	**ARL1**	**SDRL**	**ARL1**	**SDRL**
1.00	370.45	369.95	370.16	369.66	370.15	369.65
1.02	337.52	337.02	333.23	332.73	325.46	324.96
1.04	297.63	297.13	294.15	293.65	288.58	288.08
1.06	283.38	282.88	280.32	279.82	277.84	277.34
1.08	266.65	266.15	257.89	257.39	251.83	251.33
1.10	238.69	238.19	233.85	233.35	230.91	230.41
1.12	219.20	218.70	217.23	216.73	208.41	207.91
1.14	201.05	200.55	196.28	195.78	187.97	187.47

Table 22.3 *(Contd.)*

Table 22.3 (*Contd.*) The ARLs and SDRLs for the suggested chart when $\lambda = 3$, $\beta = 4$ and $ARL_0 = 370$

c	$n = 5$ $k_1 = 3.1603$ $k_2 = 2.6712$ ARL1	SDRL	$n = 7$ $k_1 = 2.8277$ $k_2 = 2.4798$ ARL1	SDRL	$n = 10$ $k_1 = 2.7475$ $k_2 = 2.4554$ ARL1	SDRL
1.16	178.91	178.41	177.08	176.58	172.69	172.19
1.18	161.90	161.40	160.80	160.30	158.77	158.27
1.20	158.98	158.48	150.88	150.38	143.63	143.13
1.22	132.63	131.83	131.79	131.29	130.45	129.95
1.24	120.18	119.68	119.25	118.75	118.00	117.50
1.26	110.45	109.95	109.03	108.53	107.08	106.58
1.28	99.31	98.81	97.18	96.67	97.35	96.85
1.30	83.60	83.10	81.79	81.29	80.85	80.34
1.32	75.72	75.22	72.05	71.55	71.20	70.69
1.34	65.60	65.09	64.32	63.82	60.31	59.81
1.36	60.55	60.05	58.79	58.29	56.73	56.22
1.38	50.50	50.00	49.66	49.15	44.70	44.20
1.40	44.74	44.23	43.04	42.54	40.66	39.16
1.42	39.42	38.91	37.93	37.42	34.84	34.33
1.44	32.23	32.72	33.73	33.22	31.86	31.35
1.46	29.77	29.27	28.30	27.79	25.56	25.06
1.48	26.71	26.20	25.92	25.41	23.41	22.90
1.50	24.65	23.75	23.37	22.86	20.21	20.71

Table 22.4 The ARLs and SDRLs for the suggested chart when $\lambda = 3$, $\beta = 4$ and $ARL_0 = 300$

c	$n = 5$ $k_1 = 2.8273$ $k_2 = 2.5219$ ARL1	SDRL	$n = 7$ $k_1 = 2.9289$ $k_2 = 2.5671$ ARL1	SDRL	$n = 10$ $k_1 = 2.9098$ $k_2 = 2.5815$ ARL1	SDRL
1.00	300.26	299.76	300.43	299.93	300.22	299.72
1.02	279.37	278.87	271.27	270.77	270.87	270.37
1.04	252.84	252.34	251.01	250.51	250.55	250.05
1.06	218.98	218.47	218.62	218.12	216.65	216.15
1.08	197.37	196.87	197.28	196.78	196.68	196.18
1.10	181.23	180.72	179.87	179.37	174.78	174.28
1.12	164.90	164.40	163.39	162.89	161.68	161.18
1.14	143.36	142.86	138.57	138.07	136.28	135.77
1.16	125.74	125.24	123.38	122.88	122.75	121.25
1.18	113.92	113.42	112.90	112.40	111.94	111.43
1.20	100.81	100.31	99.42	99.92	98.95	98.45
1.22	88.75	88.24	85.53	85.02	83.87	83.37
1.24	76.77	76.27	75.69	75.19	73.58	73.08
1.26	66.72	66.21	16.86	16.36	62.14	61.64
1.28	58.50	58.00	57.88	57.37	56.39	55.89
1.30	53.10	52.60	51.85	51.35	48.96	48.46

c	$n = 5$ $k_1 = 2.8273$ $k_2 = 2.5219$ ARL1	SDRL	$n = 7$ $k_1 = 2.9289$ $k_2 = 2.5671$ ARL1	SDRL	$n = 10$ $k_1 = 2.9098$ $k_2 = 2.5815$ ARL1	SDRL
1.32	45.10	44.60	44.29	43.79	43.02	42.52
1.34	38.93	38.43	35.53	35.03	33.58	33.08
1.36	34.35	33.85	31.24	30.73	30.75	30.24
1.38	28.72	28.21	25.01	24.51	23.31	23.80
1.40	24.69	24.18	23.56	23.06	22.50	22.00
1.42	22.27	21.77	21.20	20.69	20.84	20.34
1.44	19.14	18.64	18.87	18.37	17.11	16.61
1.46	16.80	16.29	16.09	15.58	15.73	15.22
1.48	15.51	15.00	14.96	14.45	13.65	13.14
1.50	13.82	13.31	13.31	12.81	10.53	10.03

22.5 REAL-WORLD EXAMPLES

22.5.1 Real Data

Here, we consider the data-set that represents the waiting times before service of 100 bank customers. The set was used by Merovci and Elbatal (2014) to fit the transmuted Lindley-geometric distribution. We use this dataset to illustrate the proposed control chart of N_{pkg} for WD. For easy reference, we have given data along with chart statistics in Table 22.5. The maximum likelihood estimates of WD for waiting times before service of bank customers' data are $\hat{\lambda} = 10.7847$ and $\hat{\beta} = 1.4586$. To test the goodness-of-fit for this data to WD, the Kolmogorov-Smirnov test is used. We got the maximum distance between the data and the fitted WD as 0.0582 with p-value as 0.887. Thus, the WD provides a good fit for the waiting times before service of bank customers' data. To emphasize the goodness-of-fit, we also plot a histogram and superimpose empirical density and theoretical density. Further, goodness-of-fit is stressed with Q-Q plot, empirical and theoretical CDF, and P-P plots that are displayed in Figure 22.1.

The methodology of the developed control chart is addressed with waiting times before service of bank customers' data. Table 22.5 shows reported waiting times before service of bank customers' data of 20 samples of subgroup size 5 with $\widehat{C}_{Npkg}$. The target ARL_0 has been fixed at 370 for this dataset. For $n = 5$, $\hat{\lambda} = 10.7847$, and $\hat{\beta} = 1.4586$, we obtained the following chart coefficients (during the in-control process): $k_1 = 2.9969$ and $k_2 = 2.5323$. The lower CL and upper CL are given by $LCL_2 = 0.1544$, $UCL_2 = 0.8233$; $LCL_1 = 0.0930$, $UCL_1 = 0.8847$, respectively. The proposed control chart with control limits and statistic are displayed

Table 22.5 Waiting times before service of bank customers' data (in minutes)

S. No.						$\widehat{C}_{Npkg}$
1	18.9	4.2	5.0	8.6	13.0	0.4988
2	17.3	4.2	5.3	2.6	13.1	0.4265
3	0.8	4.3	5.5	2.7	13.3	0.2669
4	18.2	8.8	5.7	2.9	13.6	0.4938
5	1.3	4.3	5.7	8.6	13.7	0.3372
6	21.3	4.4	6.1	8.8	13.9	0.5469
7	1.5	11.1	6.2	8.9	8.6	0.3643
8	12.5	31.6	6.2	8.9	14.1	0.7356
9	1.8	11.2	6.3	9.5	15.4	0.4436
10	1.9	11.2	6.7	9.6	17.3	0.4687
11	19.9	11.5	6.9	9.7	18.1	0.6634
12	1.9	4.4	7.1	9.8	18.4	0.4175
13	2.1	4.6	7.1	10.7	19.0	0.4365
14	3.1	4.7	7.1	10.9	20.6	0.4657
15	3.2	23.0	1.9	11.0	21.4	0.6072
16	3.3	27.0	7.4	11.0	21.9	0.7085
17	3.5	4.7	7.6	11.9	7.1	0.3492
18	3.6	4.8	7.7	0.8	33.1	0.5018
19	4.0	4.9	8.0	12.4	38.5	0.6804
20	4.1	4.9	8.2	12.9	6.2	0.3643

in Figure 22.2 and we noticed that all the values of $\widehat{C}_{Npkg}$ lie between the control limits. Thus, Figure 22.2 describes the in-control process. We have also compared the proposed chart with the existing chart given by Aslam et al. (2020). To develop the existing chart, employing this data when $ARL_0 = 370$, for $n = 5$, $\widehat{\lambda} = 10.7847$, and $\widehat{\beta} = 1.4586$, we obtained control chart coefficients when $k = 3.3055$. The lower and upper control limits are given by $LCL = 0.0521$ and $UCL = 0.9255$, respectively. Figure 22.3 depicts the graphical presentation of the existing control chart. In Figure 22.3, notice that all the chart statistics $\widehat{C}_{Npkg}$ values are within the control limits and near the central line, whereas the proposed control chart shows that chart statistics $\widehat{C}_{Npkg}$ values are closer to inner-control limits. Hence, the proposed chart performs better as compared to the existing chart.

22.5.2 Simulated Data

The proposed control chart methodology is exhibited using simulated data. The following is the algorithm to generate the data and depict the proposed control chart.

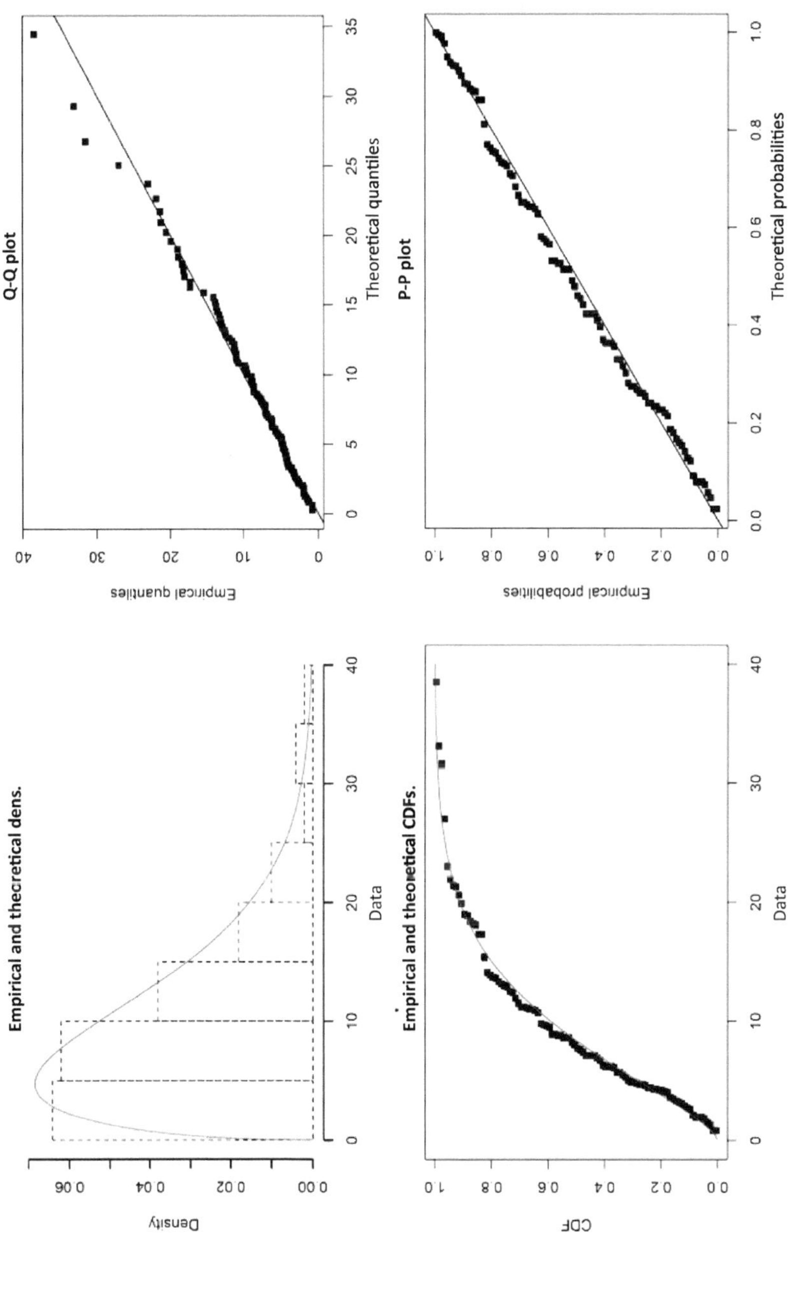

Figure 22.1 Histogram with superimposed theoretical density plot, Q-Q plot, empirical and theoretical CDFs, and P-P plots of the fitted WD for waiting times of bank customers' data.

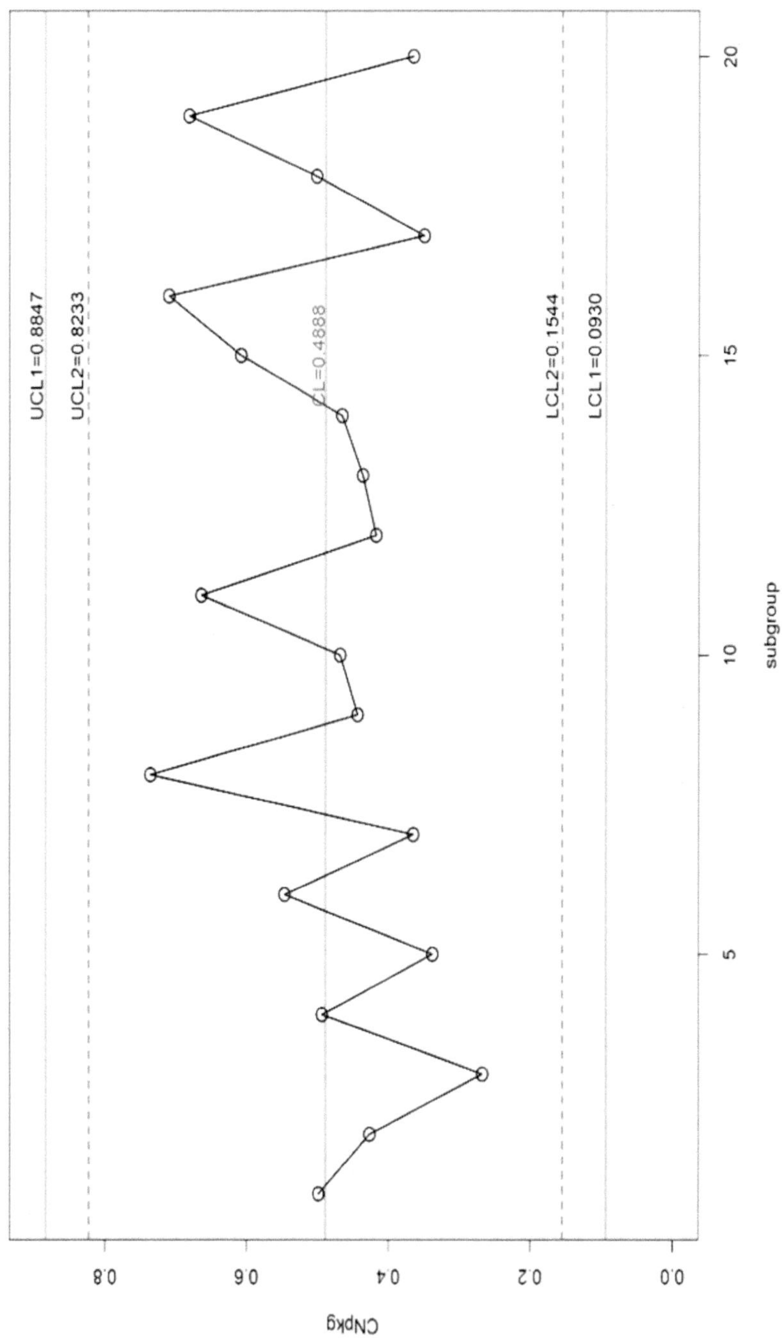

Figure 22.2 The proposed control chart for waiting times of bank customers' data.

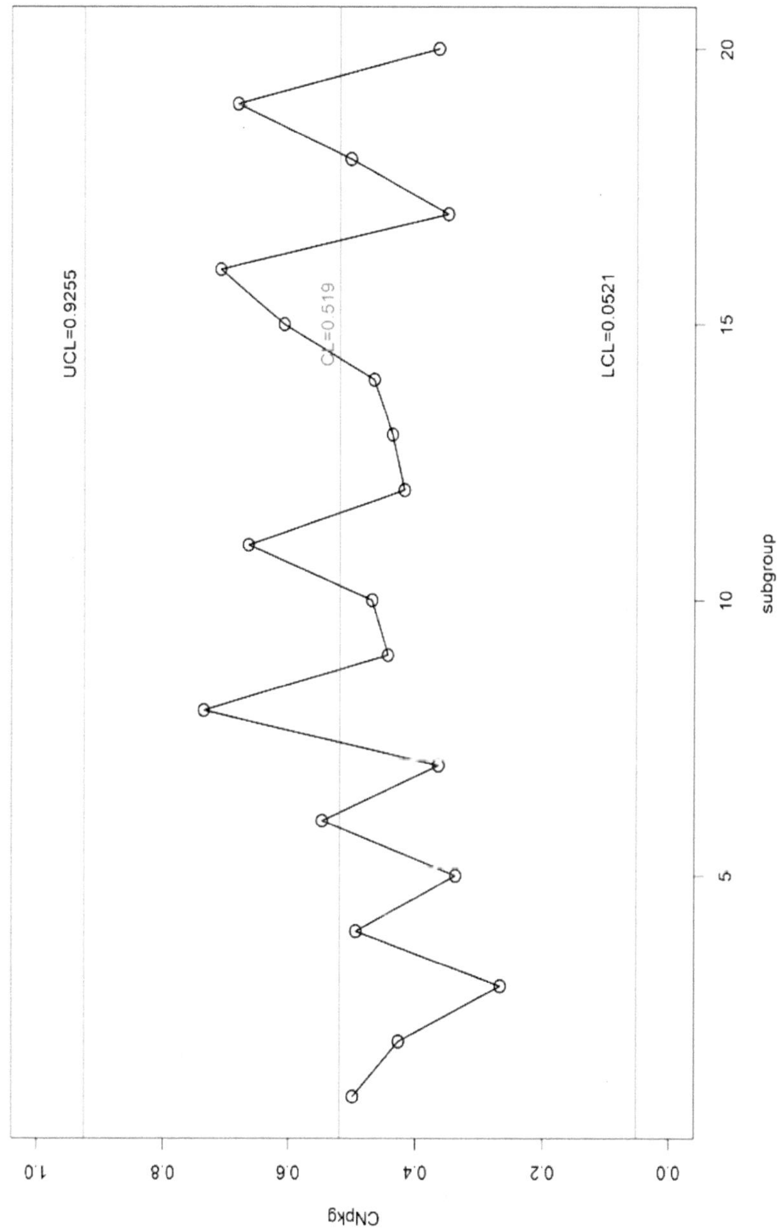

Figure 22.3 Existing control chart for waiting times of bank customers' data.

Step 1: Select a sample of subgroup size $n = 5$.

Step 2: A random variable X of each subgroup is generated from WD with scale parameter 3 and shape parameter 4.

Step 3: Compute the chart statistic $\widehat{C}_{Npkg}$ using Eq. (16).

Step 4: Generate 20 ($m = 36$) sample groups of size $n = 5$ by repeating Steps 1 to 3.

Step 5: Depict the statistics $\widehat{C}_{Npkg}$ against their sample groups and construct the control limits.

The first 20 samples of subgroup size 5 are generated using WD having in-control parameters $\lambda = 3$ and $\beta = 4$, and the second set of the 30 samples of the same subgroup size from the same WD having parameters $\lambda = 3$ and $\beta = 5.52$ (i.e., for shift of shape parameter $c = 1.38$). From Table 22.3, if $ARL_0 = 370$ and for in-control parametric values $\lambda = 3$ and $\beta = 4$, the control chart coefficients are $K_1 = 3.1603$ and $k_2 = 2.6712$ for the suggested chart when $n = 5$. In Table 22.6, the simulated $\widehat{C}_{Npkg}$ value for these 50 subgroup data is displayed. The lower and upper control limits are given by $LCL_2 = 0.1809$, $UCL_2 = 0.8571$; $LCL_1 = 0.1190$, $UCL_1 = 0.9190$, respectively. Figure 22.4 depicts the graphical presentation of the proposed control chart. From Figure 22.4, we can see that the proposed chart quickly detects the shift in the process at the 43rd, 44th, and 46th samples, which are the 13th, 14th, and 15th subgroups after the true shift, which are beyond inner-control limits and within the outer-control limits. Therefore, the decision for the process based on the simulated data is to repeat the process.

Table 22.6 Simulated $\hat{C}_{Npkg}$ values for WD data when $n = 5$, $\lambda = 3$, $\beta = 4$

S. No.	$\hat{C}_{Npkg}$	S. No.	$\hat{C}_{Npkg}$	S. No.	$\hat{C}_{Npkg}$
1	0.3641	18	0.3214	35	0.4530
2	0.5732	19	0.5213	36	0.7792
3	0.5217	20	0.5682	37	0.3665
4	0.3961	21	0.5190	38	0.5015
5	0.4066	22	0.2266	39	0.4901
6	0.8050	23	0.4935	40	0.4800
7	0.5944	24	0.4988	41	0.4324
8	0.5813	25	0.7220	42	0.8158
9	0.5712	26	0.5141	43	0.8800
10	0.4145	27	0.2628	44	0.9418
11	0.5069	28	0.7101	45	0.9513
12	0.5030	29	0.5439	46	0.8513
13	0.3845	30	0.7180	47	0.7553
14	0.4506	31	0.6661	48	0.6380
15	0.5277	32	0.5682	49	0.3329
16	0.5711	33	0.4380	50	0.4681
17	0.7970	34	0.4219		

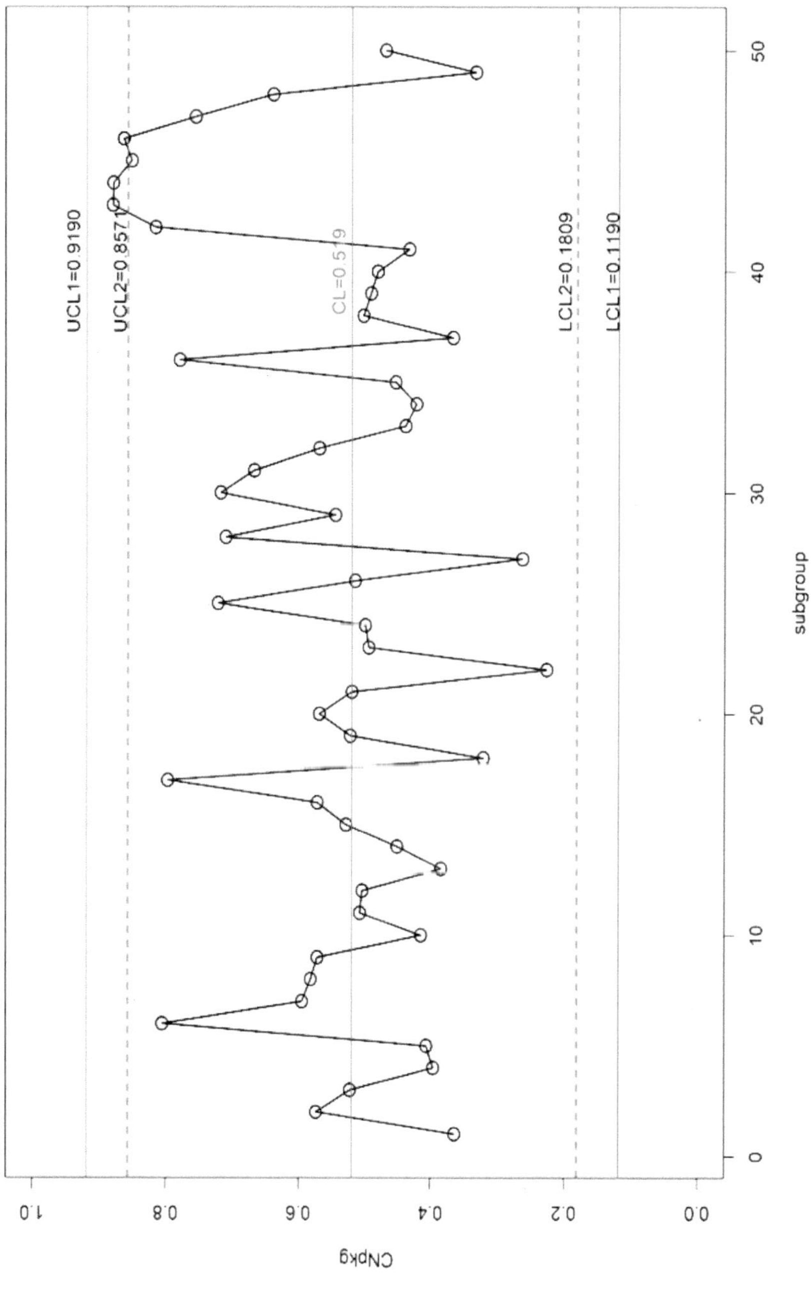

Figure 22.4 Proposed control chart for simulated data.

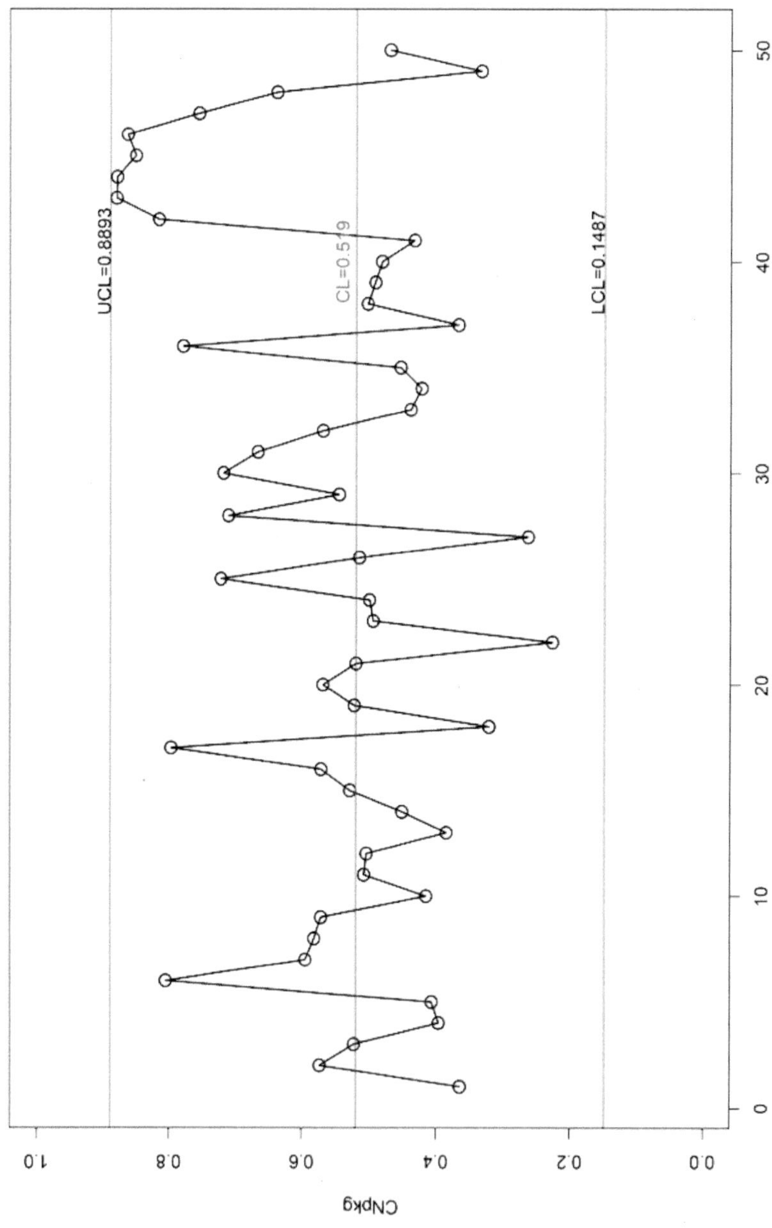

Figure 22.5 Existing control chart for simulated data.

We have also compared the proposed chart with the existing control chart presented by Aslam et al. (2020). From Aslam et al. (2020), if $ARL_0 = 370$ and for in-control the parametric values $\lambda = 3$ and $\beta = 4$, the control chart coefficients are $k = 2.9251$ when $n = 5$. The lower and upper control limits are $LCL = 0.1487$, $UCL = 0.8893$. Figure 22.5 depicts the graphical presentation of the existing control chart. In Figure 22.5, notice that all the chart statistics $\widehat{C}_{Npkg}$ values are within the control limits. Hence, we conclude that the existing control chart fails to detect a shift in the process. Hence, the proposed control chart performs better as compared with the existing control chart.

22.6 CONCLUDING REMARKS

A control chart based on repetition sampling for the PCIs C_{pk} under the Gini's mean difference using the Weibull distribution has been presented in this paper. Using the simulation method, the control chart coefficients have been determined at various parametric values of WD for specified in-control ARL values. The out-of-control ARLs have been computed for various shift values of the shape parameters to study the performance of the proposed control chart. A real data example and simulated data have been used to demonstrate the methodology of the developed control chart also included in this paper. It has been observed that the proposed chart scheme is effective in quick detection of the out-of-control process and more efficient than the existing chart.

Conflict of Interest: No conflict of interest regarding the paper.

Data Availability: The data is given in the paper.

Funding: No specific funds have been received regarding the paper.

REFERENCES

Ahmad, L., Aslam, M. and Jun, C.-H. 2014a. Coal quality monitoring with improved control charts. European Journal of Scientific Research. 125(2): 427–434.

Ahmad, L., Aslam, M. and Jun, C.-H. 2014b. Designing of X-bar control charts based on process capability index using repetitive sampling. Transactions of the Institute of Measurement and Control. 36(3): 367–374.

Ahmad, L., Aslam, M., Arif, O.H. and Jun, C.H. 2016a. Dispersion chart for some popular distributions under repetitive sampling. Journal of Advanced Mechanical Design Systems and Manufacturing. 10(4).

Ahmad, L., Aslam, M. and Jun, C.-H. 2016b. The design of a new repetitive sampling control chart based on process capability index. Transactions of the Institute of Measurement and Control. 38(8): 971–980.

Aslam, M., Mohsin, M. and Jun, C.-H. 2017. A new t-chart using process capability index. Communications in Statistics: Simulation and Computation. 46(7): 5141–5150. doi: 10.1080/03610918.2016.1146759.

Aslam, M., Rao, G.S., Ahmad, L. and Jun, C.-H. 2020. A new control chart using GINI CPK. Communications in Statistics: Theory and Methods. 1–15. doi: 10.1080/03610926.2020.1746971.

Balamurali, S. and Jun, C.H. 2006. Repetitive group sampling procedure for variables inspection. Journal of Applied Statistics. 33(3): 327–338. doi:10.1080/02664760500446010.

Chen, K., Huang, M. and Li, R. 2001. Process capability analysis for an entire product. International Journal of Production Research. 39(17): 4077–4087.

Cheng, C.-S. and Chen, P.-W. 2011. An ARL-unbiased design of time-between-events control charts with runs rules. Journal of Statistical Computation and Simulation. 81(7): 857–871.

Clements, J.A. 1989. Process capability calculations for non-normal distributions. Quality Progress. 22: 95–100.

David, H. 1968. Miscellanea: Gini's mean difference rediscovered. Biometrika. 55(3): 573–575.

Downton, F. (1966). Linear estimates with polynomial coefficient. Biometrika. 153: 129–141.

Garcia-Diaz, J.C. and Aparisi, F. 2005. Economic design of EWMA control charts using regions of maximum and minimum ARL. IIEE Transactions. 37(11): 1011–1021. doi: 10.1080/07408170500232214.

Gerstenberger, C. and Vogel, D. 2015. On the efficiency of Gini's mean difference. Statistical Methods & Applications. 24(4): 569–596.

Gini, C. 1912. Variabilità e mutabilità: Contributo allo studio delle distribuzioni e relazioni statistiche [Variability and mutability: Contribution to the study of statistical distributions and relations]. Bologna: Tipogr. di P. Cuppini.

Huang, P. and Hwang, T. 2005. The inference of Gini's mean difference. International Journal of Pure and Applied Mathematics. 25(1): 39–48.

Huang, C.-J., Chen, J.-H. and Lu, S.-L. 2019. Generally weighted moving average control charts using repetitive sampling. Communications in Statistics: Theory and Methods. doi: 10.1080/03610926.2019.1634212.

Kane, V.E. 1986. Process capability indices. Journal of Quality Technology. 18(1): 41–52.

Khan, N., Aslam, M., Ahmad, L. and Jun, C.-H. 2017. A Control chart for gamma distributed variables using repetitive sampling scheme. Pakistan Journal of Statistics and Operation Research. XIII(1). 47–61.

Knoth, S. 2007. Accurate ARL calculation for EWMA control charts monitoring normal mean and variance simultaneously. Sequential Analysis. 26(3): 251–263. doi: 10.1080/07474940701404823.

Lomnicki, Z. 1952. The standard error of Gini's mean difference. The Annals of Mathematical Statistics. 23(4): 635–637.

Maravelakis, P.E., Panaretos, J. and Psarakis, S. 2005. An examination of the robustness to non-normality of the EWMA control charts for the dispersion. Communications in Statistics: Simulation and Computation. 34(4): 1069–1079.

Merovci, F. and Elbatal, I. 2014. Transmuted Lindley-geometric distribution and its applications. Journal of Statistics Applications and Probability. 3(1): 77–91. doi: 10.12785/jsap/030107.

Montgomery, D.C. 2009. Introduction to Statistical Quality Control, 6th Edn. New York: John Wiley & Sons, Inc.

Nair, U. 1936. The standard error of Gini's mean difference. Biometrika. 28(3/4): 428–436.

Nanthakumar, D. and Vijayalakshmi, M. 2015. Construction of interquartile range (IQR) control chart using process capability for mean. International Journal OF Modern Scinence and Engineering Technology. 5(1): 114–118.

Palmer, K. and Tsui, K.-L. 1999. A review and interpretations of process capability indices. Annals of Operations Research. 87: 31–47.

Pearn, W. and Chen, K. 1997. Capability indices for non-normal distributions with an application in electrolytic capacitor manufacturing. Microelectron Reliability. 37(12): 1853–1858.

Pearn, W.L. 1998. New generalization of process capability index C_{pk}. Journal of Applied Statistics. 25(6): 801–810.

Riaz, M. and Saghirr, A. 2007. Monitoring process variability using Gini's mean difference. Quality Technology & Quantitative Management. 4(4): 439–454.

Riaz, M. and Does, R.J. 2008. An alternative to the bivariate control chart for process dispersion. Quality Engineering. 21(1): 63–71.

Riaz, M. and Abbasi, S.A. 2009. Gini's mean difference based time-varying EWMA charts. Economic Quality Control. 24(2): 269–286.

Sennaroglu, B. and Senvar, O. 2015. Performance comparison of box-cox transformation and weighted variance methods with Weibull distribution. Journal of Aeronautics and Space Technologies. 8(2): 49–55.

Sherman, R.E. 1965. Design and evaluation of repetitive group sampling plan. Technometrics. 7(1): 11–21. doi: 10.2307/1266124.

Sullivan, L.P. 1984. Reducing variability: A new approach to quality. Quality Progress. 17(7): 15–21.

Yitzhaki, S. 2003. Gini's mean difference: A superior measure of variability for non-normal distributions. Metron. 61(2): 285–316.

Chapter **23**

Robust Measures of Goodness-of-Fit and Outlier Detection in Linear Regression Models

Shalabh[1], Subhra Sankar Dhar[2] and Gaurav Garg[3]

[1,2]Department of Mathematics and Statistics
Indian Institute of Technology Kanpur, Kanpur 208016 (India).
E-mails: [1]shalab@iitk.ac.in, [2]subhra@iitk.ac.in

[3]Decision Sciences Area
Indian Institute of Management Lucknow, Noida 201307 (India).
Email: ggarg@iiml.ac.in

23.1 INTRODUCTION

The performances of standard statistical tools are based on the fundamental assumption that no outliers exist in the dataset. The presence of outliers disturbs the estimates of parameters and their statistical properties, resulting in erroneous decision-making. In the context of the multiple linear regression model, the ordinary least squares estimator (OLSE) is the best linear and unbiased estimator of the regression coefficient under the standard assumptions of the multiple linear regression model. One fundamental assumption in using the ordinary least square estimator is that the dataset used for fitting the model is free from outliers. When outliers are present in the data,

*For Correspondence: Shalabh (shalabh@iitk.ac.in)

the ordinary least squares estimator's behavior is disturbed and gives erroneous results. More precisely, the outliers in the response variables significantly affect the value of the ordinary least squares estimator of the regression coefficient and hence the slope of the line deflects significantly. For example, suppose the fitted model based on the data will be used to forecast the values of response variables. When outliers in the values of the response variable are present, the fitted model deviates and differs from the fitted model, which would have been obtained without the outliers. Consequently, forecasting the values of the response variable becomes faulty and misguiding in practical applications. Also, further applications of the fitted model heavily depend on how well the model is fitted. For example, the goodness-of-fit of the model is measured by using the estimated value of the parameter which is used in obtaining the fitted model. Various goodness-of-fit measures have been proposed in the literature, which can be utilized to determine the degree of goodness-of-fit in multiple linear regression models. One popular approach is using the coefficient of determination, known as R^2. The coefficient of determination is a function of the ordinary least squares estimator of the regression coefficient. When the outliers are present in the data, the ordinary least squares estimator's behavior is disturbed and gives erroneous results. So when the ordinary least square estimator is being incorrectly measured in the presence of outliers in the response variable, the coefficient of determination also provides erroneous results. How to measure the goodness-of-fit in a multiple linear regression model when the outliers are present in the data is an issue that needs to be addressed in the literature. As the absence or presence of outliers in the data decides the appropriate choice of various statistical measures, including the goodness-of-fit, another critical issue is how to detect the outliers in the data.

Outlier detection is an important aspect of any regression analysis and has attracted the attention of researchers in the literature. A textbook level description is well described in Chatterjee and Hadi (2013), Montgomery (2021), etc. Schweder (1976) has presented methods to detect the structural shift or outliers in regression, Dehon et al. (2009) presented a Hausman-type test to detect the presence of influential outliers in regression analysis, whereas Dünder (2021) presented a correction approach for information criteria to detect outliers in regression modeling, etc. Outlier detection in nonparametric setup in estimating production frontiers is addressed by Khezrimotlagh (2020) whereas Gumedze (2019) presented likelihood ratio tests to detect outliers under the variance shift outlier model. A clustering approach to detect multiple outliers in a linear functional relationship model for circular data is addressed in Mokhtar et al. (2018). A co-median approach to detect compositional outliers is presented in Di Palma and Gallo (2016). Bayesian methods have also been employed to detect outliers. Ekiz (2002) presented a Bayesian method to detect outliers in multivariate linear regression. Dong (2010) considered the outlier detection in ordinal data, see also Guttman and Dutter (1978). Outlier detection in time series data is considered in Embrechts et al. (1986), see also Peña and Galeano (2008).

The goodness-of-fit statistics is a sample-based measure whose value reflects the decision for the entire population. For example, the efficacy of a vaccine is judged over a sample of human beings, but the conclusions are valid for the entire human population. The experimenters and practitioners are usually interested in measuring goodness-of-fit statistics quantitatively so that the suitability of a model for using it in further applications can be numerically judged. The main objective in finding the fitted model based on the sample of data is to find a model which well captures the features of the process generating the data. The success of getting a good measure of fit depends upon the success of estimating the model parameters. Different experimental and real-life conditions decide the best choice of suitable models among different statistical and econometric models. Measuring the goodness of fit in different models has been considered in the literature. Rao et al. (2008) gave a detailed exposition about the use of coefficient of determination in multiple linear regression model. The coefficient of determination explains the explanatory power of the fitted model and is applicable under certain conditions, e.g., there are no outliers in the data, the observations are correctly recorded, the intercept term is present in the model, explanatory variables are independent of random errors, etc. Such conditions are not always satisfied, and thus the same coefficient of determination cannot be used on all types of statistical and econometric models. This promotes the search for suitable goodness-of-fit statistics suitable for a given statistical and econometric model. Cheng et al. (2014) proposed the goodness-of-fit in measurement error models, see also Cheng et al. (2016). The goodness-of-fit in parametric and nonparametric regression is discussed in Shalabh et al. (2021), see also Shalabh and Dhar (2021). The goodness-of-fit in instrumental variable models is proposed in Shalabh and Dhar (2022), see also Shalabh and Dhar (2023) and Shalabh et al. (InPress). The goodness-of-fit in generalized shrinkage estimation is addressed by Cheng et al. (2019), Tjur (2009), Hong, Ham and Kim (2005), whereas Liao and McGee (2003) considered the goodness-of-fit in the logistic regression model. The goodness-of-fit in entropy form for generalized linear models is considered in Eshima and Tabata (2010, 2011). Similarly, Hössjer (2008) suggested goodness-of-fit in the mixed regression model; Huang and Chen (2008) suggested goodness-of-fit in the local polynomial model; and Lipstiz et al. (2001) suggested goodness-of-fit in the multivariate normal repeated measures data, Knight (1980) presented the goodness-of-fit in the simultaneous equation models. A robust goodness-of-fit is proposed by Renaud and Victoria-Feser (2010). The goodness-of-fit in the case of association in a regression framework is proposed in van der Linde and Tutz (2008), and Srivastava and Shobhit (2002) proposed a family of the goodness-of-fit based on the coefficient of determination in the linear regression model. Marchand (2001) considered the coefficient of determination as point estimation; see also Marchand (1997). Tanaka and Huba (1998) suggested a goodness-of-fit for the covariance structure models under arbitrary generalized least squares. The coefficient of determination for the least absolute deviation is discussed in McKean and Sievers (1987).

It is important to note that the construction of the goodness-of-fit statistic heavily depends upon the successful estimation of the parameters. When outliers are present in the data, the ordinary least squares estimator behaves erratically and consequently, and the coefficient of determination does not provide reliable information. The outliers cannot always be removed from the sample data as many times they contain important information about the nature of the data-generating process. So the area of robust estimation plays an important role and the robust estimators of the parameters help in extracting the correct and reliable estimation of the parameters in the presence of outliers in the sample data. The interest of a practitioner lies in using the fitted model for various purposes like forecasting. Undoubtedly, the success of finding a good model depends upon the technique employed for the estimation of parameters but the practitioner is more interested in finding the correct degree of goodness-of-fit quantitatively even in the presence of outliers in the data. So robust estimation can help us in finding a good estimator of parameters in the presence of outliers but how to measure it remains unanswered. Finding the answer to such a question is attempted in this chapter.

The area of robust goodness-of-fit measures has been addressed in the literature by various researchers. Zheng et al. (2018) discussed the robust goodness-of-fit test for generalized autoregressive conditional heteroscedastic models. Brys et al. (2008) suggested the goodness-of-fit tests based on a robust measure of skewness. Bahari et al. (2017) suggested a robust goodness-of-fit test in general linear models with missing covariates. Brys et al. (2008) suggested the goodness-of-fit tests based on a robust measure of skewness and Keich and Nagarajan (2006) presented a numerically robust method for the exact multinomial goodness-of-fit test. Chan (1994) discussed goodness-of-fit tests in robust time series modeling.

The question of how to measure the goodness-of-fit in the models arises when outliers are present in the data. The goodness-of-fit measures which are robust in the sense that they can correctly measure the goodness-of-fit in the presence of outliers in the data, has not been much discussed in the literature. Before this question, another important question is how to decide if there is any outlier present in the data? In the classical multiple linear regression model, the coefficient of determination based on the ordinary least squares estimator of the regression coefficient is used to decide if any observation present in the dataset can be considered as an outlier or not, Montgomery (2021). Combining the two aspects of outlier detection and robust measuring of the goodness of fit, we propose a methodology based on the robust goodness of fit measures to decide the presence and detection of outliers in the data. Next is the construction of goodness-of-fit statistics which can measure the degree of the fitted model in the presence of outliers in the data. We have considered three different estimators: median-based estimator, least median squared regression estimator, and least trimmed squared estimator of regression coefficients in the setup of multiple linear regression models and obtain robust estimators of the regression coefficients. Then we use them for the construction of goodness-of-fit statistics. The statistical performance of the proposed measures has been evaluated through various numerical simulations. Outlier detection is another area to explore and we suggest an approach based on the robust goodness-of-fit statistics for the detection of outliers in the data on the response variable and study its performance numerically through simulation. An attempt is made to provide some guidelines for practitioners.

The plan of the paper is as follows. The description of multiple linear regression and ordinary least squares estimator is given in Section 23.2. Section 23.3 describes the goodness-of-fit measures and its construction in the multiple linear regression model without outliers. Section 23.4 presents the robust goodness-of-fit measures and its construction in the multiple linear regression model with outliers. Section 23.5 presents an approach to detect the outliers from the data. An extensive study from the Monte-Carlo simulation results is presented in Section 23.6 followed by Section 23.7 presenting the concluding remarks.

23.2 MODEL AND OLS ESTIMATOR (OLSE)

Let the n observations on the study variable be stored in y which is a $(n \times 1)$ vector and n observations on each of the p explanatory variables $X_1, X_2, \ldots, X_p$ are stored in a matrix X of order $(n \times p)$. The multiple linear regression model between y and X is expressed using β_0 as the intercept term and regression coefficients $\beta_1, \beta_2, \ldots, \beta_p$ associated with the explanatory variables $X_1, X_{2,}, \ldots, X_p$, respectively as the $(p \times 1)$ vector β and let ε be the $(n \times 1)$ vector of random disturbances in y as

$$y = \beta_0 e_n + X\beta + \varepsilon \tag{1}$$

where e_n is the $(n \times 1)$ vector with all the elements unity (**1s**) and random disturbances $\varepsilon_1, \varepsilon_2, \ldots, \varepsilon_n$ are identically and independently distributed with mean 0 and variance σ^2, i.e., ε in Eq. (1) follows a distribution with mean vector 0 and covariance matrix $\sigma^2 I$.

The ordinary least squares estimator (OLSE) of β and β_0 are obtained by minimizing

$$(y - \beta_0 e_n - X\beta)'(y - \beta_0 e_n - X\beta), \tag{2}$$

or equivalently

$$\sum_{i=1}^{n}(y_i - \beta_0 - x_{1i}\beta_1 - \cdots - x_{pi}\beta_p)^2 \tag{3}$$

which are obtained as

$$\widehat{\beta}_{OLS} = (X'PX)^{-1} X'Py = (b_{1o}, b_{2o}, \ldots, b_{po})' \tag{4}$$

and

$$\widehat{\beta_0}_{OLS} = \bar{y} - b_{1o}\bar{x}_1 - b_{2o}\bar{x}_2 - \cdots - b_{po}\bar{x}_p, \tag{5}$$

respectively, where $P = I_n - n^{-1}\mathbf{e}_n\mathbf{e}_{n'}$ and $\bar{x}_j = \frac{1}{n}\sum_{i=1}^{n} x_{ij}$.

23.3 GOODNESS-OF-FIT IN THE MODEL WITHOUT OUTLIERS

The goodness-of-fit of the model under the multiple linear regression model Eq. (1) is measured by estimating the multiple correlation coefficient between the study variable y and all the explanatory variables $X_1, X_2, \ldots, X_p$.

The square of population multiple correlation coefficient between the y and $X_1, X_2, \ldots, X_p$ is defined as

$$\theta = \frac{\beta'\Sigma\beta}{\beta'\Sigma\beta + \sigma^2}, \quad 0 \le \theta \le 1, \tag{6}$$

where $\Sigma = n^{-1}X'PX$. Note that θ involves unknown β and so cannot be used in real datasets.

The goodness-of-fit in the multiple linear regression model is judged by the coefficient of determination. The coefficient of determination is a consistent estimator of the square of population multiple correlation coefficient between the y and $X_1, X_2, \ldots, X_p$, i.e., θ in the model Eq. (1). The coefficient of determination is based on the OLSE of β and is denoted as R^2. It is given by

$$R^2_{OLS} = \frac{\hat{\beta}'_{OLS} X'PX\hat{\beta}_{OLS}}{y'Py} = \frac{y'PX(X'PX)^{-1}X'Py}{y'Py}, \; 0 \le R^2_{OLS} \le 1. \tag{7}$$

The coefficient of determination can also be obtained by partitioning the total sum of squares into two orthogonal components: the sum of squares due to regression and sum of squares due to error in the context of analysis of variance as

$$SS_{total} = SS_{regression} + SS_{error} \tag{8}$$

where $SS_{total} = y'Py$ and $SS_{regression} = y'PX\left(X'PX\right)^{-1} X'Py = \beta'_{OLS}X'PX\hat{\beta}_{OLS}$. The R^2_{OLS} can also be obtained as

$$R^2_{OLS} = \frac{SS_{regression}}{SS_{total}} = 1 - \frac{SS_{error}}{SS_{total}}, \; 0 \le R^2_{OLS} \le 1, \tag{9}$$

which indicates the proportion of the total variation being explained by the fitted model using $\hat{\beta}_{OLS}$. The advantage of using $\hat{\beta}_{OLS}$ is clear from Eq. (8) that in the context of analysis of variance, it provides an orthogonal partitioning of the total sum of squares into the sum of squares due to regression and sum of squares due to error.

The R^2_{OLS} measures the goodness-of-fit or capability of the model to explain the variation in y based on the explanatory power of the explanatory variables $X_1, X_2, \ldots, X_p$ (Rao et al. 2008). If the value of σ^2 is zero, then $\theta = 1$. This means that the variation in y is smaller and ideally zero, and under such a situation, $\theta = 1$ indicates that the model is best fitted. Similarly, when $\beta_1 = \beta_2 = \cdots = \beta_p = 0$, it means that no explanatory variable contributes to explaining the variation in y in the modeling. In such a situation, $\beta = 0$, and hence $\theta = 0$ indicates that the model is worst fitted. Any other values of $0 < \theta < 1$ indicate and measure the degree of goodness of the model quantitatively.

Since R^2_{OLS} is a consistent estimator of θ, so it is also expected to have similar interpretations about the goodness-of-fit based on OLSE. It indicates how much is the contribution of explanatory variables in explaining the variability in the values of the study variable. The values $R^2_{OLS} = 0$ and $R^2_{OLS} = 1$ indicates the worst and best fits of the model, similar to the values $\theta = 0$ and $\theta = 1$, respectively. Any other value $0 < R^2_{OLS} < 1$ measures the degree of fitted model based on OLSE. For example, if $R^2_{OLS} = 0.75$, it suggests that the model is approximately 75% good and 75% of the variation in y is explained by the involved explanatory variables. Using the assumptions $E(\varepsilon) = 0$, covariance matrix $V(\varepsilon) = \sigma^2 I$ and $\frac{X'PX}{n} = \Psi$, where Ψ is a positive definite matrix and $\frac{X'P\varepsilon}{n} = 0$ in the model (1), it can be shown that R^2 is a consistent estimator of θ in the sense that R^2_{OLS} converges to θ in probability.

Although the OLSE is the best linear unbiased estimator of β, it is sensitive to outliers. The presence of outliers in the observations deviates from the value of OLSE based on the data not having outliers and hence changes the slope of the line. This, in turn, alters the value of R^2_{OLS} providing erroneous information about the model. Consequently, all important further applications will be affected and provide incorrect statistical inferences. So an important issue arises is how to measure the goodness-of-fit when the data is influenced by the outliers.

23.4 GOODNESS OF FIT IN THE MODEL WITH OUTLIERS

It is clear from the construct of R^2_{OLS} that it is expected that any goodness-of-fit statistics is a consistent estimator of θ as long as the estimators of the unknown parameters are consistent. When $\hat{\beta}_{OLS}$ is affected by the presence of outliers, then R^2_{OLS} also gets affected. So if an estimator of β which is not affected by the outliers can be found, then an estimator of θ can be constructed using the philosophy and idea used in the construction of R^2_{OLS}. Partitioning the total sum of squares orthogonally into the sum of squares due to regression and random errors may also not always be possible with estimators other than OLSE. Thus, using the philosophy behind the construction of the coefficient of determination as the proportion of the sum of squares due to regression and the sum of squares due to random errors can be used to find a suitable form of the goodness-of-fit statistics, which is robust in the sense that it provides correct information about the model fit in the presence of outliers in the data.

The success of finding such a measure depends upon the suitable choice of a robust estimator of the regression coefficient. It is well known that the median estimator is robust in the presence of outliers in the response space. So another idea is to use the structure of R^2_{OLS} and construct a consistent estimator of θ based on the median estimator of β. The LAD estimator $\hat{\beta}_{LAD}$ of β is the minima of

$$\sum_{i=1}^{n} |y_i - \beta_0 - x_{1i}\beta_1 - \cdots - x_{pi}\beta_p| \tag{10}$$

A consistent estimator of θ based on $\hat{\beta}_{LAD}$ as an estimator of β is proposed as

$$R^2_{LAD} = \frac{\hat{\beta}'_{LAD} X'PX \hat{\beta}_{LAD}}{y'Py}, \; 0 \le R^2_{LAD} \le 1. \tag{11}$$

Other popular estimation methods of finding the robust estimators of the regression coefficients are based on using the least median squared (LMS) regression and least trimmed squared (LTS) regression. Both these estimators are robust against the outliers in the covariate and the response space. In terms of rate of convergence, the LTS estimator has usual $\sqrt{n}-$ convergence whereas the LMS estimator has slower rate of convergence which equals with $n^{1/3}$. Here n is the sample size. To gain more insight about both estimators, the readers may refer to Rousseeuw (1984).

The estimator of regression coefficient β based on the use of the least median squared (LMS) regression estimator of is defined as

$$\widehat{\beta}_{LMS} = arg\min_{\beta} median\{\varepsilon_i\}^2, \tag{12}$$

where $\varepsilon_i = (y_i - \beta_0 - x_{1i}\beta_1 - \cdots - x_{pi}\beta_p)$. The corresponding consistent estimator of θ is thus proposed as

$$R^2_{LMS} = \frac{\widehat{\beta}'_{LMS}X'PX\widehat{\beta}_{LMS}}{y'Py}, 0 \le R^2_{LMS} \le 1. \tag{13}$$

Similarly, the estimator of regression coefficient β based on the least trimmed squared (LTS) regression is defined as

$$\widehat{\beta}_{LTS,\alpha} = arg\min_{\beta} \sum_{i=[n\alpha+1]}^{n-[n\alpha]} \left(\varepsilon_{[i]}\right)^2, \tag{14}$$

where $\alpha \in \left(0, \frac{1}{2}\right)$ is the trimming proportion. The corresponding consistent estimator of θ is thus proposed as

$$R^2_{LTS} = \frac{\widehat{\beta}'_{LTS,\alpha}X'PX\widehat{\beta}_{LTS,\alpha}}{y'Py}, 0 \le R^2_{LTS} \le 1. \tag{15}$$

The interpretation of R^2_{LAD}, R^2_{LMS}, and R^2_{LTS} have the same interpretation as of R^2_{OLS}. For example, the values $R^2_{LMS} = 0$, $R^2_{LMS} = 1$, or $R^2_{LMS} = 0.85$ indicate that the model fitted by using $\widehat{\beta}_{LMS}$ is the best fitted, worst fitted, or explains only 85% of the variability in the fitted model, respectively. It is important to note that, in the context of analysis of variance, it is not possible to orthogonal partition the total sum of squares into the sum of squares due to regression and sum of squares due to error using the LAD estimator, $\widehat{\beta}_{LMS}$, and $\widehat{\beta}_{LTS,\alpha}$ of β.

Since $\widehat{\beta}_{LAD}$, $\widehat{\beta}_{LMS}$ and $\widehat{\beta}_{LTS,\alpha}$ are consistent estimators of β, so it can be proved that R^2_{LAD}, R^2_{LMS} and R^2_{LTS} are the consistent estimators of θ under the assumptions $E(\varepsilon) = 0$, covariance matrix $V(\varepsilon) = \sigma^2 I$ and $\frac{X'PX}{n} = \Psi$ (a positive definite matrix) and $\frac{X'P\varepsilon}{n} = 0$ in the model Eq. (1).

23.5 OUTLIER DETECTION

Now we address the issue of outlier detection based on the goodness-of-fit measures. In the context of the coefficient of determination, we observe that it measures the degree of the linear relationship between the dependent and independent variables. In case if there are outliers in the data, the fitted regression line deviates significantly from the line which would have been fitted without the outliers in the data. Based on this concept, we propose here to use the robust measures of goodness-of-fit and compare their values with the classical goodness-of-fit, i.e., coefficient of determination, which is based on OLSE. The values of the goodness-of-fit measures based on OLSE and robust estimates of regression coefficients will differ significantly only in the case when outliers are present in the data. So, considering the change in the values of the goodness-of-fit or the relative change occurring in the values of the goodness-of-fit, we can discover the presence of outliers in the data. The issue of how much change will indicate the presence of the data is difficult to find analytically. So, it is addressed through the Monte-Carlo simulation in Section 23.6.1. It is important to note that the outliers beyond a certain threshold will look like a part of the dataset with higher variability.

Let us denote R^2_{rob} as the value of goodness of fit statistic based on any robust estimator of the regression coefficient, e.g., R^2_{rob} in our context can be R^2_{LAD}, R^2_{LMS}, or R^2_{LTS}. We propose to compute the relative or percentage change in the values of the goodness-of-fit based on robust estimators R^2_{rob} relative to R^2_{OLS} and we term it as Outlier Detection Percentage Change (ODPC) as

$$ODPC(R^2_{rob}) = \frac{|R^2_{rob} - R^2_{OLS}|}{R^2_{OLS}} \times 100. \tag{16}$$

For example, the ODPCs based on R^2_{LAD}, R^2_{LMS}, and R^2_{LTS} are obtained as follows:

$$ODPC(R^2_{LAD}) = \frac{|R^2_{LAD} - R^2_{OLS}|}{R^2_{OLS}} \times 100, \tag{17}$$

$$ODPC(R^2_{LMS}) = \frac{|R^2_{LMS} - R^2_{OLS}|}{R^2_{OLS}} \times 100, \tag{18}$$

$$ODPC(R^2_{LTS}) = \frac{|R^2_{LTS} - R^2_{OLS}|}{R^2_{OLS}} \times 100. \quad (19)$$

If the values of $ODPC(R^2_{LAD})$, $ODPC(R^2_{LMS})$, and $ODPC(R^2_{LTS})$ are more than a cutoff value then it will indicate the presence of outliers in the data and it is recommended to use the robust measures of goodness of fit. The cutoff value is to be determined by the practitioner based on the choice of goodness-of-fit measure.

23.6 SIMULATION STUDY

It is challenging to determine the finite sample statistical properties like bias, mean squared error, etc. of R^2_{OLS}, R^2_{LAD}, R^2_{LMS}, and R^2_{LTS}. Even if these properties can be determined, their forms may be too complex to provide any insight into the performance of the estimators. The asymptotic distributions of R^2_{OLS}, R^2_{LAD}, R^2_{LMS}, and R^2_{LTS} can be derived but their expression will be complicated to reveal any useful information from them. The performance of the fitted model, superiority conditions, and instructions for their applications are typically of more relevance to any experimenter, practitioner, or applied statistician. So, we conduct a Monte-Carlo simulation experiment to study, explore, and investigate the statistical properties of R^2_{OLS}, R^2_{LAD}, R^2_{LMS}, and R^2_{LTS}.

Since the goodness-of-fit statistics are constructed based on the philosophy of coefficient of determination, some questions crop up. For example, how the proportion of outliers in the data affects the behavior of the classical coefficient of determination, how the performance of the proposed goodness-of-fit statistics changes when the distribution of random error deviates from normal or proportion of outliers increases, and what is the role of intercept term in the proposed goodness-of-fit measures, etc. Therefore, an effort is made to discover the answers to these questions through a thorough examination of the findings from a Monte Carlo simulation experiment carried out in R-software.

Different values of θ are taken into consideration in order to understand the impact of the inherent degree of linear relationship in the data and how the proposed measures respond to them. A low value of θ indicates a weaker linear relationship, indicating that there is a lack of evidence to support a linear model. On the other hand, a larger value of θ suggests a linear relationship that is too strong and can be fitted by a linear model. However, excessively high values of θ are more difficult to obtain in real-world circumstances. Clearly, $0 \leq \theta \leq 1$, so } $\theta = 0.2, 0.4, 0.6,$ and 0.8 are selected in the simulation. Practically speaking, a value of $\theta > 0.8$ may be regarded as large. The performance of the goodness-of-fit statistics will be impacted by the addition of outliers, and the performance will rely on the percentage of outliers present in the data. Data with a large proportion of outliers and a low θ value will fluctuate more and may obscure the effect of the outliers. Similar to this, a greater θ value along with a higher percentage of outliers in the data will show a higher level of intrinsic variability. It's also critical to keep in mind that even with high levels of data variability, one can still count on robust measures to produce reliable results. While fixing the percentage of outliers in the data, a balance is struck in order to assess how outliers affect the efficacy of goodness of fit measurements. Various proportions of outliers in the data are used in the simulation experiment, such as 0% (no outliers), 5%, 10%, 15%, 20%, 25%, 30%, 35%, 40%, 45%, and 50%. As the findings stabilized beyond the sample size of 200, the sample sizes of 60, 100, and 200 were selected for the simulation.

It is expected that robust measures will be able to handle the consequence of the random errors' deviation from a normal distribution. In order to verify it, we conducted the simulation and estimated the goodness-of-fit statistics using the data based on random errors from the normal and Gamma distributions. It is to be noted that the Gamma distribution has nonzero skewness and nonzero kurtosis, whereas the skewness and kurtosis of the normal distribution are both zero. We verified it by simulating the data from normal and Gamma distributions and computing the suggested goodness-of-fit measures to determine whether they are robust enough to counteract the non-normality impact and produce reliable results. The difference in the related statistical features between the data obtained from normal and Gamma distributions is anticipated to reveal the impact of deviation from normality.

The X matrix is assumed to be non-stochastic and the data on random errors is generated using normal distribution and Gamma distribution. After generating errors, the data on y vector is obtained as per the multiple linear regression model Eq. (1) using $\beta' = (-4, 2.5, 9.3, 2.1, -16.3)$ and $\beta_0 = 5$. For fixed X, the random errors are repeatedly generated and corresponding y is obtained. Considering different combinations of sample sizes, outlier proportions, θ values, and distributions, the values of absolute bias and mean squared error of R^2_{OLS}, R^2_{LAD}, R^2_{LMS}, and R^2_{LTS} are empirically computed based on 10,000 repetitions. The packages `quantreg` and `MASS` in R software are used to obtain the optimized estimates of the regression coefficients for a given set of data. The results about empirical absolute bias (AB) and empirical mean squared error (MSE) are obtained and presented in Tables 23.1–23.6 and Tables 23.7–23.12, respectively.

Table 23.1 Empirical absolute bias of R^2_{OLS}, R^2_{LAD}, R^2_{LMS}, and R^2_{LTS} for $n = 60$, when errors have normal distribution

	Proportion of Outliers	$AB(R^2_{OLS})$	$AB(R^2_{LAD})$	$AB(R^2_{LMS})$	$AB(R^2_{LTS})$
	0.00	0.062	0.088	0.287	0.290
	0.05	0.094	0.047	0.231	0.235
	0.10	0.110	0.066	0.209	0.207
	0.15	0.115	0.070	0.220	0.213
	0.20	0.116	0.071	0.203	0.210
$\theta = 0.20$	0.25	0.112	0.067	0.210	0.204
	0.30	0.114	0.072	0.218	0.215
	0.35	0.113	0.068	0.209	0.202
	0.40	0.110	0.064	0.211	0.211
	0.45	0.109	0.066	0.208	0.214
	0.50	0.108	0.066	0.215	0.214
	0.00	0.050	0.064	0.152	0.159
	0.05	0.259	0.205	0.078	0.076
	0.10	0.294	0.251	0.026	0.017
	0.15	0.312	0.267	0.013	0.009
	0.20	0.315	0.271	0.006	0.005
$\theta = 0.40$	0.25	0.310	0.265	0.010	0.003
	0.30	0.313	0.269	0.004	0.007
	0.35	0.310	0.268	0.014	0.008
	0.40	0.312	0.268	0.011	0.005
	0.45	0.302	0.255	0.014	0.009
	0.50	0.296	0.256	0.016	0.012
	0.00	0.034	0.029	0.060	0.060
	0.05	0.413	0.320	0.063	0.056
	0.10	0.480	0.436	0.169	0.162
	0.15	0.506	0.464	0.184	0.165
	0.20	0.508	0.461	0.179	0.188
$\theta = 0.60$	0.25	0.510	0.464	0.183	0.182
	0.30	0.506	0.459	0.175	0.176
	0.35	0.502	0.454	0.178	0.179
	0.40	0.507	0.464	0.183	0.185
	0.45	0.483	0.444	0.192	0.196
	0.50	0.470	0.427	0.163	0.167
	0.00	0.016	0.000	0.037	0.041
	0.05	0.550	0.251	0.129	0.121
	0.10	0.641	0.582	0.315	0.324
	0.15	0.692	0.647	0.348	0.355
	0.20	0.704	0.657	0.369	0.388
$\theta = 0.80$	0.25	0.705	0.659	0.385	0.389
	0.30	0.689	0.649	0.383	0.380
	0.35	0.686	0.642	0.377	0.386
	0.40	0.697	0.656	0.369	0.386
	0.45	0.648	0.606	0.355	0.358
	0.50	0.629	0.584	0.331	0.339

Table 23.2 Empirical absolute bias of R^2_{OLS}, R^2_{LAD}, R^2_{LMS}, and R^2_{LTS} for n = 100, when errors have normal distribution

	Proportion of Outliers	$AB(R^2_{OLS})$	$AB(R^2_{LAD})$	$AB(R^2_{LMS})$	$AB(R^2_{LTS})$
	0.00	0.037	0.061	0.219	0.249
	0.05	0.125	0.100	0.135	0.156
	0.10	0.138	0.112	0.117	0.127
	0.15	0.141	0.112	0.112	0.135
	0.20	0.146	0.118	0.105	0.121
θ = 0.20	0.25	0.144	0.118	0.108	0.122
	0.30	0.146	0.117	0.107	0.125
	0.35	0.148	0.119	0.116	0.124
	0.40	0.146	0.119	0.110	0.133
	0.45	0.145	0.119	0.122	0.133
	0.50	0.141	0.114	0.114	0.131
	0.00	0.029	0.038	0.131	0.132
	0.05	0.287	0.255	0.012	0.021
	0.10	0.319	0.294	0.073	0.049
	0.15	0.329	0.303	0.085	0.061
	0.20	0.333	0.303	0.095	0.073
θ = 0.40	0.25	0.337	0.310	0.087	0.073
	0.30	0.342	0.313	0.087	0.060
	0.35	0.340	0.313	0.083	0.080
	0.40	0.343	0.315	0.090	0.065
	0.45	0.340	0.315	0.090	0.075
	0.50	0.326	0.300	0.077	0.062
	0.00	0.019	0.018	0.051	0.056
	0.05	0.437	0.377	0.091	0.093
	0.10	0.490	0.463	0.222	0.230
	0.15	0.510	0.483	0.259	0.254
	0.20	0.521	0.494	0.296	0.260
θ = 0.60	0.25	0.525	0.497	0.276	0.261
	0.30	0.527	0.500	0.272	0.261
	0.35	0.530	0.503	0.276	0.253
	0.40	0.530	0.501	0.289	0.272
	0.45	0.531	0.505	0.276	0.271
	0.50	0.506	0.481	0.267	0.243
	0.00	0.010	0.003	0.025	0.033
	0.05	0.560	0.329	0.114	0.120
	0.10	0.632	0.588	0.362	0.362
	0.15	0.669	0.643	0.432	0.414
	0.20	0.688	0.660	0.438	0.434
θ = 0.80	0.25	0.697	0.671	0.456	0.446
	0.30	0.705	0.677	0.458	0.444
	0.35	0.711	0.686	0.477	0.467
	0.40	0.711	0.679	0.456	0.440
	0.45	0.709	0.682	0.458	0.445
	0.50	0.654	0.623	0.409	0.398

Table 23.3 Empirical absolute bias of R^2_{OLS}, R^2_{LAD}, R^2_{LMS}, and R^2_{LTS} for n = 200, when errors have normal distribution

	Proportion of Outliers	$AB(R^2_{OLS})$	$AB(R^2_{LAD})$	$AB(R^2_{LMS})$	$AB(R^2_{LTS})$
	0.00	0.016	0.029	0.156	0.174
	0.05	0.142	0.125	0.064	0.101
	0.10	0.162	0.148	0.014	0.039
	0.15	0.167	0.153	0.002	0.028
	0.20	0.169	0.157	0.004	0.018
θ = 0.20	0.25	0.171	0.157	0.002	0.023
	0.30	0.171	0.158	0.005	0.016
	0.35	0.172	0.158	0.012	0.018
	0.40	0.171	0.158	0.005	0.019
	0.45	0.169	0.155	0.000	0.023
	0.50	0.166	0.152	0.003	0.021
	0.00	0.013	0.018	0.094	0.105
	0.05	0.299	0.265	0.014	0.008
	0.10	0.346	0.328	0.157	0.119
	0.15	0.357	0.342	0.184	0.162
	0.20	0.363	0.349	0.189	0.165
θ = 0.40	0.25	0.366	0.353	0.204	0.174
	0.30	0.367	0.353	0.199	0.181
	0.35	0.365	0.351	0.201	0.171
	0.40	0.364	0.349	0.200	0.174
	0.45	0.362	0.348	0.200	0.179
	0.50	0.352	0.338	0.191	0.164
	0.00	0.008	0.008	0.042	0.047
	0.05	0.433	0.347	0.009	0.022
	0.10	0.521	0.495	0.267	0.248
	0.15	0.538	0.521	0.363	0.334
	0.20	0.551	0.535	0.380	0.352
θ = 0.60	0.25	0.557	0.544	0.391	0.376
	0.30	0.556	0.542	0.390	0.368
	0.35	0.558	0.544	0.396	0.372
	0.40	0.554	0.539	0.385	0.367
	0.45	0.548	0.534	0.382	0.354
	0.50	0.528	0.514	0.363	0.338
	0.00	0.005	0.002	0.007	0.011
	0.05	0.549	0.248	0.002	0.023
	0.10	0.676	0.600	0.245	0.249
	0.15	0.703	0.669	0.446	0.440
	0.20	0.726	0.706	0.531	0.509
θ = 0.80	0.25	0.737	0.723	0.565	0.533
	0.30	0.738	0.726	0.572	0.546
	0.35	0.737	0.723	0.576	0.556
	0.40	0.734	0.720	0.575	0.541
	0.45	0.717	0.703	0.545	0.520
	0.50	0.678	0.659	0.506	0.481

Table 23.4 Empirical absolute bias of R^2_{OLS}, R^2_{LAD}, R^2_{LMS}, and R^2_{LTS} for $n = 60$, when errors have Gamma distribution

	Proportion of Outliers	$AB(R^2_{OLS})$	$AB(R^2_{LAD})$	$AB(R^2_{LMS})$	$AB(R^2_{LTS})$
	0.00	0.066	0.092	0.291	0.285
	0.05	0.081	0.034	0.485	0.479
	0.10	0.104	0.058	0.377	0.388
	0.15	0.113	0.062	0.320	0.314
	0.20	0.114	0.063	0.300	0.297
$\theta = 0.20$	0.25	0.116	0.069	0.278	0.274
	0.30	0.114	0.068	0.290	0.278
	0.35	0.114	0.064	0.317	0.286
	0.40	0.117	0.066	0.309	0.296
	0.45	0.110	0.062	0.325	0.326
	0.50	0.109	0.065	0.291	0.290
	0.00	0.045	0.059	0.165	0.166
	0.05	0.236	0.170	0.482	0.458
	0.10	0.293	0.246	0.303	0.309
	0.15	0.304	0.254	0.196	0.184
	0.20	0.308	0.259	0.104	0.097
$\theta = 0.40$	0.25	0.310	0.262	0.082	0.090
	0.30	0.314	0.267	0.088	0.083
	0.35	0.315	0.267	0.105	0.079
	0.40	0.314	0.269	0.085	0.084
	0.45	0.313	0.264	0.117	0.120
	0.50	0.301	0.254	0.115	0.106
	0.00	0.034	0.033	0.052	0.054
	0.05	0.368	0.232	0.612	0.624
	0.10	0.476	0.420	0.530	0.531
	0.15	0.496	0.443	0.166	0.198
	0.20	0.502	0.452	0.018	0.036
$\theta = 0.60$	0.25	0.507	0.458	0.089	0.120
	0.30	0.508	0.462	0.090	0.096
	0.35	0.509	0.461	0.097	0.119
	0.40	0.514	0.465	0.083	0.088
	0.45	0.505	0.456	0.064	0.064
	0.50	0.485	0.435	0.096	0.091
	0.00	0.017	0.001	0.044	0.038
	0.05	0.486	0.037	0.975	0.963
	0.10	0.644	0.559	1.272	1.179
	0.15	0.671	0.603	0.991	0.982
	0.20	0.684	0.617	0.332	0.363
$\theta = 0.80$	0.25	0.698	0.642	0.099	0.125
	0.30	0.704	0.649	0.246	0.249
	0.35	0.709	0.654	0.270	0.278
	0.40	0.705	0.655	0.260	0.256
	0.45	0.696	0.639	0.138	0.154
	0.50	0.645	0.592	0.230	0.231

Table 23.5 Empirical absolute bias of R^2_{OLS}, R^2_{LAD}, R^2_{LMS}, and R^2_{LTS} for n = 100, when errors have Gamma distribution

	Proportion of Outliers	$AB(R^2_{OLS})$	$AB(R^2_{LAD})$	$AB(R^2_{LMS})$	$AB(R^2_{LTS})$
	0.00	0.040	0.063	0.233	0.233
	0.05	0.121	0.092	0.166	0.176
	0.10	0.131	0.105	0.113	0.120
	0.15	0.137	0.109	0.111	0.132
	0.20	0.143	0.116	0.102	0.125
θ = 0.20	0.25	0.144	0.117	0.108	0.143
	0.30	0.145	0.115	0.119	0.131
	0.35	0.145	0.118	0.107	0.127
	0.40	0.145	0.118	0.108	0.133
	0.45	0.143	0.116	0.105	0.125
	0.50	0.141	0.113	0.128	0.148
	0.00	0.024	0.033	0.129	0.136
	0.05	0.280	0.239	0.044	0.044
	0.10	0.305	0.277	0.054	0.042
	0.15	0.326	0.300	0.080	0.061
	0.20	0.334	0.309	0.085	0.060
θ = 0.40	0.25	0.338	0.311	0.083	0.058
	0.30	0.337	0.311	0.092	0.071
	0.35	0.339	0.312	0.087	0.076
	0.40	0.339	0.312	0.087	0.083
	0.45	0.336	0.313	0.084	0.070
	0.50	0.325	0.296	0.075	0.052
	0.00	0.019	0.021	0.058	0.065
	0.05	0.423	0.332	0.066	0.058
	0.10	0.467	0.433	0.212	0.201
	0.15	0.499	0.468	0.245	0.236
	0.20	0.518	0.492	0.278	0.252
θ = 0.60	0.25	0.526	0.499	0.277	0.267
	0.30	0.525	0.498	0.273	0.257
	0.35	0.525	0.499	0.269	0.269
	0.40	0.524	0.498	0.274	0.261
	0.45	0.522	0.495	0.275	0.266
	0.50	0.499	0.472	0.253	0.239
	0.00	0.011	0.002	0.020	0.023
	0.05	0.543	0.219	0.054	0.060
	0.10	0.600	0.532	0.301	0.300
	0.15	0.647	0.608	0.393	0.378
	0.20	0.689	0.663	0.442	0.434
θ = 0.80	0.25	0.697	0.667	0.446	0.427
	0.30	0.694	0.663	0.445	0.430
	0.35	0.701	0.673	0.461	0.451
	0.40	0.696	0.669	0.456	0.440
	0.45	0.691	0.665	0.444	0.428
	0.50	0.645	0.609	0.411	0.393

Table 23.6 Empirical absolute bias of R^2_{OLS}, R^2_{LAD}, R^2_{LMS}, and R^2_{LTS} for $n = 200$, when errors have Gamma distribution

	Proportion of Outliers	$AB(R^2_{OLS})$	$AB(R^2_{LAD})$	$AB(R^2_{LMS})$	$AB(R^2_{LTS})$
	0.00	0.015	0.029	0.151	0.168
	0.05	0.135	0.119	0.078	0.100
	0.10	0.152	0.136	0.026	0.052
	0.15	0.158	0.144	0.006	0.029
	0.20	0.162	0.148	0.009	0.027
$\theta = 0.20$	0.25	0.163	0.150	0.002	0.025
	0.30	0.166	0.151	0.003	0.025
	0.35	0.167	0.153	0.001	0.020
	0.40	0.168	0.154	0.007	0.022
	0.45	0.491	0.466	0.300	0.280
	0.50	0.701	0.682	0.515	0.498
	0.00	0.015	0.024	0.112	0.118
	0.05	0.278	0.248	0.008	0.014
	0.10	0.319	0.302	0.132	0.107
	0.15	0.335	0.319	0.154	0.136
	0.20	0.343	0.328	0.173	0.147
$\theta = 0.40$	0.25	0.349	0.335	0.183	0.162
	0.30	0.353	0.337	0.185	0.159
	0.35	0.355	0.342	0.192	0.160
	0.40	0.358	0.342	0.189	0.160
	0.45	0.168	0.155	0.001	0.034
	0.50	0.163	0.149	0.004	0.028
	0.00	0.010	0.011	0.057	0.057
	0.05	0.407	0.321	0.021	0.025
	0.10	0.471	0.442	0.231	0.213
	0.15	0.497	0.477	0.306	0.291
	0.20	0.515	0.498	0.341	0.317
$\theta = 0.60$	0.25	0.525	0.507	0.353	0.330
	0.30	0.529	0.513	0.368	0.346
	0.35	0.537	0.522	0.369	0.350
	0.40	0.539	0.524	0.378	0.352
	0.45	0.357	0.343	0.190	0.175
	0.50	0.348	0.334	0.170	0.154
	0.00	0.004	0.004	0.011	0.014
	0.05	0.503	0.204	0.001	0.027
	0.10	0.585	0.511	0.267	0.270
	0.15	0.623	0.579	0.392	0.366
	0.20	0.651	0.620	0.448	0.427
$\theta = 0.80$	0.25	0.671	0.647	0.488	0.463
	0.30	0.681	0.657	0.507	0.479
	0.35	0.699	0.678	0.523	0.499
	0.40	0.700	0.681	0.523	0.501
	0.45	0.540	0.527	0.371	0.347
	0.50	0.520	0.504	0.351	0.333

Table 23.7 Empirical MSE of R^2_{OLS}, R^2_{LAD}, R^2_{LMS}, and R^2_{LTS} for n = 60, when errors have normal distribution

	Proportion of Outliers	$MSE(R^2_{OLS})$	$MSE(R^2_{LAD})$	$MSE(R^2_{LMS})$	$MSE(R^2_{LTS})$
	0.00	0.011	0.022	0.137	0.140
	0.05	0.012	0.010	0.105	0.110
	0.10	0.015	0.011	0.094	0.094
	0.15	0.016	0.011	0.103	0.098
	0.20	0.016	0.011	0.086	0.093
θ = 0.20	0.25	0.015	0.010	0.095	0.090
	0.30	0.016	0.011	0.100	0.099
	0.35	0.015	0.011	0.093	0.091
	0.40	0.015	0.010	0.091	0.094
	0.45	0.015	0.011	0.094	0.095
	0.50	0.015	0.011	0.099	0.094
	0.00	0.010	0.019	0.070	0.072
	0.05	0.072	0.053	0.064	0.067
	0.10	0.090	0.070	0.058	0.052
	0.15	0.100	0.078	0.052	0.054
	0.20	0.101	0.080	0.049	0.047
θ = 0.40	0.25	0.099	0.077	0.051	0.050
	0.30	0.101	0.078	0.050	0.049
	0.35	0.099	0.077	0.049	0.051
	0.40	0.100	0.078	0.052	0.052
	0.45	0.095	0.073	0.053	0.050
	0.50	0.091	0.073	0.053	0.049
	0.00	0.005	0.012	0.040	0.040
	0.05	0.176	0.120	0.065	0.066
	0.10	0.234	0.198	0.084	0.083
	0.15	0.259	0.222	0.086	0.081
	0.20	0.261	0.219	0.083	0.085
θ = 0.60	0.25	0.263	0.222	0.085	0.083
	0.30	0.259	0.218	0.081	0.079
	0.35	0.255	0.214	0.081	0.082
	0.40	0.260	0.222	0.085	0.083
	0.45	0.237	0.204	0.086	0.086
	0.50	0.226	0.190	0.079	0.078
	0.00	0.001	0.006	0.023	0.022
	0.05	0.307	0.092	0.068	0.064
	0.10	0.415	0.351	0.160	0.164
	0.15	0.482	0.426	0.174	0.181
	0.20	0.497	0.438	0.187	0.199
θ = 0.80	0.25	0.499	0.440	0.196	0.199
	0.30	0.478	0.429	0.198	0.194
	0.35	0.474	0.420	0.191	0.198
	0.40	0.489	0.437	0.185	0.199
	0.45	0.424	0.377	0.178	0.181
	0.50	0.401	0.352	0.165	0.166

Table 23.8 Empirical MSE of R^2_{OLS}, R^2_{LAD}, R^2_{LMS}, and R^2_{LTS} for n = 100, when errors have normal distribution

	Proportion of Outliers	$MSE(R^2_{OLS})$	$MSE(R^2_{LAD})$	$MSE(R^2_{LMS})$	$MSE(R^2_{LTS})$
θ = 0.20	0.00	0.006	0.012	0.092	0.110
	0.05	0.017	0.013	0.065	0.074
	0.10	0.020	0.016	0.053	0.061
	0.15	0.021	0.016	0.050	0.058
	0.20	0.022	0.016	0.047	0.053
	0.25	0.022	0.016	0.048	0.053
	0.30	0.022	0.016	0.048	0.055
	0.35	0.023	0.017	0.049	0.053
	0.40	0.022	0.017	0.049	0.054
	0.45	0.022	0.016	0.053	0.058
	0.50	0.021	0.016	0.052	0.060
θ = 0.40	0.00	0.005	0.011	0.061	0.061
	0.05	0.085	0.070	0.058	0.062
	0.10	0.104	0.090	0.048	0.049
	0.15	0.110	0.095	0.044	0.049
	0.20	0.112	0.095	0.043	0.046
	0.25	0.115	0.099	0.042	0.043
	0.30	0.118	0.101	0.046	0.043
	0.35	0.117	0.101	0.043	0.044
	0.40	0.119	0.102	0.043	0.044
	0.45	0.117	0.102	0.045	0.045
	0.50	0.108	0.094	0.042	0.045
θ = 0.60	0.00	0.003	0.007	0.034	0.037
	0.05	0.194	0.152	0.077	0.073
	0.10	0.243	0.220	0.101	0.104
	0.15	0.262	0.237	0.109	0.107
	0.20	0.273	0.248	0.122	0.106
	0.25	0.277	0.251	0.114	0.109
	0.30	0.279	0.253	0.111	0.105
	0.35	0.283	0.257	0.113	0.104
	0.40	0.282	0.254	0.117	0.113
	0.45	0.283	0.258	0.114	0.114
	0.50	0.258	0.235	0.111	0.105
θ = 0.80	0.00	0.001	0.004	0.018	0.019
	0.05	0.316	0.129	0.061	0.064
	0.10	0.402	0.354	0.188	0.187
	0.15	0.450	0.418	0.229	0.221
	0.20	0.476	0.441	0.238	0.235
	0.25	0.487	0.455	0.246	0.244
	0.30	0.499	0.463	0.249	0.239
	0.35	0.508	0.474	0.265	0.258
	0.40	0.508	0.465	0.247	0.238
	0.45	0.505	0.469	0.250	0.241
	0.50	0.431	0.394	0.210	0.206

Table 23.9 Empirical MSE of R^2_{OLS}, R^2_{LAD}, R^2_{LMS}, and R^2_{LTS} for $n = 200$, when errors have normal distribution

	Proportion of Outliers	$MSE(R^2_{OLS})$	$MSE(R^2_{LAD})$	$MSE(R^2_{LMS})$	$MSE(R^2_{LTS})$
	0.00	0.003	0.005	0.056	0.065
	0.05	0.021	0.017	0.039	0.054
	0.10	0.027	0.023	0.020	0.026
	0.15	0.028	0.024	0.016	0.022
	0.20	0.029	0.025	0.015	0.018
$\theta = 0.20$	0.25	0.029	0.025	0.014	0.020
	0.30	0.030	0.026	0.016	0.019
	0.35	0.030	0.026	0.014	0.019
	0.40	0.030	0.026	0.014	0.021
	0.45	0.029	0.025	0.017	0.022
	0.50	0.028	0.024	0.015	0.021
	0.00	0.002	0.005	0.042	0.046
	0.05	0.090	0.073	0.058	0.061
	0.10	0.120	0.109	0.053	0.050
	0.15	0.128	0.118	0.053	0.049
	0.20	0.132	0.123	0.053	0.048
$\theta = 0.40$	0.25	0.134	0.125	0.055	0.050
	0.30	0.135	0.125	0.055	0.051
	0.35	0.134	0.124	0.057	0.050
	0.40	0.133	0.123	0.055	0.049
	0.45	0.132	0.122	0.058	0.052
	0.50	0.125	0.116	0.054	0.049
	0.00	0.001	0.004	0.027	0.029
	0.05	0.189	0.127	0.060	0.061
	0.10	0.272	0.248	0.124	0.114
	0.15	0.291	0.273	0.155	0.142
	0.20	0.304	0.288	0.164	0.149
$\theta = 0.60$	0.25	0.310	0.297	0.170	0.162
	0.30	0.310	0.295	0.170	0.156
	0.35	0.311	0.297	0.173	0.160
	0.40	0.307	0.292	0.166	0.157
	0.45	0.301	0.287	0.165	0.148
	0.50	0.280	0.266	0.153	0.141
	0.00	0.000	0.002	0.013	0.015
	0.05	0.303	0.073	0.024	0.027
	0.10	0.458	0.365	0.126	0.125
	0.15	0.496	0.450	0.249	0.242
	0.20	0.528	0.500	0.311	0.294
$\theta = 0.80$	0.25	0.545	0.524	0.343	0.313
	0.30	0.545	0.528	0.348	0.321
	0.35	0.544	0.525	0.353	0.332
	0.40	0.540	0.520	0.351	0.317
	0.45	0.516	0.495	0.324	0.300
	0.50	0.461	0.437	0.282	0.264

Table 23.10 Empirical MSE of R^2_{OLS}, R^2_{LAD}, R^2_{LMS}, and R^2_{LTS} for $n = 60$, when errors have Gamma distribution

	Proportion of Outliers	$MSE(R^2_{OLS})$	$MSE(R^2_{LAD})$	$MSE(R^2_{LMS})$	$MSE(R^2_{LTS})$
	0.00	0.012	0.023	0.140	0.132
	0.05	0.011	0.010	0.552	0.512
	0.10	0.014	0.011	0.410	0.442
	0.15	0.015	0.011	0.316	0.279
	0.20	0.016	0.011	0.209	0.207
$\theta = 0.20$	0.25	0.016	0.011	0.177	0.168
	0.30	0.015	0.011	0.209	0.175
	0.35	0.015	0.010	0.228	0.195
	0.40	0.016	0.011	0.234	0.216
	0.45	0.015	0.011	0.259	0.252
	0.50	0.015	0.010	0.203	0.197
	0.00	0.010	0.018	0.075	0.076
	0.05	0.061	0.045	0.721	0.612
	0.10	0.089	0.068	0.597	0.636
	0.15	0.096	0.072	0.476	0.422
	0.20	0.098	0.074	0.176	0.204
$\theta = 0.40$	0.25	0.099	0.076	0.148	0.116
	0.30	0.101	0.078	0.125	0.105
	0.35	0.102	0.077	0.147	0.113
	0.40	0.101	0.078	0.131	0.124
	0.45	0.100	0.077	0.187	0.160
	0.50	0.094	0.073	0.138	0.134
	0.00	0.005	0.011	0.041	0.039
	0.05	0.142	0.086	1.023	0.996
	0.10	0.231	0.186	1.464	1.414
	0.15	0.249	0.204	0.940	1.005
	0.20	0.255	0.212	0.397	0.340
$\theta = 0.60$	0.25	0.260	0.217	0.210	0.129
	0.30	0.261	0.221	0.159	0.150
	0.35	0.261	0.219	0.135	0.121
	0.40	0.266	0.222	0.158	0.138
	0.45	0.258	0.215	0.179	0.192
	0.50	0.239	0.196	0.117	0.124
	0.00	0.001	0.006	0.020	0.022
	0.05	0.240	0.060	1.484	1.467
	0.10	0.418	0.324	3.196	2.887
	0.15	0.454	0.374	4.194	3.834
	0.20	0.471	0.390	2.476	2.501
$\theta = 0.80$	0.25	0.490	0.420	0.857	0.796
	0.30	0.498	0.429	0.478	0.395
	0.35	0.505	0.434	0.189	0.215
	0.40	0.500	0.435	0.256	0.231
	0.45	0.487	0.416	0.337	0.395
	0.50	0.420	0.360	0.201	0.200

Table 23.11 Empirical MSE of R^2_{OLS}, R^2_{LAD}, R^2_{LMS}, and R^2_{LTS} for n = 100, when errors have Gamma distribution

	Proportion of Outliers	$MSE(R^2_{OLS})$	$MSE(R^2_{LAD})$	$MSE(R^2_{LMS})$	$MSE(R^2_{LTS})$
	0.00	0.005	0.010	0.058	0.066
	0.05	0.016	0.012	0.078	0.082
	0.10	0.019	0.014	0.048	0.051
	0.15	0.020	0.015	0.050	0.059
	0.20	0.022	0.016	0.044	0.053
θ = 0.20	0.25	0.022	0.016	0.047	0.058
	0.30	0.022	0.016	0.050	0.055
	0.35	0.022	0.017	0.047	0.055
	0.40	0.022	0.016	0.046	0.057
	0.45	0.022	0.016	0.045	0.054
	0.50	0.021	0.015	0.055	0.062
	0.00	0.003	0.008	0.034	0.036
	0.05	0.081	0.063	0.063	0.061
	0.10	0.095	0.081	0.045	0.046
	0.15	0.108	0.093	0.043	0.044
	0.20	0.113	0.099	0.043	0.047
θ = 0.40	0.25	0.116	0.100	0.042	0.044
	0.30	0.115	0.099	0.046	0.046
	0.35	0.116	0.100	0.042	0.043
	0.40	0.116	0.100	0.045	0.044
	0.45	0.115	0.100	0.045	0.043
	0.50	0.108	0.091	0.043	0.045
	0.00	0.001	0.004	0.017	0.019
	0.05	0.182	0.122	0.071	0.063
	0.10	0.221	0.194	0.094	0.089
	0.15	0.251	0.223	0.103	0.098
	0.20	0.271	0.246	0.113	0.105
θ = 0.60	0.25	0.278	0.252	0.114	0.112
	0.30	0.277	0.251	0.113	0.106
	0.35	0.277	0.253	0.111	0.110
	0.40	0.276	0.251	0.112	0.107
	0.45	0.274	0.249	0.110	0.109
	0.50	0.252	0.227	0.103	0.098
	0.00	0.006	0.012	0.092	0.110
	0.05	0.298	0.068	0.037	0.039
	0.10	0.363	0.292	0.149	0.148
	0.15	0.422	0.376	0.208	0.195
	0.20	0.477	0.443	0.235	0.234
θ = 0.80	0.25	0.488	0.450	0.241	0.227
	0.30	0.483	0.444	0.240	0.226
	0.35	0.493	0.457	0.249	0.243
	0.40	0.487	0.452	0.244	0.236
	0.45	0.480	0.446	0.240	0.230
	0.50	0.419	0.378	0.209	0.198

Table 23.12 Empirical MSE of R^2_{OLS}, R^2_{LAD}, R^2_{LMS}, and R^2_{LTS} for $n = 200$, when errors have Gamma distribution

	Proportion of Outliers	$MSE(R^2_{OLS})$	$MSE(R^2_{LAD})$	$MSE(R^2_{LMS})$	$MSE(R^2_{LTS})$
	0.00	0.003	0.005	0.054	0.062
	0.05	0.019	0.016	0.047	0.049
	0.10	0.024	0.020	0.022	0.028
	0.15	0.025	0.022	0.016	0.023
	0.20	0.027	0.023	0.017	0.020
$\theta = 0.20$	0.25	0.027	0.023	0.016	0.020
	0.30	0.028	0.024	0.016	0.022
	0.35	0.028	0.024	0.016	0.020
	0.40	0.029	0.025	0.017	0.020
	0.45	0.128	0.119	0.054	0.051
	0.50	0.122	0.113	0.049	0.047
	0.00	0.003	0.005	0.044	0.049
	0.05	0.079	0.065	0.058	0.058
	0.10	0.103	0.094	0.049	0.050
	0.15	0.113	0.103	0.048	0.045
	0.20	0.118	0.109	0.050	0.047
$\theta = 0.40$	0.25	0.123	0.113	0.051	0.047
	0.30	0.125	0.115	0.052	0.047
	0.35	0.127	0.118	0.054	0.048
	0.40	0.128	0.118	0.054	0.048
	0.45	0.292	0.279	0.158	0.147
	0.50	0.271	0.256	0.146	0.138
	0.00	0.002	0.004	0.029	0.032
	0.05	0.168	0.109	0.060	0.059
	0.10	0.223	0.199	0.103	0.095
	0.15	0.248	0.230	0.127	0.119
	0.20	0.266	0.250	0.140	0.131
$\theta = 0.60$	0.25	0.277	0.259	0.147	0.135
	0.30	0.280	0.265	0.153	0.144
	0.35	0.290	0.274	0.154	0.145
	0.40	0.291	0.276	0.161	0.147
	0.45	0.493	0.467	0.292	0.279
	0.50	0.438	0.394	0.233	0.215
	0.00	0.000	0.002	0.013	0.014
	0.05	0.254	0.051	0.023	0.027
	0.10	0.343	0.266	0.125	0.126
	0.15	0.389	0.339	0.195	0.177
	0.20	0.425	0.387	0.239	0.224
$\theta = 0.80$	0.25	0.451	0.421	0.269	0.250
	0.30	0.465	0.434	0.284	0.260
	0.35	0.490	0.462	0.298	0.276
	0.40	0.029	0.025	0.016	0.023
	0.45	0.027	0.023	0.016	0.021
	0.50	0.661	0.625	0.446	0.424

First we study the absolute bias of R^2_{OLS}, R^2_{LAD}, R^2_{LMS}, and R^2_{LTS} under all the aforesaid combination of θ and proportions of outliers in the data with normally distributed random errors and the results are tabulated in the Tables 23.1–23.3. It is observed that the absolute bias of all four goodness-of-fit measures is relatively very small when there are no outliers in the data. It is clear from the results that the AB of R^2_{OLS}, R^2_{LAD}, R^2_{LMS}, and R^2_{LTS} increases as the proportion of outliers increases. The impact becomes clearer when the value of θ is also high. For example, when θ is high, then the change in the values of AB can be considered with more confidence to be caused by the outliers. Such a behavior in the values of AB is true for all the considered sample sizes. It is further observed that for $\theta = 0.2$, AB(R^2_{OLS}) > AB(R^2_{LAD}) > AB(R^2_{LTS}) > AB(R^2_{LMS}). Such an observation changes as θ increases and for such values of θ, AB(R^2_{OLS}) > AB(R^2_{LAD}) but AB(R^2_{LMS}) > AB(R^2_{LTS}) when the proportion of outliers is low. But the inequality is reversed as soon as the proportion of outliers becomes high. For higher value of θ, it is observed that when the proportions of outliers is more than 10%, AB(R^2_{OLS}) and AB(R^2_{LAD}) are close to each other whereas AB(R^2_{LMS}) and AB(R^2_{LTS}) are also close to each other.

Next we observe the AB of R^2_{OLS}, R^2_{LAD}, R^2_{LMS}, and R^2_{LTS} under all the aforesaid combination of θ and proportions of outliers in the data with Gamma distributed random errors from Tables 23.4–23.6. The AB of all four goodness-of-fit measures is relatively very small when there are no outliers in the data. When outliers are present in the data, the AB increases. Among all the statistics, R^2_{LAD} performs better than others only when the proportion of outlier is low for lower values of θ whereas R^2_{LMS} as well as R^2_{LTS} performs better in all other situations. However, the difference in the AB becomes less, e.g., when there are 10% or more outliers in the data in all the sample sizes. For all other values of θ, AB(R^2_{OLS}) > AB(R^2_{LAD}) but AB(R^2_{LTS}) and AB(R^2_{LMS}) are practically close to each other though AB(R^2_{LMS}) is marginally lower than AB(R^2_{LTS}) in most of the cases. When θ increases, the AB of R^2_{OLS}, R^2_{LAD}, R^2_{LMS}, and R^2_{LTS} increase.

Comparing the AB of R^2_{OLS}, R^2_{LAD}, R^2_{LMS}, and R^2_{LTS} under normal and Gamma distributed random errors from Tables 23.1–23.3 and Tables 23.4–23.6, it is observed that the respective values of AB differ. It clearly indicates that the departure from the normality affects the AB of R^2_{OLS}, R^2_{LAD}, R^2_{LMS}, and R^2_{LTS}. The proposed goodness-of-fit measures are capable of handling the presence of outliers in the data and provide dependable results with respect to AB.

Next we observe the variability in R^2_{OLS}, R^2_{LAD}, R^2_{LMS}, and R^2_{LTS}. The empirical mean squared errors of R^2_{OLS}, R^2_{LAD}, R^2_{LMS}, and R^2_{LTS} under all the aforesaid combination of θ and proportions of outliers in the data are presented in Tables 23.7–23.12. First we observe the MSEs of

R^2_{OLS}, R^2_{LAD}, R^2_{LMS}, and R^2_{LTS} under all the aforesaid combination of θ and proportions of outliers in the data under normally distributed random errors from Tables 23.7–23.9. The MSEs of all four goodness-of-fit measures are relatively very small when there are no outliers in the data. As the proportion of outliers increases, for all values of θ, the MSE(R^2_{OLS}) and MSE(R^2_{LAD}) also increased. On the other hand, the MSE(R^2_{LTS}) and MSE(R^2_{LMS}) decreases when the value of θ is lower but increases steadily for higher values of θ. The values of MSE(R^2_{OLS}) are close to the values of MSE(R^2_{LAD}) whereas the values of MSE(R^2_{LTS}) are close to MSE(R^2_{LMS}). With a significant proportion of outliers, the statistics R^2_{LTS} and R^2_{LMS} perform better than R^2_{OLS} and R^2_{LAD} under the criterion of MSE.

Now we consider the MSEs of R^2_{OLS}, R^2_{LAD}, R^2_{LMS}, and R^2_{LTS} under Gamma distributed random errors from the Tables 23.10–23.12 under all the aforesaid combination of θ and proportions of outliers in the data. The MSEs of all four goodness of fit measures are relatively very small when there are no outliers in the data. When the linearity in the model is ensured, the MSEs of all the measures R^2_{OLS} R^2_{LAD}, R^2_{LMS}, and R^2_{LTS} increase with an increase in the proportion of outliers in the data as well as increase in the value of θ. Similar to the case of normally distributed random errors, the values of MSE(R^2_{OLS}) are close to the MSE(R^2_{LAD}) whereas the values of MSE(R^2_{LTS}) are close to MSE(R^2_{LMS}) in the case of Gamma distributed random errors.

Comparing the results from Tables 23.7–23.12 under normally distributed random errors and Tables 23.10–23.12 under Gamma distributed random errors, it is observed that the results differ under the respective settings. Such a difference can be attributed as arising from the departure from normality. An important and useful finding is that the proposed measures of goodness of fit also perform well when there is a departure from the normality of random errors. Hence, the proposed goodness-of-fit measures can also handle the departure from non-normality in the data.

The coefficient of determination R^2_{OLS} in the standard multiple linear regression model is defined only when there is an intercept term in the model. Such an assumption is analytically verifiable with the OLSE of the regression coefficient. It is difficult to verify analytically that the presence of an intercept term in the model is also required or essential for R^2_{LAD}, R^2_{LMS}, and R^2_{LTS}. So an attempt is made to verify it numerically with the simulated results. The value of R^2_{LAD}, R^2_{LMS}, and R^2_{LTS} are computed with a model without intercept terms. Tremendous fluctuations were noticed in the estimated values of R^2_{LAD}, R^2_{LMS}, and R^2_{LTS}, their ABs and MSEs. Such an experiment was conducted in all the settings where the model with intercept term is considered. It is concluded that R^2_{OLS}, R^2_{LAD}, R^2_{LMS}, and R^2_{LTS} are not well defined in multiple linear regression models without an intercept term.

23.6.1 Simulation on Outlier Detection

As discussed in Section 23.5, we compute the *ODPC*s based on R^2_{LAD}, R^2_{LMS}, and R^2_{LTS} to detect the presence outliers in the data as follows:

$$ODPC(R^2_{LAD}) = \frac{|R^2_{LAD} - R^2_{OLS}|}{R^2_{OLS}} \times 100,$$

$$ODPC(R^2_{LMS}) = \frac{|R^2_{LMS} - R^2_{OLS}|}{R^2_{OLS}} \times 100,$$

$$ODPC(R^2_{LTS}) = \frac{|R^2_{LTS} - R^2_{OLS}|}{R^2_{OLS}} \times 100.$$

The results have been obtained from the same data used in the simulation reported in Section 23.6 where R^2_{OLS}, R^2_{LAD}, R^2_{LMS}, and R^2_{LTS} were computed and their absolute bias and mean squared errors were empirically computed under the normal and Gamma distributed random errors. We choose $\alpha = 0.5$ to compute R^2_{LTS} in numerical studies.

First, we observe the *ODPC* values presented in Tables 23.13–23.18 computed under the normal and Gamma distributed random errors. It is clearly noticed that $ODPC(R^2_{LAD}) < ODPC(R^2_{LMS}) < ODPC(R^2_{LTS})$ for all the sample sizes and the distributions of random errors. We observe that $ODPC(R^2_{LAD})$, $ODPC(R^2_{LMS})$, and $ODPC(R^2_{LTS})$ values are very small, sometimes negligible, when there are no outliers in the data. As the proportion of outliers increases, the *ODPC* values initially increase and then decrease. The *ODPC*s increase significantly in the data with 5% outliers. For $\theta = 0.2$ and $\theta = 0.4$, the *ODPC*s increase moderately after that. For $\theta = 0.6$ and $\theta = 0.8$, the *ODPC*s are very high for 5% outliers in the data. The values of *ODPC*s are smaller when there are more outliers. Such an observation is explained as follows. When the percentage of outliers is low, the outliers are easily distinguishable from the rest of the data. As the proportion of outliers increases, they get mixed up with the data and it becomes difficult to distinguish between the data and the outliers. Consequently, the data appears as the data with a lower coefficient of determination, i.e., a reduced degree of linearity between the study and explanatory variables. The maximum values of *ODPC* are observed in the data with 5% outliers and it can be concluded that the number of outliers more than 5% starts merging with the data. As R^2_{LMS} and R^2_{LTS} are based on robust estimators, they exhibit higher changes and, consequently, higher values of *ODPC*s.

Table 23.13 OPDC values of R^2_{OLS}, R^2_{LAD}, R^2_{LMS}, and R^2_{LTS} for n = 60, when errors have normal distribution

	Proportion of Outliers	$OPDC(R^2_{LAD})$	$OPDC(R^2_{LMS})$	$OPDC(R^2_{LTS})$
	0.00	14	124	122
	0.05	41	506	511
	0.10	45	520	512
	0.15	48	461	456
	0.20	52	461	449
θ = 0.20	0.25	51	453	440
	0.30	51	437	427
	0.35	51	442	431
	0.40	50	457	445
	0.45	51	457	446
	0.50	49	441	432
	0.00	6	57	53
	0.05	29	475	467
	0.10	38	525	509
	0.15	44	449	443
	0.20	47	438	444
θ = 0.40	0.25	46	407	397
	0.30	47	396	395
	0.35	47	390	388
	0.40	46	394	384
	0.45	43	407	401
	0.50	45	387	376
	0.00	3	26	27
	0.05	33	497	494
	0.10	31	545	551
	0.15	35	444	440
	0.20	41	424	419
θ = 0.60	0.25	41	344	337
	0.30	40	347	341
	0.35	38	318	304
	0.40	39	329	333
	0.45	34	323	317
	0.50	37	322	320
	0.00	1	10	10
	0.05	92	544	541
	0.10	24	768	779
	0.15	26	673	672
	0.20	32	567	571
θ = 0.80	0.25	29	302	289
	0.30	30	262	254
	0.35	28	224	224
	0.40	28	236	234
	0.45	23	227	225
	0.50	29	233	231

Table 23.14 OPDC values of R^2_{OLS}, R^2_{LAD}, R^2_{LMS}, and R^2_{LTS} for $n = 100$, when errors have normal distribution

	Proportion of Outliers	$OPDC(R^2_{LAD})$	$OPDC(R^2_{LMS})$	$OPDC(R^2_{LTS})$
	0.00	10	100	102
	0.05	40	489	543
	0.10	42	446	479
	0.15	47	450	456
	0.20	47	458	473
$\theta = 0.20$	0.25	46	451	488
	0.30	50	461	507
	0.35	50	486	519
	0.40	55	512	542
	0.45	50	514	541
	0.50	48	437	467
	0.00	4	39	40
	0.05	35	481	523
	0.10	38	440	457
	0.15	37	359	384
	0.20	41	396	415
$\theta = 0.40$	0.25	42	393	423
	0.30	47	411	452
	0.35	52	446	472
	0.40	45	417	464
	0.45	43	433	491
	0.50	40	368	398
	0.00	2	17	18
	0.05	57	596	605
	0.10	30	487	542
	0.15	25	312	376
	0.20	35	326	362
$\theta = 0.60$	0.25	40	330	364
	0.30	35	357	382
	0.35	42	382	420
	0.40	38	374	387
	0.45	39	377	391
	0.50	37	306	319
	0.00	1	8	9
	0.05	152	635	653
	0.10	41	913	909
	0.15	28	573	619
	0.20	31	374	429
$\theta = 0.80$	0.25	33	290	305
	0.30	30	275	299
	0.35	32	272	304
	0.40	33	269	294
	0.45	30	282	303
	0.50	24	191	196

Table 23.15 OPDC values of R^2_{OLS}, R^2_{LAD}, R^2_{LMS}, and R^2_{LTS} for $n = 200$, when errors have normal distribution

	Proportion of Outliers	$OPDC(R^2_{LAD})$	$OPDC(R^2_{LMS})$	$OPDC(R^2_{LTS})$
	0.00	6	65	68
	0.05	26	351	399
	0.10	31	401	467
	0.15	37	439	498
	0.20	41	496	548
$\theta = 0.20$	0.25	43	495	598
	0.30	45	509	597
	0.35	47	525	608
	0.40	43	529	636
	0.45	51	590	657
	0.50	40	504	551
	0.00	2	26	29
	0.05	27	303	330
	0.10	24	315	388
	0.15	26	291	350
	0.20	29	356	368
$\theta = 0.40$	0.25	33	382	442
	0.30	28	346	396
	0.35	33	414	479
	0.40	34	408	483
	0.45	36	420	495
	0.50	32	359	412
	0.00	1	11	13
	0.05	48	368	362
	0.10	25	513	574
	0.15	21	258	318
	0.20	22	256	277
$\theta = 0.60$	0.25	26	267	304
	0.30	22	248	285
	0.35	24	279	314
	0.40	22	289	321
	0.45	22	308	382
	0.50	22	230	261
	0.00	0	4	5
	0.05	122	386	392
	0.10	44	968	985
	0.15	25	844	924
	0.20	22	357	451
$\theta = 0.80$	0.25	19	199	240
	0.30	17	154	172
	0.35	17	168	182
	0.40	18	190	216
	0.45	16	219	259
	0.50	18	154	171

Table 23.16 OPDC values of R^2_{OLS}, R^2_{LAD}, R^2_{LMS}, and R^2_{LTS} for $n = 60$, when errors have Gamma distribution

	Proportion of Outliers	$OPDC(R^2_{LAD})$	$OPDC(R^2_{LMS})$	$OPDC(R^2_{LTS})$
$\theta = 0.20$	0.00	15	132	138
	0.05	38	408	411
	0.10	41	427	421
	0.15	43	420	409
	0.20	48	407	399
	0.25	49	397	405
	0.30	45	414	409
	0.35	53	447	431
	0.40	51	438	416
	0.45	53	480	481
	0.50	46	404	392
$\theta = 0.40$	0.00	5	52	51
	0.05	39	341	339
	0.10	32	425	390
	0.15	33	331	337
	0.20	39	341	338
	0.25	37	333	347
	0.30	41	377	381
	0.35	48	356	350
	0.40	41	359	348
	0.45	42	412	405
	0.50	39	332	340
$\theta = 0.60$	0.00	3	25	26
	0.05	53	325	330
	0.10	27	377	382
	0.15	23	298	297
	0.20	30	275	298
	0.25	33	299	289
	0.30	36	297	304
	0.35	37	291	294
	0.40	37	311	301
	0.45	39	363	366
	0.50	31	269	259
$\theta = 0.80$	0.00	1	9	9
	0.05	111	321	324
	0.10	34	484	500
	0.15	25	378	400
	0.20	26	328	341
	0.25	26	247	237
	0.30	29	220	219
	0.35	30	212	211
	0.40	27	221	208
	0.45	34	280	268
	0.50	32	189	176

Table 23.17 OPDC values of R^2_{OLS}, R^2_{LAD}, R^2_{LMS}, and R^2_{LTS} for n = 100, when errors have Gamma distribution

	Proportion of Outliers	$OPDC(R^2_{LAD})$	$OPDC(R^2_{LMS})$	$OPDC(R^2_{LTS})$
	0.00	10	95	103
	0.05	33	404	450
	0.10	42	446	510
	0.15	46	437	479
	0.20	50	458	496
θ = 0.20	0.25	51	497	530
	0.30	50	488	509
	0.35	51	460	522
	0.40	50	496	524
	0.45	51	517	557
	0.50	47	436	455
	0.00	3	35	41
	0.05	27	401	408
	0.10	31	403	469
	0.15	36	428	456
	0.20	41	393	425
θ = 0.40	0.25	44	399	427
	0.30	45	417	441
	0.35	43	422	448
	0.40	51	444	485
	0.45	48	435	470
	0.50	36	296	331
	0.00	2	17	19
	0.05	38	412	414
	0.10	27	477	494
	0.15	31	427	449
	0.20	32	370	393
θ = 0.60	0.25	36	332	379
	0.30	35	337	349
	0.35	40	364	374
	0.40	39	338	373
	0.45	38	375	414
	0.50	24	210	230
	0.00	0	7	7
	0.05	105	431	432
	0.10	36	836	834
	0.15	28	951	1001
	0.20	27	672	718
θ = 0.80	0.25	27	362	435
	0.30	25	264	308
	0.35	27	269	274
	0.40	29	282	301
	0.45	24	316	369
	0.50	17	139	141

Table 23.18 OPDC values of R^2_{OLS}, R^2_{LAD}, R^2_{LMS}, and R^2_{LTS} for n = 200, when errors have Gamma distribution

	Proportion of Outliers	$OPDC(R^2_{LAD})$	$OPDC(R^2_{LMS})$	$OPDC(R^2_{LTS})$
	0.00	5	60	70
	0.05	26	407	450
	0.10	33	417	471
	0.15	40	477	539
	0.20	42	536	601
θ = 0.20	0.25	45	545	587
	0.30	46	521	603
	0.35	49	550	618
	0.40	48	534	605
	0.45	40	522	601
	0.50	38	496	558
	0.00	2	22	27
	0.05	31	395	438
	0.10	23	351	404
	0.15	29	345	415
	0.20	32	362	426
θ = 0.40	0.25	35	402	464
	0.30	35	388	453
	0.35	35	410	470
	0.40	28	401	447
	0.45	33	424	469
	0.50	30	340	376
	0.00	1	11	13
	0.05	51	459	478
	0.10	22	522	585
	0.15	21	326	417
	0.20	22	255	313
θ = 0.60	0.25	26	284	327
	0.30	23	273	310
	0.35	24	283	331
	0.40	26	281	325
	0.45	22	284	331
	0.50	21	222	260
	0.00	0	5	5
	0.05	136	501	496
	0.10	35	961	995
	0.15	21	1099	1213
	0.20	17	591	741
θ = 0.80	0.25	18	207	276
	0.30	19	191	221
	0.35	17	186	219
	0.40	17	185	200
	0.45	17	185	219
	0.50	15	138	165

Next, we illustrate how to find the cutoff values of *ODPC* such that any value of *ODPC* above it will indicate the presence of outliers in the data. Looking at the *ODPC* values based on the simulation results when $\theta = 0.8$, a possible cutoff can be defined as follows: Based on R^2_{LMS} and R^2_{LTS}, if $ODPC \geq 200$ then it indicates the presence of outliers in the data. Such a cutoff value based on ${R_{med}}^2$ is $ODPC \geq 20$. So $ODPC \geq 200$ in case of R^2_{LMS} and R^2_{LTS} and $ODPC \geq 20$ for R^2_{LAD} will indicate the presence of outliers in the data and recommends the use of robust measures of goodness-of-fit. There are some minor differences in the *ODPCs* values of normal and Gamma distributed random errors, but the conclusions are the same in both cases. The results obtained by increasing the sample sizes differ little under normal and Gamma distributed random errors but the conclusions again remain the same, and the cutoff values also remain almost the same. The effect of non-normal random errors on the values of *ODPC* is also clearly observed by comparing the *ODPC* values under normal and Gamma distributed random errors with the same sample sizes. It is important to note that such cutoff values obtained here will work like a rule of thumb and will change depending upon the nature of the data as well as the choice of the robust estimator of regression coefficient used in the construction of goodness-of-fit statistics.

23.7 CONCLUSIONS

Outliers in the data cause regression coefficient estimators to perform poorly, which causes the goodness of fit statistics to not accurately reflect the reliable values indicating the quality of fit of the model. Outliers frequently represent the extreme values in the experimental design and cannot be disregarded. In such a case, the issue of how to fit the model and assess the goodness of fit emerges. In these circumstances, robust estimation techniques can be used to estimate the regression coefficients, but it is not yet known how to assess the quality of fit based on those estimates. The robust method of estimation based on the median-based estimator, least median squared regression estimator, and least trimmed squared estimator of regression coefficients are used to estimate the regression coefficients, which are then used to construct the goodness-of-fit statistics following the philosophy of constructing the coefficient of determination in the multiple linear regression model. Such goodness-of-fit statistics can also be used to diagnose the presence of outliers in the data. The significant difference between the goodness-of-fit values based on the robust estimator and OLSE of regression coefficient is measured by ODPC, which indicates the presence of outliers in the data. It is challenging to analytically derive the statistical properties of the derived goodness-of-fit statistics. In order to examine the effectiveness of the proposed goodness-of-fit measures, we conducted a thorough Monte Carlo simulation. The simulated results show that the proposed goodness-of-fit measures continue to perform well and provide us with a trustworthy value for the goodness-of-fit statistics even when the fraction of outliers in the data is raised. They measure the goodness-of-fit effectively even when there are a significant number of outliers in the data, e.g., nearly 50% of the data is considered as outlier. In the case of a multiple linear regression model, whenever there is a departure from the normality of the random errors, the robust estimates are supposed to take care of the departure from the non-normality. We have taken into account the random errors generated from the two distributions: the normal distribution and the Gamma distributions, and the goodness-of-fit measures are evaluated in order to determine whether the suggested measures are robust with respect to the deviation from normality. Even in the presence of data that deviates from normalcy, the proposed goodness-of-fit methods are effective and deliver reliable findings. The ODPC values are computed for the robust goodness-of-fit measures with respect to the coefficient of determination. The simulated results on ODPC confirms that they are capable of diagnosing the presence of outliers in the data. The cutoff value of ODPC differ based on the choice of robust estimator.

REFERENCES

Bahari, F., Parsi, S. and Ganjali, M. 2017. A doubly robust goodness-of-fit test in general linear models with missing covariates. Communication Statistical Simulation Computational. 46(10): 79097923.

Brys, G., Hubert, M. and Struyf, A. 2008. Goodness-of-fit tests based on a robust measure of skewness. Computational Simulation. 23(3): 429–442.

Chatterjee, S. and Hadi, A.S. 2013. Regression Analysis by Example, 5th Edn. Wiley.

Chan, W.S. 1994. On portmanteau goodness-of-fit tests in robust time series modeling. Computational Simulation. 9(4): 301–310.

Cheng, C.L., Shalabh and Garg, G. 2014. Coefficient of determination for multiple measurement error models. Journal of Multivariate Analysis. 123: 137–152.

Cheng, C.L., Shalabh and Garg, G. 2016. Goodness of fit in restricted measurement error models. Journal of Multivariate Analysis. 145: 101–116.

Cheng, C.L., Shalabh and Chaturvedi, A. 2019. Goodness of fit for generalized shrinkage estimation. Theory of Probability and Mathematical Statistics (American Mathematical Society). 100: 177–197.

Dehon, C., Gassner, M. and Verardi, V. 2009. A Hausman-type test to detect the presence of influential outliers in regression analysis. Economic Letters. 105(1): 64–67.

Di Palma, M.A. and Gallo, M. 2016. A co-median approach to detect compositional outliers. Journal of Applied Statistics. 43(13): 2348–2362.

Dong, F. 2010. Bayesian method to detect outliers for ordinal data. Communication Statistical Simulation Computational. 39(7): 1470–1484.

Dünder, Emre. 2021. A new correction approach for information criteria to detect outliers in regression modeling. Communication Statistical Theory Methods. 50(10): 2451–2465.

Ekiz, Ufuk. 2002. A Bayesian method to detect outliers in multivariate linear regression. Hacet.British Journal of Mathematical and Statistical. 31: 77–82.

Embrechts, P., Herzberg, A.M. and Ng, Allen, C.K. 1986. An investigation of Andrews' plots to detect period and outliers in time series data. Communication Statistical. B: Simulation Computational. 15(4): 1027–1051.

Eshima, N. and Tabata, M. 2010. Entropy coefficient of determination for generalized linear models. Computational Statistics and Data Analysis. 54(5): 1381–1389.

Eshima, N. and Tabata, M. 2011. Three predictive power measures for generalized linear models: The entropy coefficient of determination, the entropy correlation coefficient and the regression correlation coefficient. Computational Statistics and Data Analysis. 55(11): 3049–3058.

Gumedze. 2019. Freedom N. Use of likelihood ratio tests to detect outliers under the variance shift outlier model. Journal of Applied Statistics. 46(4): 598–620.

Guttman, I., Dutter, R. and Freeman, P.R. 1978. Care and handling of univariate outliers in the general linear model to detect spuriosity a Bayesian approach. Technometrics. 20(2): 187–193.

Hong, C.S., Ham, J.H. and Kim, H.I. 2005. Variable selection for logistic regression model using adjusted coefficients of determination. Korean Journal of Applied Statistics. 18(2): 435–443.

Hössjer, O. 2008. On the coefficient of determination for mixed regression models. Journal of Statistical Planning and Inference. 138(10): 3022–3038.

Huang, L.-S. and Chen, J. (2008). Analysis of variance, coefficient of determination and F-test for local polynomial regression. The Annals of Statistics. 36(5): 2085–2109.

Keich, U. and Nagarajan, N. 2006. A fast and numerically robust method for exact multinomial goodness-of-fit test. Journal of Computing Graph Statistical. 15(4): 779–802.

Khezrimotlagh, D., Cook, Wade D. and Zhu, J. 2020. A nonparametric framework to detect outliers in estimating production frontiers. European Journal of Operational Research. 286(1): 375–388.

Knight, J.L. 1980. The coefficient of determination and simultaneous equation systems. Journal of Econometrics. 14(2): 265–270.

Liao, J.G. and McGee, D. (2003). Adjusted coefficients of determination for logistic regression. The American Statistician. 57(3): 161–165.

Lipsitz, S.R., Leong, T., Ibrahim, J. and Lipshultz, S. 2001. A partial correlation coefficient and coefficient of determination for multivariate normal repeated measures data. The Statistician. 50(1): 87–95.

Marchand, E. 1997. On moments of beta mixtures, the noncentral beta distribution and the coefficient of determination. Journal of Statistical Computation and Simulation. 59(2): 161–178.

Marchand, E. 2001. Point estimation of the coefficient of determination. Statistics and Decisions. 19(2): 137–154.

McKean, J.W. and Sievers G.L. 1987. Coefficients of determination for least absolute deviation analysis. Statistics and Probability Letters. 5(1): 49–54.

Mokhtar, N.A., Zubairi, Y.Z. and Hussin, A.G. (2018). A clustering approach to detect multiple outliers in linear functional relationship model for circular data. Journal of Applied Statistics. 45(6): 1041–1051.

Montgomery, D.C., Peck, E.A. and Vining, G.G. 2021. Introduction to Linear Regression Analysis. 6th Edn. Wiley.

Peña, D. and Galeano, P. 2008. A unified approach to model selection, discrimination, goodness of fit and outliers in time series. In: Advances in Mathematical and Statistical Modeling. 267–278. Statistics of Industrial Technology, Birkhuser, Boston, MA.

Rao, C.R., Toutenburg, H., Shalabh and Heumann, C. 2008. Linear Models and Generalizations - Least Squares and Alternatives. Springer.

Renaud, O. and Victoria-Feser, M.-P. 2010. A robust coefficient of determination for regression. Journal of Statistical Planning and Inference. 140(7): 1852–1862.

Rousseeuw, P. 1984. Least median of squares regression. Journal of the American Statistical Association. 79: 871–880.

Schweder, Tore. 1976. Some 'optimal' methods to detect structural shift or outliers in regression. Journal of the American Statistical Association. 71(354): 491–501. (Reviewer: A. Hedayat).

Shalabh, Dhar, S.S., Chakroborty, C and Jha, P. (Inpress). Goodness of Fit Based and Variable Selection in Non-parametric Measurement Error Model. In: Statistical Modeling and Applications on Real-Time Problems. Chandra Shekhar and Raghaw Raman Sinha (Eds.) Taylor's & Francis, CRC Press (Accepted for publication).

Shalabh, Dhar, S.S. and Balakrishna, N. 2021. Goodness of fit in parametric and nonparametric econometric models in optimal decision making. In: Operations Research & Statistics: Methodologies and Applications, (Eds.) Leopoldo Eduardo Ca'rdenas-Barro'n, Aquil Ahmed, Irfan Ali and Ali Akbar Shaikh (Eds.). pp. 68–91. Taylor's & Francis, Routledge, CRC Press.

Shalabh and Dhar, S.S. 2021. Goodness of Fit in Nonparametric Regression Modelling. Journal of Statistical Theory and Practice (Invited paper for the special issue dedicated to Professor C.R. Rao on "Celebrating the Centenary of Professor C.R. Rao"). Journal of Statistical Theory and Practice. 15: 18.

Shalabh and Dhar, S.S. 2022. GIVE statistic for goodness of fit in instrumental variables models with application to COVID Data. Nature Scientific Reports

(Nature Publication), 12, 9472. DOI: https://doi.org/10.1038/s41598-022-13240-y..

Shalabh and Dhar, S.S. 2023. Testing the goodness of fit in instrumental variables models in recent advances. In: G Families of Probability Distributions, Mir Masoom Al, Irfan Ali, Haitham M. Youso and Mohamed Ibrahim (Eds.). Taylor's & Francis, CRC Press (Accepted for publication).

Srivastava, A.K. and Shobhit. 2002. A family of estimators for the coefficient of determination in linear regression models. Journal of Applied Statistical Science. 11(2): 133–144.

Tanaka, J.S. and Huba, G.J. 1998. A general coefficient of determination for covariance structure models under arbitrary GLS estimation. British Journal of Mathematical and Statistical Psychology. 42(2): 233–239.

Tjur, T. 2009. Coefficients of determination in logistic regression models—A new proposal: the coefficient of discrimination. The American Statistician. 63(4): 366–372.

van der Linde, A. and Tutz, G. 2008. On association in regression: The coefficient of determination revisited. Statistics. 42(1): 1–24.

Zheng, Y., Li, W.K. and Li, G.A. 2018. A robust goodness-of-fit test for generalized autoregressive conditional heteroscedastic models. Biometrika. 105(1): 73–89.

Index

W

X